PROGRESS IN

Molecular Biology and Translational Science

Volume 95

PROGRESS IN

Molecular Biology and Translational Science

Molecular Biology of Cancer: Translation to the Clinic

edited by

Raymond W. Ruddon, M.D., Ph.D.

Department of Pharmacology,
University of Michigan Medical School,
Ann Arbor, Michigan, USA

Volume 95

ELSEVIER

AMSTERDAM • BOSTON • HEIDELBERG • LONDON
NEW YORK • OXFORD • PARIS • SAN DIEGO
SAN FRANCISCO • SINGAPORE • SYDNEY • TOKYO
Academic Press is an imprint of Elsevier

Academic Press is an imprint of Elsevier
32 Jamestown Road, London, NW1 7BY, UK
Radarweg 29, PO Box 211, 1000 AE Amsterdam, The Netherlands
30 Corporate Drive, Suite 400, Burlington, MA 01803, USA
525 B Street, Suite 1900, San Diego, CA 92101-4495, USA

Library of Congress Cataloging-in-Publication Data
A catalog record for this book is available from the Library of Congress

British Library Cataloguing in Publication Data
A catalogue record for this book is available from the British Library

ISBN: 978-0-12-385071-3
ISSN: 1877-1173

For information on all Academic Press publications
visit our website at elsevierdirect.com

Printed and bound by CPI Group (UK) Ltd, Croydon, CR0 4YY
Transferred to Digital Printing, 2013

Contents

Contributors

Numbers in parentheses indicate the pages on which the authors' contributions begin.

James R. Baker, Michigan Nanotechnology Institute for Medicine and Biological Sciences, University of Michigan, Ann Arbor, Michigan, USA (193)

Mahaveer S. Bhojani, Department of Radiology, University of Michigan, Center for Molecular Imaging, Ann Arbor, Michigan, USA (237)

Dean E. Brenner, University of Michigan Medical Center and VA Medical Center, Ann Arbor, Michigan, USA (351)

Arul M. Chinnaiyan, Michigan Center for Translational Pathology, and Department of Pathology, and Howard Hughes Medical Institute, Maryland, and Department of Urology; and Comprehensive Cancer Center, University of Michigan, Ann Arbor, Michigan, USA (55)

Seok Ki Choi, Michigan Nanotechnology Institute for Medicine and Biological Sciences, University of Michigan, Ann Arbor, Michigan, USA (193)

Rosemarie Chirco D'Angelo, Department of Internal Medicine, Division of Hematology and Oncology, University of Michigan Comprehensive Cancer Center, University of Michigan, Ann Arbor, Michigan, USA (113)

Craig J. Galbán, Department of Radiology, University of Michigan, Center for Molecular Imaging, Ann Arbor, Michigan, USA (237)

Stefanie Galbán, Department of Radiation Oncology, University of Michigan, Center for Molecular Imaging, Ann Arbor, Michigan, USA (237)

Daniel F. Hayes, Breast Oncology Program, University of Michigan Comprehensive Cancer Center, Ann Arbor, Michigan, USA (95)

Baohua Huang, Michigan Nanotechnology Institute for Medicine and Biological Sciences, University of Michigan, Ann Arbor, Michigan, USA (193)

Madhuri Kakarala, University of Michigan Medical Center and VA Medical Center, Ann Arbor, Michigan, USA (351)

John S. Lazo, Department of Pharmacology and Chemical Biology, University of Pittsburgh Drug Discovery Institute and Cancer Institute, University of Pittsburgh, Pittsburgh, Pennsylvania, USA (9)

Kyung-Hoon Lee, Michigan Nanotechnology Institute for Medicine and Biological Sciences and Department of Chemistry, University of Michigan, Ann Arbor, Michigan, USA (193)

Gary D. Luker, Department of Radiology, University of Michigan, Center for Molecular Imaging, Ann Arbor, Michigan, USA (237)

Istvan J. Majoros, Michigan Nanotechnology Institute for Medicine and Biological Sciences, University of Michigan, Ann Arbor, Michigan, USA (193)

Andrzej Myc, Michigan Nanotechnology Institute for Medicine and Biological Sciences, University of Michigan, Ann Arbor, Michigan, USA (193)

Gilbert S. Omenn, Department of Internal Medicine, Department of Human Genetics, School of Public Health, Center for Computational Medicine and Bioinformatics, University of Michigan, Ann Arbor, Michigan, USA (159)

Nallasivam Palanisamy, Michigan Center for Translational Pathology, and Department of Pathology; and Comprehensive Cancer Center, University of Michigan, Ann Arbor, Michigan, USA (55)

Lalit Patel, Department of Medicine; and Department of Urology, Michigan Center for Translational Pathology and the University of Michigan Comprehensive Cancer Center, University of Michigan, Ann Arbor, Michigan, USA (31)

Kenneth J. Pienta, Department of Medicine; and Department of Urology, Michigan Center for Translational Pathology and the University of Michigan Comprehensive Cancer Center, University of Michigan, Ann Arbor, Michigan, USA (31)

Alnawaz Rehemtulla, Department of Radiation Oncology, University of Michigan, Center for Molecular Imaging, Ann Arbor, Michigan, USA (237)

Brian D. Ross, Department of Radiology, University of Michigan, Center for Molecular Imaging, Ann Arbor, Michigan, USA (237)

Raymond W. Ruddon, Department of Pharmacology, University of Michigan Medical School, Ann Arbor, Michigan, USA (1)

Jeffrey B. Smerage, Breast Oncology Program, University of Michigan Comprehensive Cancer Center, Ann Arbor, Michigan, USA (95)

Marcian E. Van Dort, Department of Radiology, University of Michigan, Center for Molecular Imaging, Ann Arbor, Michigan, USA (237)

Brent B. Ward, Oral and Maxillofacial Surgery, University of Michigan Hospitals, Ann Arbor, Michigan, USA (193)

Wendell Weber, Department of Pharmacology, University of Michigan Medical School, Ann Arbor, Michigan, USA (299)

Max S. Wicha, Department of Internal Medicine, Division of Hematology and Oncology, University of Michigan Comprehensive Cancer Center, University of Michigan, Ann Arbor, Michigan, USA (113)

Jian Zhang, Department of Medicine; and Department of Urology, Michigan Center for Translational Pathology and the University of Michigan Comprehensive Cancer Center, University of Michigan, Ann Arbor, Michigan, USA (31)

Preface

The purpose of this volume in the Progress in Molecular Biology and Translational Science series is to explore some of the most exciting recent advances in basic research on the molecular biology of cancer and how this knowledge leads to advances in the diagnosis, treatment, and prevention of cancer.

The chapter topics include introduction to the molecular biology of cancer (Ruddon), molecularly targeted approaches to the development of anticancer drugs (Lazo), targeting chemokine ligands and their role in cancer metastasis (Pienta), discovery of cancer cell fusion genes in solid cancers (Chinnaiyan), role of circulating tumor cells in cancer diagnosis, disease progression and response to therapy (Hayes), cancer stem cells (WIcha), bioinformatics and systems biology of cancers(Omenn), progress in cancer nanotechnology (Baker), molecular imaging (Ross), cancer epigenetics (Weber), and cancer prevention (Brenner).

The senior investigators represented here are all University of Michigan faculty, with the exception of John Lazo, from the University of Pittsburgh, but even he has a U of M genealogy, having received his Ph.D. in Pharmacology from the Ruddon lab at U of M. However, all of these scientists are leaders in their fields of research. Thus, this is not just a parochial concatenation of narrowly focused local research, but it is, in fact, representative of the most advanced research in the fields discussed here.

R. W. Ruddon, Ann Arbor

Introduction to the Molecular Biology of Cancer: Translation to the Clinic

Raymond W. Ruddon, M.D., Ph.D.

Department of Pharmacology, University of Michigan Medical School, Ann Arbor, Michigan, USA

Advances in molecular biology over the last several decades are being steadily applied to our understanding of the molecular biology of cancer, and these advances in knowledge are being translated into the clinical practice of oncology.

Many examples can be cited to demonstrate this. Some of them are listed below. Everyone has their favorite list, of course; however, everyone would probably agree that the items on this list should be included in any such list of advances.[1]

1. *Techniques of modern molecular biology*: These include PCR, DNA microarrays, proteomics, molecular imaging, identification of cancer stem cells, and the analysis of DNA methylation in cancer epigenetics. Other advances are in methods for metabolomic measurements, nanotechnology, systems biology, cancer immunology and monoclonal antibody production, and bioinformatics. Many of these advances will be discussed in this volume.
2. *Cancer susceptibility genes*: In the early 1900s, it was known that familial clustering of some cancers occurred, for example, with colon cancer and breast cancer, but the genes involved in this were not known. The APC, BRCA-1, BRCA-2, and p53-inherited mutations, for example, were not known until more recently. Research in this area has identified a number of genes involved in cancer susceptibility, and with modern cloning techniques, more are identified every few months.
3. *Genes involved in cancer initiation and promotion*: It has been known for a long time that chemicals and irradiation could damage DNA and initiate cancer in animals and humans, but what genes were altered was almost completely unknown until the advances in molecular biology were applied. We now know a lot about what genes are involved at various stages of a number of cancers. For example, the work of Bert Vogelstein and his colleagues have defined a pathway, sometimes called the "Vogelgram," for the progression of colon cancer.[2] We knew that

Progress in Molecular Biology
and Translational Science, Vol. 95
DOI: 10.1016/S1877-1173(10)95001-1

1877-1173/10 $35.00

DNA repair was important and that heritable conditions of defective DNA repair (e.g., xeroderma pigmentosum) could lead to cancer, but the ideas about the mechanisms of DNA repair were primitive until fairly recently.[3]

4. *The identification of oncogenes*: This did not really take off until the early 1980s. The *src* gene was identified in 1976 by Stehelin *et al.* and *erb*, *myc*, and *myb* oncogenes in the late 1970s, but this was about the limit of our knowledge (reviewed in Ref. 4).
5. *Tumor suppressor genes*: The term "tumor suppressor gene" was not even coined until the early 1980s, although their existence had been implied from the cell fusion experiments of Henry Harris, who showed that if you fused a normal cell with a malignant cell, the phenotype was usually nonmalignant (reviewed in Ref. 5). The RB gene was the first one cloned, in 1986 by Friend *et al.*[6] P53 was originally thought of as an oncogene. It was not realized until 1989 that wild-type p53 could actually suppress malignant transformation. A number of tumor suppressor genes have, of course, been identified since.
6. *Cell cycle checkpoints*: These were identified in yeast starting in the 1970s by Lee Hartwell and colleagues, but the identification of many of the human homologs of these genes did not occur until the late 1980s.
7. *Tumor immunology*: The mechanism of the immune response and the ability to manipulate it with cytokines, activated dendritic cells, vaccines, and drugs was not in the treatment armamentarium until recently. Advances in identification of tumor antigens and in the techniques to produce monoclonal antibodies are now leading to newer treatment modalities.
8. *The viral etiology of cancer*: This was still being widely debated in the 1980s. The involvement of Epstein–Barr virus in Burkitt's lymphoma and of hepatitis B virus in liver cancer was becoming accepted, but the role of viruses in these diseases and in cervical cancer, Kaposis' sarcoma, and in certain T-cell lymphomas became clearer much later.
9. *Growth factors that affect cancer*: Even though growth factors that stimulate cell replication, such as IGF-1 and 2, FGF, NGF, PDGF, and EGF, have been known for a long time, knowledge about their receptors and signal transduction mechanisms have been greatly expanded. Importantly, it is now known that the signal transduction mechanisms that cancer cells use are overlapping and redundant. Thus, cancer cells can become resistant to anticancer drugs by upregulating alternate pathways.

The explosion of knowledge about signal transduction mechanisms and how these pathways interact have also been a tremendous boon to our understanding of how cells respond to signals in their microenvironment and communicate with one another.

10. *Regulation of gene expression*: Current information on the packaging of chromatin, transcription factors, coinducers and corepressors, inhibitory RNA (siRNA), and micro RNA is expanding our knowledge of how gene expression is regulated in cancer.

Several decades of advances in cancer cell molecular biology have led to a rich pipeline of anticancer agents aimed at membrane-bound receptor protein kinases, intracellular signaling kinases, epigenetic abnormalities, as well as to agents that affect protein folding and degradation, tumor vasculature, and the tumor cell microenvironment (reviewed in Ref. 7).

The history of chemotherapy has seen its ups and downs since the introduction of nitrogen mustard after World War II.[8] The focus until recently has been almost exclusively on toxic chemicals that somewhat nonselectively kill dividing cells. Now, however, there is a large focus on what has been called "molecularly targeted agents." Two successful examples are trastuzumab (Herceptin) and imatinib (Gleevec). The number of successes for this approach to date is small, however, and the success of imatinib in treatment of chronic myelogenous leukemia (CML) may be the exception rather than the rule, arguing for the need to continue to search for agents that are toxic to dividing cells (albeit, hopefully, more selectively toxic ones).[9]

The other school of thought argues that targeted therapies can be developed against specific targets that are selectively overexpressed or mutated in cancer cells and thus be less toxic to normal tissues.[10] To this end, a detailed understanding of the target's structure and function will be necessary to identify a targeted drug's mechanism of action and the mechanisms of resistance development to the drug. In addition, the use of cytotoxic drugs in combination with targeted drugs will most likely continue to be the most efficacious approach to treating cancer.

In Chapter 2 of this volume, John Lazo expands on these concepts.

Of course, all successful approaches to new cancer treatment depend on the adaptation of molecular findings into the clinic, so-called translational research. In Chapter 3, Ken Pienta *et al.* describe a beautiful example of how this is done. They have shown how the monocyte chemoattractant protein MCP-1 (CCL2) is involved in the lethal phenotype of prostate cancer cells, that it is increased in prostate cancer bone metastases, and that it is a pro-survival and pro-angiogenic factor that leads to metastasis.[11] They have also shown that monoclonal antibody targeted to CCL-2 inhibits bone metastasis and increases survival.

In Chapter 4, Nallasivam Palanisamy and Arul Chinnaiyan discuss the recent discovery of gene fusions in "solid cancers" such as prostate, breast, and lung cancers. Up until this research, it was generally believed that gene fusion events occurred typically in hematologic malignancies and rare bone and soft-tissue tumors, but not in tumors such as prostate, lung, breast, and colon. Novel gene fusions were discovered utilizing integrative analysis of high throughput long- and short-read transcriptome sequencing of cancer cells. The finding in human prostate cancers that a fusion (TMPRSS2-ERG) occurs between an oncogene and an androgen regulated gene was a seminal observation. Subsequently, gene fusion events have been found in a variety of other human solid cancers, including breast, lung, and pancreatic cancers.[12]

In the Chapter 5, Dan Hayes and Jeffrey Smerage discuss the technique to isolate cancer cells from the blood of cancer patients and how the monitoring of serial changes in circulatory breast cancer cells can be used as an index of cancer progression in both murine xenograft models of human breast cancer[13] and in patients with breast cancer.[14]

In this method, an iron-tagged (ferrofluid) monoclonal antibody to epithelial cell surface markers is used to isolate epithelial cells from red and white blood cells. Since epithelial cells do not normally circulate, these cells are often of tumor origin. Hayes *et al.*[14] have shown that the number of circulatory tumor cells (CTCs) obtained before treatment is an independent predictor of progression-free survival and overall survival in patients with metastatic breast cancer.

The concept that cancer stem cells (discussed in chapter 6 by D'Angelo and Wicha) are the most aggressive cell type in a tumor cell population is an area of evolving research. The concept is based on studies that indicate that only a very small subset of cells (often less than 0.1%) of cells in a tumor have the ability to generate a new tumor from an implanted human cancer cell population in immunodeficient (SCID) mice.[15] The stem cells in a human breast cancer cell population have a specific stem cell population phenotype: $CD44^+/CD24^-/$ aldehyde dehydrogenase 1^+. These findings have important diagnostic and therapeutic implications. For example, as noted earlier, currently available chemotherapeutic drugs were developed largely on the basis of their ability to shrink a tumor mass in experimental models and clinical trials. This essentially predicts the ability of a drug to kill the bulk of cells in a cancer population, potentially leaving the more aggressive, drug-resistant cells behind. Thus, drugs more specifically targeted to the cancer stem cell population would most likely result in more effective and durable responses.

As more is learned about the phenotype and genotype of cancer cells, it is becoming apparent that cancer cells are extremely complex. They not only utilize many of the same genes and proteins that normal cells express but can up- or downregulate functions in order to survive extraordinary environmental threats such as toxic drugs or radiation therapy. They have redundant and overlapping signal transduction pathways that enable them to circumvent many challenges to their viability. The relatively new science of systems biology is beginning to unravel much of this complexity.

Systems biology is a conceptual framework to quantify and integrate the types of biological information contained in cells, tissues, organisms, and populations of individuals. It is these interacting networks that modulate and regulate life. This includes the study of genomics, transcriptomics, proteomics, metabolomics, and most other "-omics" that have yet to be so named. Systems biology attempts to delineate and integrate the dynamic relationships between DNA as it is packaged in chromalin, RNA (including mRNA, rRNA, siRNA, and microRNA), gene regulatory networks, protein–protein interactions, and cellular communication systems as well as interactions at the tissue and organ levels.

This is a tremendously complicated business and requires the analysis and integration of enormous data sets. Thus, the science of bioinformatics is playing a crucial role in this new way to look at cancer cell biology. For example, even in lower organisms, such as yeast, *Caenorhabditis elegans*, and *Drosophila*, the interactions of gene and protein networks are high (reviewed in Ref. 16). The global mapping of a yeast genetic interaction network containing 1000 genes revealed over 4000 interactions. A single large network of 1548 proteins showed 2538 interactions. In *C. elegans*, more than 5500 protein–protein interactions were identified. In Drosophila, a total of 10,623 gene transcripts produced a map of 7048 proteins with 20,405 predicted interactions.

Some of the issues of determining the complicated interactions in cells and the use of bioinformatics in solving these puzzles are described by Gil Omenn in Chapter 7.

Studies in the relatively new field of metabolomics are beginning to reveal a lot about the cancer cell phenotype. Metabolomics is at the same time less complicated and more complicated than its "-omics" cousins genomics, transcriptomics, and proteomics. It is less complicated in the sense that whereas there are 25,000–30,000 genes in the human genome, 100,000 transcripts, and 1,000,000 proteins, there are only about 1800 compounds that constitute the metabolome. It is more complicated in that these 1800 compounds are in a constant, rapid state of flux depending on absorption of dietary substances, hormone levels, drug intake, exercise, body temperature, and the presence of disease states such as diabetes, infections, or cancer. NMR and mass spectrometry are the typical tools used in the study of metabolomics.[17]

Multiple complex metabolic events characterize cancer development and progression. Recently, sarcosine, an *N*-methyl derivative of the amino acid glycine, was identified in urine as a metabolite that differentiates prostate cancer from benign prostate tissue and as a biological marker that was greatly increased during prostate cancer progression.[18]

The molecular profile of cancer cells can also be used to target delivery of drugs to cancer cells. One way to take advantage of this is the use of nanoparticles bound to anticancer drugs to improve a chemotherapeutic response (see Baker *et al.*, Chapter 8 in this volume).

Nano- is a prefix for something that is a one billionth part (10^{-9} of a specified unit, e.g., nanometer, nanosecond, etc.). Baker and colleagues have designed dendrimer nanoparticles that target the intracellular folate receptor in cancer cells (that use an uptake mechanism for folate different from normal cells) to selectively deliver the anticancer drug methotrexate. This kills 100-fold more cultured cancer cells than free, unbound methotrexate.[19]

Brian Ross *et al.* (Chapter 9) discuss the use of molecular imaging as an early biomarker of cancer treatment response. There is a critical need for such biomarkers because it is very difficult for a clinical oncologist to know with any high degree of accuracy if and when a patient is responding to a therapeutic regimen. In this context, a biomarker is defined as a biochemical entity that can be measured in plasma, urine, ascites fluid, or other body fluids or in tissue by molecular imaging. These biomarkers can be used as an indicator of pathological processes or as an indicator of therapeutic response. Ross and colleagues have used an innovative technique of measuring diffusion coefficients in tissue, which is a measure of the degree of water Brownian motion (and hence, tissue "flexibility") as a measure of drug response. The diffusion coefficient varies between an untreated solid cancer (e.g., a glioblastoma) and one undergoing necrosis in response to treatment.

Not all characteristics of cells, including cancer cells, are totally regulated by the DNA sequence of genes; far from it. A number of regulatory events are modulated by DNA modifications such as DNA methylation and by modification of histones bound to chromatin such as methylation and acetylation. This phenomenon is called epigenetics. DNA methylation status and level of histone acetylation, for example, determine which genes may be expressed or repressed. Increased DNA methylation of tumor suppressor genes is frequently observed in cancer cells that have lost or diminished expression of these genes.

In Chapter 10, Wendell Weber discusses how epigenetics affects the normal function of cells and what epigenetic changes are seen during carcinogenesis and that occur in already transformed cancer cells. The chapter also describes the influence of epigenetic profiling in diagnosis, therapy response, and prognosis.

In spite of the fact that remarkable progress has been made over the past few decades in the understanding of the molecular, cellular, and tissue processes involved in precancer and during cancer progression, the development of effective and safe modalities for prevention of cancer remains slow, inefficient, and expensive. One of the problems is that treatment with cancer preventative agents usually means that individuals need to take a chemopreventive agent for many years, if not for a lifetime. This means that such treatments must be extremely nontoxic and safe. Secondly, pharmaceutical companies are loath to fund such clinical trials because the trials must go on for years if not decades and involve very large numbers of individuals. All this costs a lot of money. Of course, initially such clinical trials can be focused on high-risk individuals, which would cut down on the scale of the trial. Also, if accurate surrogate markers for effectiveness could be developed, the amount of time and number of patients could be reduced significantly.[20]

Similarly, the application of epidemiology to cancer prevention can provide significant information on cancer risk and the usefulness of preventive strategies.[21] For example, the elucidation of clues to cancer causation flows from observed associations of population exposures to tobacco, diet, environmental chemicals, and other exogenous factors with the development of cancer in patients. Indeed, the one real success story for cancer prevention has come about through smoking cessation.

Nevertheless, everyone wants the "magic pill" that will allow them to keep their life styles and at the same time enable them to avoid getting cancer. As numerous studies have shown, alterations in diet or ingestion of mega doses of antioxidants and vitamins have so far proved to be of little avail. As better molecular models of cancer initiation and promotion become available, the hope is that effective chemopreventive agents may be developed. The history of cancer prevention for the past 100 years has been recently reviewed.[22] Current clinical data for "cancer risk reductive interventions" (also previously known as "chemoprevention") are discussed by Kakarala and Brenner in Chapter 11.

References

1. Ruddon RW. *Cancer Biology*. 4th ed. London: Oxford University Press; 2007. p. 1–16.
2. Fearon ER, Vogelstein BA. A genetic model for colorectal tumorigenesis. *Cell* 1990;**61**:759–67.
3. Hoeïjmakers JHJ. Genome maintenance mechanisms for preventing cancer. *Nature* 2001;**411**:366–74.
4. Ruddon RW. *Cancer Biology*. 4th ed. London: Oxford University Press; 2007. p. 321–352.
5. Ruddon RW. *Cancer Biology*. 4th ed. London: Oxford University Press; 2007. p. 352–367.

6. Friend SH, Bernards R, Rogelj S, Weinberg RA, Rapaport JM, Albert DM, et al. A human DNA segment with properties of the gene that predisposes to retinoblastoma and osteosarcoma. *Nature* 1986;**323**:643–6.
7. Ma WW, Adjei AA. Novel agents on the horizon for cancer therapy. *CA Cancer J Clin* 2009;**59**:111–37.
8. DeVita Jr. VT, Chu E. A history of cancer chemotherapy. *Cancer Res* 2008;**68**:8643–53.
9. Hambley TW. Is anticancer drug development heading in the right direction? *Cancer Res* 2009;**69**:1259–62.
10. Hait WN. Targeted cancer therapeutics. *Cancer Res* 2009;**69**:1263–7.
11. Li X, Loberg R, Liao J, Ying C, Snyder LA, Pienta KJ, et al. A destructive cascade mediated by CCL2 facilitates prostate cancer growth in bone. *Cancer Res* 2009;**69**:1685–92.
12. Maher CA, Kumar-Sinha C, Cao X, Kalyana-Sundaram S, Han B, Jing X, et al. Transcriptome sequencing to detect gene fusions in cancer. *Nature* 2009;**458**:97–101.
13. Eliane J-P, Repollet M, Luker KE, Brown M, Rae JM, Dontu G, et al. Monitoring serial changes in circulating human breast cancer cells in murine xenograft models. *Cancer Res* 2008;**68**:5529–32.
14. Hayes DF, Cristofanilli M, Budd GT, Ellis MJ, Stopeck A, Miller MC, et al. Circulating tumor cells at each follow-up time point during therapy of metastatic breast cancer patients predict progression-free and overall survival. *Clin Cancer Res* 2006;**12**:4218–24.
15. Al-Hajj M, Wicha MS, Benito-Hernandez A, Morrison SJ, Clarke MF. Prospective identification of tumorigenic breast cancer cells. *Proc Natl Acad Sci USA* 2003;**100**:3983–8.
16. Ruddon RW. *Cancer Biology*. 4th ed. London: Oxford University Press; 2007. p. 376–379.
17. Nicholson JK, Lindon JC. Metabonomics. *Nature* 2008;**455**:1054–6.
18. Sreekumar A, Poisson LM, Rajendiran TM, Khan AP, Cao Q, Yu J, et al. Metabolomic profiles delineate potential role for sarcosine in prostate cancer progression. *Nature* 2009;**457**:910–4.
19. Quintana A, Raczka E, Piehler L, Lee I, Mhyc A, Majoros I, et al. Design and function of a dendrimer-based therapeutic nano-device targeted to tumor cells through the folate receptor. *Pharmacol Res* 2002;**19**:1310–6.
20. Kelloff GJ, Bast Jr. RC, Coffey DS, D'Amico AV, Kerbel RS, Park JW, et al. Biomarkers, surrogate end points, and the acceleration of drug development for cancer prevention and treatment: an update. *Clin Cancer Res* 2004;**10**:3881–4.
21. Greenwald P, Dunn BK. Landmarks in the history of cancer epidemiology. *Cancer Res* 2009;**69**:2151–62.
22. Lippman SM, Hawk ET. Cancer prevention: from 1727 to milestones of the past 100 years. *Cancer Res* 2009;**69**:5269–84.

Molecular Biology and Anticancer Drug Discovery

John S. Lazo

Department of Pharmacology and Chemical Biology, University of Pittsburgh Drug Discovery Institute and Cancer Institute, University of Pittsburgh, Pittsburgh, Pennsylvania, USA

The profound impact of molecular biology on the philosophy of how one should seek new cancer therapeutics cannot be overstated. It has enabled the discovery of unique drugs as well as the identification of new drug targets and biomarkers and the creation of powerful animal models. Nevertheless, the process of cancer drug discovery remains inherently complex and inefficient. This is partially a consequence of the requirement of any successful therapy to show differential effects toward tumor cells relative to nonmalignant cells. The goal of this chapter is to outline the impact of molecular biology on modern approaches to anticancer drug discovery and to highlight the continuing challenges.

I. Introduction

Contemporary cancer drug discovery largely focuses on identifying new therapies by leveraging advances in our understanding of the molecular biology of cancer. This has been facilitated by the enormous increase in our knowledge concerning the molecular pathogenesis of cancer in the past two decades, which

Progress in Molecular Biology
and Translational Science, Vol. 95
DOI: 10.1016/S1877-1173(10)95002-3

was stimulated by the profound investment in research funding from both private and public sources. In addition to great improvements in diagnostics and imaging, there are now more than 200 approved drugs for cancer; hundreds are in clinical development. Nonetheless, our ability to control, much less cure, cancer has been disappointing. Oncology has among the worst success rates (5%) for clinical drug development compared to all other major disease sectors.[1] Drug discovery and development in cancer remains a challenging, high-risk, inefficient, and complex process. Why is that so?

There are several reasons for the poor success in anticancer drug discovery and development. First, as mentioned in Chapter 1, tumors evolve through a multistep process in which cancer cells frequently acquire a large number of low-frequency genetic or epigenetic changes that can contribute to oncogenesis. These mutations affect the expression or functionality of oncogenes or tumor suppressors. The substantial number of alterations makes it challenging to distinguish the critical changes that are worthy of being targeted with drugs from the nonessential ones or "passengers" that are a legacy of the disease process. Some changes may have been important for the initial stages of oncogenesis but now are completely dispensable for the oncogenic phenotype and, therefore, unworthy of being drug targets. A second problem is that many of the oncogenes represent overexpressed versions of normal proteins or they contain mutations that are only subtly different from the normal gene product. Thus, selective targeting of cancer cells may be considered problematic. Generally, drugs affect gain-of-function abnormalities, making it at least conceptually challenging to design drugs that will replace the loss of a tumor suppressor. Finally, the fundamental genetic instability of cancer produces inherent plasticity in the disease that promotes rapid drug resistance.

There is no doubt that the cancer drug discovery landscape has benefited from developments in a host of disciplines and technologies. Advances in organic chemistry have facilitated the complete synthesis of complex natural products with powerful pharmacological activities.[2] Combinatorial chemistry and diversity-oriented synthesis have expanded greatly the numbers of drug-like compounds that now can be regularly probed for anticancer activities. Publically accessible databases of compounds and biological actions, such as PubChem, PharmGKB, ChemSpider, and ChemExper, provide information on tens of millions of compounds.[3] Advances in automated liquid handling platforms have enabled high throughput screening of hundreds of thousands of compounds for bioactivity. Similarly, high content platforms have empowered investigators to directly test compounds for cellular actions. Computational and structural biology have generated tools for molecular docking and visualization of drug or ligand interactions with putative targets.[4] Innovation in drug delivery and nanotechnology offer hope for better pharmacokinetics and drug targeting to tumors.[2] The emergence of systems biology as a discipline has created new tools to visualize signaling pathways controlling attributes that are vital for the neoplastic phenotype. How are these developments in various disciplines integrated into the process of discovering a new anticancer drug?

Fundamentally, there are two broad approaches used to identify new drugs: Forward Pharmacology and Reverse Pharmacology.[5] These approaches are not unique to anticancer drug discovery, but there is some merit to the argument that the field of oncology drug discovery has uniquely benefited from the advances of molecular biology with the cataloging of vast numbers of oncogenes and tumor suppressors, which has accelerated the transition from Forward Pharmacology to Reverse Pharmacology. The oldest strategy, Forward Pharmacology, relies on observing phenotypic changes of cells, organs, or organisms by chemical substances. Even before the acquisition of any fundamental understanding of the regulation of physiological systems or the nature of pathological processes, one could readily measure the inhibition of cancer cell proliferation in a culture dish. This molecular target-unbiased approach merely focused on the desired biology without any knowledge of how the process was regulated. There are at least six phenotypic attributes associated with tumor cells that can be examined in culture (Table I). Indeed, a majority of our existing clinically used anticancer drugs were discovered using a phenotypic assay and with little knowledge of how they might be selectively toxic to cancer cells. In contrast, Reverse Pharmacology is practiced with the goal of finding a modulator, usually an inhibitor, of a molecular target that is believed to be critical for the cancer phenotype.[5] Clearly, molecular biology has profoundly influenced this branch of pharmacology and anticancer

TABLE I
EXAMPLES OF PHENOTYPIC ATTRIBUTES OF CANCER CELLS AND COMPOUNDS THAT ALTER THE PROCESS

Phenotypic target	Proposed mechanism of action	Drug or compound	References
Proliferation	DNA damage, Cyclin-dependent kinase, Dihydrofolate reductase	Cisplatin, AT7519, P276-00, Methotrexate	6,7
Angiogenesis	Vascular endothelial growth factor A (VEGF-A), VEGF receptor	Bevacizumab, Sunitinib	8
Metastasis/invasion	Increased activating transcription factor 3 (ATF3)	Sulindac	9
Senescence	DNA damage	Doxorubicin	10
Stem cell	Potassium ionophore (?)	Salinomycin	11
Differentiation	Induction of CCAAT/enhancer binding protein	All-*trans* retinoic acid (Tretinoin)	12

drug discovery. Molecular biology and molecular genetics have been critical in identifying potential causal factors in neoplasia. Once identified, the molecular target (generally an enzyme) can be produced in a recombinant form permitting *in vitro* interrogation for inhibitors. This class of assays has the distinct advantage of generally being precise, rapid, and easier to conduct than assays requiring animal, organs, or cells. One can sometimes even determine at an atomic level the interactions between the compound and the target, making the generation of a chemical structure–activity relationship practical and the synthesis of more potent analogs rational. Target specificity and modes of inhibition assays are also conceptually easier to perform with a known target than with phenotypic assays. For molecular targeted anticancer drugs, it is desirable that most, if not all, vital cell types in the human body not depend on the target for survival. Otherwise, the drug is likely to have a narrow therapeutic window owing to the requirement for the target in normal cells. Although currently there is enormous enthusiasm for target-driven anticancer drug discovery, phenotypic screening approaches are regaining popularity in part because they can be designed to ensure compound entry into cells and stability, the failure of which can lead to the demise of compounds identified by more reductionist assays.

II. Phenotypic Targets

As mentioned above, efforts to identify compounds that kill cancer cells in culture have existed for more than half a century and have produced what are now commonly called the "cytotoxic drugs." Murine cell lines, such as P388 leukemia, L1210 leukemia, and B16 melanoma, dominated the early years of cancer cell testing both in culture and in mice, but with the successful culturing of human cancer cells, such as HeLa, and the development of immunosuppressed mice, such as nude mice, there was a gradual movement to the use of human-derived cancer cells and xenografts. The establishment by the National Cancer Institute in 1990 of a panel of 60 human cell lines (NCI60) for compound interrogation marked a major innovation in anticancer high throughput drug screening and profiling. The NCI60 comprises cell lines from nine cancer types: six leukemias, nine melanomas, nine non-small-cell lung, seven colorectal, six CNS, seven ovarian, six breast, two prostate, and eight renal. Despite this diversity, the NCI60 has not been valuable in predicting which specific human tumor types will be responsive to an experimental drug. Some have argued that a much larger cell panel, perhaps with thousands of cultured cell types, would be required to achieve that goal.[13] Others[14] have suggested that human tumor cell lines, which have been adapted to grow under the extremely artificial conditions of serum, culture medium, and high oxygen, acquire new mutations resulting in a population that is inherently different from primary tumors. After all, most tumors isolated from patients do not yield cells that can grow productively in culture. Moreover, cell lines adapt distinctive phenotypes after multiple passages in different laboratories. Thus, a HeLa cell that has been extensively passaged in one laboratory is not necessarily the same as that studied in another laboratory. Nonetheless, the US National Cancer Institute has established a large publicly available annotated database, which can be mined with programs, such as the COMPARE algorithm (see http://dtp.nci.nih.gov/docs/compare/compare.html), to examine the profile of drug sensitivity to the panel. Studies using this algorithm have clustered compounds with common mechanisms of action, identified compounds with new mechanisms of action, and exposed agents that are substrates for multiple drug resistance.[13] The landmark work of the National Cancer Institute has also spawned other tumor cell line panels in an effort to develop an ideal tumor cell screening panel.[13]

All cell proliferation or cell death assays depend on the assay protocol but most importantly on the endpoint that is being measured. Some assays simply measure the total number of cells on the plate after a fixed time point. Others determine the number of viable cells using, for example, Trypan blue exclusion or Alamar blue reduction. In most cases, however, there is little distinction between cell growth inhibition and cell death. There is a growing interest in developing assays that will differentiate between apoptosis, necrosis, necroptosis, anoikis, autophagy, and other processes associated with cell death. Obviously, it is desirable to establish preferential toxicity against cancer cells versus normal cells, but the question of selecting the appropriate normal cell type for the comparison often is difficult to answer.

Phenotypic cytotoxicity assays clearly are only biological models: the composition of the culture medium, the presence of fetal bovine serum, and the absence of other stromal cells make the environment extremely artificial. Additionally, the nonphysiological oxygen concentrations and the lack of a normal extracellular matrix and stromal cells prevent a precise recapitulation of the environment in which tumor cells find themselves *in vivo*. The importance of the tumor microenvironment in response to therapy is well established.[13] Studies with three-dimensional cultures, tumor spheroid systems, and mixed stromal cell substrates may provide improved platforms for future drug testing systems.

Although drugs identified with phenotypic assays or Forward Pharmacology are generally regarded in the current parlance as "cytotoxics" and not targeted, all probably function by influencing a molecular target. Potent inhibitors of dihydrofolate reductase, topoisomerase I, topoisomerase II, ribonucleotide reductase, microtubule stabilizers, and tubulin disruptors were first identified as inhibitors of cancer cell proliferation and only later were their mechanisms of action, namely molecular targets, determined (often with the assistance of molecular biology tools). For example, methotrexate binds to its target dihydrofolate reductase at picomolar concentrations and has orders of magnitude preference for this target over its secondary target, thymidylate synthase. Conversely, many drugs identified with molecular targets in mind, such as proteosome inhibitors (bortezomib) or histone deacetylase inhibitors (vorinostatin), are not easily distinguished from the traditional cytotoxics and might reasonably be referred to as "neocytotoxics." Nonetheless, the term "targeted therapy" has become closely associated with drugs that are identified with Reverse Pharmacology processes aimed at attacking enzymes thought to be essential for neoplasia.

In addition to cell proliferation as a phenotypic endpoint, there are a number of other cancer-associated properties that have been employed to identify potentially new and useful anticancer agents. These include assays to measure inhibition of tumor cell invasion and metastasis,[15] angiogenesis,[8] differentiation,[16,17] and senescence[10,18] (Table I). Successful metastasis only occurs if positive and negative regulatory mechanisms, governing angiogenesis, proteolysis, motility, host defense systems, and cellular adhesion events are concomitantly deregulated. Because metastasis is a complex multifactorial event, it is extremely difficult to evaluate *in vitro*. Therefore, there are few available robust assays and most of the research effort has focused on identifying agents that alter cell migration and invasion. One common cell migration and invasion assay uses a Boyden chamber or a Transwell device, which either lack or contain a coating of an extracellular matrix emulator.[15] Cells are placed in an upper chamber and the lower chamber is loaded with medium containing serum. The migration of cells into the lower chamber is then quantified at various times. There are a growing number of molecular targets associated with tumor cell invasion and metastases, which have been studied. These include enzymes involved in the cleavage of collagens, namely the metalloproteinases. The cell culture invasion assays have been quite useful to validate these compounds.

The recognition of the critical role of angiogenesis in cancer and the advanced knowledge concerning biological factors regulating tumor-stimulated angiogenesis and the emergence of clinically approved drugs have stimulated the development of a number of innovative assays aimed at discovering new antiangiogenic agents.[8] One of the most sophisticated automated *in vivo* assays exploits transgenic zebrafish to quantify the microscopic effects of drugs on blood vessel formation.[19]

Differentiation therapy is generally defined in oncology as the use of small or large molecular entities that induce the reversion of malignancy with the restoration of mature cells of the same histological lineage. Clinically, the poster child for differentiation therapy has been all-*trans* retinoic acid (ATRA) for the treatment of acute promyelocytic leukemia (APL), which has dramatically improved the prognosis for patients with this disease. Exposure of APL cells in culture to ATRA leads to differentiation into mature neutrophils, most likely due to the induction of CCAAT/enhancer binding proteins.[12] There is preclinical evidence that methyltransferase inhibitors, such as azacytidine, and the histone deactylase inhibitor sodium phenylbutrate can also induce differentiation in APL.[17] In addition, arsenic trioxide has been documented to induce differentiation in APL cells, leading to clinical remissions, although an incontrovertible mechanism of action is lacking.

In contrast to differentiation, tumor cell senescence is the loss of replicative ability that is accompanied by cellular enlargement, flattened morphology with an increased cytoplasmic area, increased granularity, extensive cytoplasmic vacuoles, and multinucleation. In addition, senescent cells have changes in gene expression that is largely independent of cell type, including an increase in the expression senescence-associated β-galactosidase, cyclin D, and cyclin-dependent kinase inhibitors, p21 and p16. Senescent cells also have a reduction in the tumor suppressor protein Rb. A surprisingly large number of clinically used, mechanistically distinct anticancer agents, including DNA alkylators, topoisomerase poisons, antimetabolites, and microtubule stabilizers, as well as irradiation, have been reported to induce senescence.[10] Typically, senescence is seen with cultured human tumor cells at low drug concentrations, while apoptosis is observed with higher drug concentrations, presumably because of the greater cell damage. For example, treatment of human hepatoma cells with a low concentration of the indirect DNA-damaging agent doxorubicin (< 100 nM) causes senescence, whereas higher concentrations (20 μM) produces frank apoptosis. It should be emphasized, however, that the importance of the phenomenon of senescence for the clinical activity of any drug remains to be determined.

Interest also exists in identifying drug-like compounds that can disrupt the essential dialogue between tumor cells and stromal cells, but currently, there are no clinically used agents that faithfully act in this manner. As will be mentioned in Chapter 6, there is considerable enthusiasm for identifying drugs that selectively kill cancer stem cells. Unfortunately, useful *in vitro* assays for identifying agents that specifically kill epithelial cancer stem cells have not been available because of the paucity of these cancer stem cells within tumor cell populations and their instability in culture. The recent development of an innovative assay by Gupta *et al.*[11] may have overcome this barrier, however. Using this assay, Gupta *et al.*[11] discovered several compounds, including salinomycin, with selective toxicity for breast cancer stem cells. Treatment of mice with salinomycin inhibited mammary tumor growth *in vivo* and induced increased epithelial differentiation of tumor cells. Thus, it may be possible to target what is thought to be a critical cancer cell population normally resistant to existing therapies.

III. Molecular Targets

A. General Issues

Remarkable advances in our fundamental knowledge of the oncogenic process catalyzed by the tools of molecular biology have enabled the current so-called "targeted" therapy approach. There is no doubt that a majority of all current anticancer drug discovery programs are now driven by molecular targets. Not only does this reductionist approach provide a pleasant feeling of rationality that is frequently absent in the phenotypic approach mentioned above but it also facilitates ligand–target interaction studies. The approach has also been propelled by the availability of enormous databases annotating gene expression profiles, protein expression, metabolomic patterns, cancer mutations, and other changes in human tumors. Finally, the clinical success of the tyrosine kinase inhibitor, imatinib, has been a primal stimulus for molecular targeted drug discovery.

The tools of molecular biology have armed investigators with many approaches to identify potential molecular targets. Perhaps the most powerful has been small interference RNA (siRNA) screens. MacKeigen *et al.*,[20] using siRNA libraries aimed at suppressing all of the known cellular kinases and phosphatases, demonstrated that 73 of the 650 known or putative kinases (11%) were essential for the survival of HeLa cells and 72 of the 222 known or putative protein phosphatases or their regulatory protein (32%) were essential for cell survival. Each of these enzymes represent at potential molecular targets for novel anticancer therapies. Clearly, there are many hurdles that must be overcome for the individual kinase or phosphatase to be fully validated as a cancer drug target, perhaps most importantly, the need to document selective toxicity against malignant cells versus nonmalignant cells. In some cases, it appears that the actions of even highly selective kinase inhibitors, such as the RAF inhibitors, are exquisitely dependent on the cellular context in which the kinase target is expressed.[21]

Before considering potential anticancer drug discovery targets, it might be instructive for the oncology community to review the lessons learned from another genomic-derived, target-based drug discovery program, namely the infectious disease field. The sequencing of the first complete bacterial genome in 1995 heralded a new era of hope for antibacterial drug discoverers, who now had powerful tools to search entire genomes for new antibacterial targets. Armed with knowledge about pathogen genomes, several companies launched genomics-derived, target-based approaches to screen for new classes of drugs with unique modes of action. One company identified 160 genes that were vital for pathogenic bacterial survival and, thus, were considered ideal drug targets. They cleverly narrowed the list further by excluding those genes that encoded proteins that had close human homologs. Over a 7-year period, they systematically screened many of these seemingly ideal targets with large 260,000–530,000 compound libraries for selective inhibitors.[22] Despite completing 67 high throughput screening campaigns on these highly attractive and novel antibacterial targets, no credible development candidate emerged. It is sobering to recognize that the disappointing results from that well-conceived genome-derived, target-based approach in the infectious disease sector by one company have been shared by other pharmaceutical and biotechnology companies. This reflects the significant challenges of target-based drug discovery even with seemingly "simple" diseases. How can failures be avoided or at least be reduced in our target-based strategies in oncology? The answer is not necessarily immediately obvious. Certainly, the composition of the chemical library is critical.[22] It is also clear from this exercise that selection of tractable pharmacologically validated targets is critical. Extracellular targets, such as receptors, transporters, exoenzymes, cell surface antigens, proteoglycans, and extracellular matrix components, are much more likely to be attractive than intracellular molecular targets. Being fully validated is more important than being new. Finally, whole cell assays probably provide better discriminators of drug-like compounds than simple *in vitro* assays for intracellular molecular targets because the compound must easily enter cells in order to be efficacious.

It is axiomatic that any successful anticancer therapy, whether designed to cure or to just control cancer, must demonstrate preferential actions on tumor cells relative to normal cells in the patient. This forms the basis of the therapeutic index. Almost all of the cytotoxic and most of the current molecularly targeted drugs have remarkably narrow therapeutic indices limiting the amount of drug that can be administered.

B. Drug Resistance

Soon after the first clinical trials with cytotoxic anticancer drugs in the middle of the last century, it became apparent that human tumors rapidly acquired resistance to chemotherapy. Indeed, modern combination chemotherapy, which uses multiple agents with distinct mechanisms of actions, is structured on the concept of deterring drug resistance. Despite considerable preclinical and clinical efforts, however, we have been remarkably unsuccessful in identifying drugs that reverse drug resistance. Drugs, such as verapamil, ZNRD1, biricodar, and INF271, which inhibit the multidrug resistance ABC cassette protein ABCB1 (also known as MDR1 or *p*-glycoprotein), have not emerged as viable clinical supplements for the existing chemotherapeutic substrates of this membrane transporter. This could be due to the lack of efficacy of the agents tested, unacceptable toxicity associated with inhibition of ABCB1 in critical normal tissues, or a relatively minor role for ABCB1 in human drug resistance. In contrast, an understanding of the mechanism of drug resistance has enabled the successful design and development of second generation anticancer drugs that retain efficacy in tumor cells with acquired or innate resistance to the first generation agents. For example, soratinib is effective against tumor cells resistant to imatinib. Premetrexate inhibits multiple folate-related targets, some of which are involved in resistance to methotrexate. Many of the second generation compounds now have embedded in them multiple targets, some of which are considered resistance mechanisms or participants in the oncogenic phenotype.

C. Oncogene and Nononcogene Addiction

If cancer-causing genes products are not unique to tumor cells but are exploited normal proteins required for development or normal cell signaling, then how can one target them? This is the conundrum facing the medical oncologist. Surprisingly, basic cell biological studies have revealed a remarkable dependency of cancer cells on the expression of certain oncogenes or the lack of expression of some tumor suppressors. Weinstein[23] first coined the term "oncogene addiction" as the enhanced state of dependency on the presence of an oncogene; he later extended the term to include tumor suppressor gene hypersensitivity.[24] The Weinstein hypothesis is that the malignant phenotype of a cell with one or more oncogenes or with the absence of a tumor suppressor protein mandates an extensive alteration in cellular signaling processes. Once the adaptation occurs, tumor maintenance depends on the continued activity of an oncogene or the absence of the tumor suppressor. Tumor addiction to oncogenes could help explain the sensitivity of cancer cells to some targeted therapies (Table II). This anthropomorphic notion of "tumor addiction" has been expanded to a dependency on existing intracellular pathways that include participating proteins that may themselves not be oncogenes or tumor suppressors.[6] For example, elevation in a proximal tyrosine kinase activity in an oncogenic signaling pathway, such as the mitogen-activating protein kinase pathway, could lead to making other members of the pathway, which are not themselves oncogenes, required because they now are rate-limiting in the pathway. Alternatively, cells with an oncogene-mediated hyperactive mitogen-activating protein kinase pathway may become more dependent on normal protein phosphatases to regulate the hyperactivity. These alterations, which are required for the maintenance of the oncogenic state, become very logical and attractive molecular targets.

Many oncogenes paradoxically induce both pro-mitogenic signals and anti-mitogenic or pro-apoptotic signals. Oncogene activation presumably stimulates growth because the former is dominant. As long as the oncogene signal is sustained, the proliferative signal, which promotes mitogenesis, survival, or both, dominates. According to the theory, acute inactivation of the oncogene

TABLE II
EXAMPLES OF ONCOGENE AND NONONCOGENE ADDICTION TARGETS AND AGENTS

Target	Agent
Oncogene targets	
BCR-Abl, c-Kit	Imatinib, Dasatinib
Retinoic acid receptor, Retinoic X receptor	ATRA (Tretinoin)
ERBB2	Trastuzumab
Epidermal growth factor receptor	Erlotinib, Gefitinib
MDM2	Nutlin-3
BCL-2	Oblimersen, ABT-737
PI3K	GDC-0941, BEZ235
Nononcogene targets	
mTOR	Temsirolimus, Everolimus
Topoisomerase I	Topotecan, Irinotecan
Dihydrofolate reductase	Methotrexate, Pemetrexed
Vascular endothelial growth factor	Bevacizumab
PARP1	AZD2281, ABT-888, AG014699
Heat shock protein-90	Geldanamycin, Alvespimycin
Mitotic Spindles	Vinblastine, Vincristine, Paclitaxel, Epothilone B
Chk1	PF-00477736

causes growth cessation or death when the antimitogenic or pro-apoptotic signals decay more slowly than the mitogenic signals, perhaps due to differences in mRNA and/or protein half-life.

There are now multiple examples supporting the notion that cancer cells *in vitro* are dependent on or addicted to certain activated oncogenes; similar evidence is mounting for tumor cells becoming addicted to the loss of the functionality of tumor suppressor genes. Oncogene addiction is thought to be the basis of the success of the tyrosine kinase inhibitor imatinib for CML in which BCR–ABL is the oncogene and gastrointestinal stromal tumors in which KIT is the oncogene.

Considering the large number of identified oncogenes and tumor suppressors, how does one determine what to target? This is particularly important when one remembers that tumor onset and progression frequently mandate the accumulation of multiple genetic and epigenetic lesions. Computational and experimental methods are emerging that permit the identification of the so-called "dominant" and "recessive" changes.[25] Using the Met receptor as a model system, Bertotti *et al.*[25] combined multiplex phosphoproteomics, genome-wide expression profiling, and functional assays to deconvolute signatures that are required for sustained oncogene addiction, and these would seem to represent attractive drug targets. This is one of what will no doubt be many attempts to deconvolute the profile associated with oncogene-addicted human tumors.

The concept of oncogene addiction occurring *in vivo* has been established with several oncogenes. For example, lymphomas, osteosarcomas, and papillomas formed with a chemically induced MYC oncogene were shown to regress when MYC expression is terminated by removal of the inducing chemical.[26] A comparable phenomenon has been documented with HRAS, BCL-ABL, and KRAS oncogenes. Similarly, reintroduction of a tumor suppressor, such as p53, to tumors in which there is loss of p53 through gene deletion, mutation, or epigenetic silencing, causes tumor regression.[27] It is critical for the oncology community to identify the complete constellation of oncogene and tumor suppressors to which human tumors are addicted. Some might consider the term "addiction" to be too vague and not sufficiently quantitative to ensure therapeutic success. Indeed, the term is not necessarily favored in the field of drug abuse, because it describes a human behavior rather than a physiologically measurable phenomenon. In the context of cancer, however, the long-term regression of tumors after the withdrawal of an oncogene or the reexpression of a tumor suppressor appears to provide a sound quantitative method to validate an addiction participant.

How can we pharmacologically exploit the concept of tumor suppressor addiction? Loss of the tumor suppressors Rb, p16, p21, or p27—all cause an increase in cyclin-dependent kinase activity, which permits cell cycle progression. Therefore, one could theorize that such tumors should be hypersensitive to inhibitors of cyclin-dependent kinases. Tumors that have lost the tumor suppressor and lipid phosphatase PTEN, which normally acts to restrain PI3K signaling, should theoretically be hypersensitive to PI3K inhibitors.

D. Synthetic Lethal

If oncogene expression and tumor suppressor loss alter the networks upon which tumor cells depend, is there a strategy to uncover the proteins and pathways? An unbiased experimental approach termed synthetic lethal for identifying molecular targets or chemosensitivity nodes provides a powerful conceptual framework. The synthetic lethal strategy evolved from yeast genetic studies in which investigators were interested in identifying genes that collaborated with yeast functionality. Two genes were labeled to be "synthetic lethal" if deletion or mutation of either gene alone did not result in death but loss of both genes caused death (Table III). Yeast synthetic lethal interactions have most commonly been described for loss-of-function alleles but they also can involve gain-of-function genes. In *Saccharomyces cerevisiae*, approximately 20% of genes are individually essential, but synthetic lethal screens reveal

TABLE III
Genetic Synthetic Lethality

	Gene A	Gene B	Phenotype
Deletion/Mutation	–	–	Viable
Deletion/Mutation	+	–	Viable
Deletion/Mutation	–	+	Viable
Deletion/Mutation	+	+	Death

TABLE IV
Mixed Drug and Genetic Synthetic Lethality

	Drug A	Gene A	Phenotype
Treatment/Deletion/Mutation	–	–	Viable
Treatment/Deletion/Mutation	+	–	Viable
Treatment/Deletion/Mutation	–	+	Viable
Treatment/Deletion/Mutation	+	+	Death

that a large fraction of the remaining genes can collaborate in death processes.[28,29] Many of the targets for existing cytotoxic agents are essential gene products such as tubulin or topoisomerase. First principles would teach us that they are not attractive because they are essential for both normal and malignant cells; this is why existing compounds have such a narrow therapeutic index. If the synthetic lethal concept from yeast genetic screens could be extended to human tumor cells, which have preexisting mutations, deletions, and epigenetic abnormalities, it might be possible to identify innovative drug targets. It might even be possible that some of our current cytotoxic drugs are efficacious because they inadvertently exploit synthetic lethality. In this model, one of the gene deletions or mutations is replaced with a drug (Table IV). One attractive example is the sensitivity of cancers carrying BRCA1 mutations, which is responsible for homologous DNA recombination repair. These tumors are exceptionally sensitive to DNA-damaging and cross-linking agents such as temozolomide and cisplatin.[6] These tumors are heavily reliant on orthogonal forms of DNA repair, such base-excision repair that is mediated by poly-ADP-ribose polumerase (PARP1). PARP1 facilitates repair of single-strand breaks. In nonmalignant cells, endogenous DNA damage generated by PARP1 inhibition by agents, such as ABT 888, is well tolerated because of functional compensation from homologous recombination-mediated repairs. This is an

TABLE V
DRUG–DRUG SYNTHETIC LETHALITY

	Drug A	Drug B	Phenotype
Treatment	−	−	Viable
Treatment	+	−	Viable
Treatment	−	+	Viable
Treatment	+	+	Death

excellent example of nononcogene addiction of tumor cells with BRCA1 mutations on PARP1. The synthetic lethality model can also be extended by replacing a gene deletion with a second drug (Table V). This could obviously be a powerful strategy for identifying novel drug combinations in an unbiased manner (see below).

Molecular biology has provided a long list of human oncogenes and tumor suppressors. A number of groups have established isogenic pairs of normal and malignant cell lines suitable for screening of compound libraries or plant, soil, and marine extracts, which are complex mixtures of Natural Products, for agents that selectively inhibit cells with cancer-relevant genetic alterations in a high throughput manner. For example, compounds have been identified in marine sponge extracts that preferentially inhibited $Trp53^{-/-}$ mouse embryonic fibroblast proliferation relative to wild-type mouse embryonic fibroblasts.[30] Colon cancer cells with the KRAS mutant alleles have been created and engineered to produce a blue fluorescent protein reporter. An isogenic counterpart was also created in which the mutant KRAS allele was eliminated by homologous recombination; these latter cells were engineered to produce yellow fluorescent protein. This allowed the investigators to monitor the drug treated pair of cells for differential killing using the ratio of blue/yellow fluorescence.[31] The availability of the isogenic pair enabled the investigators to identify a novel cytosine nucleoside, which was selectively toxic to cells containing mutant KRAS. The general synthetic lethal approach has been expanded further based on the recent recognition of the minimal genes required for transformation. Thus, primary human cells have been transformed with human telomerase reverse transcriptase, RAS, and oncoproteins that affect pRB, p53, and/or protein phosphatase 2A (PP2A).[32] A number of compounds that were identified included several clinically useful inhibitors of topoisomerase I and II, which help validate the notion that existing cytotoxic agents found by Forward Pharmacology were likely discovered unknowingly with a synthetic lethal approach. Obviously, it is also possible to exploit the synthetic lethal strategy for new drug combinations.

E. Combination Chemotherapy

Almost all regimens for the treatment of cancer rely on drug combinations. They have been created almost exclusively by empirically amalgamating drugs with different mechanisms of action, untoward effects, and mechanisms of resistance. Molecular biology has provided new information about the interconnections of the anticancer targets that could provide some theoretical guidance for the formulation of drug combinations. Nonetheless, even with the introduction of targeted therapies where there is a quantifiable pharmacological effect, it is most common to develop drug combinations in the clinic by escalating the individual agents to the maximum tolerated dose until the aggregate effect of toxicity is considered to be excessive. This approach presumes, perhaps incorrectly, that the maximum tolerated dose is the maximum effective dose. There is no incontrovertible clinical evidence to support this notion. It has been extremely challenging to provide any *in vitro* guidance for this important question, especially as we see the use of combinations containing multiple agents. We know each of the individual drugs is likely to exert multiple pharmacological effects at different concentrations and each has a unique pharmacokinetic profile that alters the plasma and tumor concentration over time. Thus, it is important to address this issue, because there is compelling evidence that the anticancer effect of drug combinations can be profoundly dependent on the ratio of the individual drugs.[33]

F. Nontraditional Targets

As mentioned above, the primary targets for almost all current drugs are the catalytic site of enzymes or ligand binding sites. Many enzymes, however, depend on allosteric regulation to be fully active, and there is a growing interest in disrupting this process. We know that the intracellular spatial, kinetic, and substrate specificity regulation of enzymes can be controlled by multiple protein–protein interactions (PPIs). These are widely believed to control all major cellular functions, including the maintenance of DNA topology, DNA replication, mRNA transcription, protein translation and its proper folding, the assembly and maintenance of morphological structures, and the regulation of cellular metabolism and signaling pathways. The protein interactions comprise identical or homotypic binding and nonidentical or heterotypic binding between two or more polypeptides; they can be characterized thermodynamically and kinetically as anything from high affinity stable contacts to low affinity transient interactions. Some are critical for functionality and others are gratuitous.

It is easy to imagine how the critical PPIs might provide valuable potential molecular targets that would be mechanistically distinct from the common active sites or ligand binding drug discovery targets. While PPIs might superficially seem attractive, until recently it was widely believed that PPI surfaces were not druggable because they were large and amorphous.[34] Refined structural elucidation of many protein–protein complexes indicate that the protein-binding interfaces can be dissected into discrete patches with critical residues, termed "hot spots," that are vital for binding.[34] Biochemical PPI assays exploiting methods, such as co-precipitation, co-purification, affinity chromatography, ultracentrifugation, nuclear magnetic resonance, surface plasmon resonance, mass spectrometry, and isothermal titration calorimetry, have been developed.[34,35] Additionally, higher throughput biochemical PPI screening assay formats have been developed, including capture ELISAs, cell surface binding, fluorescence polarization, time-resolved fluorescence, ligand-induced changes in thermal stability, bead-based technologies, such as AlphaScreen and Luminex, and resonance energy transfer assays.[34,35] Cell-based PPIs assays exploiting two-hybrid transcriptional reporter systems have also been developed in mammalian cells. Several imaging-based high content screening assay formats have also been employed to study PPIs, such as the co-localization of fluorescently labeled protein partners, fluorescent resonance energy transfer measurements between PPI partners bearing donor and acceptor fluorescently labeled protein, protein fragment complementation assays, and positional biosensors that measure the PPI-induced redistribution of fluorescently labeled protein partners.

Perhaps one of the most successful discovery examples of PPI inhibitors has been with p53 and MDM2. The p53 tumor suppressor is a transcriptional activator that regulates the expression of target genes involved in processes that serve to restrict the initiation, progression, or survival of cancer cells. p53 controls cell cycle arrest, DNA damage repair, apoptosis, senescence, metastasis, and angiogenesis. In more than half of all human cancers, p53 is inactivated by single-point missense mutation in the DNA-binding domain, resulting in deficient regulation of p53 target genes. In a significant proportion of the remaining cancers where wild-type p53 is functional, MDM2 is overexpressed and blocks the tumor suppressor activity of p53. In this case, MDM2 binds to the N-terminal transactivation domain of p53, thereby inhibiting activation of p53 target genes. Because MDM2 contains a E3-ubiquitin-ligase activity, it also tags p53 for degradation by the proteasome.[36] The structure of the protein–protein binding interfaces between p53 and MDM2 has been comprehensively mapped and characterized; there are three amino acid residues in the p53 N-terminus (Phe19, Trp23, and Leu26) that bind to a small hydrophobic pocket on the surface of MDM2. This protein–protein binding interface was targeted using a combination of structure-based drug design and an SPR assay that measured the disruption of the p53-MDM2 binding.[37] A *cis*-imidazoline compound, called nutlin, was identified as a small molecule that occupied the hydrophobic pocket on the surface of MDM2 disrupting the p53-MDM2 PPI. Nutlin caused stabilization of p53 in cells and suppressed tumor growth in xenograft models.[37] The initial success of nutlin has spawned several other groups to investigate and identify small molecules that focus on disrupting the p53–MDM2 interaction. Benzodiazepene analogs such as TDP665759 have been shown to disrupt p53–MDM2 interactions *in vitro* with low micromolar concentrations.[36] They have also been found to suppress the growth of p53 wild-type tumor cells in culture. Treatment of normal mice with TDP665759 produces an increase in the p53-dependent transcript, cyclin-dependent kinase inhibitor p21 in liver. It is therapeutically interesting that TDP665759 had a synergistic growth-inhibitory effect with doxorubicin both in culture and in xenografts. Potent spirooxindoles such as MI-215 have been synthesized that inhibit the p53–MDM2 interaction *in vitro* in multiple tumor cell lines. MI-219 displays good pharmacokinetic and bioavailability properties in mice and induces tumor regression without obvious reported untoward effects.

There is a vast array of potential PPI targets. It seems likely that the initial success with p53-MDM2 PPI inhibitors will stimulate the field. There are already efforts to generate PPI inhibitors for MYC-MAX[38] and BCL2-BAX[39] heterodimers. One would expect that some of these will enter advance preclinical development in the near future. In addition, others are attempting to produce inhibitors of protein–DNA interactions using analogs of natural antibiotics like netropsin and distamycin.[40] These compounds might be able to regulate transcription in tumor cells but the preclinical development of these has been quite slow probably due to poor uptake of the compounds and the lack of the requisite cancer selectivity.

Additional strategies are being pursued to reactivate p53 tumor suppressor activity as potential cancer therapeutics. These include searching for compounds that inhibit MDM2 ubiquitin E3 ligase activity or restore the thermal stability and DNA-binding activity of p53 DNA-binding mutants. The core domain of wild-type p53 is somewhat unstable, with a melting temperature of 44 °C and a short half-life of nine minutes. Some of the known p53 mutations add to the thermal instability of the protein, leading to the loss of DNA binding and the p53 response. Small molecules that could stabilize p53 in its active biological conformation and, thus, restore its DNA-binding functionality, could potentially rescue wild-type p53 function. Proof-of-principle studies were provided with antibodies that bind the C-terminus of p53 and with synthetic peptides derived from the C-terminal domain.[41] The antibodies and peptides showed a stimulatory effect on the DNA-binding ability of p53. Because these antibodies and peptides lack ideal pharmacological attributes, drug discovery campaigns have been conducted to identify small molecules that would emulate the antibodies and peptides.

One compound, CP31398, was identified using an *in vitro* assay for conformational refolding of mutant p53 to a wild-type shape.[41] CP31398 induced the expression of both reporter and endogenous p53 target genes. More recent results suggest that CP31398 did not act as a PPI inhibitor but rather intercalated with DNA, altering and destabilizing the p53-DNA core domain complex. CP31398 also decreased sequence-specific DNA binding of wild-type p53 and the His273 mutant p53. It also produced cellular toxicity that was independent of mutant p53. Therefore, there continues to be a challenge with respect to reactivating mutant p53 with small molecules. If such compounds were discovered, they will probably need to target a large fraction of the known mutation observed in p53 to be clinically interesting.

IV. Other Contemporary Issues in Anticancer Drug Discovery

In addition to the complexities of anticancer drug discovery mentioned above, it is useful to consider some of the other aspects that will affect how future drugs will be identified and developed. There is enormous interest in the potential role of fusing diagnostics with therapeutics leading to personalized medicines. There is considerable debate about the statistical metrics that should be used to distinguish a driver from a noncontributing passenger mutation. Clearly, there is enormous complexity in the patterns of mutations in tumors among individuals and even within an individual.

Every tumor harbors a complex combination of low-frequency mutations thought to drive the cancer phenotype. Does this mean one will require thousands of drugs to successfully treated tumors or are there chemosensitivity nodes that can be exploited? Should we be seeking highly potent and selective agents or multitargeted and promiscuous inhibitors? Highly potent and selective enzyme inhibitors have been the mantra for cancer drug discoverers for the last two decades. With the recognition that the complex signaling pathways often have collateral participants that can function as redundant elements when the targeted protein is inactivated, multitargeted drugs become more attractive.

What are the appropriate animal models for studying anticancer drugs? The lack of reliable animal models for predicting the clinical efficacy of new anticancer drugs has long plagued oncology.[13,42,43] The most promising solution remains highly controversial. While xenografts of human cancer cell lines have been useful for determining the pharmacological properties of new agents, including their pharmacokinetics, they have not been very successful in predicting drug efficacy.[44] It is possible that orthotopic models will improve the predictability of human xenografts. An extracellular matrix-embedded hollow fiber assay system has been promoted as one convenient approach to emulate the three-dimensional tumor microenvironment *in vivo*.[45] Recent reports have argued in favor of genetically engineered mouse models as being superior to xenograft approaches.[46–48] Genetically engineered mouse models of cancer appear to recapitulate many aspects of sporadic human tumors. Nevertheless, the research community has been slow to adopt genetically engineered mouse models of cancer for drug discovery. There are several practical and philosophical reasons for this reluctance. One concern with the genetically engineered mouse models is that the resulting tumors are thought to be genetically too homogenous. While this uniformity may conceptually be beneficial for preclinical evaluation of drugs, some believe that the tumors lack the complex genetic heterogeneity seen in human tumors, which may be an important factor determining the responsiveness to therapy. Using genetically engineered mouse models of cancer required the establishment of a substantial infrastructure and commitment of resources. In general, there is the need to use a larger number of mice than with xenograft models to generate the appropriate preclinical dataset. Recent studies with conditional mutant KRAS-driven non-small-cell lung cancer and pancreatic adenocarcinoma models illustrate the power and perhaps the superiority of genetically engineered mouse models.[48] The investigators examined standard-of-care agents both alone and in combination with clinically used agents that block the Epidermal Growth Factor receptor and the Vascular Endothelial Growth Factor using noninvasive imaging methods and conventional survival endpoints. They found a concordance between the preclinical results and existing clinical data leading them to conclude that the genetically engineered mouse platform model human disease well and that they had predictive value for anti-cancer drug development.

V. Conclusions

Molecular biology has irreversibly altered the manner in which new anti-cancer drugs are discovered. It has provided numerous validated molecular targets that have been the subject of high throughput drug screens. It has also provided reagents to conduct imaging-based cellular screens and to generate new animal models of cancer. Although the inherent complexities of human cancer remain, the new knowledge obtained about the processes permitting the development of cancer should enable the creation of more efficacious drugs for the treatment of hematological and solid tumors.

References

1. Kamb A, Wee S, Lengauer C. Why is cancer drug discovery so difficult? *Nat Rev Drug Discov* 2007;**6**:115–20.
2. Pors K, Goldberg FW, Leamon CP, Rigby AC, Snyder SA, Falconer RA. The changing landscape of cancer drug discovery: a challenge to the medicinal chemist of tomorrow. *Drug Discov Today* 2009;**14**:1045–50.
3. Lazo JS, Brady LS, Dingledine RJ. Building a pharmacological lexicon: small molecule discovery in academia. *Mol Pharmacol* 2007;**72**:1–7.

4. Bakan A, Lazo JS, Wipf P, Brummond KM, Bahar I. Toward a molecular understanding of the interaction of dual specificity phosphatases with substrates: insights from structure-based modeling and high throughput screening. *Curr Med Chem* 2008;**15**:2536–44.
5. Lazo JS. Rear-view mirrors and crystal balls: a brief reflection on drug discovery. *Mol Interv* 2008;**8**:60–3.
6. Luo J, Solimini NL, Elledge SJ. Principles of cancer therapy: oncogene and non-oncogene addiction. *Cell* 2009;**136**:823–37.
7. Lapenna S, Giordano A. Cell cycle kinases as therapeutic targets for cancer. *Nat Rev Drug Discov* 2009;**8**:547–66.
8. Kerbel RS. Antiangiogenic therapy: a universal chemosensitization strategy for cancer? *Science* 2006;**312**:1171–5.
9. Bottone Jr. FG, Moon Y, Kim JS, Alston-Mills B, Ishibashi M, et al. The anti-invasive activity of cyclooxygenase inhibitors is regulated by the transcription factor ATF3 (activating transcription factor 3). *Mol Cancer Ther* 2005;**4**:693–703.
10. Schmitt CA. Cellular senescence and cancer treatment. *Biochim Biophys Acta* 2007;**1775**:5–20.
11. Gupta PB, Onder TT, Jiang G, Tao K, Kuperwasser C, Weinberg RA, et al. Identification of selective inhibitors of cancer stem cells by high-throughput screening. *Cell* 2009;**138**:645–59.
12. Truong BT, Lee YJ, Lodie TA, Park DJ, Perrotti D, Watanabe N, et al. CCAAT/enhancer binding proteins repress the leukemic phenotype of acute myeloid leukemia. *Blood* 2003;**101**:1141–8.
13. Sharma SV, Haber DA, Settleman J. Cell line-based platforms to evaluate the therapeutic efficacy of candidate anticancer agents. *Nat Rev Cancer* 2010;**10**:241–53.
14. Borrell B. How accurate are cancer cell lines? *Nature* 2010;**463**:245.
15. Achiwa H, Lazo JS. PRL-1 tyrosine phosphatase regulates c-Src levels, adherence, and invasion in human lung cancer cells. *Cancer Res* 2007;**67**:643–50.
16. Lazo JS, Ruddon RW. Neurite extensions and malignancy of murine neuroblastoma cells after treatment with prostaglandin E1 and papaverine. *J Natl Cancer Inst* 1977;**59**:137–43.
17. Spira AJ, Carducci MA. Differentiation therapy. *Curr Opin Pharmacol* 2003;**3**:338–43.
18. Tierno MB, Kitchens CA, Petrik B, Graham TH, Wipf P, Xu F, et al. Microtubule binding and disruption and induction of premature senescence by Disorazole C1. *J Pharmacol Exp Ther* 2009;**328**:715–22.
19. Vogt A, Cholewinski A, Shen S, Nelson S, Lazo JS, Tsang M, et al. Automated image-based phenotypic analysis in zebrafish embryos. *Dev Dynamics* 2009;**238**:656–63.
20. MacKeigan JP, Murphy LO, Blenis J. Sensitized RNAi screen of human kinases and phosphatases identified new regulators of apoptosis and chemoresistance. *Nat Cell Biol* 2005;**7**:591–600.
21. Poulikakos PI, Zhang C, Bollag G, Shokat KM, Rosen N. RAF inhibitors transactivate RAF dimers and ERK signalling in cells with wild-type BRAF. *Nature* 2010;**464**:427–31.
22. Payne DJ, Gwynn MN, Holmes DJ, Pompliano DL. Drugs for bad bugs: confronting the challenges of antibacterial discovery. *Nat Rev Drug Discov* 2007;**6**:29–40.
23. Weinstein IB. Cancer. Addition to oncogenes-the Achilles heal of cancer. *Science* 2002;**297**:63–4.
24. Weinstein IB, Joe A. Oncogene addiction. *Cancer Res* 2008;**68**:3077–80.
25. Bertotti A, Burbridge MF, Gastaldi S, Galimi F, Torti D, Medico E, et al. Only a subset of met-activated pathways are required to sustain oncogene addiction. *Sci Signal* 2009;**2**:er11.
26. Felsher DW, Bishop JM. Transient excess of MYC activity can elicit genomic instability and tumorigenesis. *Proc Natl Acad Sci USA* 1999;**96**:3940–4.
27. Ventura A, Kirsch DG, McLaughlin ME, Tuveson DA, Grimm J, Lintault L, et al. Restoration of p53 function leads to tumor regression in vivo. *Nature* 2007;**445**:661–5.

28. Tong AH, Evangelista M, Parsons AB, Xu H, Bader GD, Pagé N, et al. Systematic genetic analysis with ordered arrays of yeast deletion mutants. *Science* 2001;**294**:2364–8.
29. Kaelin J, Kaelin WG. The concept of synthetic lethality in the context of anticancer therapy. *Nat Rev Cancer* 2005;**5**:689–98.
30. Stockwell BR, Haggarty SJ, Schreiber SL. High-throughput screening of small molecules in miniaturized mammalian cell-based assays involving post-translational modifications. *Chem Biol* 1999;**6**:71–83.
31. Torrance CJ, Agrawal V, Vogelstein B, Kinzler KW. Use of isogenic human cancer cells for high-throughput screening and drug discovery. *Nat Biotech* 2001;**19**:940–5.
32. Dolma S, Lessnick SL, Hahn WC, Stockwell BR. Identification of genotype-selective anti-tumor agents using synthetic lethal chemical screening in engineered human tumor cells. *Cancer Cell* 2003;**3**:285–96.
33. Tardi PG, Dos Santos N, Harasym TO, Johnstone SA, Zisman N, Tsang AW, et al. Drug ratio-dependent antitumor activity of irinotecan and cisplatin combinations in vitro and in vivo. *Mol Cancer Ther* 2009;**8**:2266–75.
34. Colas P. High-throughput screening assays to discover small-molecule inhibitors of protein interactions. *Curr Drug Discov Technol* 2008;**5**:190–9.
35. Pagliaro L, Felding J, Audouze K, Nielsen SJ, Terry RB, Krog-Jensen C, et al. Emerging classes of protein–protein interaction inhibitors and new tools for their development. *Curr Opin Chem Biol* 2004;**8**:442–9.
36. Brown CJ, Lain S, Verma CS, Fersht AR, Lane DP. Awakening guardian angels: drugging the p53 pathway. *Nat Rev Cancer* 2009;**9**:862–73.
37. Vassilev LT, Vu BT, Graves B, Carvajal D, Podlaski F, Filipovic Z, et al. In vivo activation of the p53 pathway by small-molecule antagonists of MDM2. *Science* 2004;**303**:844–8.
38. Yin XY, Giap C, Lazo JS, Prochownik EV. Low molecular weight inhibitors of Myc-Max interaction and function. *Oncogene* 2003;**22**:6151–9.
39. Oltersdorf T, Elmore SW, Shoemaker AR, Armstrong RC, Augeri DJ, Belli BA, et al. An inhibitor of Bcl-2 family proteins induces regression of solid tumours. *Nature* 2005;**435**:677–81.
40. Reddy BS, Sharma SK, Lown JW. Recent developments in sequence selective minor groove DNA effectors. *Curr Med Chem* 2001;**8**:475–508.
41. Foster BA, Coffey HA, Morin MJ, Rastinejad F. Pharmacological rescue of mutant p53 conformation and function. *Science* 1999;**286**:2507–10.
42. Kelland LR. Of mice and men: values and liabilities of the athymic nude mouse model in anticancer drug development. *Eur J Cancer* 2004;**40**:827–36.
43. Klein S, Levitzki A. Targeted cancer therapy: promise and reality. *Adv Cancer Res* 2007;**97**:295–319.
44. Suggitt M, Bibby MC. 50 years of preclinical anticancer drug screening: empirical to target-driven approaches. *Clin Cancer Res* 2005;**11**:971–81.
45. Hall LA, Krauthauser CM, Wexler RS, Hollingshead MG, Slee AM, Kerr JS. The hollow fiber assay: continued characterization with novel approaches. *Anticancer Res* 2000;**20**:903–11.
46. Olive KP, Jacobetz MA, Davidson CJ, Gopinathan A, McIntyre D, Honess D, et al. Inhibition of Hedgehog signaling enhances delivery of chemotherapy in a mouse model of pancreatic cancer. *Science* 2009;**324**:1457–61.
47. Olive KP, Tuveson DA. The use of targeted mouse models for preclinical testing of novel cancer therapeutics. *Clin Cancer Res* 2006;**12**:5277–87.
48. Singh M, Lima A, Molina R, Hamilton P, Clermont AC, Devasthali V, et al. Assessing therapeutic responses in Kras mutant cancers using genetically engineered mouse models. *Nat Biotech* 2010;**28**:585–93.

Targeting Chemokine (C-C motif) Ligand 2 (CCL2) as an Example of Translation of Cancer Molecular Biology to the Clinic

Jian Zhang,[*,†] Lalit Patel,[*,†] and Kenneth J. Pienta[*,†]

[*]*Department of Medicine, Michigan Center for Translational Pathology and the University of Michigan Comprehensive Cancer Center, University of Michigan, Ann Arbor, Michigan, USA*

[†]*Department of Urology, Michigan Center for Translational Pathology and the University of Michigan Comprehensive Cancer Center, University of Michigan, Ann Arbor, Michigan, USA*

Chemokines are a family of small and secreted proteins that play pleiotropic roles in inflammation-related pathological diseases, including cancer. Among the identified 50 human chemokines, chemokine (C-C motif) ligand 2 (CCL2) is of particular importance in cancer development since it serves as one of the key mediators of interactions between tumor and host cells. CCL2 is produced

Progress in Molecular Biology
and Translational Science, Vol. 95
DOI: 10.1016/S1877-1173(10)95003-5

1877-1173/10 $35.00

by cancer cells and multiple different host cells within the tumor microenvironment. CCL2 mediates tumorigenesis in many different cancer types. For example, CCL2 has been reported to promote prostate cancer cell proliferation, migration, invasion, and survival, via binding to its functional receptor CCR2. Furthermore, CCL2 induces the recruitment of macrophages and induces angiogenesis and matrix remodeling. Targeting CCL2 has been demonstrated as an effective therapeutic approach in preclinical prostate cancer models, and currently, neutralizing monoclonal antibody against CCL2 has entered into clinical trials in prostate cancer. In this chapter, targeting CCL2 in prostate cancer will be used as an example to show translation of laboratory findings from cancer molecular biology to the clinic.

I. Biology of CCL2

A. CCL2 Basics

Chemokines, a family of chemoattractant cytokines, are classified into four sub-families as CXC, CC, CX3C, and C based upon the number and location of the cysteine residues at the N-terminus of the protein. Chemokine (C-C motif) ligand 2 (CCL2), also known as monocyte chemotactic protein-1 (MCP-1), is a small, secreted protein that belongs to the CC chemokine family. CCL2 was purified and cloned in 1989 from human gliomas and myelomonocytic cells by two independent research groups based on its ability to chemoattract monocytes.[1,2] Subsequent to its cloning, it was confirmed that this protein was also identical to the product of the human *JE* gene. The *JE* gene, originally identified in mouse fibroblasts, is a platelet-derived growth factor (PDGF)-inducible gene. Since then, CCL2 has been shown to display chemoattractic activity for not only monocytes, but also memory T cells, natural killer (NK) cells, and perhaps dendritic cells, resulting in recruiting of these cells to sites of tissue injury and inflammatory responses.[3,4] The human *CCL2* cDNA encodes a 99 amino acid residue precursor protein with a hydrophobic signal peptide of 23 amino acids and a mature peptide of 76 amino acids.[5,6] The *CCL2* gene is located on the chromosome 17 where many of the genes of the CC chemokine family are located. The mouse or rat CCL2 gene has about 75% homology to humans. CCL2 functions through binding to a functional chemokine receptor CCR2, although it also binds to CCR4.[7] The roles of CCL2 have been implicated in the pathogenesis of various diseases that associate with monocyte infiltration, for example, rheumatoid arthritis, atherosclerosis, and multiple types of cancer [see review in Ref. 8].

B. CCL2 Expression

CCL2 is expressed in a wide array of tissues. It can be produced by multiple cell types, including fibroblasts, macrophages, lymphocytes, astrocytes, mast cells, endothelial cells, and osteoblasts.[1,9–14] In addition, CCL2 can also be produced by a variety of human and murine malignant cells [see review in Refs. 15–17]. In prostate cancer, determined by immunohistochemical staining on a human tissue microarray, CCL2 positive staining was located mostly in epithelial and fibromuscular stromal cells[18]; however, CCL2 positive staining was also observed in the extracellular areas surrounding neoplastic glands and epithelial cells,[18] suggesting both autocrine and paracrine production of CCL2 in the tumor microenvironment. In at least one report, CCL2 expression levels positively correlated with Gleason score (a measure of tumor aggressiveness) and pathologic stages.[18–20] CCL2 production has been determined by enzyme-linked immunosorbent assay (ELISA) in conditioned media collected from prostate cancer cell lines and compared to primary prostate epithelial cells.[18] Prostate cancer cells produce higher amounts of CCL2 as compared to non-malignant prostate epithelial cells.[18] Higher production of CCL2 at the sites of bone metastases was demonstrated by a clinically related report in which Loberg *et al.* collected tumor bone metastatic and normal bone specimens from vertebral lesions in three patients with prostate cancer.[21] Total protein lysates were isolated and analyzed by cytokine array. Elevated CCL2 production was identified in the tumor-bone microenvironment compared to normal bone microenvironment, suggesting that CCL2 plays a critical role in prostate tumorigenesis in bone metastases.[21] *In vitro*, it has been further reported that human osteoblasts and bone marrow endothelial cells produced higher amount of CCL2 compared to prostate epithelial cells.[13] CCL2 can also be produced by osteoclasts.[22–25] Its production can be induced by Receptor Activator of NF-κB ligand (RANKL) and tumor necrosis factor alpha (TNFα).[22–25] To date, it remains unclear which cell type(s) play a major role in the production of CCL2 in the tumor microenvironment.

C. CCL2 Functions

CCL2 functions as a chemoattractant through binding to its receptor on monocytes, macrophages, and lymphocytes [see review in Ref. 26]. The existence of CCL2/CCR2 axis has been validated using CCR2 knockout animals.[27] In acute inflammatory response, CCL2 has been shown to actively recruit monocytes to the site of inflammation.[28,29] CCL2 also plays important roles in T-cell immunity, and CCL2 expression is associated with Th2 response.[30–32] For example, CCL2 is overproduced in an animal model of Th2 immune-mediated asthma.[33] CCL2 is a potent factor in the polarization of Th0 cells toward a Th2 phenotype.[34] It has been demonstrated that CCL2 induced interleukin-4 (IL-4) production through direct activation of IL-4 promoter in T cells.[35] CCL2 knockout mice demonstrated impaired Th2 immunity, but intact Th1 immune response.[36] There is a large body of evidence showing the crucial roles of CCL2/CCR2 axis in various chronic inflammatory conditions that are associated with macrophage infiltration such as atherosclerosis,[37,38] multiple sclerosis,[39] rheumatoid arthritis,[40] glomerulonephritis,[41] pulmonary hypertension,[42] and pulmonary fibrosis.[43] It has been reported that CCL2 is expressed at the site of tooth eruption and bone resorption.[44] It has been shown that CCL2 can induce osteoclast maturation and formation as represented by the formation of Tartrate Resistant Acid Phosphatase (TRAP)-positive, multinuclear cells in the absence of RANKL, but the produced osteoclasts lack the ability to cause bone resorption.[25,45] In addition, CCL2 has been reported to cause the degranulation of basophiles and migration of mast cells.[46,47] This effect can be enhanced by pretreatment with IL-3 and other cytokines.[48] In tumor development, stimulation of infiltrating macrophages has been shown to augment antitumor activity or promote tumor development, depending on the type of cancer [see review in Ref. 49].

D. CCR2, The Functional Receptor for CCL2

CCR2, a G protein-coupled receptor, is the key functional receptor for CCL2. The activation of the ligand-receptor binding leads to the activation of intracellular signaling cascades that mediate chemotactic response. CCR2 has both pro-inflammatory (mediated by APC and T cells) and anti-inflammatory (mediated by regulatory T cells) effects [see review in Ref. 8]. CCR2-deficient mice have been shown to have altered inflammatory responses in an allergic asthma model.[50]

CCR2 can be expressed by both hematopoietic cells such as macrophages and nonhematopoietic cells such as endothelial cells,[51] fibroblasts,[52] and mesenchymal stem cells.[53] In prostate cancer, it has been shown that CCR2 expression correlates with prostate cancer progression and metastasis as determined by *in situ* immunohistochemical staining.[54,55] Specifically, CCR2 expression correlated with Gleason score and pathological stages; however, these published reports were not able to distinguish which cell type(s) may produce CCR2 transcripts in the metastatic sites. In prostate cancer cell lines, differential expression of CCR2 has been reported.[21,54] In particular, more aggressive cancer cells express greater levels of CCR2 compared with the less aggressive cancer cells or nonneoplastic cells.[54] These findings suggest that the CCL2/CCR2 axis may be a target for prostate cancer treatment. It has been recognized that CCR2 antagonists are potential therapeutic agents in preventing, treating, or ameliorating a CCR2-mediated inflammatory syndrome or disease such as psoriasis, uveitis, rheumatoid arthritis, multiple sclerosis, asthma, obesity, and chronic obstructive pulmonary disease [see review in Ref. 56]. In a prostate cancer study, a CCR2 antagonist has been shown to diminish the prostate cancer cell proliferation and invasion *in vitro*.[18]

Recently, CCR2 was reported as a key factor in balancing the bone remodeling process.[57] It was shown that CCR2 knockout mice had high bone mass and stability (biomechanical properties by compression) due to a decrease in number, size, and function of osteoclasts.[57] RANK expression is diminished in CCR2 knockout mice, and CCL2 enhances RANK expression via NFkB and ERK1/2 pathways[57]; therefore, CCR2 could become a therapeutic target in postmenopausal bone loss.

E. Regulation of CCL2 and CCR2

1. CCL2 Regulation

CCL2 production is elevated in various diseases that are associated with chronic inflammation and macrophage infiltration. Similar to other inflammation-associated soluble factors, CCL2 production can be induced by oxidative stress, cytokines, and growth factors. In prostate cancer, it has been reported that serum CCL2 levels are elevated in patients with skeletal metastases compared to localized prostate cancer.[58] One of the key regulators has been suggested to be parathyroid hormone-related protein (PTHrP), a 141-amino acid protein that has limited homology to PTH, but binds the same receptor as PTH with similar biological activity.[59,60] It was also reported that PTHrP is highly expressed in metastatic bone lesions, compared to a moderate expression on localized prostate cancer tissues and cell lines.[61–63] PTHrP has been shown to enhance bone metastases in animal models of both prostate cancer[64] and breast cancer.[65–68] It has been demonstrated that PTHrP treatment of osteoblastic cells upregulates CCL2[13,69] which can be blocked by a PTHrP antagonist,[70,71] suggesting that prostate cancer cell-derived PTHrP plays an important role in elevation of osteoblast-derived CCL2. It was further demonstrated that PTHrP induces CCL2 production in human bone marrow endothelial (HBME) cells.[13] Investigation of the mechanisms through which CCL2 is upregulated in osteoblasts and HBME cells is needed to provide a better understanding of the roles of tumor microenvironment in skeletal metastasis.

The *CCL2* gene is regulated in a tissue-specific and stimulus-specific manner.[13] In its promoter region, there are a pair of C/EBP-binding sites (− 2591 to − 2579; − 3118 to − 3107) that are important for the response to insulin activation of PI3K,[72] a pair of NFkB-binding sites (− 2639 to − 2630; − 2612 to − 2603) that are important for interleukin-1 (IL-1) and TNFα stimulation, and a GC box (− 64 to − 59) that binds Sp1 and is important for CCL2 basal expression.[73] Using CCL2 reporters that were transfected into human fetal osteoblast (hFOB) cells, PTHrP induced CCL2 promoter activity in hFOB cells through NFkB and C/EBP activation[13]; however, it remains unknown (a) whether PTHrP can also upregulate the CCL2 promoter-luciferase reporter set in the HBME cells, as well as in prostate cancer cell lines, (b) whether PTHrP does so via NFkB and C/EBP activations as with hFOB, (c) whether other stimuli such as IL-6 or TNFα or RANKL can also upregulate this promoter in these cells, and, (d) if so, via which transcription factor, NFkB and/or C/EBP. It is worthy to note that TNFa has been reported to induce CCL2 expression in sensory neurons[74–76] and RANKL induces CCL2 expression in osteoclasts at transcriptional levels.[25,57]

Recently, a gene expression profile in individual human prostate cancer specimens before and after exposure to chemotherapy (docetaxel treatment) was determined.[77] In that study, several genes, including CCL2, were upregulated after chemotherapy treatment. In addition, docetaxel was shown to induce CCL2 expression in prostate cancer cell lines *in vitro*. CCL2-specific siRNA inhibited prostate cancer cell proliferation and enhanced the growth-inhibitory effect of low-dose docetaxel. This protective effect of CCL2 was associated with activation of the ERK/MAP kinase and PI3K/AKT.[77] These findings suggest a mechanism of chemotherapy resistance mediated by cellular stress responses involving the induction of CCL2 expression and indicate that inhibiting CCL2 activity could enhance therapeutic responses to taxane-based therapy. To date, elevated serum CCL2 levels in cancer patients have not been demonstrated to be specific enough to correlate with disease activity, tumor burden, or response to therapy.

2. CCR2 Regulation

Little is known about the regulation of the *CCR2* gene in normal or cancerous tissues. It is downregulated as monocytes move down the macrophage differentiation pathway while other related chemokine receptors are not.[78] IFNγ+ M-CSF or PMA + ionomycin downregulate CCR2 expression in monocytes, and this can be replicated with a − 1220/+115 hCCR2 promoter-pGL3 luciferase reporter.[78] Peroxisome proliferator-activated receptor-gamma (PPAR-γ) ligands (i.e., Rosiglitazone) also downregulate CCR2 in circulating monocytes, while cholesterol slightly upregulated CCR2.[79] While proinflammatory cytokines rapidly reduce CCR2 expression in monocytes, they upregulate CCR2 expression in the brain.[80] Constitutive tissue-specific expression of CCR2 in THP-1 monocyte cells has been shown to be dependent upon a 31-bp region (− 89 to − 59) adjacent to the TATA box that contains an Oct-1 binding site and a pair of tandem C/EBP binding sites located in the 5′UTR (+ 50 to + 77 bp).[81] In addition to the Oct-1 and C/EBP binding sites that function in monocyte CCR2 expression, the hCCR2 5′ flank and UTR contains an array of possible binding sites for PPAR/RXR, SREBP, GATA, STAT, NFAT, and AP-1. It remains unknown whether these sequences are sufficient for positive regulation in prostate cancer cells, but the monocyte expression of CCR2 suggests that they should function in osteoclast precursor cells, for example, RAW264.7 cells.

II. CCL2 in Prostate Cancer

A. Proliferation and Survival

CCL2 has been shown to promote prostate cancer cell proliferation and invasion *in vitro* via the phophatidylinositol 3-kinase (PI3K)/AKT signaling pathway.[18,21] CCL2 induces Akt phosphorylation in prostate cancer cells. In addition, CCL2 stimulates p70-S6 kinase phosphorylation, a downstream target of Akt, resulting in actin rearrangement, a critical step in the formation of the migratory phenotype of the tumor cells.[21] Activation of p70-S6 kinase alters the actin cytoskeleton microstructure[82] and the binding of CCL2, and CCR2 has been linked to the actin skeleton through interactions with FOUNT, a novel activator of C-C chemokines.[83–85] Constitutive activation of this PI3K/AKT pathway has been implicated in prostate cancer progression,[86–89] and activation of AKT pathway further induces survival benefits for the tumor cells.[90] The later protective roles of CCL2 have shown to upregulate survivin gene expression; therefore, CCL2 plays an important role for the survival of tumor cells, possibly through reduction of autophagosome formation.[90,91] Survivin has been demonstrated to serve as a key molecule that protects the tumor cells from autophagic death.[90,92]

B. Angiogenesis

Chemokines play an important role in the maintenance of hematopoietic homeostasis, regulation of cell proliferation, tissue morphogenesis, and angiogenesis.[93] In human breast cancer, it was reported that CCL2 levels in the excised breast cancer tissue were correlated significantly with the levels of angiogenic factors, including vascular endothelial growth factor (VEGF), thymidine phosphorylase, TNFa, and IL-8.[94] Transfection of colon cancer cells with the CCL2 gene induces angiogenesis in a murine model.[95] It has been demonstrated that both CCL2 and VEGF expression positively correlates with TAM infiltration and angiogenesis in breast cancer.[96] In prostate cancer, it has been reported that CCL2 induces tumor cells to produce the pro-angiogenic factor VEGF-A, which indirectly induces sprout formation in human bone marrow endothelial cells.[71] *In vivo*, it has been shown that administration of neutralizing antibody against CCL2 significantly reduces tumor blood vessel density and decreases the prostate cancer tumor burden (Fig. 1)[71,97]; therefore, CCL2 is a key mediator of tumor angiogenesis.

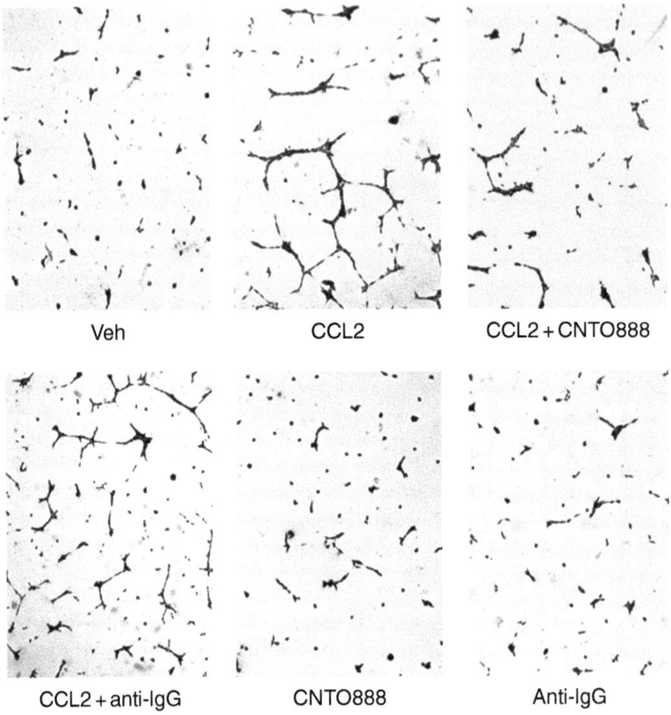

FIG. 1. CCL2 has pro-angiogenic effects on endothelial cells.[71]

C. Migration, Invasion, and Metastasis

Metastasis is a multi-step process that begins with a cell that has a phenotype which allows higher motility, invasion through tissue layers in the primary tumor, survival in the circulation, as well as establishment, expansion, and growth in a "hostile" microenvironment at the distant target organ. In this process, the interaction between the tumor cells and the tumor microenvironment has become the focus of therapeutic opportunities.[98–104] There is growing evidence to suggest that CCL2 may act directly on the cancer epithelial cells and regulate the migration and invasion, thus enhancing metastatic potential. In particular, CCL2 has been shown to be a potent chemotactic factor, in both autocrine and paracrine manners, for prostate cancer cells *in vitro*.[18,21] It has been also reported that CCL2 acts as a chemoattractant for myeloma cell migration.[24,105,106] In addition, CCL2 has been suggested as a predictor of colorectal cancer hepatic metastasis and poor survival.[107]

D. Macrophage Infiltration

Chronic inflammation has been recognized as a risk factor in a variety of cancer types, including prostate cancer.[108,109] In the tumor microenvironment, inflammatory components present as a large number of infiltrating macrophages.[110] These macrophages are most likely derived from circulating monocytic lineage[111] and have been termed tumor-associated macrophages (TAMs). It is well accepted that TAMs provide a direct link between inflammation and malignancy.[49,112–119] TAMs are increasingly recognized as important regulators to cancer progression and metastasis, both in positive or negative ways.[49,116–119] TAMs can be stimulated to inhibit tumor growth,[49,117–119] but on the other hand, they can produce soluble mediators, such as CCL2, to directly or indirectly promote cancer epithelial cell proliferation in the tumor microenvironment.[110] CCL2 has been suggested to be one of the crucial determinants of human tumor macrophage content,[120,121] and a large number of TAMs have been identified in prostate cancer tissues compared to nonneoplastic tissues.[122] In human breast cancer, CCL2 concentration from the excised tumor was associated with TAM accumulation.[94,123] TAM infiltration has been demonstrated in preclinical animal models in prostate,[97,122,124] breast,[123,125] cervix,[126] and pancreatic carcinoma.[127,128] Recently, CCL2-overexpressing breast cancer cells were shown to promote macrophage chemotaxis *in vivo* in a mouse model.[125] In colon cancer, it has been shown that blocking TNFa/TNF receptor axis reduces colorectal carcinogenesis, intracolonic macrophage infiltration, and CCL2 mRNA expression.[129]

Macrophages are classified as M1-type, the "classically activated macrophages" and M2-type, the "alternatively activated macrophages" that contribute to the tumor progression.[110,114] Furthermore, CCL2, in concert with IL-6, has been shown to promote survival of human CD11b positive peripheral blood mononuclear cells and induce M2-type macrophage polarization.[122] The mechanistic studies have shown that both cytokines inhibit the apoptotic cleavage of caspase-8 and promote enhanced autophagic activity to protect the monocyte recruited to the tumor (Fig. 2).[122] CCL2's antitumor activity has been demonstrated *in vitro* by its ability to augment cytostatic activity against tumor cells upon addition to macrophages in tissue culture, and by its ability to induce FAS ligand expression in cultured endometrial stromal cells, thus driving the cells to apoptosis.[130–133] These findings suggest that targeting both CCL2 and IL-6 could become an optional therapeutic approach in prostate cancer treatment.

E. Osteoclast Recruitment and Activation

Several cancer types, including prostate, breast, and lung cancer, preferentially metastasize to the skeleton. Unlike breast cancer that usually cause bone resorptive osteolysis, prostate cancer bone metastatic lesions usually represent

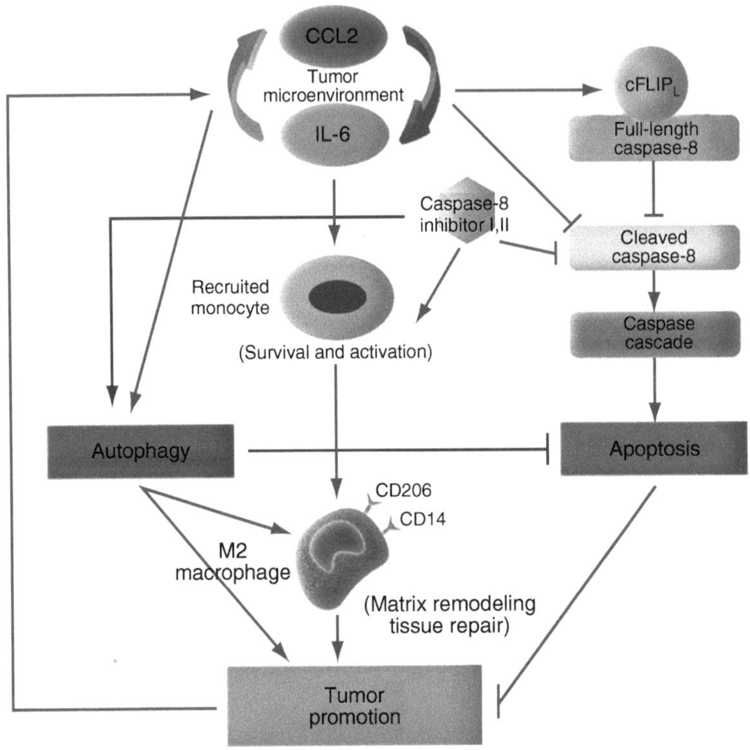

FIG. 2. Proposed mechanism by which CCL2 and IL-6 potentiate tumor progression by protecting tumor infiltrating monocytes and inducing their differentiation toward M2-type macrophages. CCL2 and IL-6 induce each other and boost their expression in the tumor microenvironment. Both cytokines inhibit the apoptotic cleavage of caspase-8 and promote enhanced autophagic activity to protect the monocytes recruited to the tumor and, at the same time, induce their differentiation toward M2-type macrophages.[122]

a mixture of predominant osteoblastic response (woven bone formation) and osteolytic (bone resorptive) activity. It is well documented that the tumor-induced osteoclast activity may be a prerequisite for prostate cancer establishment as micrometastases in the bone microenvironment [see review in Ref. 134]. CCL2 participates in the recruitment of osteoclast precursor cells, osteoclast activation, and maturation.[25,45,135]

Bone is a dynamic tissue, being continuously remodeled by the coordinated actions of osteoclasts and osteoblasts. Osteoblasts, the bone-forming cells, are derived originally from pluripotent mesenchymal stem cells. Osteoblasts express protease-activated receptor-1 and VEGF.[136] Osteoclasts arise from monocytic precursor cells. Cytokines and hormones regulate osteoclast formation and activity. Most osteotropic factors, such as PTH, 1,25-hydroxy vitamin D3, TNFα, and prostaglandins, promote osteoclast formation mediated by induction of RANKL on marrow stromal cells and osteoblasts. Osteoprotegerin (OPG), a decoy receptor for RANKL, inhibits osteoclast formation and activity. In prostate cancer bone metastasis, the number of osteoclasts is increased because of cytokines, and chemokines produced or induced by tumor cells increase the ratio of RANKL to OPG, and thereby increase osteoclast formation.[137] Tumors metastatic to bone increase osteoclast numbers, which in turn, increase bone destruction and create space for the tumor growth.

In prostate cancer animal models, CCL2 has been shown to induce osteoclast differentiation and maturation using human bone marrow monocytes[138] and peripheral blood mononuclear cells.[139] It has been shown that *in vivo*, CCL2 mediates prostate cancer cell-induced osteoclast activity.[71,140] It has been reported that CCL2 knockdown in prostate cancer cells by shRNA methodology significantly reduced the conditioned media (collected from the CCL2 knockdown cells compared to the scramble control knockdown cells)-induced osteoclast maturation *in vitro* and diminished partially prostate cancer growth in bone in an intratibial-injection mouse model.[58] Similar inhibition of tumor growth in bone was demonstrated in other cancer types as in lung cancer[141] and breast cancer.[125] Another chemokine, RANTES, can also stimulate the differentiation of preosteoclasts into mature osteoclasts.[45]

III. CCL2 Development as a Therapeutic Target

A. Preclinical Animal Models

Due to the bench findings that CCL2 directly stimulate the tumor cell growth, survival, invasion, and migration, and indirectly promotes macrophage infiltration and osteoclast maturation and activity [reviewed in Refs. 15,16,142], anti-CCL2 has been tested as a therapeutic option in preclinical animal models in prostate cancer[55,71] and breast cancer.[125] In one study, using neutralizing antibodies against human CCL2 (CNTO888) and/or the mouse CCL2 (C1142), it was shown that treatment with C1142 attenuated prostate cancer PC3 cell-mediated overall tumor burden in an intracardiac injection model by 96% at weeks post the tumor cells injection, although targeting the human CCL2 derived from the human tumor cells only modestly inhibited the tumor growth.[97] This suggests that host-derived CCL2 plays a prominent role in tumor progression and metastasis.[21,97] In addition, it was shown that the combination of the chemotherapy drug docetaxel with the neutralizing antibodies against CCL2 further reduced the tumor growth compared to using either treatment alone.[97] Future work is needed to delineate the role of host-derived and the tumor-derived CCL2 in prostate cancer tumorigenesis and metastasis.

The effects of CCR2 deficiency from host cells have been initially tested in a preclinical colon cancer model.[143] In that study, murine colon adenocarcinoma colon 26 cells were intraportally injected into wild-type and CCR2 knockout mice.[143] After 10 days, the number and size of tumor foci were significantly reduced in CCR2-deficient mice, with a concomitant reduction in the macrophage accumulation in the tumor, compared to wild-type mice, although tumor formation occurred at similar rates in wild-type and CCR2-deficient mice up to 10 days after tumor cell injection. Further evaluation is still needed to determine the effects of CCL2 in this model.

TAMs are a pivotal component of stromal cells in the tumor microenvironment, releasing a variety of growth factors, proteolytic enzymes, cytokines, and inflammatory mediators; therefore, TAMs have been implicated as therapeutic targets. It was demonstrated, for example, that extratumoral macrophages promote tumor and vascular growth in an orthotopic rat prostate tumor model.[144] This study was performed using Dunning R-3327 AT-1 rat prostate tumor cells that have been demonstrated to produce CCL2 *in vivo*. Recently, clodronate- or other bisphosphonate liposome-mediated macrophage depletion regimens have been tested in preclinical models.[144–147] It has been shown that combined with antibodies against VEGF, depletion of TAMs was accompanied by significant inhibition of tumor growth in tumor models.[146] In a human melanoma xenograft model, it has been shown that targeting TAMs by clodronate liposomes reduced the tumor growth associated with less angiogenesis and macrophage infiltration.[145]

In a preclinical lung cancer model, it was very recently shown that the combination of neutralizing antibody against CCL2 with a tumor vaccine effectively augmented efficacy with enhanced reduction in tumor volume and cures of approximately 50% of the tumors.[148] The combined therapy generated more total intratumoral CD8+ T cells that were more activated and more antitumor antigen-specific, as measured by tetramer evaluation. A potential mechanism is suggested by the reduction in intratumoral T-regulatory cells in this model. These findings suggest that CCL2 is indeed a key chemokine that mediates immune suppression in the tumors.

B. Clinical Studies

Neutralizing monoclonal antibody against CCL2 (CNTO888) has entered into Phase I trials for safety and Phase II clinical tests in prostate cancer to test efficacy. In the near future, a combination of therapeutic approaches such as neutralizing antibody against CCL2/CCR2 axis or small molecule CCR2 antagonist with other therapeutic approaches such as chemotherapy or other immune modulators should provide new therapeutic approaches for prostate and other cancers.

IV. Conflicting Reports on the Roles of CCL2 in Cancer

There are a few conflicting reports on the role of CCL2 in tumor progression and metastasis. Specifically, it was reported that in 4T1E breast cancer parental cells, CCL2 is highly expressed but shows low bone metastasis based on incidence of metastasis and histology from a group of 10–11 animals.[149] The 4T1E-derived 4T1E/M3 cells have extremely low levels of CCL2 expression, but have a high incidence of metastasis.[149] Transfection of the *CCL2* gene into a highly metastatic murine colon carcinoma CT-26 cells reduced tumorigenicity and suppressed metastatic potential.[150] The same group presented a similar report in the renal adenocarcinoma cell line RENCA.[151] TAM-associated modulation of tumor growth *in vivo* in Panc-1, a human pancreatic carcinoma cell line, and WM115, a human melanoma cell line, was reported using antibodies.[152] Addition of CCL2 did not have effects on cancer cell proliferation and apoptosis.[153] Monocyte recruitment was blocked using a rat monoclonal antibody against murine CD11b, and CCL2 was blocked using a mouse monoclonal antibody against human CCL2. In another study, the CCL2 gene was introduced into Chinese hamster ovary (CHO) cells, and the ability of transfected cells to form tumors *in vivo* was evaluated.[154] Clones transfected with human CCL2 or murine CCL2, via mammalian expression vector did not show significant differences in growth rate *in vivo* compared with clones transfected with vectors. Finally, it was demonstrated that when nontumorigenic melanoma cells were transfected with CCL2 expression vector and injected *in vivo*, high levels of CCL2 production resulted in extensive monocyte invasion and elimination of the tumor growth, and low-level CCL2 production resulted in a low level of monocyte recruitment and promotion of tumor angiogenesis.[155] Based on the aforementioned reports, CCL2 needs further investigation in different tumor types. Like CCL2, CCR2 has low sequence homology between human and lower species which could raise the question of whether blocking the CCL2/CCR2 axis could generate sufficient clinical efficacy in certain diseases, as predicted by many preclinical animal modes.[56]

V. Conclusions

It has been demonstrated that CCL2 promotes prostate cancer tumorigenesis and metastasis via (1) direct promotional effects on tumor cell growth and survival, and (2) indirect modulatory effects on macrophage infiltration and

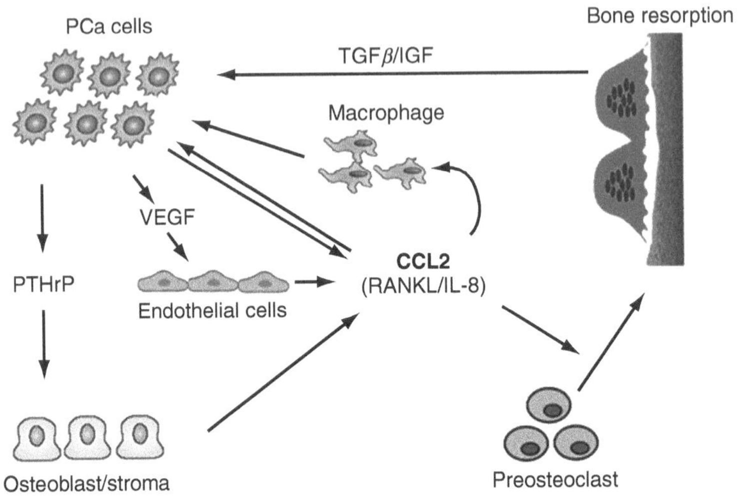

FIG. 3. Roles of chemokine (C-C motif) ligand 2 (CCL2) in prostate cancer cells and the bone microenvironment. The CCL2/CCR2 axis has been identified as an important contributor to prostate tumorigenesis. CCL2, by binding to its receptor CCR2, directly stimulates prostate cancer cell proliferation, survival, and migration. In addition, CCL2 contributes to the development of metastases in the bone microenvironment by stimulating macrophage recruitment and education, angiogenesis, and activation of osteoclastogenesis. Prostate cancer cells produce parathyroid hormone-related peptide (PTHrP), which stimulates CCL2 expression from osteoblasts. CCL2 appears to mediate the interactions between tumor-derived factors, such as PTHrP, and host-derived chemokines and cytokines, which act together to promote metastatic tumor growth in bone. IGF = insulin-like growth factor; IL-8 = interleukin 8; PCa = prostate cancer; RANKL = receptor activator of NF-κB ligand; TGF-β = transforming growth factor β; VEGF = vascular endothelial growth factor.[16]

osteoclast activation[16] (Fig. 3). The laboratory investigations of CCL2 have been successfully translated to the clinic: (1) studies of CCL2 were initiated by the discovery of high production of CCL2 in bone metastasis compared to primary prostate cancer; (2) the functional roles of CCL2 in the tumor development *in vitro* were investigated; (3) the roles of CCL2 *in vivo* in preclinical animal models were confirmed; and (4) neutralizing antibodies against CCL2 are currently being evaluated in clinical trials. CCL2 can serve as an example of other chemokines and cytokines for therapeutic development in cancer.

ACKNOWLEDGMENTS

We thank Rhonda Hotchkin for her editing.

References

1. Yoshimura T, Robinson EA, Tanaka S, Appella E, Kuratsu J, Leonard EJ. Purification and amino acid analysis of two human glioma-derived monocyte chemoattractants. *J Exp Med* 1989;**169**:1449–59.
2. Matsushima K, Larsen CG, DuBois GC, Oppenheim JJ. Purification and characterization of a novel monocyte chemotactic and activating factor produced by a human myelomonocytic cell line. *J Exp Med* 1989;**169**:1485–90.
3. Balkwill F. Cancer and the chemokine network. *Nat Rev Cancer* 2004;**4**:540–50.
4. Charo IF, Taubman MB. Chemokines in the pathogenesis of vascular disease. *Circ Res* 2004;**95**:858–66.
5. Cochran BH, Reffel AC, Stiles CD. Molecular cloning of gene sequences regulated by platelet-derived growth factor. *Cell* 1983;**33**:939–47.
6. Van Coillie E, Van Damme J, Opdenakker G. The MCP/eotaxin subfamily of CC chemokines. *Cytokine Growth Factor Rev* 1999;**10**:61–86.
7. Graves DT, Jiang Y, Valente AJ. Regulated expression of MCP-1 by osteoblastic cells in vitro and in vivo. *Histol Histopathol* 1999;**14**:1347–54.
8. Deshmane SL, Kremlev S, Amini S, Sawaya BE. Monocyte chemoattractant protein-1 (MCP-1): an overview. *J Interferon Cytokine Res* 2009;**29**:313–26.
9. Barna BP, Pettay J, Barnett GH, Zhou P, Iwasaki K, Estes ML. Regulation of monocyte chemoattractant protein-1 expression in adult human non-neoplastic astrocytes is sensitive to tumor necrosis factor (TNF) or antibody to the 55-kDa TNF receptor. *J Neuroimmunol* 1994;**50**:101–7.
10. Brown Z, Strieter RM, Neild GH, Thompson RC, Kunkel SL, Westwick J. IL-1 receptor antagonist inhibits monocyte chemotactic peptide 1 generation by human mesangial cells. *Kidney Int* 1992;**42**:95–101.
11. Cushing SD, Berliner JA, Valente AJ, Territo MC, Navab M, Parhami F, et al. Minimally modified low density lipoprotein induces monocyte chemotactic protein 1 in human endothelial cells and smooth muscle cells. *Proc Natl Acad Sci USA* 1990;**87**:5134–8.
12. Standiford TJ, Kunkel SL, Phan SH, Rollins BJ, Strieter RM. Alveolar macrophage-derived cytokines induce monocyte chemoattractant protein-1 expression from human pulmonary type II-like epithelial cells. *J Biol Chem* 1991;**266**:9912–8.
13. Lu Y, Xiao G, Galson DL, Nishio Y, Mizokami A, Keller ET, et al. PTHrP-induced MCP-1 production by human bone marrow endothelial cells and osteoblasts promotes osteoclast differentiation and prostate cancer cell proliferation and invasion in vitro. *Int J Cancer* 2007;**121**:724–33.
14. Yoshimura T, Yuhki N, Moore SK, Appella E, Lerman MI, Leonard EJ. Human monocyte chemoattractant protein-1 (MCP-1). Full-length cDNA cloning, expression in mitogen-stimulated blood mononuclear leukocytes, and sequence similarity to mouse competence gene JE. *FEBS Lett* 1989;**244**:487–93.
15. Zhang J, Patel L, Pienta KJ. CC chemokine ligand 2 (CCL2) promotes prostate cancer tumorigenesis and metastasis. *Cytokine Growth Factor Rev* 2010;**21**:41–8.
16. Zhang J, Lu Y, Pienta KJ. Multiple roles of chemokine (C-C motif) ligand 2 in promoting prostate cancer growth. *J Natl Cancer Inst* 2010;**102**:522–8.
17. Craig MJ, Loberg RD. CCL2 (Monocyte Chemoattractant Protein-1) in cancer bone metastases. *Cancer Metastasis Rev* 2006;**25**:611–9.
18. Lu Y, Cai Z, Galson DL, Xiao G, Liu Y, George DE, et al. Monocyte chemotactic protein-1 (MCP-1) acts as a paracrine and autocrine factor for prostate cancer growth and invasion. *Prostate* 2006;**66**:1311–8.

19. Mazzucchelli L, Loetscher P, Kappeler A, Uguccioni M, Baggiolini M, Laissue JA, et al. Monocyte chemoattractant protein-1 gene expression in prostatic hyperplasia and prostate adenocarcinoma. *Am J Pathol* 1996;**149**:501–9.
20. Chetcuti A, Margan S, Mann S, Russell P, Handelsman D, Rogers J, et al. Identification of differentially expressed genes in organ-confined prostate cancer by gene expression array. *Prostate* 2001;**47**:132–40.
21. Loberg RD, Day LL, Harwood J, Ying C, St John LN, Giles R, et al. CCL2 is a potent regulator of prostate cancer cell migration and proliferation. *Neoplasia* 2006;**8**:578–86.
22. Stifter S. The role of nuclear factor kappaB on angiogenesis regulation through monocyte chemotactic protein-1 in myeloma. *Med Hypotheses* 2006;**66**:384–6.
23. Hurwitz AA, Lyman WD, Berman JW. Tumor necrosis factor alpha and transforming growth factor beta upregulate astrocyte expression of monocyte chemoattractant protein-1. *J Neuroimmunol* 1995;**57**:193–8.
24. Johrer K, Janke K, Krugmann J, Fiegl M, Greil R. Transendothelial migration of myeloma cells is increased by tumor necrosis factor (TNF)-alpha via TNF receptor 2 and autocrine up-regulation of MCP-1. *Clin Cancer Res* 2004;**10**:1901–10.
25. Kim MS, Day CJ, Morrison NA. MCP-1 is induced by receptor activator of nuclear factor-{kappa}B ligand, promotes human osteoclast fusion, and rescues granulocyte macrophage colony-stimulating factor suppression of osteoclast formation. *J Biol Chem* 2005;**280**:16163–9.
26. Balkwill F. Chemokine biology in cancer. *Semin Immunol* 2003;**15**:49–55.
27. Kurihara T, Warr G, Loy J, Bravo R. Defects in macrophage recruitment and host defense in mice lacking the CCR2 chemokine receptor. *J Exp Med* 1997;**186**:1757–62.
28. Ajuebor MN, Flower RJ, Hannon R, Christie M, Bowers K, Verity A, et al. Endogenous monocyte chemoattractant protein-1 recruits monocytes in the zymosan peritonitis model. *J Leukoc Biol* 1998;**63**:108–16.
29. Kurth I, Willimann K, Schaerli P, Hunziker T, Clark-Lewis I, Moser B. Monocyte selectivity and tissue localization suggests a role for breast and kidney-expressed chemokine (BRAK) in macrophage development. *J Exp Med* 2001;**194**:855–61.
30. Chensue SW, Warmington KS, Lukacs NW, Lincoln PM, Burdick MD, Strieter RM, et al. Monocyte chemotactic protein expression during schistosome egg granuloma formation. Sequence of production, localization, contribution, and regulation. *Am J Pathol* 1995;**146**:130–8.
31. Chensue SW, Warmington KS, Ruth JH, Sanghi PS, Lincoln P, Kunkel SL. Role of monocyte chemoattractant protein-1 (MCP-1) in Th1 (mycobacterial) and Th2 (schistosomal) antigen-induced granuloma formation: relationship to local inflammation, Th cell expression, and IL-12 production. *J Immunol* 1996;**157**:4602–8.
32. Handel TM, Domaille PJ. Heteronuclear (1H, 13C, 15N) NMR assignments and solution structure of the monocyte chemoattractant protein-1 (MCP-1) dimer. *Biochemistry* 1996;**35**:6569–84.
33. Gonzalo JA, Lloyd CM, Wen D, Albar JP, Wells TN, Proudfoot A, et al. The coordinated action of CC chemokines in the lung orchestrates allergic inflammation and airway hyperresponsiveness. *J Exp Med* 1998;**188**:157–67.
34. Gu L, Tseng S, Horner RM, Tam C, Loda M, Rollins BJ. Control of TH2 polarization by the chemokine monocyte chemoattractant protein-1. *Nature* 2000;**404**:407–11.
35. Karpus WJ, Lukacs NW, Kennedy KJ, Smith WS, Hurst SD, Barrett TA. Differential CC chemokine-induced enhancement of T helper cell cytokine production. *J Immunol* 1997;**158**:4129–36.

36. Dewald O, Zymek P, Winkelmann K, Koerting A, Ren G, Abou-Khamis T, et al. CCL2/monocyte chemoattractant protein-1 regulates inflammatory responses critical to healing myocardial infarcts. *Circ Res* 2005;**96**:881–9.
37. Boring L, Gosling J, Cleary M, Charo IF. Decreased lesion formation in CCR2−/− mice reveals a role for chemokines in the initiation of atherosclerosis. *Nature* 1998;**394**:894–7.
38. Namiki M, Kawashima S, Yamashita T, Ozaki M, Hirase T, Ishida T, et al. Local overexpression of monocyte chemoattractant protein-1 at vessel wall induces infiltration of macrophages and formation of atherosclerotic lesion: synergism with hypercholesterolemia. *Arterioscler Thromb Vasc Biol* 2002;**22**:115–20.
39. Izikson L, Klein RS, Charo IF, Weiner HL, Luster AD. Resistance to experimental autoimmune encephalomyelitis in mice lacking the CC chemokine receptor (CCR)2. *J Exp Med* 2000;**192**:1075–80.
40. Shahrara S, Proudfoot AE, Park CC, Volin MV, Haines GK, Woods JM, et al. Inhibition of monocyte chemoattractant protein-1 ameliorates rat adjuvant-induced arthritis. *J Immunol* 2008;**180**:3447–56.
41. Wada T, Yokoyama H, Furuichi K, Kobayashi KI, Harada K, Naruto M, et al. Intervention of crescentic glomerulonephritis by antibodies to monocyte chemotactic and activating factor (MCAF/MCP-1). *FASEB J* 1996;**10**:1418–25.
42. Kasahara Y, Kimura H, Kurosu K, Sugito K, Mukaida N, Matsushima K, et al. MCAF/MCP-1 protein expression in a rat model for pulmonary hypertension induced by monocrotaline. *Chest* 1998;**114**:67S.
43. Moore BB, Paine 3rd R, Christensen PJ, Moore TA, Sitterding S, Ngan R, et al. Protection from pulmonary fibrosis in the absence of CCR2 signaling. *J Immunol* 2001;**167**:4368–77.
44. Volejnikova S, Laskari M, Marks Jr. SC, Graves DT. Monocyte recruitment and expression of monocyte chemoattractant protein-1 are developmentally regulated in remodeling bone in the mouse. *Am J Pathol* 1997;**150**:1711–21.
45. Kim MS, Day CJ, Selinger CI, Magno CL, Stephens SR, Morrison NA. MCP-1-induced human osteoclast-like cells are tartrate-resistant acid phosphatase, NFATc1, and calcitonin receptor-positive but require receptor activator of NFkappaB ligand for bone resorption. *J Biol Chem* 2006;**281**:1274–85.
46. Alam R, Kumar D, Anderson-Walters D, Forsythe PA. Macrophage inflammatory protein-1 alpha and monocyte chemoattractant peptide-1 elicit immediate and late cutaneous reactions and activate murine mast cells in vivo. *J Immunol* 1994;**152**:1298–303.
47. Taub D, Dastych J, Inamura N, Upton J, Kelvin D, Metcalfe D, et al. Bone marrow-derived murine mast cells migrate, but do not degranulate, in response to chemokines. *J Immunol* 1995;**154**:2393–402.
48. Valent P. Cytokines involved in growth and differentiation of human basophils and mast cells. *Exp Dermatol* 1995;**4**:255–9.
49. Pollard JW. Tumour-educated macrophages promote tumour progression and metastasis. *Nat Rev Cancer* 2004;**4**:71–8.
50. Kim Y, Sung S, Kuziel WA, Feldman S, Fu SM, Rose Jr. CE. Enhanced airway Th2 response after allergen challenge in mice deficient in CC chemokine receptor-2 (CCR2). *J Immunol* 2001;**166**:5183–92.
51. Salcedo R, Ponce ML, Young HA, Wasserman K, Ward JM, Kleinman HK, et al. Human endothelial cells express CCR2 and respond to MCP-1: direct role of MCP-1 in angiogenesis and tumor progression. *Blood* 2000;**96**:34–40.
52. Carulli MT, Ong VH, Ponticos M, Shiwen X, Abraham DJ, Black CM, et al. Chemokine receptor CCR2 expression by systemic sclerosis fibroblasts: evidence for autocrine regulation of myofibroblast differentiation. *Arthritis Rheum* 2005;**52**:3772–82.

53. Klopp AH, Spaeth EL, Dembinski JL, Woodward WA, Munshi A, Meyn RE, et al. Tumor irradiation increases the recruitment of circulating mesenchymal stem cells into the tumor microenvironment. *Cancer Res* 2007;**67**:11687–95.
54. Lu Y, Cai Z, Xiao G, Liu Y, Keller ET, Yao Z, et al. CCR2 expression correlates with prostate cancer progression. *J Cell Biochem* 2007;**101**:676–85.
55. Loberg RD, Ying C, Craig M, Day LL, Sargent E, Neeley C, et al. Targeting CCL2 with systemic delivery of neutralizing antibodies induces prostate cancer tumor regression in vivo. *Cancer Res* 2007;**67**:9417–24.
56. Xia M, Sui Z. Recent developments in CCR2 antagonists. *Expert Opin Ther Pat* 2009;**19**:295–303.
57. Binder NB, Niederreiter B, Hoffmann O, Stange R, Pap T, Stulnig TM, et al. Estrogen-dependent and C-C chemokine receptor-2-dependent pathways determine osteoclast behavior in osteoporosis. *Nat Med* 2009;**15**:417–24.
58. Lu Y, Chen Q, Corey E, Xie W, Fan J, Mizokami A, et al. Activation of MCP-1/CCR2 axis promotes prostate cancer growth in bone. *Clin Exp Metastasis* 2009;**26**:161–9.
59. Juppner H, Abou-Samra AB, Freeman M, Kong XF, Schipani E, Richards J, et al. A G protein-linked receptor for parathyroid hormone and parathyroid hormone-related peptide. *Science* 1991;**254**:1024–6.
60. Burtis WJ. Parathyroid hormone-related protein: structure, function, and measurement. *Clin Chem* 1992;**38**:2171–83.
61. Dougherty KM, Blomme EA, Koh AJ, Henderson JE, Pienta KJ, Rosol TJ, et al. Parathyroid hormone-related protein as a growth regulator of prostate carcinoma. *Cancer Res* 1999;**59**:6015–22.
62. Iwamura M, di Sant'Agnese PA, Wu G, Benning CM, Cockett AT, Deftos LJ, et al. Immunohistochemical localization of parathyroid hormone-related protein in human prostate cancer. *Cancer Res* 1993;**53**:1724–6.
63. Iwamura M, Gershagen S, Lapets O, Moynes R, Abrahamsson PA, Cockett AT, et al. Immunohistochemical localization of parathyroid hormone-related protein in prostatic intraepithelial neoplasia. *Hum Pathol* 1995;**26**:797–801.
64. Deftos LJ, Barken I, Burton DW, Hoffman RM, Geller J. Direct evidence that PTHrP expression promotes prostate cancer progression in bone. *Biochem Biophys Res Commun* 2005;**327**:468–72.
65. Guise TA, Yin JJ, Thomas RJ, Dallas M, Cui Y, Gillespie MT. Parathyroid hormone-related protein (PTHrP)-(1-139) isoform is efficiently secreted in vitro and enhances breast cancer metastasis to bone in vivo. *Bone* 2002;**30**:670–6.
66. Yin JJ, Selander K, Chirgwin JM, Dallas M, Grubbs BG, Wieser R, et al. TGF-beta signaling blockade inhibits PTHrP secretion by breast cancer cells and bone metastases development. *J Clin Invest* 1999;**103**:197–206.
67. Dunbar ME, Wysolmerski JJ. Parathyroid hormone-related protein: a developmental regulatory molecule necessary for mammary gland development. *J Mammary Gland Biol Neoplasia* 1999;**4**:21–34.
68. Bouizar Z, Spyratos F, De vernejoul MC. The parathyroid hormone-related protein (PTHrP) gene: use of downstream TATA promotor and PTHrP 1-139 coding pathways in primary breast cancers vary with the occurrence of bone metastasis. *J Bone Miner Res* 1999;**14**:406–14.
69. Li X, Qin L, Bergenstock M, Bevelock LM, Novack DV, Partridge NC. Parathyroid hormone stimulates osteoblastic expression of MCP-1 to recruit and increase the fusion of pre/osteoclasts. *J Biol Chem* 2007;**282**:33098–106.
70. Liao J, Li X, Koh AJ, Berry JE, Thudi N, Rosol TJ, et al. Tumor expressed PTHrP facilitates prostate cancer-induced osteoblastic lesions. *Int J Cancer* 2008;**123**:2267–78.

71. Li X, Loberg R, Liao J, Ying C, Snyder LA, Pienta KJ, et al. A destructive cascade mediated by CCL2 facilitates prostate cancer growth in bone. *Cancer Res* 2009;**69**:1685–92.
72. Sekine O, Nishio Y, Egawa K, Nakamura T, Maegawa H, Kashiwagi A. Insulin activates CCAAT/enhancer binding proteins and proinflammatory gene expression through the phosphatidylinositol 3-kinase pathway in vascular smooth muscle cells. *J Biol Chem* 2002;**277**:36631–9.
73. Ueda A, Okuda K, Ohno S, Shirai A, Igarashi T, Matsunaga K, et al. NF-kappa B and Sp1 regulate transcription of the human monocyte chemoattractant protein-1 gene. *J Immunol* 1994;**153**:2052–63.
74. Milligan ED, Twining C, Chacur M, Biedenkapp J, O'Connor K, Poole S, et al. Spinal glia and proinflammatory cytokines mediate mirror-image neuropathic pain in rats. *J Neurosci* 2003;**23**:1026–40.
75. Twining CM, Sloane EM, Milligan ED, Chacur M, Martin D, Poole S, et al. Peri-sciatic proinflammatory cytokines, reactive oxygen species, and complement induce mirror-image neuropathic pain in rats. *Pain* 2004;**110**:299–309.
76. Jung H, Miller RJ. Activation of the nuclear factor of activated T-cells (NFAT) mediates upregulation of CCR2 chemokine receptors in dorsal root ganglion (DRG) neurons: a possible mechanism for activity-dependent transcription in DRG neurons in association with neuropathic pain. *Mol Cell Neurosci* 2008;**37**:170–7.
77. Qian DZ, Rademacher BL, Pittsenbarger J, Huang CY, Myrthue A, Higano CS, et al. CCL2 is induced by chemotherapy and protects prostate cancer cells from docetaxel-induced cytotoxicity. *Prostate* 2010;**70**:433–42.
78. Phillips RJ, Lutz M, Premack B. Differential signaling mechanisms regulate expression of CC chemokine receptor-2 during monocyte maturation. *J Inflamm (Lond)* 2005;**2**:14.
79. Chen Y, Green SR, Ho J, Li A, Almazan F, Quehenberger O. The mouse CCR2 gene is regulated by two promoters that are responsive to plasma cholesterol and peroxisome proliferator-activated receptor gamma ligands. *Biochem Biophys Res Commun* 2005;**332**:188–93.
80. Glabinski AR, Bielecki B, Kolodziejski P, Han Y, Selmaj K, Ransohoff RM. TNF-alpha microinjection upregulates chemokines and chemokine receptors in the central nervous system without inducing leukocyte infiltration. *J Interferon Cytokine Res* 2003;**23**:457–66.
81. Yamamoto K, Takeshima H, Hamada K, Nakao M, Kino T, Nishi T, et al. Cloning and functional characterization of the 5′-flanking region of the human monocyte chemoattractant protein-1 receptor (CCR2) gene. Essential role of 5′-untranslated region in tissue-specific expression. *J Biol Chem* 1999;**274**:4646–54.
82. Raymond CR, Redman SJ, Crouch MF. The phosphoinositide 3-kinase and p70 S6 kinase regulate long-term potentiation in hippocampal neurons. *Neuroscience* 2002;**109**:531–6.
83. Gavrilin MA, Gulina IV, Kawano T, Dragan S, Chakravarti L, Kolattukudy PE. Site-directed mutagenesis of CCR2 identified amino acid residues in transmembrane helices 1, 2, and 7 important for MCP-1 binding and biological functions. *Biochem Biophys Res Commun* 2005;**327**:533–40.
84. Terashima Y, Onai N, Murai M, Enomoto M, Poonpiriya V, Hamada T, et al. Pivotal function for cytoplasmic protein FROUNT in CCR2-mediated monocyte chemotaxis. *Nat Immunol* 2005;**6**:827–35.
85. Satoh M, Akatsu T, Ishkawa Y, Minami Y, Nakamura M. A novel activator of C-C chemokine, FROUNT, is expressed with C-C chemokine receptor 2 and its ligand in failing human heart. *J Card Fail* 2007;**13**:114–9.
86. Majumder PK, Sellers WR. Akt-regulated pathways in prostate cancer. *Oncogene* 2005;**24**:7465–74.

87. Zi X, Singh RP, Agarwal R. Impairment of erbB1 receptor and fluid-phase endocytosis and associated mitogenic signaling by inositol hexaphosphate in human prostate carcinoma DU145 cells. *Carcinogenesis* 2000;**21**:2225–35.
88. Miyake H, Nelson C, Rennie PS, Gleave ME. Overexpression of insulin-like growth factor binding protein-5 helps accelerate progression to androgen-independence in the human prostate LNCaP tumor model through activation of phosphatidylinositol 3′-kinase pathway. *Endocrinology* 2000;**141**:2257–65.
89. Murillo H, Huang H, Schmidt LJ, Smith DI, Tindall DJ. Role of PI3K signaling in survival and progression of LNCaP prostate cancer cells to the androgen refractory state. *Endocrinology* 2001;**142**:4795–805.
90. Roca H, Varsos Z, Pienta KJ. CCL2 protects prostate cancer PC3 cells from autophagic death via phosphatidylinositol 3-kinase/AKT-dependent survivin up-regulation. *J Biol Chem* 2008;**283**:25057–73.
91. Roca H, Varsos ZS, Mizutani K, Pienta KJ. CCL2, survivin and autophagy: new links with implications in human cancer. *Autophagy* 2008;**4**:969–71.
92. Roca H, Varsos ZS, Pienta KJ. CCL2 is a negative regulator of AMP-activated protein kinase to sustain mTOR complex-1 activation, survivin expression, and cell survival in human prostate cancer PC3 cells. *Neoplasia* 2009;**11**:1309–17.
93. Karnoub AE, Weinberg RA. Chemokine networks and breast cancer metastasis. *Breast Dis* 2006;**26**:75–85.
94. Ueno T, Toi M, Saji H, Muta M, Bando H, Kuroi K, et al. Significance of macrophage chemoattractant protein-1 in macrophage recruitment, angiogenesis, and survival in human breast cancer. *Clin Cancer Res* 2000;**6**:3282–9.
95. Nakashima E, Mukaida N, Kubota Y, Kuno K, Yasumoto K, Ichimura F, et al. Human MCAF gene transfer enhances the metastatic capacity of a mouse cachectic adenocarcinoma cell line in vivo. *Pharm Res* 1995;**12**:1598–604.
96. Valkovic T, Dobrila F, Melato M, Sasso F, Rizzardi C, Jonjic N. Correlation between vascular endothelial growth factor, angiogenesis, and tumor-associated macrophages in invasive ductal breast carcinoma. *Virchows Arch* 2002;**440**:583–8.
97. Loberg RD, Ying C, Craig M, Yan L, Snyder LA, Pienta KJ. CCL2 as an important mediator of prostate cancer growth in vivo through the regulation of macrophage infiltration. *Neoplasia* 2007;**9**:556–62.
98. Roodman GD. Mechanisms of bone metastasis. *N Engl J Med* 2004;**350**:1655–64.
99. Mundy GR. Metastasis to bone: causes, consequences and therapeutic opportunities. *Nat Rev Cancer* 2002;**2**:584–93.
100. Taichman RS, Loberg RD, Mehra R, Pienta KJ. The evolving biology and treatment of prostate cancer. *J Clin Invest* 2007;**117**:2351–61.
101. Keller ET, Zhang J, Cooper CR, Smith PC, McCauley LK, Pienta KJ, et al. Prostate carcinoma skeletal metastases: cross-talk between tumor and bone. *Cancer Metastasis Rev* 2001;**20**:333–49.
102. Lu X, Kang Y. Organotropism of breast cancer metastasis. *J Mammary Gland Biol Neoplasia* 2007;**12**:153–62.
103. Akhtari M, Mansuri J, Newman KA, Guise TM, Seth P. Biology of breast cancer bone metastasis. *Cancer Biol Ther* 2008;**7**:3–9.
104. Morrissey C, Vessella RL. The role of tumor microenvironment in prostate cancer bone metastasis. *J Cell Biochem* 2007;**101**:873–86.
105. Vanderkerken K, Vande Broek I, Eizirik DL, Van Valckenborgh E, Asosingh K, Van Riet I, et al. Monocyte chemoattractant protein-1 (MCP-1), secreted by bone marrow endothelial cells, induces chemoattraction of 5T multiple myeloma cells. *Clin Exp Metastasis* 2002;**19**:87–90.

106. Vande Broek I, Asosingh K, Vanderkerken K, Straetmans N, Van Camp B, Van Riet I. Chemokine receptor CCR2 is expressed by human multiple myeloma cells and mediates migration to bone marrow stromal cell-produced monocyte chemotactic proteins MCP-1, -2 and -3. *Br J Cancer* 2003;**88**:855–62.
107. Hu H, Sun L, Guo C, Liu Q, Zhou Z, Peng L, et al. Tumor cell-microenvironment interaction models coupled with clinical validation reveal CCL2 and SNCG as two predictors of colorectal cancer hepatic metastasis. *Clin Cancer Res* 2009;**15**:5485–93.
108. Coussens LM, Werb Z. Inflammation and cancer. *Nature* 2002;**420**:860–7.
109. DeMarzo AM, Nelson WG, Isaacs WB, Epstein JI. Pathological and molecular aspects of prostate cancer. *Lancet* 2003;**361**:955–64.
110. Sica A, Schioppa T, Mantovani A, Allavena P. Tumour-associated macrophages are a distinct M2 polarised population promoting tumour progression: potential targets of anti-cancer therapy. *Eur J Cancer* 2006;**42**:717–27.
111. Mantovani A, Bottazzi B, Colotta F, Sozzani S, Ruco L. The origin and function of tumor-associated macrophages. *Immunol Today* 1992;**13**:265–70.
112. Balkwill F, Charles KA, Mantovani A. Smoldering and polarized inflammation in the initiation and promotion of malignant disease. *Cancer Cell* 2005;**7**:211–7.
113. Allavena P, Sica A, Solinas G, Porta C, Mantovani A. The inflammatory micro-environment in tumor progression: the role of tumor-associated macrophages. *Crit Rev Oncol Hematol* 2008;**66**:1–9.
114. Murdoch C, Muthana M, Coffelt SB, Lewis CE. The role of myeloid cells in the promotion of tumour angiogenesis. *Nat Rev Cancer* 2008;**8**:618–31.
115. Mantovani A. La mala educacion of tumor-associated macrophages: diverse pathways and new players. *Cancer Cell* 2010;**17**:111–2.
116. Solinas G, Germano G, Mantovani A, Allavena P. Tumor-associated macrophages (TAM) as major players of the cancer-related inflammation. *J Leukoc Biol* 2009;**86**:1065–73.
117. Balkwill F, Mantovani A. Inflammation and cancer: back to Virchow? *Lancet* 2001;**357**:539–45.
118. Bingle L, Brown NJ, Lewis CE. The role of tumour-associated macrophages in tumour progression: implications for new anticancer therapies. *J Pathol* 2002;**196**:254–65.
119. Brigati C, Noonan DM, Albini A, Benelli R. Tumors and inflammatory infiltrates: friends or foes? *Clin Exp Metastasis* 2002;**19**:247–58.
120. Mantovani A, Sozzani S, Locati M, Allavena P, Sica A. Macrophage polarization: tumor-associated macrophages as a paradigm for polarized M2 mononuclear phagocytes. *Trends Immunol* 2002;**23**:549–55.
121. Mantovani A. Tumor-associated macrophages in neoplastic progression: a paradigm for the in vivo function of chemokines. *Lab Invest* 1994;**71**:5–16.
122. Roca H, Varsos ZS, Sud S, Craig MJ, Ying C, Pienta KJ. CCL2 and interleukin-6 promote survival of human CD11b+ peripheral blood mononuclear cells and induce M2-type macrophage polarization. *J Biol Chem* 2009;**284**:34342–54.
123. Lin EY, Gouon-Evans V, Nguyen AV, Pollard JW. The macrophage growth factor CSF-1 in mammary gland development and tumor progression. *J Mammary Gland Biol Neoplasia* 2002;**7**:147–62.
124. Mizutani K, Roca H, Varsos Z, Pienta KJ. Possible mechanism of CCL2-induced Akt activation in prostate cancer cells. *Anticancer Res* 2009;**29**:3109–13.
125. Lu X, Kang Y. Chemokine (C-C motif) ligand 2 engages CCR2+ stromal cells of monocytic origin to promote breast cancer metastasis to lung and bone. *J Biol Chem* 2009;**284**:29087–96.
126. Pahler JC, Tazzyman S, Erez N, Chen YY, Murdoch C, Nozawa H, et al. Plasticity in tumor-promoting inflammation: impairment of macrophage recruitment evokes a compensatory neutrophil response. *Neoplasia* 2008;**10**:329–40.

127. Eisold S, Schmidt J, Ryschich E, Gock M, Klar E, von Knebel Doeberitz M, et al. Induction of an antitumoral immune response by wild-type adeno-associated virus type 2 in an in vivo model of pancreatic carcinoma. *Pancreas* 2007;**35**:63–72.
128. Gocheva V, Wang HW, Gadea BB, Shree T, Hunter KE, Garfall AL, et al. IL-4 induces cathepsin protease activity in tumor-associated macrophages to promote cancer growth and invasion. *Genes Dev* 2010;**24**:241–55.
129. Popivanova BK, Kitamura K, Wu Y, Kondo T, Kagaya T, Kaneko S, et al. Blocking TNF-alpha in mice reduces colorectal carcinogenesis associated with chronic colitis. *J Clin Invest* 2008;**118**:560–70.
130. Palframan RT, Jung S, Cheng G, Weninger W, Luo Y, Dorf M, et al. Inflammatory chemokine transport and presentation in HEV: a remote control mechanism for monocyte recruitment to lymph nodes in inflamed tissues. *J Exp Med* 2001;**194**:1361–73.
131. Brown Z, Robson RL, Westwick J. Regulation and expression of chemokines: potential role in glomerulonephritis. *J Leukoc Biol* 1996;**59**:75–80.
132. Haberstroh U, Stilo K, Pocock J, Wolf G, Helmchen U, Wenzel U, et al. L-arginine suppresses lipopolysaccharide-induced expression of RANTES in glomeruli. *J Am Soc Nephrol* 1998;**9**:203–10.
133. Zachariae CO, Anderson AO, Thompson HL, Appella E, Mantovani A, Oppenheim JJ, et al. Properties of monocyte chemotactic and activating factor (MCAF) purified from a human fibrosarcoma cell line. *J Exp Med* 1990;**171**:2177–82.
134. Martin TJ, Mundy GR. Bone metastasis: can osteoclasts be excluded? *Nature* 2007;**445**:E19, [Discussion E-20].
135. Kyriakides TR, Foster MJ, Keeney GE, Tsai A, Giachelli CM, Clark-Lewis I, et al. The CC chemokine ligand, CCL2/MCP1, participates in macrophage fusion and foreign body giant cell formation. *Am J Pathol* 2004;**165**:2157–66.
136. Steinbrech DS, Mehrara BJ, Saadeh PB, Greenwald JA, Spector JA, Gittes GK, et al. VEGF expression in an osteoblast-like cell line is regulated by a hypoxia response mechanism. *Am J Physiol Cell Physiol* 2000;**278**:C853–60.
137. Dovio A, Data V, Angeli A. Circulating osteoprotegerin and soluble RANKL: do they have a future in clinical practice? *J Endocrinol Invest* 2005;**28**:14–22.
138. Lu Y, Cai Z, Xiao G, Keller ET, Mizokami A, Yao Z, et al. Monocyte chemotactic protein-1 mediates prostate cancer-induced bone resorption. *Cancer Res* 2007;**67**:3646–53.
139. Mizutani K, Sud S, Pienta KJ. Prostate cancer promotes CD11b positive cells to differentiate into osteoclasts. *J Cell Biochem* 2009;**106**:563–9.
140. Mizutani K, Sud S, McGregor NA, Martinovski G, Rice BT, Craig MJ, et al. The chemokine CCL2 increases prostate tumor growth and bone metastasis through macrophage and osteoclast recruitment. *Neoplasia* 2009;**11**:1235–42.
141. Cai Z, Chen Q, Chen J, Lu Y, Xiao G, Wu Z, et al. Monocyte chemotactic protein 1 promotes lung cancer-induced bone resorptive lesions in vivo. *Neoplasia* 2009;**11**:228–36.
142. Zhang J, Patel L, Pienta KJ. CC chemokine ligand 2 (CCL2) promotes prostate cancer tumorigenesis and metastasis. *Cytokine Growth Factor Rev* 2009;**21**:41–8.
143. Yang X, Lu P, Ishida Y, Kuziel WA, Fujii C, Mukaida N. Attenuated liver tumor formation in the absence of CCR2 with a concomitant reduction in the accumulation of hepatic stellate cells, macrophages and neovascularization. *Int J Cancer* 2006;**118**:335–45.
144. Halin S, Rudolfsson SH, Van Rooijen N, Bergh A. Extratumoral macrophages promote tumor and vascular growth in an orthotopic rat prostate tumor model. *Neoplasia* 2009;**11**:177–86.
145. Gazzaniga S, Bravo AI, Guglielmotti A, van Rooijen N, Maschi F, Vecchi A, et al. Targeting tumor-associated macrophages and inhibition of MCP-1 reduce angiogenesis and tumor growth in a human melanoma xenograft. *J Invest Dermatol* 2007;**127**:2031–41.

146. Zeisberger SM, Odermatt B, Marty C, Zehnder-Fjallman AH, Ballmer-Hofer K, Schwendener RA. Clodronate-liposome-mediated depletion of tumour-associated macrophages: a new and highly effective antiangiogenic therapy approach. *Br J Cancer* 2006;**95**:272–81.
147. Kimura YN, Watari K, Fotovati A, Hosoi F, Yasumoto K, Izumi H, et al. Inflammatory stimuli from macrophages and cancer cells synergistically promote tumor growth and angiogenesis. *Cancer Sci* 2007;**98**:2009–18.
148. Fridlender ZG, Buchlis G, Kapoor V, Cheng G, Sun J, Singhal S, et al. CCL2 blockade augments cancer immunotherapy. *Cancer Res* 2009;**70**:109–18.
149. Takahashi M, Miyazaki H, Furihata M, Sakai H, Konakahara T, Watanabe M, et al. Chemokine CCL2/MCP-1 negatively regulates metastasis in a highly bone marrow-metastatic mouse breast cancer model. *Clin Exp Metastasis* 2009;**26**:817–28.
150. Huang S, Singh RK, Xie K, Gutman M, Berry KK, Bucana CD, et al. Expression of the JE/MCP-1 gene suppresses metastatic potential in murine colon carcinoma cells. *Cancer Immunol Immunother* 1994;**39**:231–8.
151. Huang S, Xie K, Singh RK, Gutman M, Bar-Eli M. Suppression of tumor growth and metastasis of murine renal adenocarcinoma by syngeneic fibroblasts genetically engineered to secrete the JE/MCP-1 cytokine. *J Interferon Cytokine Res* 1995;**15**:655–65.
152. Zhang L, Yoshimura T, Graves DT. Antibody to Mac-1 or monocyte chemoattractant protein-1 inhibits monocyte recruitment and promotes tumor growth. *J Immunol* 1997;**158**:4855–61.
153. Monti P, Leone BE, Marchesi F, Balzano G, Zerbi A, Scaltrini F, et al. The CC chemokine MCP-1/CCL2 in pancreatic cancer progression: regulation of expression and potential mechanisms of antimalignant activity. *Cancer Res* 2003;**63**:7451–61.
154. Hirose K, Hakozaki M, Nyunoya Y, Kobayashi Y, Matsushita K, Takenouchi T, et al. Chemokine gene transfection into tumour cells reduced tumorigenicity in nude mice in association with neutrophilic infiltration. *Br J Cancer* 1995;**72**:708–14.
155. Nesbit M, Schaider H, Miller TH, Herlyn M. Low-level monocyte chemoattractant protein-1 stimulation of monocytes leads to tumor formation in nontumorigenic melanoma cells. *J Immunol* 2001;**166**:6483–90.

Chromosomal Aberrations in Solid Tumors

Arul M. Chinnaiyan[*,†,‡,§,¶] and Nallasivam Palanisamy[*,†,¶]

[*]*Michigan Center for Translational Pathology, University of Michigan, Ann Arbor, Michigan, USA*

[†]*Department of Pathology, University of Michigan, Ann Arbor, Michigan, USA*

[‡]*Howard Hughes Medical Institute, Maryland, USA*

[§]*Department of Urology, University of Michigan, Ann Arbor, Michigan, USA*

[¶]*Comprehensive Cancer Center, University of Michigan, Ann Arbor, Michigan, USA*

Ever since the identification of the exact number of human chromosomes in 1956, several cancer-specific chromosomal abnormalities have been identified in different tumors. Among the various genetic changes, such as alterations in oncogenes, tumor suppressor genes, and microRNA genes, recurrent chromosome translocations have been identified as an important class of mutations in hematological malignancies, soft tissue sarcomas, and more recently in prostate cancer and lung cancer. Recurrent gene fusions are used for cancer classification and as diagnostic markers, and some have been successfully targeted for drug development. Recent advances in high-throughput sequencing technology and the ambitious undertaking of "The Cancer Genome Atlas" (TCCA) project will help drive the identification of the underlying genetic aberrations in most of the

Progress in Molecular Biology
and Translational Science, Vol. 95
DOI: 10.1016/S1877-1173(10)95004-7

1877-1173/10 $35.00

solid cancers. This chapter presents an overview on the current status of the knowledge on chromosome aberrations in solid cancers, cytogenetic and non-cytogenetic methods for the characterization of changes at the DNA and RNA levels, technological advancements in high-throughput characterization of the cancer genome and transcriptome, and the current understanding of the molecular mechanism involved in the formation of gene fusions in solid cancer.

I. Introduction

Given the worldwide increase in the incidence of cancer and the urgent need to find ways for prevention and cure, cancer research is progressing at a rapid rate toward understanding the mechanism of the transformation of a normal cell into cancer and eventually spreading to different organs (metastasis). The evolution of cancer occurs through stepwise accumulation of specific genetic changes at each stage over a period of many years.[1] A wealth of knowledge, accumulated through sustained research, on the incidence and genetic heterogeneity of different types of cancer indicates the overwhelming task ahead to unravel the genetic complexity before specific therapies are developed. Although, there is vast information on many of the various causative factors of cancer (genetic and environmental), the mechanism behind the development of cancer is not well understood. Cancer at its earliest stage of development with few genetic abnormalities does not present with any symptoms. The accumulation of more complex aberrations during disease progression results in various clinical symptoms. In order to find the exact mechanism behind cancer development, it is important to understand the most common genetic alterations, occurring in a large number of cases for each cancer type, which trigger the transformation processes and leads eventually to metastasis. Analysis of the whole genome (DNA level) and transcriptome (RNA level) at different stages of each cancer in an unbiased manner is a powerful and rational approach to gain insights into their behavior, facilitating the identification of biomarkers, as well as the development of targeted methods for diagnosis and treatment.[2]

Due to the heterogeneity and complexity in genetic changes in each type of cancer, it is difficult to pinpoint the earliest event responsible for the transformation of a normal cell into cancer cell. Based on the insights gleaned from several decades of research, it is important to note that the fundamental changes have to happen at the gene level, which could be a gain of function mutation in the gene(s), genomic amplification, loss or deletion of a whole chromosome or segments, formation of fusion genes by interchromosomal or intrachromosomal rearrangements, and/or general genomic instability due to high level aneuploidy and associated complex chromosome aberrations.[3–5] The functional consequences of gross chromosomal changes occurring at the DNA level can be assessed by systematic characterization of the transcriptome for deregulated genes, mutated genes, and fusion genes. The recent discovery of microRNA,[6,7] noncoding RNA,[8–10] and expressed pseudogenes,[11] along with their role in the regulation of key pathway genes in cancer, adds a new dimension to the complex genetic makeup of cancer.

Chromosomal aberrations in hematological malignancies (leukemia and lymphoma) are simple and disease specific and are well characterized by karyotype analysis. Recurrent intrachromosomal and interchromosomal balanced reciprocal translocations and associated gene fusions have been identified for most types of these malignancies and are used for disease classification, routine diagnosis, disease follow-up, response assessment, and drug development. Detection of cancer-specific recurrent chromosome aberrations in solid cancers is rare, mainly due to technical limitations rather than any fundamental genetic differences between hematological malignancies and solid cancers. Given the technological improvements and development of new analytical approaches, recurrent gene fusions in prostate cancer,[12,13] lung cancer,[13] and other rare solid cancers have come to light, which has changed our view that recurrent gene fusions once thought to be occurring only in hematological malignancies and sarcomas are now recognized as a common type of mutation detectable in some of the solid cancers. Due to the inherent complexity of the solid cancer genome, established approaches have failed to identify similar aberrations in other solid cancers. Emerging technologies and approaches (discussed below), however, are poised to unravel the genetic complexity of other solid cancer types.

In this chapter, we discuss the nature of various chromosome aberrations in solid cancers, conventional and advanced methods for the identification and characterization of chromosome translocations and gene fusions, as well as their use in clinical practice. We also discuss the recent developments on our understanding on the mechanisms of the formation of gene fusions in cancer.

II. Historical Background: Discovery of Chromosome Aberrations in Cancer

The concept of errors in cell division resulting in abnormal chromosome complements responsible for cancer development was first proposed by Boveri in 1914.[14] The landmark observation by Tjio and Levan in 1956[15] confirmed the presence of 46 chromosomes in normal human cells. In 1960, Nowell and Hungerford[16,17] identified a recurrent deleted chromosome in patients with chronic myeloid leukemia (CML). The advent of chromosome banding techniques in the early 1970s by Caspersson[18] led to the confirmation of the recurrent deleted chromosome as a translocation between chromosomes 9 and 22, which was later characterized to carry the fusion between the *ABL* and *BCR* genes.[19–21] Since then Boveri's concept has gained wide acceptance in the scientific community leading to the identification and characterization of a series of cancer-specific chromosome translocations and associated gene fusions in many types of leukemia and lymphoma, and recognition of more complex chromosome complements in solid cancers.[22,23] The rapid identification of recurrent translocations and gene fusions in hematological malignancies are due to the ease in the analytical methods for studying a large number of cases in each cancer type to establish disease-specific recurrent chromosome aberrations. Despite the presence of multiple complex aberrations, identification of cancer-specific recurrent chromosome aberrations in solid cancers, however, is limited due to technical challenges in the study of a large number of cases in each cancer type with abnormal karyotypes.[23,24] New methods have been developed that provide an unprecedented view of the cancer genome (Fig. 1). These remarkable discoveries represent a paradigm shift in our view of the solid cancer genome.

The exact molecular events underlying the formation of lineage or tissue-specific translocations in different cancer types are not clearly understood; however, the products of gene fusions have been aggressively studied for the development of drugs for treatment. A successful outcome of targeting gene fusions for drug development was the development of a new drug, imatinib mesylate (Gleevec®),[25] targeting the abnormal gene fusion product in CML, and many other drugs target mutant[26] and overexpressed genes.[27]

III. Discovery of Gene Fusions in Cancer

In the last 50 years, since the discovery of *BCR-ABL* gene fusion in CML, many recurring nonrandom chromosome aberrations have been identified in different cancer types. The accumulated data from cancer cytogenetics[23] of every cancer type studied with a sufficient number of cases allowed tumor stratification based on specific chromosome abnormalities. Among the various chromosome aberrations (e.g., deletions, duplications, and aneuploidy), balanced reciprocal translocations have been identified with remarkable specificity for hematological malignancies and soft tissue sarcomas. The incidence of cancer-specific translocations is heavily biased in favor of hematological malignancies as compared to solid tumors. About 1400 breakpoints[28] of reciprocal translocations and 500 tumor-specific translocations have been reported in the literature (http://cgap.nci.nih.gov/Chromosomes/Mitelman; http://www.unav.es/genetica/TICdb[28]), of which about 75% of the aberrations were identified in hematological malignancies.[23,29]

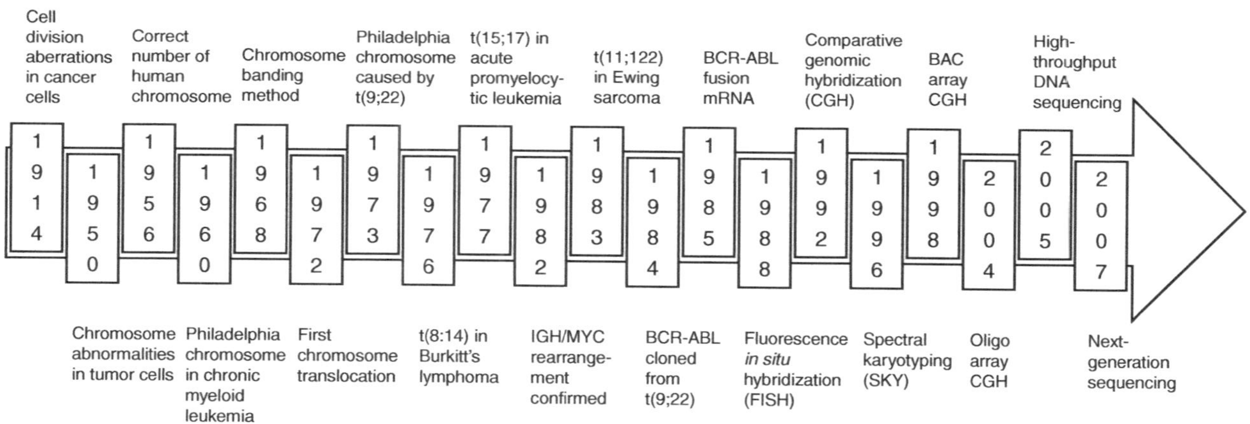

FIG. 1. History of the identification of chromosome aberrations and recent advances in technology for the identification of gene fusions in cancer.

Given the fact that more than 80% of cancer deaths are due to solid tumors, as opposed to only 10% die due to hematological malignancies, our knowledge of recurrent chromosomal aberrations, as well as the significance of the high incidence of complex rearrangements in solid tumors, is rudimentary. The lack of cytogenetic and molecular data from solid tumors makes it difficult to understand the tissue-specific differences in the mechanisms of tumor development. Until recently, methods were developed based on the available information from the human genome and were focused mainly on mutation analysis in cancer-causing genes, differential gene expression patterns of known genes (gene expression microarray methods), and detection of copy number changes at low resolution using low-throughput technologies (comparative genomic hybridization, CGH). Each method provided knowledge about the genetic heterogeneity and complexity of solid cancers; however, due to poor resolution and several other technical limitations, none of them were suitable for the identification of tumor-specific markers or recurrent gene fusions within various translocations and other complex aberrations.

Studies on hematological malignancies and sarcomas have provided important clues for the existence of recurrent chromosome translocations across many cancer types. It is rational to pursue our search for recurrent translocations in all cancer types because of the identification and usefulness of specific gene products (gene fusions) associated with each recurrent translocation. The molecular consequences of nonrandom reciprocal translocations result either in juxtaposition of a normal gene under the regulation of a new gene, resulting mostly in an abnormal level of expression of a normal gene from one of the rearranged chromosomes (e.g., *IGH-BCL2* in follicular lymphoma; *IGH-MYC* in Burkitt's lymphoma) or in the formation of a fusion gene from the breakpoints in the introns of two different genes (e.g., *BCR-ABL* in CML; *PML-RARA* in acute promyelocytic leukemia, APL) (Fig. 2) and many others. Most of the well-known fusion genes described in soft tissue sarcomas and other solid cancers are functionally important and are either transcription control genes (transcription

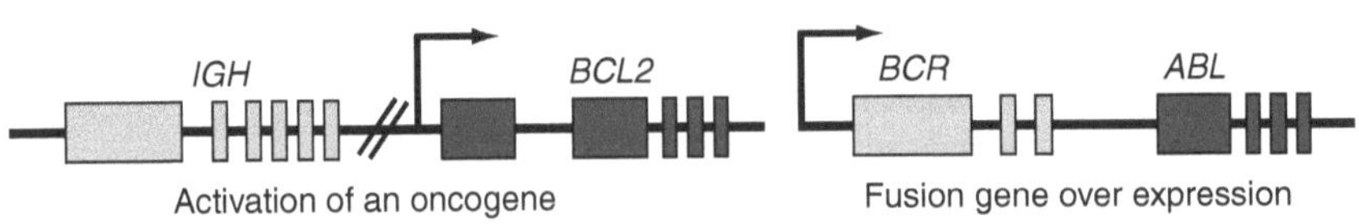

FIG. 2. Molecular consequences of chromosomal translocations. Activation of an oncogene by the juxtaposition of a normal gene under the control of 5′ regulatory elements of a different gene (left). Formation of a fusion between two different genes located at a distance on the same chromosome or from two different chromosomes due to the breakage and joining of introns resulting in the production of an abnormal level of functional chimeric RNA with exons derived from two different genes (right).

factors, TF) or tyrosine kinases (TK), which are potential targets for drug development (Table I). TFs, in general, are considered to be poor therapeutic targets when compared to the kinase family of genes[122,123]; however, recent advances in cancer research have found ways to target the genes regulated by the TFs, rather than using the fusion gene product as direct targets.[124–126]

IV. New Approaches for Gene Fusion Identification

Identification of gene fusions in hematological malignancies were made possible by the painstaking observations of recurrent chromosomal aberrations resolved at the level of G-band karyotype analysis. Due to the near diploid nature of the genome in hematological malignancies, and rarely with the presence of only one or a few secondary aberrations, and involvement of large chromosomal segments in translocations, it was easy to recognize the abnormal chromosomes. However, a few hidden aberrations such as inv(16)(p13q22) in acute myeloid leukemia and t(4;14)(p16;q32) in multiple myeloma can be easily missed even by trained cytogeneticists because the banding pattern of the abnormal chromosomes appear similar to normal chromosomes and involve small chromosome segments near the terminal regions of the chromosomes. Molecular characterization of breakpoint junctions were time consuming and laborious. However, these findings are a strong scientific foundation for the search for gene fusions across many cancer types. The complexity of aberrations in solid cancers compared to hematological malignancies was beyond the resolving power of the then available methods. Even with successful chromosome preparations from primary tumors and established cell lines of solid cancers, recurrent chromosome translocations were not identified by karyotype analysis and other methods, suggesting that balanced reciprocal translocations were either nonexistent or difficult to resolve in solid cancers. Although technological limitations were considered to be the reason for the failure of finding recurrent aberrations in solid cancers, recent discoveries using advanced noncytogenetic methods have proved otherwise by revealing the cryptic nature of these aberrations, as discussed below.

The nature of genomic aberrations (DNA level) in the recently identified recurrent gene fusions in prostate cancer and lung cancer are confined to a small genomic region on the same chromosome (intrachromosomal aberrations) and are impossible to identify using conventional cytogenetic methods. More importantly, the prostate cancer (*TMPRSS2-ERG*)[12] and lung cancer gene fusions (*EML4-ALK*)[127] were identified without any knowledge or understanding of the nature of genomic aberrations. The *TMPRSS2-ERG* gene fusion was revealed using a new bioinformatics approach called cancer outlier profile

TABLE I
LIST OF GENE FUSIONS IN DIFFERENT TYPES OF SOLID CANCER

Tumor	Translocation	Fusion gene	Function
Sarcoma solid tumor translocations			
Ewing sarcoma, primitive neuroectodermal tumor	t(11;22)(q24.3;q12.2)	FLI1;EWSR1[30]	TF
	t(21;22)(q22.2;q12.2)	ERG;EWSR1[31]	TF
	t(17;22)(q21;q12)	ETV4;EWSR1[32]	TF
	t(2;22)(q33;q12)	FEV;EWSR1[33]	TF
	t(16;21)(p11;q12)	ERG;FUS[34,35]	TF
	t(7;22)(p22;q12)	ETV1;EWSR1[35]	TF
Ewing-like soft tissue sarcoma	t(4;19)(q35;q13)	CIC-DUX4[36]	
Round cell sarcoma	t(1;22)(p36;q12)	EWSR1-ZNF278[37]	TF
Desmoplastic small round cell tumor	t(11;22)(p13;q12.2)	WT1;EWSR1[38]	TF
Inflammatory myofibroblastic tumor	t(2;19)(2p23;p13.12)	TPM4-ALK[39]	TK
	t(2;4)(p23;q21)	SEC31L1-ALK[40]	TK
	inv(2)(p23q35)	ATIC-ALK[41]	TK
	t(1;2)(q21.3;p23)	TPM3-ALK[39]	TK
	inv(2;2)(p23q13)	RANBP2-ALK[42]	TK
	t(2;17)(p23;q23),	CLTC-ALK[43]	TK
	t(2;11)(p23;p15.4)	CARS-ALK[44]	TK
Clear cell sarcoma	t(12;22)(q13;q12)	ATF1;EWSR1[45]	TF
Extraskeletal myxoid chondro	t(9;22)(q31.1;q12.2)	NR4A3;EWSR1[46]	TF
Sarcoma	t(9.17)(q22;q11)	NR4A3;TAF15[47]	TF
	t(9;15)(q22;q21)	NR4A3;TCF12[48]	TF
	t(3;9)(q11-12;22.33)	TFG-NR4A3[49]	TF
Alveolar rhabdomyosarcoma	t(2;13)(q35;q14)	PAX3;FKHR[50]	TF
	t(X;2)(q13.1;q36.1)	PAX3-FOXO4[51]	TF
	t(X;2)(q13.1;p23.3)	PAX3-NCOA1[51]	TF
	t(1;13)(p36.3;q14)	PAX7;FKHR[52]	TF
Synovial sarcoma	t(X;18)(p11.2;q11.2)	SS18;SSX1[53]	TF
	t(X;18)(p11.2;q11.2)	SS18;SSX2[53]	TF
	t(X;18)(p11.2;q11.2)	SS18;SSX4[54]	TF
Myxoid liposarcoma	t(12;16)(q13.3;p11.2)	CHOP;TLS[55]	TF
	t(12;22)(q13.3;q12.2)	CHOP;EWSR1[56]	TF
Dermatofibro-sarcoma protuberans	t(17;22)(q22;q13)	COL1A1-PDGFB[57]	GF

(*Continues*)

TABLE I (*Continued*)

Tumor	Translocation	Fusion gene	Function
Congenital fibrosarcoma	t(12;15)(p13;q25)	ETV6-NTRK3[58]	TK
Soft tissue chondrosarcoma	t(9;22)(q31;q12)	EWSR1-NR4A3[46]	TF
Angiomatoid fibrous histiocytoma	t(12;16)(q13;p11)	FUS-ATF1[59]	TF
Alveolar soft-part sarcoma	t(X;17)(p11.23;q25.3)	ASPSCR1-TFE3[60,61]	TF
Fibromyxoid soft tissue sarcoma	t(7;16)(q33;p11)	FUS-CREB3L2[62]	TF
	t(11;16)(p11;p11)	FUS-CREB3L1[63]	TF
Endometrial stromal sarcoma	t(7;17)(p15;q11)	JAZF1-SUZ12[64]	TF
	t(6;10)(21.32;p11.22)	EPC1-PHF1[65]	TF
	t(6;7)(p21;p22)	JAZF1-PHF1[64]	TF
Bone sarcoma	t(6;22)(p21;q12)	EWSR1-POU5F1[66]	TF
Non-sarcoma solid tumor translocations			
Papillary thyroid carcinoma	inv(1)(q21q22)	TPM3-NTRK1[67]	TK
	inv(1)(q21q25)	TPR-NTRK1[68]	TK
	i(1;3)(q21;q11)	TFG-NTRK1[68]	TK
	inv(10)(q11.2;q21)	HRH4-RET[69]	TK
	inv(10)(q11;q22)	NCOA4-RET[70]	TK
	t(7;10)(q32–34;q11.2)	TRIM24-RET[71]	TK
	t(1;10)(p13;q11.2)	TRIM33-RET[71]	TK
	t(10;12)(q11.2;p13.3)	ERC1-RET[72]	TK
	t(10;14)(q11.2;q22.1)	KTN1-RET[73]	TK
	t(10;18)(q11.2;q21–22)	RFG9-RET[74]	TK
	t(8;10)(p21–22;q11.2)	PCM1-RET[75]	TK
	t(6;10)(p21;q11.2)	TRIM27-RET[76]	TK
	t(10;14)(q32.12;q11.2)	GOLGA5-RET[73]	TK
	t(8;10)(p11.21;q11.2)	HOOK3-RET[77]	TK
Follicular thyroid carcinoma	t(2;3)(q13;p25)	PAX8-PPARg[78]	NR
Radiation induced thyroid cancer	t(7;7)(q21–22;q34)	AKAP9-BRAF[79]	TK
Pilocytic astrocytoma	dup(7)(q34)	KIAA1549-BRAF[80]	TK
Congenital melanocytic levi	t(2;7)(q24q33)	FCHSD1-BRAF[81]	TK
Gastric cancer	t(1;7)(p36.22;q34)	AGTRAP-BRAF[82]	TK

(*Continues*)

TABLE I (*Continued*)

Tumor	Translocation	Fusion gene	Function
Melanoma	t(?;3)(?;p25.1)	?-RAF1[82]	TK
	t(?;7)(?;q34)	?-BRAF[82]	TK
Prostate	t(1;7)(q32.1;q34)	SLC45A3-BRAF[82]	TK
	t(3;8)(p25.1;q22.1)	ESRP1-RAF1[82]	TK
	del/ins(21)(q22.2)	TMPRSS2-ERG[12]	TF
	t(16;21)(q13;q22.2)	HERPUD1-ERG[83]	TF
	t(1;21)(q32.1;q22.2)	SLC45A3-ERG[84]	TF
	t(8;21)(q24.22;q22.2)	NDRG1-ERG[85]	TF
	t(7;22)(p21.2;q11.23)	HERVK-ETV1[86]	TF
	?(1q32.1)	SLC45A3-ELK4[87]	TF
	t(7;21)(p21.2;q22.3)	TMPRSS2-ETV1[88,89]	TF
	t(7;15)(p21.3;q21)	C15orf21-ETV1[88]	TF
	t(7;7)(p21.2;p15)	HNRPA2B1-ETV1[88]	TF
	(1;7)(q32;p21.2)	SLC45A3-ETV1[90]	TF
	t(2;7)(q36.1p21.2)	ACSL3-ETV1[89]	TF
	t(7;17)(p21.2;p13.1)	AX747630-ETV1[90]	TF
	t(17;21)(q21;q22.3)	TMPRSS2-ETV4[91]	TF
	t(17;19)(q21;q13)	KLK2-ETV4[92]	TF
	inv(17;17)(q22;q25)	CANT1-ETV4[90]	TF
	t(17;17)(q21;q21)	DDX5-ETV4[90]	TF
	t(3;21)(q27;q22.3)	TMPRSS2-ETV5[93]	TF
	t(1;3)(q32;q27)	SLC45A3-ETV5[93]	TF
	t(7;14)(p21.2;q21)	EST14-ETV1[86]	TF
Non-small cell lung cancer, lung adenocarcinoma	inv(2)(p23;p21)or t(2;2)(p23;p21)	EML4-ALK[13]	TK
	t(2;3)(p23;q12.2)	TFG-ALK[94]	TK
Melanoma of the soft parts	t(12;22)(q13.12;q12.2)	EWSR1-ATF1[95]	TF
Meningioma	t(12;22)(p13;q11)	MN1-ETV6[96]	TF
Glioblastoma	del(6)(q21;q21)	GOPC-ROS1[97]	TK

(*Continues*)

TABLE I (*Continued*)

Tumor	Translocation	Fusion gene	Function
Secretory breast cancer	t(12;15)(p13;q25)	ETV6-NTRK3[98]	TK
Salivary gland carcinoma	t(12;15)(p13;q25)	ETV6-NTRK3[99]	TK
Congenital mesoblastic nephroma	t(12;15)(p13;q25)	ETV6-NTRK3[100]	TK
Hamartoma of the liver	t(11;19)(q11;q13.4)	MALAT-1/ MHLB1[101]	NC
Renal cell carcinoma	t(X;1)(p11;q21)	PRCC-TFE3[102]	TF
	t(X;17)(p11;q25)	ASPSCR1-TFE3[103]	TF
	t(6;11)(p21;q13)	MALAT1-TFEB[104]	TF
	t(6;11)(p21.1;q13)	Alpha-TFEB[105]	TF
	t(X;1)(p11;p34)	SFPQ-TFE3[106]	TF
	inv(X)(p11;q12)	NonO-TFE3[107]	TF
	(X;17)(p11.2;q23)	CLTC-TFE3[108]	TF
	t(X;17)(p11.2;q25.3)	RCC17-TFE3[109]	TF
Breast carcinoma	t(12;15)(p13;q25)	ETV6-NTRK3[98]	TK
	dic(8;11)(p12;q14)	ODZ4-NRG1[110]	TK
	t(8;11)(p12;q14)	ODZ4-NRG1[111]	TK
	t(3;6)(q26;q25)	TBL1XR1-RGS17[112]	TF
Aggressive midline carcinoma	t(15;19)(q13;p13.1)	BRD3-NUT[113]	NP
	t(9;15)(q34;q13)	BRD3-NUT[113]	NP
Pleomorphic adenoma	t(3;8)(p21;q12)	CTNNB1-PLAG1[114]	ZP
	t(5;8)(p13;q12)	LIFR-PLAG1[115]	ZP
	t(8;8)(q12;q11.2)	TCEA1-PLAG1[116]	ZP
	t(8;8)(q12;q11.2)	CHCHD7-PLAG1[116]	ZP
	t(3;13)(p14.2;q13–15)	HMGA2-FHIT[117]	TS
	t(9;12)(p12–22;q13–15)	HMGA2-NFIB[118,119]	TF
Mucoepidermoid carcinoma	t(11;19)(q21-22;p13.11)	CRTC1-MAML2[120]	TF
	t(11;15)(q21;q26.1)	CRTC3-MAML2[121]	TF

TK, tyrosine kinase; TF, transcription factor; ZP, zinc finger protein; NR, nuclear receptor; GF, growth factor; NC, noncoding RNA; NP, nuclear protein.

analysis (COPA)[128] using prostate cancer gene expression microarray data. The *EML4-ALK* gene fusion was found by screening a retroviral cDNA expression library constructed from non-small cell lung cancer (NSCLC) samples.[127]

The nature of genomic aberrations were later characterized by fluorescent *in situ* hybridization (FISH) analysis using probes based on the structure of the fusion gene transcript (messenger RNA). The *TMPRSS2-ERG* gene fusion, as revealed by FISH analysis, was produced by both balanced (translocation/insertion) and an unbalanced (deletion) aberration. It is rare that a same gene fusion is created both by a deletion and translocation/insertion mechanism within a solid cancer type.

Involvement of two aberrations to create the same gene fusion has been reported in leukemia. For example, inversion and balanced reciprocal translocation have been associated with *CBFB-MYH11* gene fusion in acute myeloid leukemia M4.[129] The two mechanisms involved in the formation of prostate cancer gene fusions have clinical implications with altered prognosis; particularly, *TMPRSS2-ERG* gene fusions formed by deletions were found to be associated in the majority of patients with androgen-independent metastatic prostate cancer.[130] Deletions associated with *BCR-ABL* translocations have been reported with altered prognosis; however, these deletions are identified as a secondary aberration formed during the progression of the disease, and are not primarily responsible for fusion gene formation.[131] Further, the *TMPRSS2* and *ERG* genes are located within a small genomic distance (3 MB) on the long arm of chromosome 21, which is beyond the resolution limit of the cytogenetic and noncytogenetic methods. Deletion or insertion of the intervening 3 MB genomic fragment results in the fusion of the first exon of the 5′ androgen-regulated *TMPRSS2* gene with mostly the exon 4 of *ERG*. Similarly, *EML4-ALK* gene fusion occurs in 1–5% of NSCLC and is formed by a cryptic inversion on the short arm of chromosome 2. These two seminal discoveries realigned our thinking about the nature of causal genetic aberrations in solid cancer. Furthermore, genomic aberrations not indicative of the presence of gene fusions have led to the discovery of recurrent gene fusions involving the RAF family of genes in thyroid cancer (*AKAP9-BRAF*)[79] and pilocytic astrocytoma (*KIAA1549-BRAF*).[80]

Recurrent aberrations involving cryptic inversions and duplications, in thyroid cancer and brain cancer, respectively, were discovered using FISH and high-density array CGH (a-CGH) (see discussion below). Identification of the RAF genes, along with other genes within the small genomic aberration, led to the identification of the activation of the fusion gene by the truncated *BRAF* gene. Fusions of *BRAF* with the 5′ gene result in the loss of the *N*-terminus autoregulatory domain retaining the *C*-terminus kinase domain in the fusion gene. Furthermore, the expression of two reciprocal transcripts from both rearranged chromosomes in a balanced reciprocal translocation is being observed more frequently due to the unbiased nature of paired-end transcriptome sequence analysis (see below). A couple of examples are *ESRP1-RAF1*[82] in prostate cancer and *CLDN18-ARHGAP26* in gastric cancer (Palanisamy *et al.*, unpublished data). Unlike the hematological malignancies, identification of these four important fusion genes in four different solid cancers and rare balanced reciprocal translocations by noncytogenetic methods revealed the hidden and complex nature of genomic aberrations and justify the long delay in the discovery of recurrent gene fusions in solid cancer. These findings suggest the need for the development of an unbiased and high-resolution method for whole genome analysis.

Furthermore, it is important to note that as the *TMPRSS2-ERG* gene fusion was identified based on the outlier expression of *ERG* in prostate cancer; extension of this approach to outlier genes identified in other cancers has found that all the outlier expression genes do not necessarily form a gene fusion. Involvement of alternate molecular mechanisms such as altered methylation patterns and loss of microRNA has been implicated for the overexpression of genes. For example, overexpression of the enhancer of *zeste 2 isoform a* (*EZH2*) in several solid cancers is due to the loss of microRNA-101 by genomic deletion.[132] Therefore, attempts to identify gene fusions targeting the outlier genes in a cancer may not yield the expected outcome. A recent study using next-generation sequencing (NGS) of ETS (erythroblastosis virus E26 transformation-specific) negative prostate cancer provides an alternate view that recurrent gene fusions can be found among the nonoutlier genes but as an outlier gene, not necessarily always, at the level of an individual tumor.[82] While G-band karyotyping, gene expression profiling, and copy number profiling have been used to identify gene fusions in cancer, they have been considered biased approaches because prior knowledge as to the genes or the locations of their aberrations in the genome was needed to initiate follow-up investigations. Gene expression microarrays and oligo-based CGH arrays for copy number profiling were constructed based on the information about annotated genes only, therefore these approaches are considered biased and they do not provide a complete view or understanding of the complex events that occur in the solid cancer genome.

There has been a gradual improvement in the cytogenetic methods that offer a clear perspective on the genomic complexity and have helped to identify a long list of gene fusions (Table I). These methods have also identified many regions with recurrent amplifications and deletions that are associated with disease progression and clinical outcome.[133,134] These observations indicate that multiple recurrent aberrations may exist in a given cancer type and that each identifies only a small subtype of each cancer. Unlike hematological malignancies, there is no single aberration/gene fusion that can hold the answer for all the cases in a given cancer type. A single aberration may identify a small subset in different cancer types, as evident from the genes belonging to the family of kinases and TFs in subsets of more than one cancer type. For example, identification of gene fusions involving the ETS family TFs in different sarcoma and solid cancers, as well as the RAF family kinase in small subsets of prostate cancer, thyroid cancer, brain cancer, melanoma, and gastric cancer support this view. A recent study on the characterization of mutations in genes across multiple solid cancer types indicated the involvement of all mutated genes affecting the MAPK pathway irrespective of the cancer type.[135] These studies warrant an in-depth investigation of a large number of cases in each cancer type in a case by case manner rather than as an entity based on histological and morphological classification. Identification of a biologically important gene fusion recurrent in a small subset of a cancer type justifies an extensive search for similar aberrations across several cancer types.

With the identification of recurrent gene fusions in a subset of prostate cancer, lung cancer, brain cancer, gastric cancer, and thyroid cancer, it is reasonable to believe that other solid tumors may also carry specific gene fusions, thereby emphasizing gene fusion as a dominant class of mutations across all cancer types. All the tested methods and approaches applied to prostate cancer and lung cancer may not yield expected results. Hence, an alternative approach using high-throughput sequencing is being extensively used, and is yielding promising results thus far. Studies aimed at sequencing the genomic DNA of solid cancers, not surprisingly, has identified a large number of genomic breakpoints due to multiple complex rearrangements. However, classification of these aberrations into tumor-causing (driver) and tumor-maintaining (passenger) aberrations remains a challenge because many of these aberrations either occur at gene desert (no gene mapped) regions in the genome or do not produce a functional RNA transcript. Alternatively, in order to directly understand the molecular consequences of genomic aberrations, sequencing expressed sequences (RNA) in an unbiased manner is a rational approach. Recent studies have shown that sequencing RNA from tumor samples is the best alternative approach. RNA sequencing has identified rare druggable gene fusions[82] in prostate cancers that do not carry the known *TMPRSS2-ERG* gene fusion, melanoma and gastric cancer. RNA sequencing has also revealed the presence of multiple fusion genes in a single tumor, posing a new challenge to classify the driver and passenger gene fusions.

Given the unbiased nature of the sequencing approach, multiple gene fusions are identified from each sample and classified into different groups based on the genomic organization of the genes.[136] Besides interchromosomal gene fusions (genes from two different chromosomes), the presence of a large number of intrachromosomal (genes from the same chromosome within a small genomic distance) gene fusions with or without evidence for genomic rearrangements are the new class of gene fusions detected by sequencing methods. While a vast number of gene fusions have been identified, a systematic analysis of a large number of tumors with matching normal controls will lead to the identification of all the functionally important fusions in cancer.

It is interesting to note that the gene fusions in hematological malignancies and soft tissue sarcomas were first identified based on the balanced reciprocal translocations identified by cytogenetic methods, whereas for *TMPRSS2-ERG* and *EML4-ALK*, gene fusion transcripts (RNA) were identified first without any knowledge about the genomic aberration. Similarly, the *KIAA1549-BRAF* and *AKAP9-BRAF* fusions were identified by noncytogenetic methods such as FISH and a-CGH. Identification of gene fusions within small amplifications indicates that the unbalanced aberrations in solid cancer produce many intrachromosomal rather than interchromosomal cryptic aberrations not detectable by cytogenetic methods. Furthermore, the extension of the informatics approach used in the identification of the *TMPRSS2-ERG* gene fusion in prostate cancer to other solid cancers did not reveal the presence of gene fusions even for the outlier genes in other cancer types. This underscores the need for high-resolution methods for characterization of the solid cancer genome.

NGS is emerging as an alternative approach to current methods. To harness the power of NGS technology, an ambitious cancer genome characterization project (http://cancergenome.nih.gov/) has been initiated by the National Institutes of Health (NIH) with the goal of providing the complete characterization of a large number of cancer samples from each one of the most common cancer types. Data generated from this project will be made available to the scientific community to conduct an integrated analysis to get a better understanding on the genomic complexities of cancer and possibly identify important molecular markers for diagnosis and drug development.

Taken together, the knowledge gained over 35 years starting from the discovery of the *BCR-ABL* to *TMPRSS2-ERG* gene fusions and with new technological advancements, the identification of gene fusions in solid cancer is no longer a challenging task. However, new obstacles emerge to understand the biological significance of the multiple gene fusions within one sample and also the many variant types within a cancer type. Nonrecurrent but functionally important gene fusions present only in one sample of a given cancer type are hard to ignore. Now with the massively parallel sequencing approach, the cancer research community has the opportunity to classify the innumerable gene fusions and understand their role in cancer development and progression. The questions to address include: Among the many fusion genes in a sample, how many of them are causal aberrations? How many of them are responsible for tumor growth? How do so many abnormal gene fusion products cooperate in a tumor environment? At what stage of the cancer do they occur? Emerging evidence suggests that the occurrence of the same gene fusion across multiple tumors may facilitate reclassification of a cancer based on molecular aberrations rather than its morphological and histological subtypes. This line of inquiry lays the foundation for treatment strategies targeting gene fusion products.

V. Methods for the Characterization of Chromosome Aberrations in Solid Tumors

Cancer cells are characterized by uncontrolled cell division, which is believed to have its roots in various random and nonrandom chromosomal abnormalities. Cytogenetics is the study of chromosome abnormalities in genetic diseases including cancer. A rather new discipline, molecular cytogenetics, is the study of complex chromosome abnormalities at a much higher resolution. Many noncytogenetic methods based on polymerase chain reaction (PCR) were developed mostly for the diagnosis and confirmation of cytogenetic aberrations. Conventional cytogenetic analysis by chromosome banding and advanced molecular cytogenetic analysis using FISH, spectral karyotyping (SKY), multicolor FISH (M-FISH), CGH, and a-CGH have been extensively applied to almost all cancer types. These investigations have identified an array of novel chromosome abnormalities that cause the deregulation of genes associated with favorable and unfavorable clinical outcomes (Fig. 3). In the following sections, we present an overview of the utilization of conventional cytogenetic and noncytogenetic methods for the analysis of chromosome aberrations, and discuss newly recognized gene fusions using noncytogenetic methods.

The most commonly ignored problem in solid tumor genome analysis is the characterization of complex rearrangements and translocation breakpoints and their effect at the transcriptome level. Gene expression microarray and low-resolution copy number analysis methods do not provide information on genomic rearrangements. Conventional cytogenetic karyotyping analysis on hematological malignancies and solid tumors identified 58,819 (http://cgap.nci.nih.gov/Chromosomes/Mitelman) abnormal karyotypes as of August 10, 2010. Complete molecular characterization of various abnormalities resulted in the identification of more than 275 genes involved in chromosome rearrangements.[23] The specificity of chromosome translocations has led to the subclassification of tumors solely based on chromosome aberrations. To date, about 500 such tumor-specific translocations have been identified. There is a higher incidence of cancer deaths due to solid tumors (80%) as compared with hematological malignancies (10%) but the proportion of available cytogenetics data is biased in favor of hematological malignancies. The cytogenetic changes in hematological malignancies are very few even in advanced stage cancers, and the types of chromosome changes are specific to a particular histological type as well. Chromosome aberrations in solid tumors are highly complex even at the early stage or at diagnosis. Among the various changes, distinction between tumor-associated primary abnormality and progression-associated changes are not yet possible. Additional complexities are due to clonal heterogeneity, which is present in less than 5% of hematological cancers and very common in solid tumors.

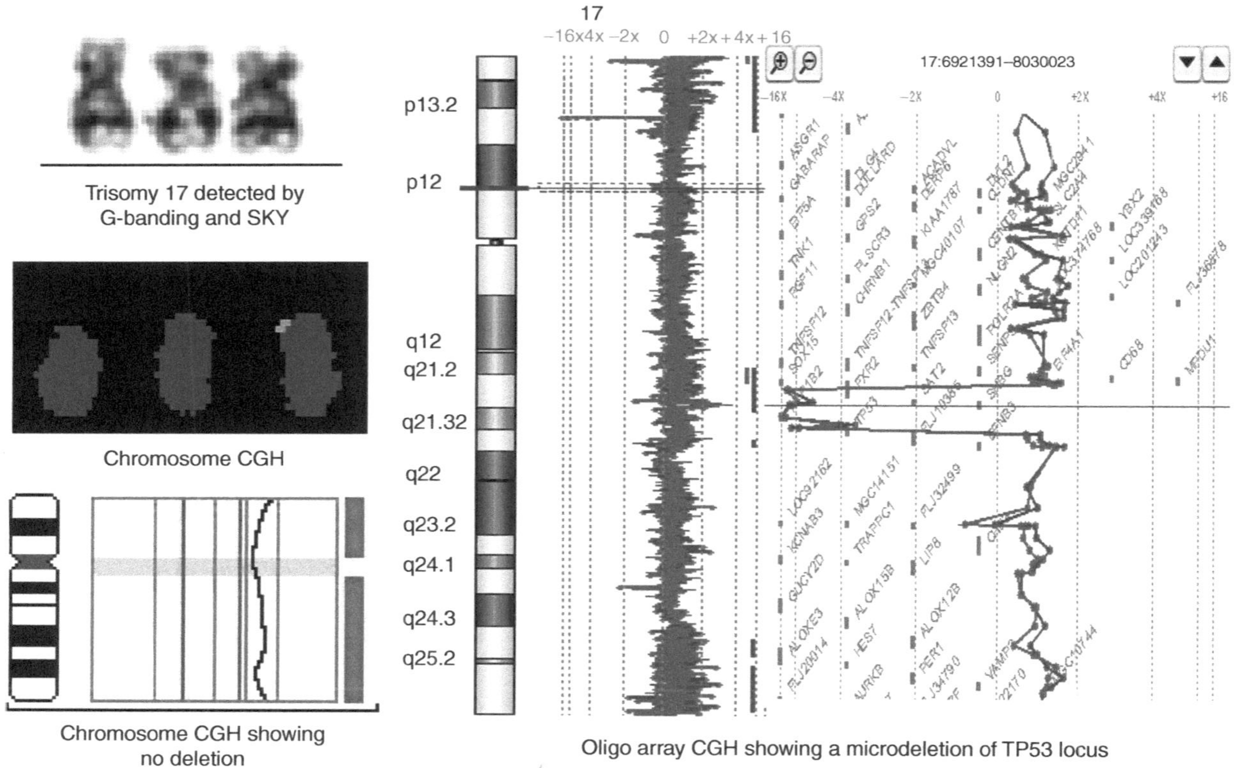

FIG. 3. Evolution of cytogenetic methods for the study of chromosome aberrations in cancer. (A) Partial karyotype showing trisomy of chromosome 17 confirmed by G-band karyotype analysis in HEPG2, a liver cancer cell line. (B) Spectral karyotyping (SKY) analysis also confirmed that all the three copies of chromosome 17 do not carry any small structural rearrangements. (C) Chromosome-based comparative genomic hybridization analysis (C-CGH) showing the gain of chromosome 17. (D) Oligonucleotide-based array comparative genomic hybridization (a-CGH) revealed the presence of a homozygous cryptic deletion of TP53 gene in all the three copies of chromosome 17 showing the resolving power of the high-resolution technologies to bring out hidden chromosomal aberrations.

New approaches have been developed to analyze the primary and advanced solid tumor genome to obtain additional cytogenetics information, which may help to identify the tissue-specific mechanisms of chromosome aberrations in tumor formation. The number of tests offered in a clinical cytogenetics laboratory has been increasing due to the identification of clinically useful cytogenetic and molecular markers. The molecular cytogenetic methods such as FISH, CGH, SKY, and microarray technology combined with bioinformatics tools are being used to identify molecular biomarkers. These markers are used for routine diagnosis, to predict the outcome of targeted therapy, and to monitor minimal residual disease.

A. Cytogenetic Methods

1. Chromosome Banding Analysis

Cytogenetic analysis, which involves culturing tumor cells from bone marrow aspirate, lymph node, or other tissue biopsies, is precise, if dividing cells are available. The methodology involves accumulation of metaphase cells by treatment with colchicine (used to synchronize cells at the mitotic stage of cell division) and fixation on glass slides. Slides containing adequate numbers of cells in the metaphase/prometaphase stage are "banded" by a number of techniques; the most widely used is G-banding. Viewed microscopically, the dark and pale bands along the length of the chromosome are consistent and reproducible for normal chromosomes (banding pattern). Consequently, the dark and light bands along the length of the chromosomes serve as landmarks for chromosome identification, as well as for assignment of breakpoints at the sites of rearrangement in abnormal chromosomes (Fig. 4). The drawbacks of this method are that it requires dividing cells and cell-by-cell analysis, which is labor intensive and time consuming, and even with a good karyotype, many chromosomes are difficult to classify due to complex banding patterns, resulting in a separate class known as marker chromosomes.

The current rate of successful cytogenetic analysis of newly diagnosed cases ranges from 60% to 80% for hematologic malignancies and $<40\%$ of solid tumors. This percentage falls dramatically in posttreatment bone marrow/blood samples because of the lack of adequate specimen or poor proliferation. Thus, cytogenetics is a less useful tool for patient follow-up. The identification of deletions less than the size of a small cytogenetic band is difficult or impossible using this method. Although cytogenetic analysis is not a practical method of choice for chromosome analysis of solid tumors, earlier studies in the pre- and postbanding era were used to identify recurrent balanced reciprocal translocations particularly in soft tissue sarcomas such as lipomas and

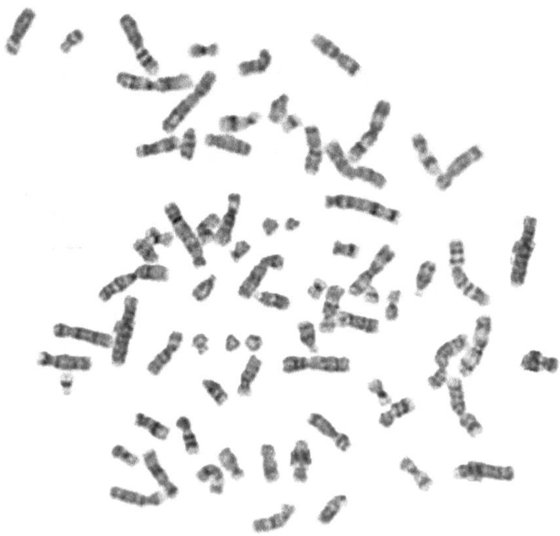

FIG. 4. Metaphase cell from a breast cancer cell line showing high level of aneuploidy and complex chromosome rearrangement.

leiomyosarcomas.[137–139] Analysis of high level of complexity, unbalanced rearrangements, deletions, amplifications, and isochromosomes[140] led to the development of more advanced methods discussed below.

2. Spectral Karyotyping

SKY[141] is a multicolor probe hybridization-based method for the analysis of chromosome rearrangements involving the whole or a portion of a chromosome(s), translocations involving two or more chromosomes, and to identify marker chromosomes not resolvable by G-band karyotype analysis. Metaphase chromosomes from a specimen of interest are required for this analysis. Probes for hybridization are generated by labeling each normal chromosomal DNA with different fluorochromes in different combinations to generate spectrally different colors for each chromosome (Fig. 5). Hybridization signals are captured using a fluorescent microscope equipped with special devices and image processing software. A total of 20–50 metaphase cells are examined to interpret the overall chromosomal abnormalities. Two or more cells with the same pattern of chromosomal aberration are considered a clonal abnormality. Deletions below 20 MB are not detected by this method. SKY techniques reveal the clonal heterogeneity of cancer cells. No specific diagnostic markers have been

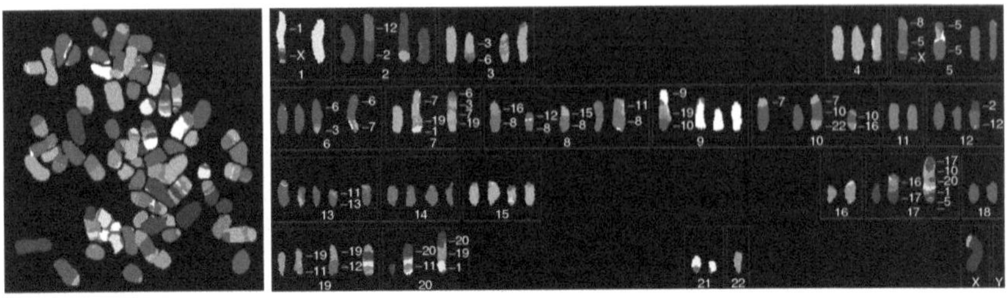

FIG. 5. Spectral karyotype analysis of solid cancer. Metaphase (left) and karyotype analysis (right) of a breast cancer cell line (MCF7) show high level of aneuploidy and complex rearrangements. Identification of tumor-causing aberrations from the large number of complex aberrations is beyond the resolving power of this technology. Normal chromosomes are recognized based on the uniform staining of entire chromosome compared with morphology using DAPI staining. Abnormal chromosomes are readily identified by the presence of more than one color in a chromosome. Unique color pattern for each chromosome identify the different chromosomes involved in rearrangement.

developed for routine use detectable only by this method; however, it is a powerful tool to study complex chromosome aberrations, mostly in human and mouse cancer cell lines.

B. Noncytogenetic Methods

Several noncytogenetic methods have been used as screening and validation tools rather than a discovery method under research settings. FISH screening of a known aberration or a gene fusion in a large number of specimens with high sensitivity and specificity, and different probe development approaches and labeling methods (dual color or multicolor) using appropriate probes or reagents derived from the known aberrations are utilized. Some of these methods find wider application under clinical settings for routine diagnosis, treatment response follow-up. The applications of some of these methods are discussed below.

1. Southern Blotting

Southern blotting is a method routinely used in molecular biology for detection of a specific DNA sequence in DNA samples. It has been used in the discovery and validation of several hematological gene fusions.[142,143] DNA isolated from the tumor samples are subjected to restriction endonuclease digestion and separation on agarose gels by electrophoresis. The digested DNA fragments transferred to nylon membrane are probed by hybridization with known DNA sequences adjacent to the breakpoint. A novel rearranged band, in addition to the germ line band derived from the normal allele, indicates the presence of rearrangement. Probes in the vicinity of the suspected breakpoints are always used. Moreover, detection of rearrangements depends on clustering of the breakpoints in one of the rearranging chromosomes within a small region. For example, the immunoglobulin heavy chain (*IGH*)-associated gene fusions in lymphomas were first identified with the confirmation of rearrangement in the *IGH* gene using Southern blotting, followed by genomic screening of phage libraries with selected *IGH* probe, resulting in the identification of a clone containing the rearranged segment from the tumor genome.[144,145] Translocations such as *BCR/ABL* with multiple breakpoints require multiple probes and multiple enzyme digestions of DNA, which is costly and time consuming. Another significant disadvantage is that this method can only detect translocations in tissues with at least 5–10% tumor cells, and hence it is an insensitive method, especially for monitoring relapse. With the development of PCR-based methods, Southern blotting is often not a method of choice.

2. Polymerase Chain Reaction

Discovery of gene fusions by DNA- and RNA-based PCR methods requires information about at least one of the partner genes, preferably the exon close to the breakpoint involved in the fusion; however, this is not a strict requirement. A modified PCR method such as rapid amplification of cDNA ends (RACE) is a method of choice for cloning fusion partners. Depending on the available information, the direction of the PCR (5′ or 3′ RACE) will be determined. Upon confirmation of both genes by RACE-PCR, primers from the exons flanking the breakpoint are used for a second PCR for confirmation. PCR amplification will occur only from tumor cells, not normal cells. Reverse transcription PCR (RT-PCR) to identify fusion RNA can also be employed in cases with widely scattered breakpoints, but successful amplification depends on consistent breakpoints within the same intronic regions of the two genes. In the RT-PCR method, primers are selected from the sequences of the exons near the breakpoints from each of the two genes. The PCR method is widely used to screen a large number of samples in a short time with high specificity and sensitivity and is used for diagnosis and follow-up.

C. Molecular Cytogenetic Methods

1. Fluorescent *In Situ* Hybridization

FISH is a powerful method with wide application in cancer for discovery, diagnosis, and treatment follow-up. *De novo* assessment of rearrangement in a gene can be accomplished with FISH using a dual- or multicolor approach. Simultaneous assessment of more than one gene is accomplished with the multicolor probe cocktail. FISH is accepted as the gold standard for the confirmation of DNA-level abnormalities in cancer such as copy number aberrations (chromosome enumeration probes), amplification, deletion, inversion, rearrangement, or gene fusion. The constant expansion of the list of recurrent gene fusions in solid cancer underscores the need for developing appropriate diagnostic reagents for routine screening. Given the technical and sample requirement limitations associated with other molecular methods (discussed above), FISH is an appropriate tool for routine screening of gene fusions and other aberrations as it has high sensitivity and specificity. FISH is well suited for detecting even small numbers of tumor cells because: (1) the method is rapid and easy to perform, (2) dividing cells are not necessary, (3) a large number of cells can be scored rapidly, and (4) hybridization can be performed on tissues preserved in virtually any form. In addition, recent advances in molecular genetics and genome analysis have offered a variety of reagents for use as DNA probes, that is, yeast artificial chromosome (YACs), bacterial artificial chromosomes (BACs), and P1 artificial chromosomes (PACs). Well-established YAC/BAC/PAC maps are available for most of the chromosomal regions involved in cancer translocations. Although FISH is a simple and straightforward method, a thorough knowledge and understanding of the gene and associated aberrations, experience, and knowledge in interpreting atypical signal patterns are important to developing a highly specific and sensitive FISH assay.

a. Approaches for Probe Development. The power of interphase FISH analysis has been well recognized. Traditionally, probes have been developed to demonstrate rearrangements within a gene using a single- or dual-color approach. Due to the high rate of false positive signals from random colocalization in normal cells, the single-color approach is not suitable for interphase FISH analysis. Metaphase cells are necessary to interpret a complex or atypical signal pattern, which is not a choice when analyzing solid cancer. Highly sensitive probes using the dual-color approach have been described for translocation detection in hematological malignancies and sarcoma translocations[146] (Fig. 6). Currently, commercial probes are available in the two and multicolor format for the simultaneous detection of more than two aberrations in a single hybridization experiment. Identification of a specific type of translocation at the time of diagnosis will be useful for clinicians determining appropriate treatment plans.

FISH is a highly suitable pilot study tool for solid cancer gene rearrangement identification, diagnosis, and follow-up over conventional karyotyping, RT-PCR, and other molecular methods. Metaphase cells are not necessary for interpretation of signals, thus eliminating culture setup. The FISH method completely eliminates the need to collect intact fresh specimens from an operation theater, which is necessary to preserve RNA for RT-PCR. RT-PCR requires designing primers from the exact sequence region to get reliable

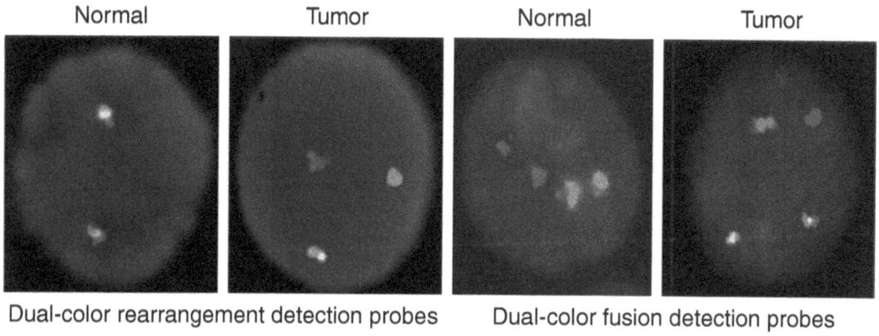

FIG. 6. Dual-color FISH analysis of chromosome rearrangement and gene fusion detection. For rearrangement detection, probes are generated from the immediately flanking regions (5′ and 3′ side) of the gene of interest and are differentially labeled. In normal cells, due to the close proximity of the 5′ and 3′ probes, signals will colocalize; in cancer cells, individual color signals indicate rearrangement. For gene fusion detection, one probe each from the 5′ end of the 5′ gene and from the 3′ end of the 3′ gene are labeled differentially. In normal cells, two sets of individual color signals corresponding to the two copies of normal chromosomes and in the tumors cells with balanced reciprocal translocations two fusion signal corresponding to two rearranged chromosomes will be identified. Signal patterns may vary based on the nature and complexity of aberration in a given sample.

results. Any deviation from the expected fragment size amplification requires further sequencing and other molecular characterization, which is time consuming and laborious, and moreover adds to the cost of the test. Complex rearrangements other than the known breakpoints, and associated deletions, lead to false negatives by RT-PCR and genomic PCR-based assays. Utilization of probes from about the 1 MB region from the flanking regions of the genes involved in the breakpoint provides a robust signal pattern.

2. Comparative Genomic Hybridization

Given the complexity of chromosome aberrations in solid cancer, high-throughput whole genome analysis methods have gained wide acceptance; however, the resolution limit is not sensitive enough for gene fusion analysis. CGH is used to study unbalanced chromosomal aberrations only. Balanced chromosomal aberrations such as reciprocal translocations are not detectable by this method.

The CGH method identifies the regions of copy number changes involving the whole chromosome or part of a chromosome for all the chromosomes in a single hybridization experiment. High-molecular weight genomic DNA from the specimen is required. DNA from formalin-fixed paraffin-tissue sections can also be used. Hybridization probes are generated by differentially labeling equal amounts of sample (tumor green) and reference DNA (normal red).

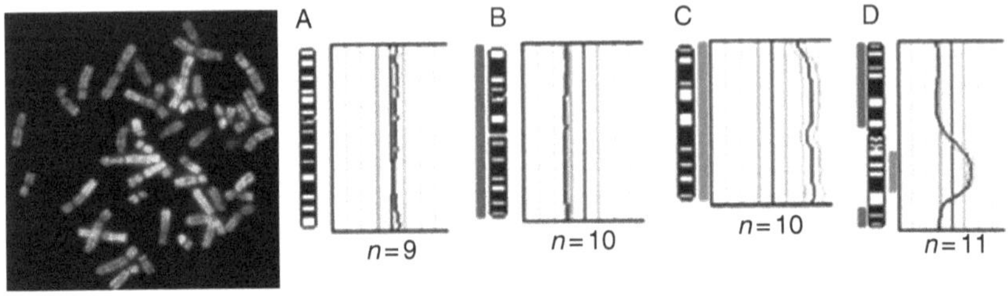

FIG. 7. Chromosome-based comparative genomic hybridization analysis of solid cancer. Differentially labeled normal and tumor DNA hybridized to normal human metaphase chromosomes (left). Fluorescent intensity across the entire length of the chromosome measured to assess the copy number changes for normal (A), loss (B), or gain (C) of whole chromosome or partial gain and loss of a region of a chromosome (D).

Labeled DNA is mixed with a 10-fold excess of human cot-1 DNA (to suppress repeat sequences) and hybridized to normal human chromosomes. A fluorescent microscope is used to view the hybridization signal patterns. Changes in the intensity patterns are compared with standard ideograms to identify the regions of gains and losses (Fig. 7). Regions with a high level of amplification will be demarked by intense green fluorescent spots. The resolution limit for this assay is 30–50 MB, which is not suitable for identification of boundaries of amplifications and deletions within a small genomic interval; thus, this methodology is not suitable for gene fusion detection.

3. Array Comparative Genomic Hybridization

Oligonucleotide-based a-CGH is a new emerging technology designed for high-precision mapping of copy number gains and losses and genomic rearrangements in cancer and genetic diseases.[147] The high-resolution analysis of copy number changes allows mapping of boundaries of amplifications and deletions with high precision in known genes, enabling the identification of novel gene fus ions at breakpoint junctions.[148] RACE-PCR methods can be used to identify the gene fusion from the breakpoint of a gene. Earlier CGH methods (e.g., chromosome-based CGH, cDNA arrays, and BAC clone arrays) suffer from poor resolution. Although these low-resolution methods identify the copy number changes, mapping of breakpoints is always within a large genomic distance of more than 100 KB to several megabases. The high-density arrays with >1 million features include probes from the well-known and cancer-related genes, along with a minimal number of probes derived from intergenic regions. Tiling arrays with features derived at fixed genomic intervals are not suitable for gene fusion testing as they miss breakpoints in important genes. Given the unique design and reproducibility of the high-density arrays, high-precision mapping of genomic rearrangements and copy number changes are obtained with remarkable specificity. The major limitation of this method is that only unbalanced copy number aberrations can be analyzed. Diagnostic tests based on known recurrent aberrations can be performed on a routine basis under clinical settings. An international cytogenomic array consortium (https://isca.genetics.emory.Browser/edu/iscaconsortiumMembers.action) has been established to set standards for the clinical utility and interpretation of data for diagnosis. Gene fusions cloned from focal genomic amplifications in thyroid and brain cancers are good examples of the clinical utility of array-based tests.

4. Characterization of Balanced Rearrangements

High aneuploidy and complex rearrangements in solid tumors create a large genome size and cause a heavy imbalance in the copy number at multiple locations in the genome when compared to the diploid genome. Boundaries of amplification, deletion, gains, and losses are a rich source of rearrangements to identify putative fusion and deregulated genes. Despite the highly unbalanced karyotypes, many regions in solid tumor genomes are not affected by copy number changes; however, they may harbor balanced rearrangements including pericentric and paracentric inversions and cryptic deletions in the size range of a few kilobases, which are not detectable by conventional banding, SKY methods, or even with high-resolution a-CGH. Recurrent chromosomal rearrangements in solid tumors might be hidden in such regions, as noted for the *EML4-ALK* gene fusion in lung cancer.

In addition to CGH, several noncytogenetic methods have been developed for the characterization of balanced rearrangements, including representational difference analysis,[149] restriction landmark genome scanning,[150,151] loss of heterozygosity analysis, Gene Identification Signature analysis using Paired-End diTagging (GIS-PET),[152] end sequence profiling,[153] digital karyotyping,[154] and representational oligonucleotide microarray analysis (ROMA).[155] These methods have provided a unique view of the cancer genome; however, only GIS-PET is suitable for the direct identification of fusion genes.

VI. Next-Generation Sequencing Technology

The cytogenetic and noncytogenetic methods discussed above are uniquely powerful but their approaches and applications are considered biased. Most of these methods are based on the known human genome information; therefore, most of them are not suitable for gene fusion discovery. Except for the oligonucleotide-based gene expression microarray, all other methods were developed to analyze the genome (DNA) rather than transcriptome (RNA). Gene expression array methods are focused on the analysis of differential gene expression patterns and detection of recurrent copy number changes at a very low resolution. Therefore, a multidisciplinary approach utilizing low-cost, high-throughput, and genome-wide characterization methods has been developed to interrogate the cancer genome on a larger scale. The current initiative of "The Cancer Genome Atlas" (TCGA; http://cancergenome.nih.gov/) was launched to address the fundamental issues of identifying cancer-causing aberrations in at least 25 different cancer types and to provide a catalog of genomic alterations (DNA/RNA-based) to accelerate prognostic biomarker identification, as well as drug development for therapeutic targets. The integrated analysis of the data, generated through this initiative, will help to unravel the genomic complexity of cancer and enhance our understanding of this process at the transcriptome level.

The associated weaknesses with regard to the resolving power and throughput in all the cytogenetic and noncytogenetic methods can be easily overcome with the application of NGS technology. Although NGS is a moving target with rapid technological improvements, it is a unique and powerful approach for the interrogation of the complex cancer genome in an "unbiased" manner. High sample throughput and in-depth sequencing platforms gained wide acceptance for cancer genome characterization through a variety of applications, which include target capture and resequencing, deep sequencing, *de novo* sequencing, transcriptome analysis, whole genome analysis, gene expression analysis, epigenetic analysis, structural variation, interaction of DNA sequences, small RNA discovery, copy number variation (CNV) discovery, ChIP-Seq, comparative genomics, metagenomics, population genomics, complex disease, personal medicine, human microbiome, and viral sequencing (Fig. 8). These methods provide an unprecedented view of both the genome and/or transcriptome. Improvements in the sequencing technology to overcome some of the technical limitations associated with some sequencing platforms such as "single molecule sequencing technology" is emerging to reduce the nonspecific sequences (e.g., those yielding false positive results).

NGS technology has been used to sequence tumor genomic DNA[156] for identification of genomic rearrangements and DNA binding sites by ChIP-Seq[157] at a very high resolution. RNA-Seq is performed using fragmented DNA from cDNA libraries, and can be used for both the "single long read" (sequence obtained from one end of the cDNA fragment) and "paired-end" (sequence obtained from both ends of the fragmented cDNA fragments) approaches. Sequences up to 100 bp can be obtained by "single read" and "paired-end" sequencing. This method overcomes many of the limitations associated with 454 sequencing technology[158] with regard to coverage, as well as the number of false positive reads found in gene fusion analyses.[136]

In RNA-Seq, long read sequences and deep coverage facilitate accurate mapping to the reference genome for the detection of most, if not all, normal and variant transcripts. Further, it provides an accurate quantization of the

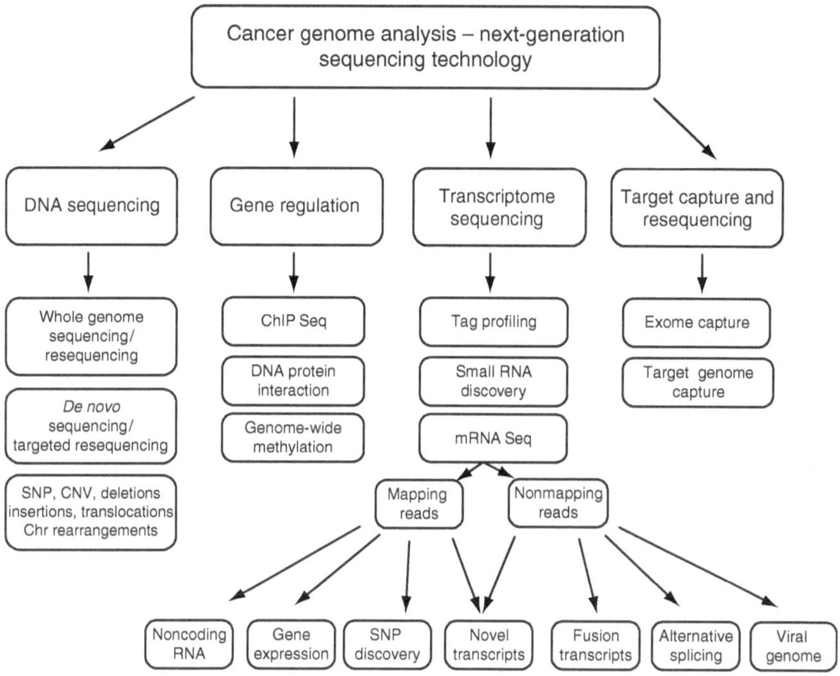

FIG. 8. Applications of next-generation sequencing technology.

expressed transcripts similar to expression array data. Importantly, the RNA-Seq approach readily overcomes many limitations associated with hybridization-based gene expression studies, which require prior knowledge of the genes in a genome of interest, background signals, and utilization of complex normalization methods for data interpretation. Moreover, due to the nature of the transcripts, it is impossible to distinguish between normal, splice variant, and chimeric transcripts generated from the microarray data. Low-throughput and expensive methods, such as serial analysis of gene expression (SAGE),[159] cap analysis of gene expression (CAGE),[160] massively parallel signature sequencing (MPSS),[161] and gene identification signature analysis of paired-end tag analysis (GIS-PET),[152] generate tag sequences from only a portion of a transcript. Using RNA-Seq analysis, however, information for the entire transcript (5′–3′) can be obtained. Recent studies on initial validation and application in cancer genome analysis employing RNA-Seq have confirmed the usefulness of this new technology for the discovery of gene fusions and analysis of variant transcripts, identification of noncoding RNA, and SNP detection.[82,83,136]

RNA-Seq is a valid approach for gene fusion identification as it will provide a direct assessment of all expressed transcripts in a sample. Moreover, it significantly reduces the complexity of the cancer genome because coding sequences from the human genome comprise only 1.5% of the human genome.[162] RNA-Seq analysis of prostate cancer and many other solid cancer transcriptomes has revealed the presence of multiple gene fusions in a sample, posing a bioinformatics challenge to identify the cancer-causing gene fusion. Pioneering studies on the application of this technology firmly established that despite the presence of multiple gene fusions in a sample, the cancer-causing and biologically important gene fusions are always expressed at a higher level. It has been shown that *TMPRSS2-ERG* gene fusion was identified as an outlier gene fusion in the fusion-positive samples based on the ranking of the fusion reads for all of the fusion genes.[136] This approach can identify the cancer-causing (driver) gene fusions within a population of many nonspecific (passenger) or cooperating gene fusions. The likelihood of identifying a recurrent gene fusion for a given cancer type is by the systematic characterization and ranking of all expressed fusion genes across multiple samples. Given the high cost of NGS (expected to be reduced significantly in the near future), sample selection is an important prerequisite to increasing the chances of identifying a recurrent gene fusion, at least in a subgroup of the cancer of interest. Rapid improvements in sequencing technology will allow the generation of longer sequence reads (currently 80–100 bp) with more coverage, which will reduce the informatics challenge to eliminate the nonspecific fusion genes.

VII. Structural Classification of Gene Fusions

Gene fusions in general are produced through interchromosomal and intrachromosomal rearrangements. Interchromosomal gene fusions are mainly due to the simultaneous occurrence of double-strand breaks at the intronic regions. The exchange of DNA segments results in either a balanced (no loss) or unbalanced (loss of one of the rearranged chromosome) translocation. In general, functional fusion genes form at the junction of one of the rearranged chromosomes; however, there are exceptions to two different reciprocal transcripts expressed from both rearranged chromosomes.[82] The functional consequences and biological significance of either or both the fusion genes are determined by constitutive activation of the gene with intact functional domains, irrespective of the level of expression. Interchromosomal aberrations are formed mostly by pericentric (including centromere) or paracentric inversions (without centromere) in a chromosome between genes located in opposite orientations at close or distant locations on the same chromosome. The NGS approach has identified transcription-mediated chimeras (TMCs) or commonly "read-through transcripts," which are a type of gene fusion. Read-through transcripts are generated between two neighboring genes located in the same direction and with or without obvious genomic rearrangements. Due to the poor resolution of the available genomic profiling methods (oligo-based a-CGH and SNP arrays), genomic aberrations, possibly micro deletions, occurring within small genomic intervals (< 20–30 KB) are not detected. For example, the *SLC45A3-ELK4* gene fusion was identified in a subset of prostate cancer that did not carry the well-known ETS family gene rearrangement.[136] Due to the high prevalence of read-through gene fusions in both normal and cancer tissues, the biological significance of such fusions is unknown. Based on the current evidence, these read-through transcripts are treated as a new class of genes. Like normal genes, read-through fusion genes may be regulated in a tissue- or cancer-specific manner. Identification of such fusion genes will help us to understand their true biological significance.

VIII. Functional Classification of Gene Fusions

Based on function, the majority of gene fusions can be classified into "TF" or "TK" genes. Among the various gene fusions identified thus far, given the diverse nature and molecular complexity in solid cancers, identification of majority of the functionally important gene fusions that belong to the TF and TK family underscores the unknown molecular mechanism operating in the selection of these specific aberrations potentially involved in tumorigenesis. Although TFs are functionally similar, the selection of different cancer-specific TF genes indicates a fundamental molecular event necessary in each cancer type to cooperate with the TFs in initiating tumor development. Moreover, it is important to note that TF fusion genes exhibit a remarkable promiscuity in selection of the partner genes, whose regulatory elements are selectively regulated in a tissue-specific manner. For example, the ETS family of genes that are associated gene fusions have been identified both in soft tissue sarcomas and prostate cancer, but the fusion partners are different in each and are regulated in a tissue-specific manner. A majority of the fusion partners of *ERG*,[12] *ETV1*,[88] *ETV4*,[91] and *ETV5*[93] in prostate cancer are androgen-regulated genes. Such selective mechanisms provide clues that may help in the development of treatment strategies targeting the regulator, as well as the regulated genes, particularly the TFs that are difficult to target by conventional therapeutic approaches.

Mutations and rearrangement in the kinase family of genes are well established in different cancers. While TFs regulate the expression of downstream genes by binding to the regulatory elements of the target genes at the DNA level, the expressed kinases regulate by chemical modification (phosphorylation) of the genes in the pathways responsible for growth signaling, differentiation, adhesion, and motility. For example, the RAF family of kinases has been known to be activated both by mutation and rearrangement. Mutations in the *BRAF* gene were found in different solid cancers; for example, there was a high incidence in melanoma.[163] Rearrangements of the RAF family of genes have been identified in many cancer types including prostate, brain, melanoma, thyroid, and gastric[82] at varying frequencies. The prevalence of kinase gene fusions in a solid cancer type mostly identifies a subset of each cancer. Given the prevalence of the same family of genes across different cancer types, perhaps cancer classification should be based on molecular aberrations rather than on the tissue of origin. This new classification system of integrating common molecular subtypes of different cancers would pave the way for research integrating patient groups in order to understand the common molecular etiology of cancer irrespective of the tissue of origin.

Based on the important discoveries, it is promising that many such recurrent gene fusions in other solid cancers may come to light in the near future. These observations underscore gene fusions as a dominant class of mutations across many cancer types, which can be used to develop diagnostic tests targeting the molecular aberrations and potential therapy for the affected individuals. The difficulties associated with characterization of genomic amplifications, deletions, inversions, duplications, and cryptic chromosome translocations can now be easily overcome by employing sequencing technology. An overview of the technologies used in the characterization of chromosome aberrations is presented in the following sections.

IX. Mechanism of the Formation of Gene Fusions in Cancer

With the growing list of gene fusions from a variety of cancers, many fundamental questions remain. What is the mechanism of the formation of gene fusions in cancer? What is the earliest event or the upstream events that trigger the gene fusion formation? Are there any tissue-specific predisposing factors that determine the formation of tissue- or lineage-specific gene fusion? Finding answers to these questions will enhance our understanding to develop ways to prevent rather than cure cancer. The prevailing school of thought has been that (1) cells undergoing genotoxic stress may suffer from random double-strand chromosomal breakages, (2) fusion occurs at random, and (3) selection and fixation of some of the aberrations confers a growth advantage that eventually leads to the formation of a cancer cell. Recently, two independent studies,[164,165] using prostate cancer as a model, have shown that the breakages occur in a nonrandom manner at selective sites dictated by androgen signaling, which brings distant chromosomal segments into proximity. In addition, secondary insult from genotoxic stress (radiation) causes breakage and translocation involving specific genes. The fusion of the ETS family of genes to the androgen-regulated 5′ genes offers an important clue toward understanding the mechanism of the formation of a gene fusion. The requirement of androgen signaling and the simultaneous occurrence of a secondary insult are essential for the formation of gene fusions in prostate cancer. These studies have made a conceptual jump in our understanding on the mechanism of the formation of gene fusions in cancer.

In summary, it has been well established that among various genomic aberrations in cancer, recurrent gene fusions have been identified as a dominant class of mutation in hematological malignancies and solid cancers, and they follow a distinct pattern of occurrence based on their origin, lineage, and tissue specificity, structure, and function. Gene fusions in lymphomas are mostly associated with an *IGH* gene by balanced reciprocal translocation, resulting in the overexpression of apparently normal transcripts at an abnormal level driven by the *IGH* gene regulatory elements. Chromosome rearrangements in leukemia, however, result mostly in the formation of a fusion gene. Promiscuous rearrangement is common in some of the gene fusions, notably *MLL*, with more than 60 3′ partner genes.[166] Cancer-specific translocations and associated gene fusions have been identified in distinct types of sarcomas involving the ETS family of genes and other genes. Promiscuous rearrangement of *EWSR1* is notable, suggesting lineage- and tissue-specific selection of regulatory elements.

The recent discovery of the ETS family rearrangement with the androgen-regulated 5′ partner genes in prostate cancer, the *EML4-ALK* gene fusion in lung cancer, and the RAF family gene rearrangement in a subset of at least five different solid cancers are true testimony for the need to develop more alternate approaches for gene fusion identification. The read-through transcripts discussed above, produced mostly without genomic rearrangements, exist in solid tumors. The majority of these read-through fusions, present both in normal and tumor samples, indicates the widespread occurrence of gene fusions in a variety of cancer types; the biological significance remains to be established for many of them. This is a remarkable and a rewarding achievement for the cancer genomics research community and all the patients who have benefited from this research. With the application of NGS, most, if not all, of the technical limitations associated with cytogenetic and noncytogenetic methods, discussed above, have been reduced or eliminated due to the unbiased nature of the sequencing approach.

A long list of cancer-specific gene fusions from a variety of solid cancers poses a challenge to the research community who seek to understand the significance of the coexistence of multiple gene fusions in a cancer cell. Current and future research is needed to (1) classify the fusions based on their molecular origin; (2) determine the regulation of expression, functional role in cancer, and association with specific pathways; (3) ascertain cancer specificity; (4) identify the biologically important fusion; (5) develop new approaches for development drugs targeting either the gene fusion product or the gene(s) regulated by the fusion genes; and (6) use the developed drugs in treatment strategies targeted to the individual patient.

References

1. Hanahan D, Weinberg RA. The hallmarks of cancer. *Cell* 2000;**100**:57–70.
2. Fearon ER, Vogelstein B. A genetic model for colorectal tumorigenesis. *Cell* 1990;**61**:759–67.
3. Rajagopalan H, Lengauer C. Aneuploidy and cancer. *Nature* 2004;**432**:338–41.
4. Sugimura T, Terada M, Yokota J, Hirohashi S, Wakabayashi K. Multiple genetic alterations in human carcinogenesis. *Environ Health Perspect* 1992;**98**:5–12.
5. Ried T, Heselmeyer-Haddad K, Blegen H, Schrock E, Auer G. Genomic changes defining the genesis, progression, and malignancy potential in solid human tumors: a phenotype/genotype correlation. *Genes Chromosom Cancer* 1999;**25**:195–204.
6. Bartel DP. MicroRNAs: genomics, biogenesis, mechanism, and function. *Cell* 2004;**116**:281–97.
7. Bartel DP. MicroRNAs: target recognition and regulatory functions. *Cell* 2009;**136**:215–33.
8. Cheng J, Kapranov P, Drenkow J, Dike S, Brubaker S, Patel S, et al. Transcriptional maps of 10 human chromosomes at 5-nucleotide resolution. *Science* 2005;**308**:1149–54.
9. Birney E, Stamatoyannopoulos JA, Dutta A, Guigo R, Gingeras TR, Margulies EH, et al. Identification and analysis of functional elements in 1% of the human genome by the ENCODE pilot project. *Nature* 2007;**447**:799–816.
10. Washietl S, Pedersen JS, Korbel JO, Stocsits C, Gruber AR, Hackermuller J, et al. Structured RNAs in the ENCODE selected regions of the human genome. *Genome Res* 2007;**17**:852–64.
11. Poliseno L, Salmena L, Zhang J, Carver B, Haveman WJ, Pandolfi PP. A coding-independent function of gene and pseudogene mRNAs regulates tumour biology. *Nature* 2010;**465**:1033–8.
12. Tomlins SA, Rhodes DR, Perner S, Dhanasekaran SM, Mehra R, Sun XW, et al. Recurrent fusion of TMPRSS2 and ETS transcription factor genes in prostate cancer. *Science* 2005;**310**:644–8.
13. Mano H. Novel EML4-ALK fusion oncogene in lung cancer. *Rinsho Ketsueki* 2009;**50**:468–73.
14. Boveri T. Anton Dohbn. *Science* 1912;**36**:453–68.
15. Tjio JH, Levan A. The chromosome number of man. *Hereditas* 1956;**42**:1–6.
16. Nowell PC, Hungerford DA. A minute chromosome in human chronic granulocytic leukemia. *Science* 1960;**132**:1497–501.
17. Nowell PC, Hungerford DA. Chromosome studies on normal and leukemic human leukocytes. *J Natl Cancer Inst* 1960;**25**:85–109.
18. Caspersson T, Lomakka G, Zech L. The 24 fluorescence patterns of human metaphase chromosomes- distinguishing characters and variability. *Hereditas* 1972;**67**(1):89–102.
19. Rowley JD. Letter: a new consistent chromosomal abnormality in chronic myelogenous leukaemia identified by quinacrine fluorescence and Giemsa staining. *Nature* 1973;**243**:290–3.
20. Ren R. Mechanisms of BCR-ABL in the pathogenesis of chronic myelogenous leukaemia. *Nat Rev Cancer* 2005;**5**:172–83.

21. Li S, Ilaria Jr. RL, Million RP, Daley GQ, Van Etten RA. The P190, P210, and P230 forms of the BCR/ABL oncogene induce a similar chronic myeloid leukemia-like syndrome in mice but have different lymphoid leukemogenic activity. *J Exp Med* 1999;**189**:1399–412.
22. Futreal PA, Coin L, Marshall M, Down T, Hubbard T, Wooster R, et al. A census of human cancer genes. *Nat Rev Cancer* 2004;**4**:177–83.
23. Mitelman F, Johansson B, Mertens F. Fusion genes and rearranged genes as a linear function of chromosome aberrations in cancer. *Nat Genet* 2004;**36**:331–4.
24. Mitelman F, Mertens F, Johansson B. Prevalence estimates of recurrent balanced cytogenetic aberrations and gene fusions in unselected patients with neoplastic disorders. *Genes Chromosom Cancer* 2005;**43**:350–66.
25. Druker BJ, Lydon NB. Lessons learned from the development of an abl tyrosine kinase inhibitor for chronic myelogenous leukemia. *J Clin Invest* 2000;**105**:3–7.
26. Tsai J, Lee JT, Wang W, Zhang J, Cho H, Mamo S, et al. Discovery of a selective inhibitor of oncogenic B-Raf kinase with potent antimelanoma activity. *Proc Natl Acad Sci USA* 2008;**105**:3041–6.
27. McAlpine JN, Wiegand KC, Vang R, Ronnett BM, Adamiak A, Kobel M, et al. HER2 overexpression and amplification is present in a subset of ovarian mucinous carcinomas and can be targeted with trastuzumab therapy. *BMC Cancer* 2009;**9**:433.
28. Novo FJ, de Mendibil IO, Vizmanos JL. TICdb: a collection of gene-mapped translocation breakpoints in cancer. *BMC Genomics* 2007;**8**:33.
29. Mitelman F, Johansson B, Mertens F. The impact of translocations and gene fusions on cancer causation. *Nat Rev Cancer* 2007;**7**:233–45.
30. Delattre O, Zucman J, Plougastel B, Desmaze C, Melot T, Peter M, et al. Gene fusion with an ETS DNA-binding domain caused by chromosome translocation in human tumours. *Nature* 1992;**359**:162–5.
31. Desmaze C, Brizard F, Turc-Carel C, Melot T, Delattre O, Thomas G, et al. Multiple chromosomal mechanisms generate an EWS/FLI1 or an EWS/ERG fusion gene in Ewing tumors. *Cancer Genet Cytogenet* 1997;**97**:12–9.
32. Ishida S, Yoshida K, Kaneko Y, Tanaka Y, Sasaki Y, Urano F, et al. The genomic breakpoint and chimeric transcripts in the EWSR1-ETV4/E1AF gene fusion in Ewing sarcoma. *Cytogenet Cell Genet* 1998;**82**:278–83.
33. Peter M, Couturier J, Pacquement H, Michon J, Thomas G, Magdelenat H, et al. A new member of the ETS family fused to EWS in Ewing tumors. *Oncogene* 1997;**14**:1159–64.
34. Shing DC, McMullan DJ, Roberts P, Smith K, Chin SF, Nicholson J, et al. FUS/ERG gene fusions in Ewing's tumors. *Cancer Res* 2003;**63**:4568–76.
35. Jeon IS, Davis JN, Braun BS, Sublett JE, Roussel MF, Denny CT, et al. A variant Ewing's sarcoma translocation (7;22) fuses the EWS gene to the ETS gene ETV1. *Oncogene* 1995;**10**:1229–34.
36. Kawamura-Saito M, Yamazaki Y, Kaneko K, Kawaguchi N, Kanda H, Mukai H, et al. Fusion between CIC and DUX4 up-regulates PEA3 family genes in Ewing-like sarcomas with t(4;19)(q35;q13) translocation. *Hum Mol Genet* 2006;**15**:2125–37.
37. Mastrangelo T, Modena P, Tornielli S, Bullrich F, Testi MA, Mezzelani A, et al. A novel zinc finger gene is fused to EWS in small round cell tumor. *Oncogene* 2000;**19**:3799–804.
38. Kim J, Lee K, Pelletier J. The DNA binding domains of the WT1 tumor suppressor gene product and chimeric EWS/WT1 oncoprotein are functionally distinct. *Oncogene* 1998;**16**:1021–30.
39. Lawrence B, Perez-Atayde A, Hibbard MK, Rubin BP, Dal Cin P, Pinkus JL, et al. TPM3-ALK and TPM4-ALK oncogenes in inflammatory myofibroblastic tumors. *Am J Pathol* 2000;**157**:377–84.
40. Panagopoulos I, Nilsson T, Domanski HA, Isaksson M, Lindblom P, Mertens F, et al. Fusion of the SEC31L1 and ALK genes in an inflammatory myofibroblastic tumor. *Int J Cancer* 2006;**118**:1181–6.

41. Colleoni GW, Bridge JA, Garicochea B, Liu J, Filippa DA, Ladanyi M. ATIC-ALK: a novel variant ALK gene fusion in anaplastic large cell lymphoma resulting from the recurrent cryptic chromosomal inversion, inv(2)(p23q35). *Am J Pathol* 2000;**156**:781–9.
42. Ma Z, Hill DA, Collins MH, Morris SW, Sumegi J, Zhou M, et al. Fusion of ALK to the Ran-binding protein 2 (RANBP2) gene in inflammatory myofibroblastic tumor. *Genes Chromosom Cancer* 2003;**37**:98–105.
43. Bridge JA, Kanamori M, Ma Z, Pickering D, Hill DA, Lydiatt W, et al. Fusion of the ALK gene to the clathrin heavy chain gene, CLTC, in inflammatory myofibroblastic tumor. *Am J Pathol* 2001;**159**:411–5.
44. Debelenko LV, Arthur DC, Pack SD, Helman LJ, Schrump DS, Tsokos M. Identification of CARS-ALK fusion in primary and metastatic lesions of an inflammatory myofibroblastic tumor. *Lab Invest* 2003;**83**:1255–65.
45. Comin CE, Novelli L, Tornaboni D, Messerini L. Clear cell sarcoma of the ileum: report of a case and review of literature. *Virchows Arch* 2007;**451**:839–45.
46. Filion C, Labelle Y. The oncogenic fusion protein EWS/NOR-1 induces transformation of CFK2 chondrogenic cells. *Exp Cell Res* 2004;**297**:585–92.
47. Panagopoulos I, Mencinger M, Dietrich CU, Bjerkehagen B, Saeter G, Mertens F, et al. Fusion of the RBP56 and CHN genes in extraskeletal myxoid chondrosarcomas with translocation t(9;17)(q22;q11). *Oncogene* 1999;**18**:7594–8.
48. Sjogren H, Wedell B, Meis-Kindblom JM, Kindblom LG, Stenman G. Fusion of the NH2-terminal domain of the basic helix-loop-helix protein TCF12 to TEC in extraskeletal myxoid chondrosarcoma with translocation t(9;15)(q22;q21). *Cancer Res* 2000;**60**:6832–5.
49. Hisaoka M, Ishida T, Imamura T, Hashimoto H. TFG is a novel fusion partner of NOR1 in extraskeletal myxoid chondrosarcoma. *Genes Chromosom Cancer* 2004;**40**:325–8.
50. Sorensen PH, Lynch JC, Qualman SJ, Tirabosco R, Lim JF, Maurer HM, et al. PAX3-FKHR and PAX7-FKHR gene fusions are prognostic indicators in alveolar rhabdomyosarcoma: a report from the children's oncology group. *J Clin Oncol* 2002;**20**:2672–9.
51. Wachtel M, Dettling M, Koscielniak E, Stegmaier S, Treuner J, Simon-Klingenstein K, et al. Gene expression signatures identify rhabdomyosarcoma subtypes and detect a novel t(2;2)(q35;p23) translocation fusing PAX3 to NCOA1. *Cancer Res* 2004;**64**:5539–45.
52. Collins MH, Zhao H, Womer RB, Barr FG. Proliferative and apoptotic differences between alveolar rhabdomyosarcoma subtypes: a comparative study of tumors containing PAX3-FKHR or PAX7-FKHR gene fusions. *Med Pediatr Oncol* 2001;**37**:83–9.
53. de Leeuw B, Balemans M, Olde Weghuis D, Geurts van Kessel A. Identification of two alternative fusion genes, SYT-SSX1 and SYT-SSX2, in t(X;18)(p11.2;q11.2)-positive synovial sarcomas. *Hum Mol Genet* 1995;**4**:1097–9.
54. Skytting B, Nilsson G, Brodin B, Xie Y, Lundeberg J, Uhlen M, et al. A novel fusion gene, SYT-SSX4, in synovial sarcoma. *J Natl Cancer Inst* 1999;**91**:974–5.
55. Kanoe H, Nakayama T, Hosaka T, Murakami H, Yamamoto H, Nakashima Y, et al. Characteristics of genomic breakpoints in TLS-CHOP translocations in liposarcomas suggest the involvement of Translin and topoisomerase II in the process of translocation. *Oncogene* 1999;**18**:721–9.
56. Alaggio R, Coffin CM, Weiss SW, Bridge JA, Issakov J, Oliveira AM, et al. Liposarcomas in young patients: a study of 82 cases occurring in patients younger than 22 years of age. *Am J Surg Pathol* 2009;**33**:645–58.
57. Wang J, Morimitsu Y, Okamoto S, Hisaoka M, Ishida T, Sheng W, et al. COL1A1-PDGFB fusion transcripts in fibrosarcomatous areas of six dermatofibrosarcomas protuberans. *J Mol Diagn* 2000;**2**:47–52.
58. Knezevich SR, McFadden DE, Tao W, Lim JF, Sorensen PH. A novel ETV6-NTRK3 gene fusion in congenital fibrosarcoma. *Nat Genet* 1998;**18**:184–7.

59. Waters BL, Panagopoulos I, Allen EF. Genetic characterization of angiomatoid fibrous histiocytoma identifies fusion of the FUS and ATF-1 genes induced by a chromosomal translocation involving bands 12q13 and 16p11. *Cancer Genet Cytogenet* 2000;**121**:109–16.
60. Argani P, Antonescu CR, Illei PB, Lui MY, Timmons CF, Newbury R, et al. Primary renal neoplasms with the ASPL-TFE3 gene fusion of alveolar soft part sarcoma: a distinctive tumor entity previously included among renal cell carcinomas of children and adolescents. *Am J Pathol* 2001;**159**:179–92.
61. Hernandez-Marti MJ, Orellana-Alonso C, Badia-Garrabou L, Verdeguer Miralles A, Paradis-Alos A. Renal adenocarcinoma in an 8-year-old child, with a t(X;17)(p11.2;q25). *Cancer Genet Cytogenet* 1995;**83**:82–3.
62. Matsuyama A, Hisaoka M, Shimajiri S, Hashimoto H. DNA-based polymerase chain reaction for detecting FUS-CREB3L2 in low-grade fibromyxoid sarcoma using formalin-fixed, paraffin-embedded tissue specimens. *Diagn Mol Pathol* 2008;**17**:237–40.
63. Mertens F, Fletcher CD, Antonescu CR, Coindre JM, Colecchia M, Domanski HA, et al. Clinicopathologic and molecular genetic characterization of low-grade fibromyxoid sarcoma, and cloning of a novel FUS/CREB3L1 fusion gene. *Lab Invest* 2005;**85**:408–15.
64. Panagopoulos I, Mertens F, Griffin CA. An endometrial stromal sarcoma cell line with the JAZF1/PHF1 chimera. *Cancer Genet Cytogenet* 2008;**185**:74–7.
65. Micci F, Panagopoulos I, Bjerkehagen B, Heim S. Consistent rearrangement of chromosomal band 6p21 with generation of fusion genes JAZF1/PHF1 and EPC1/PHF1 in endometrial stromal sarcoma. *Cancer Res* 2006;**66**:107–12.
66. Yamaguchi S, Yamazaki Y, Ishikawa Y, Kawaguchi N, Mukai H, Nakamura T. EWSR1 is fused to POU5F1 in a bone tumor with translocation t(6;22)(p21;q12). *Genes Chromosom Cancer* 2005;**43**:217–22.
67. Butti MG, Bongarzone I, Ferraresi G, Mondellini P, Borrello MG, Pierotti MA. A sequence analysis of the genomic regions involved in the rearrangements between TPM3 and NTRK1 genes producing TRK oncogenes in papillary thyroid carcinomas. *Genomics* 1995;**28**:15–24.
68. Greco A, Miranda C, Pagliardini S, Fusetti L, Bongarzone I, Pierotti MA. Chromosome 1 rearrangements involving the genes TPR and NTRK1 produce structurally different thyroid-specific TRK oncogenes. *Genes Chromosom Cancer* 1997;**19**:112–23.
69. Giannini R, Salvatore G, Monaco C, Sferratore F, Pollina L, Pacini F, et al. Identification of a novel subtype of H4-RET rearrangement in a thyroid papillary carcinoma and lymph node metastasis. *Int J Oncol* 2000;**16**:485–9.
70. Elisei R, Romei C, Castagna MG, Lisi S, Vivaldi A, Faviana P, et al. RET/PTC3 rearrangement and thyroid differentiation gene analysis in a struma ovarii fortuitously revealed by elevated serum thyroglobulin concentration. *Thyroid* 2005;**15**:1355–61.
71. Klugbauer S, Rabes HM. The transcription coactivator HTIF1 and a related protein are fused to the RET receptor tyrosine kinase in childhood papillary thyroid carcinomas. *Oncogene* 1999;**18**:4388–93.
72. Nakata T, Kitamura Y, Shimizu K, Tanaka S, Fujimori M, Yokoyama S, et al. Fusion of a novel gene, ELKS, to RET due to translocation t(10;12)(q11;p13) in a papillary thyroid carcinoma. *Genes Chromosom Cancer* 1999;**25**:97–103.
73. Salassidis K, Bruch J, Zitzelsberger H, Lengfelder E, Kellerer AM, Bauchinger M. Translocation t(10;14)(q11.2:q22.1) fusing the kinetin to the RET gene creates a novel rearranged form (PTC8) of the RET proto-oncogene in radiation-induced childhood papillary thyroid carcinoma. *Cancer Res* 2000;**60**:2786–9.
74. Klugbauer S, Demidchik EP, Lengfelder E, Rabes HM. Detection of a novel type of RET rearrangement (PTC5) in thyroid carcinomas after Chernobyl and analysis of the involved RET-fused gene RFG5. *Cancer Res* 1998;**58**:198–203.
75. Corvi R, Berger N, Balczon R, Romeo G. RET/PCM-1: a novel fusion gene in papillary thyroid carcinoma. *Oncogene* 2000;**19**:4236–42.

76. Saenko V, Rogounovitch T, Shimizu-Yoshida Y, Abrosimov A, Lushnikov E, Roumiantsev P, et al. Novel tumorigenic rearrangement, Delta rfp/ret, in a papillary thyroid carcinoma from externally irradiated patient. *Mutat Res* 2003;**527**:81–90.
77. Ciampi R, Giordano TJ, Wikenheiser-Brokamp K, Koenig RJ, Nikiforov YE. HOOK3-RET: a novel type of RET/PTC rearrangement in papillary thyroid carcinoma. *Endocr Relat Cancer* 2007;**14**:445–52.
78. Kroll TG, Sarraf P, Pecciarini L, Chen CJ, Mueller E, Spiegelman BM, et al. PAX8-PPAR-gamma1 fusion oncogene in human thyroid carcinoma [corrected]. *Science* 2000;**289**:1357–60.
79. Ciampi R, Knauf JA, Kerler R, Gandhi M, Zhu Z, Nikiforova MN, et al. Oncogenic AKAP9-BRAF fusion is a novel mechanism of MAPK pathway activation in thyroid cancer. *J Clin Invest* 2005;**115**:94–101.
80. Forshew T, Tatevossian RG, Lawson AR, Ma J, Neale G, Ogunkolade BW, et al. Activation of the ERK/MAPK pathway: a signature genetic defect in posterior fossa pilocytic astrocytomas. *J Pathol* 2009;**218**:172–81.
81. Dessars B, De Raeve LE, El Housni H, Debouck CJ, Sidon PJ, Morandini R, et al. Chromosomal translocations as a mechanism of BRAF activation in two cases of large congenital melanocytic nevi. *J Invest Dermatol* 2007;**127**:1468–70.
82. Palanisamy N, Ateeq B, Kalyana-Sundaram S, Pflueger D, Ramnarayanan K, Shankar S, et al. Rearrangements of the RAF kinase pathway in prostate cancer, gastric cancer and melanoma. *Nat Med* 2010;**16**(7):793–8.
83. Maher CA, Palanisamy N, Brenner JC, Cao X, Kalyana-Sundaram S, Luo S, et al. Chimeric transcript discovery by paired-end transcriptome sequencing. *Proc Natl Acad Sci USA* 2009;**106**:12353–8.
84. Esgueva R, Perner S, J LaFargue C, Scheble V, Stephan C, Lein M, et al. Prevalence of TMPRSS2-ERG and SLC45A3-ERG gene fusions in a large prostatectomy cohort. *Mod Pathol* 2010;**23**(4):539–46.
85. Pflueger D, Rickman DS, Sboner A, Perner S, LaFargue CJ, Svensson MA, et al. N-myc downstream regulated gene 1 (NDRG1) is fused to ERG in prostate cancer. *Neoplasia* 2009;**11**:804–11.
86. Hermans KG, van der Korput HA, van Marion R, van de Wijngaart DJ, Ziel-van der Made A, Dits NF, et al. Truncated ETV1, fused to novel tissue-specific genes, and full-length ETV1 in prostate cancer. *Cancer Res* 2008;**68**:7541–9.
87. Rickman DS, Pflueger D, Moss B, VanDoren VE, Chen CX, de la Taille A, et al. SLC45A3-ELK4 is a novel and frequent erythroblast transformation-specific fusion transcript in prostate cancer. *Cancer Res* 2009;**69**:2734–8.
88. Tomlins SA, Laxman B, Dhanasekaran SM, Helgeson BE, Cao X, Morris DS, et al. Distinct classes of chromosomal rearrangements create oncogenic ETS gene fusions in prostate cancer. *Nature* 2007;**448**:595–9.
89. Attard G, Clark J, Ambroisine L, Mills IG, Fisher G, Flohr P, et al. Heterogeneity and clinical significance of ETV1 translocations in human prostate cancer. *Br J Cancer* 2008;**99**:314–20.
90. Han B, Mehra R, Dhanasekaran SM, Yu J, Menon A, Lonigro RJ, et al. A fluorescence in situ hybridization screen for E26 transformation-specific aberrations: identification of DDX5-ETV4 fusion protein in prostate cancer. *Cancer Res* 2008;**68**:7629–37.
91. Tomlins SA, Mehra R, Rhodes DR, Smith LR, Roulston D, Helgeson BE, et al. TMPRSS2:ETV4 gene fusions define a third molecular subtype of prostate cancer. *Cancer Res* 2006;**66**:3396–400.
92. Hermans KG, Bressers AA, van der Korput HA, Dits NF, Jenster G, Trapman J. Two unique novel prostate-specific and androgen-regulated fusion partners of ETV4 in prostate cancer. *Cancer Res* 2008;**68**:3094–8.

93. Helgeson BE, Tomlins SA, Shah N, Laxman B, Cao Q, Prensner JR, et al. Characterization of TMPRSS2:ETV5 and SLC45A3:ETV5 gene fusions in prostate cancer. *Cancer Res* 2008;**68**:73–80.
94. Shinmura K, Kageyama S, Tao H, Bunai T, Suzuki M, Kamo T, et al. EML4-ALK fusion transcripts, but no NPM-, TPM3-, CLTC-, ATIC-, or TFG-ALK fusion transcripts, in non-small cell lung carcinomas. *Lung Cancer* 2008;**61**:163–9.
95. Hallor KH, Mertens F, Jin Y, Meis-Kindblom JM, Kindblom LG, Behrendtz M, et al. Fusion of the EWSR1 and ATF1 genes without expression of the MITF-M transcript in angiomatoid fibrous histiocytoma. *Genes Chromosom Cancer* 2005;**44**:97–102.
96. Buijs A, van Rompaey L, Molijn AC, Davis JN, Vertegaal AC, Potter MD, et al. The MN1-TEL fusion protein, encoded by the translocation (12;22)(p13;q11) in myeloid leukemia, is a transcription factor with transforming activity. *Mol Cell Biol* 2000;**20**:9281–93.
97. Charest A, Lane K, McMahon K, Park J, Preisinger E, Conroy H, et al. Fusion of FIG to the receptor tyrosine kinase ROS in a glioblastoma with an interstitial del(6)(q21q21). *Genes Chromosom Cancer* 2003;**37**:58–71.
98. Tognon C, Knezevich SR, Huntsman D, Roskelley CD, Melnyk N, Mathers JA, et al. Expression of the ETV6-NTRK3 gene fusion as a primary event in human secretory breast carcinoma. *Cancer Cell* 2002;**2**:367–76.
99. Skalova A, Vanecek T, Sima R, Laco J, Weinreb I, Perez-Ordonez B, et al. Mammary analogue secretory carcinoma of salivary glands, containing the ETV6-NTRK3 fusion gene: a hitherto undescribed salivary gland tumor entity. *Am J Surg Pathol* 2010;**34**:599–608.
100. Anderson J, Gibson S, Sebire NJ. Expression of ETV6-NTRK in classical, cellular and mixed subtypes of congenital mesoblastic nephroma. *Histopathology* 2006;**48**:748–53.
101. Rajaram V, Knezevich S, Bove KE, Perry A, Pfeifer JD. DNA sequence of the translocation breakpoints in undifferentiated embryonal sarcoma arising in mesenchymal hamartoma of the liver harboring the t(11;19)(q11;q13.4) translocation. *Genes Chromosom Cancer* 2007;**46**:508–13.
102. Sidhar SK, Clark J, Gill S, Hamoudi R, Crew AJ, Gwilliam R, et al. The t(X;1)(p11.2;q21.2) translocation in papillary renal cell carcinoma fuses a novel gene PRCC to the TFE3 transcription factor gene. *Hum Mol Genet* 1996;**5**:1333–8.
103. Aulmann S, Longerich T, Schirmacher P, Mechtersheimer G, Penzel R. Detection of the ASPSCR1-TFE3 gene fusion in paraffin-embedded alveolar soft part sarcomas. *Histopathology* 2007;**50**:881–6.
104. Kuiper RP, Schepens M, Thijssen J, van Asseldonk M, van den Berg E, Bridge J, et al. Upregulation of the transcription factor TFEB in t(6;11)(p21;q13)-positive renal cell carcinomas due to promoter substitution. *Hum Mol Genet* 2003;**12**:1661–9.
105. Davis IJ, Hsi BL, Arroyo JD, Vargas SO, Yeh YA, Motyckova G, et al. Cloning of an Alpha-TFEB fusion in renal tumors harboring the t(6;11)(p21;q13) chromosome translocation. *Proc Natl Acad Sci USA* 2003;**100**:6051–6.
106. Tanaka M, Kato K, Gomi K, Matsumoto M, Kudo H, Shinkai M, et al. Perivascular epithelioid cell tumor with SFPQ/PSF-TFE3 gene fusion in a patient with advanced neuroblastoma. *Am J Surg Pathol* 2009;**33**:1416–20.
107. Kuroda N, Tamura M, Tanaka Y, Hes O, Michal M, Inoue K, et al. Adult-onset renal cell carcinoma associated with Xp11.2 translocations/TFE3 gene fusion with smooth muscle stroma and abnormal vessels. *Pathol Int* 2009;**59**:486–91.
108. Argani P, Lui MY, Couturier J, Bouvier R, Fournet JC, Ladanyi M. A novel CLTC-TFE3 gene fusion in pediatric renal adenocarcinoma with t(X;17)(p11.2;q23). *Oncogene* 2003;**22**:5374–8.
109. Heimann P, El Housni H, Ogur G, Weterman MA, Petty EM, Vassart G. Fusion of a novel gene, RCC17, to the TFE3 gene in t(X;17)(p11.2;q25.3)-bearing papillary renal cell carcinomas. *Cancer Res* 2001;**61**:4130–5.
110. Adelaide J, Chaffanet M, Mozziconacci MJ, Popovici C, Conte N, Fernandez F, et al. Translocation and coamplification of loci from chromosome arms 8p and 11q in the MDA-MB-175 mammary carcinoma cell line. *Int J Oncol* 2000;**16**:683–8.

111. Wang XZ, Jolicoeur EM, Conte N, Chaffanet M, Zhang Y, Mozziconacci MJ, et al. gamma-heregulin is the product of a chromosomal translocation fusing the DOC4 and HGL/NRG1 genes in the MDA-MB-175 breast cancer cell line. *Oncogene* 1999;**18**:5718–21.
112. Hahn Y, Bera TK, Gehlhaus K, Kirsch IR, Pastan IH, Lee B. Finding fusion genes resulting from chromosome rearrangement by analyzing the expressed sequence databases. *Proc Natl Acad Sci USA* 2004;**101**:13257–61.
113. French CA, Ramirez CL, Kolmakova J, Hickman TT, Cameron MJ, Thyne ME, et al. BRD-NUT oncoproteins: a family of closely related nuclear proteins that block epithelial differentiation and maintain the growth of carcinoma cells. *Oncogene* 2008;**27**:2237–42.
114. Kas K, Voz ML, Roijer E, Astrom AK, Meyen E, Stenman G, et al. Promoter swapping between the genes for a novel zinc finger protein and beta-catenin in pleiomorphic adenomas with t(3;8)(p21;q12) translocations. *Nat Genet* 1997;**15**:170–4.
115. Voz ML, Astrom AK, Kas K, Mark J, Stenman G, Van de Ven WJ. The recurrent translocation t (5;8)(p13;q12) in pleomorphic adenomas results in upregulation of PLAG1 gene expression under control of the LIFR promoter. *Oncogene* 1998;**16**:1409–16.
116. Asp J, Persson F, Kost-Alimova M, Stenman G. CHCHD7-PLAG1 and TCEA1-PLAG1 gene fusions resulting from cryptic, intrachromosomal 8q rearrangements in pleomorphic salivary gland adenomas. *Genes Chromosom Cancer* 2006;**45**:820–8.
117. Geurts JM, Schoenmakers EF, Roijer E, Stenman G, Van de Ven WJ. Expression of reciprocal hybrid transcripts of HMGIC and FHIT in a pleomorphic adenoma of the parotid gland. *Cancer Res* 1997;**57**:13–7.
118. Nilsson M, Panagopoulos I, Mertens F, Mandahl N. Fusion of the HMGA2 and NFIB genes in lipoma. *Virchows Arch* 2005;**447**:855–8.
119. Pierron A, Fernandez C, Saada E, Keslair F, Hery G, Zattara H, et al. HMGA2-NFIB fusion in a pediatric intramuscular lipoma: a novel case of NFIB alteration in a large deep-seated adipocytic tumor. *Cancer Genet Cytogenet* 2009;**195**:66–70.
120. Winnes M, Molne L, Suurkula M, Andren Y, Persson F, Enlund F, et al. Frequent fusion of the CRTC1 and MAML2 genes in clear cell variants of cutaneous hidradenomas. *Genes Chromosom Cancer* 2007;**46**:559–63.
121. Nakayama T, Miyabe S, Okabe M, Sakuma H, Ijichi K, Hasegawa Y, et al. Clinicopathological significance of the CRTC3-MAML2 fusion transcript in mucoepidermoid carcinoma. *Mod Pathol* 2009;**22**:1575–81.
122. Pearson MA, Fabbro D. Targeting protein kinases in cancer therapy: a success? *Expert Rev Anticancer Ther* 2004;**4**:1113–24.
123. Grant SK. Therapeutic protein kinase inhibitors. *Cell Mol Life Sci* 2009;**66**:1163–77.
124. Brennan P, Donev R, Hewamana S. Targeting transcription factors for therapeutic benefit. *Mol Biosyst* 2008;**4**:909–19.
125. Frank DA. Targeting transcription factors for cancer therapy. *IDrugs* 2009;**12**:29–33.
126. Hurst HC. Transcription factors as drug targets in cancer. *Eur J Cancer* 1996;**32A**:1857–63.
127. Soda M, Choi YL, Enomoto M, Takada S, Yamashita Y, Ishikawa S, et al. Identification of the transforming EML4-ALK fusion gene in non-small-cell lung cancer. *Nature* 2007;**448**:561–6.
128. MacDonald JW, Ghosh D. COPA—cancer outlier profile analysis. *Bioinformatics* 2006;**22**:2950–1.
129. Claxton DF, Liu P, Hsu HB, Marlton P, Hester J, Collins F, et al. Detection of fusion transcripts generated by the inversion 16 chromosome in acute myelogenous leukemia. *Blood* 1994;**83**:1750–6.
130. Mehra R, Tomlins SA, Yu J, Cao X, Wang L, Menon A, et al. Characterization of TMPRSS2-ETS gene aberrations in androgen-independent metastatic prostate cancer. *Cancer Res* 2008;**68**:3584–90.

131. Sinclair PB, Nacheva EP, Leversha M, Telford N, Chang J, Reid A, et al. Large deletions at the t(9;22) breakpoint are common and may identify a poor-prognosis subgroup of patients with chronic myeloid leukemia. *Blood* 2000;**95**:738–43.
132. Varambally S, Cao Q, Mani RS, Shankar S, Wang X, Ateeq B, et al. Genomic loss of microRNA-101 leads to overexpression of histone methyltransferase EZH2 in cancer. *Science* 2008;**322**:1695–9.
133. Tanner MM, Tirkkonen M, Kallioniemi A, Holli K, Collins C, Kowbel D, et al. Amplification of chromosomal region 20q13 in invasive breast cancer: prognostic implications. *Clin Cancer Res* 1995;**1**:1455–61.
134. Haverty PM, Fridlyand J, Li L, Getz G, Beroukhim R, Lohr S, et al. High-resolution genomic and expression analyses of copy number alterations in breast tumors. *Genes Chromosom Cancer* 2008;**47**:530–42.
135. Kan Z, Jaiswal BS, Stinson J, Janakiraman V, Bhatt D, Stern HM, et al. Diverse somatic mutation patterns and pathway alterations in human cancers. *Nature* 2010; doi:10.1038/nature09208.
136. Maher CA, Kumar-Sinha C, Cao X, Kalyana-Sundaram S, Han B, Jing X, et al. Transcriptome sequencing to detect gene fusions in cancer. *Nature* 2009;**458**:97–101.
137. Sandberg AA, Turc-Carel C, Gemmill RM. Chromosomes in solid tumors and beyond. *Cancer Res* 1988;**48**:1049–59.
138. Mandahl N, Heim S, Arheden K, Rydholm A, Willen H, Mitelman F. Three major cytogenetic subgroups can be identified among chromosomally abnormal solitary lipomas. *Hum Genet* 1988;**79**:203–8.
139. Mark J, Havel G, Grepp C, Dahlenfors R, Wedell B. Cytogenetical observations in human benign uterine leiomyomas. *Anticancer Res* 1988;**8**:621–6.
140. Atkin NB. Lack of reciprocal translocations in carcinomas. *Cancer Genet Cytogenet* 1986;**21**:275–8.
141. Schrock E, du Manoir S, Veldman T, Schoell B, Wienberg J, Ferguson-Smith MA, et al. Multicolor spectral karyotyping of human chromosomes. *Science* 1996;**273**:494–7.
142. Iida S, Rao PH, Nallasivam P, Hibshoosh H, Butler M, Louie DC, et al. The t(9;14)(p13;q32) chromosomal translocation associated with lymphoplasmacytoid lymphoma involves the PAX-5 gene. *Blood* 1996;**88**:4110–7.
143. Dyomin VG, Palanisamy N, Lloyd KO, Dyomina K, Jhanwar SC, Houldsworth J, et al. MUC1 is activated in a B-cell lymphoma by the t(1;14)(q21;q32) translocation and is rearranged and amplified in B-cell lymphoma subsets. *Blood* 2000;**95**:2666–71.
144. Chen W, Palanisamy N, Schmidt H, Teruya-Feldstein J, Jhanwar SC, Zelenetz AD, et al. Deregulation of FCGR2B expression by 1q21 rearrangements in follicular lymphomas. *Oncogene* 2001;**20**:7686–93.
145. Hatzivassiliou G, Miller I, Takizawa J, Palanisamy N, Rao PH, Iida S, et al. IRTA1 and IRTA2, novel immunoglobulin superfamily receptors expressed in B cells and involved in chromosome 1q21 abnormalities in B cell malignancy. *Immunity* 2001;**14**:277–89.
146. Dewald GW, Wyatt WA, Juneau AL, Carlson RO, Zinsmeister AR, Jalal SM, et al. Highly sensitive fluorescence in situ hybridization method to detect double BCR/ABL fusion and monitor response to therapy in chronic myeloid leukemia. *Blood* 1998;**91**:3357–65.
147. Barrett MT, Scheffer A, Ben-Dor A, Sampas N, Lipson D, Kincaid R, et al. Comparative genomic hybridization using oligonucleotide microarrays and total genomic DNA. *Proc Natl Acad Sci USA* 2004;**101**:17765–70.
148. Kawamata N, Ogawa S, Zimmermann M, Niebuhr B, Stocking C, Sanada M, et al. Cloning of genes involved in chromosomal translocations by high-resolution single nucleotide polymorphism genomic microarray. *Proc Natl Acad Sci USA* 2008;**105**:11921–6.

149. Lisitsyn N, Wigler M. Cloning the differences between two complex genomes. *Science* 1993;**259**:946–51.
150. Hatada I, Hayashizaki Y, Hirotsune S, Komatsubara H, Mukai T. A genomic scanning method for higher organisms using restriction sites as landmarks. *Proc Natl Acad Sci USA* 1991;**88**:9523–7.
151. Hirotsune S, Hatada I, Komatsubara H, Nagai H, Kuma K, Kobayakawa K, et al. New approach for detection of amplification in cancer DNA using restriction landmark genomic scanning. *Cancer Res* 1992;**52**:3642–7.
152. Ng P, Wei CL, Sung WK, Chiu KP, Lipovich L, Ang CC, et al. Gene identification signature (GIS) analysis for transcriptome characterization and genome annotation. *Nat Meth* 2005;**2**:105–11.
153. Volik S, Zhao S, Chin K, Brebner JH, Herndon DR, Tao Q, et al. End-sequence profiling: sequence-based analysis of aberrant genomes. *Proc Natl Acad Sci USA* 2003;**100**:7696–701.
154. Wang TL, Maierhofer C, Speicher MR, Lengauer C, Vogelstein B, Kinzler KW, et al. Digital karyotyping. *Proc Natl Acad Sci USA* 2002;**99**:16156–61.
155. Jobanputra V, Sebat J, Troge J, Chung W, Anyane-Yeboa K, Wigler M, et al. Application of ROMA (representational oligonucleotide microarray analysis) to patients with cytogenetic rearrangements. *Genet Med* 2005;**7**:111–8.
156. Quail MA, Kozarewa I, Smith F, Scally A, Stephens PJ, Durbin R, et al. A large genome center's improvements to the Illumina sequencing system. *Nat Meth* 2008;**5**:1005–10.
157. Jothi R, Cuddapah S, Barski A, Cui K, Zhao K. Genome-wide identification of in vivo protein-DNA binding sites from ChIP-Seq data. *Nucleic Acids Res* 2008;**36**:5221–31.
158. Emrich SJ, Barbazuk WB, Li L, Schnable PS. Gene discovery and annotation using LCM-454 transcriptome sequencing. *Genome Res* 2007;**17**:69–73.
159. Velculescu VE, Zhang L, Vogelstein B, Kinzler KW. Serial analysis of gene expression. *Science* 1995;**270**:484–7.
160. Kodzius R, Kojima M, Nishiyori H, Nakamura M, Fukuda S, Tagami M, et al. CAGE: cap analysis of gene expression. *Nat Meth* 2006;**3**:211–22.
161. Brenner S, Johnson M, Bridgham J, Golda G, Lloyd DH, Johnson D, et al. Gene expression analysis by massively parallel signature sequencing (MPSS) on microbead arrays. *Nat Biotechnol* 2000;**18**:630–4.
162. Lander ES, Linton LM, Birren B, Nusbaum C, Zody MC, Baldwin J, et al. Initial sequencing and analysis of the human genome. *Nature* 2001;**409**:860–921.
163. Davies H, Bignell GR, Cox C, Stephens P, Edkins S, Clegg S, et al. Mutations of the BRAF gene in human cancer. *Nature* 2002;**417**:949–54.
164. Mani RS, Tomlins SA, Callahan K, Ghosh A, Nyati MK, Varambally S, et al. Induced chromosomal proximity and gene fusions in prostate cancer. *Science* 2009;**326**:1230.
165. Lin C, Yang L, Tanasa B, Hutt K, Ju BG, Ohgi K, et al. Nuclear receptor-induced chromosomal proximity and DNA breaks underlie specific translocations in cancer. *Cell* 2009;**139**:1069–83.
166. Collins EC, Rabbitts TH. The promiscuous MLL gene links chromosomal translocations to cellular differentiation and tumour tropism. *Trends Mol Med* 2002;**8**:436–42.

Circulating Tumor Cells

Daniel F. Hayes and
Jeffrey B. Smerage

Breast Oncology Program, University of Michigan Comprehensive Cancer Center, Ann Arbor, Michigan, USA

Circulating tumor cells (CTCs) can be separated and characterized from normal hematopoietic cellular constituents by a variety of methods. Different strategies have included separation by physical characteristics, such as size or weight, or by biological characteristics, such as expression of epithelial or cancer-specific markers. Of the latter, rtPCR for epithelial-related gene message, such as cytokeratin, and immunoseparation techniques using monoclonal antibodies against epithelial cellular adhesion molecule, have gained the most widespread use in investigational and standard clinical application to date. Detection and monitoring of CTCs might be useful for screening, prognosis, prediction of response to therapy, or monitoring clinical course in patients with primary or metastatic cancer. Currently, monitoring patients with metastatic disease is the most practical application of CTCs. In this regard, several studies have demonstrated that approximately 50–70% of patients with metastatic breast, colon, and prostate cancers have elevated CTC levels, when evaluated using a highly automated immunomagnetic CTC assay system, designated CellSearch®. These studies demonstrate that elevated CTC levels prior to initiation of a new systemic therapy are associated with a worse prognosis than those that do not, and that persistently elevated or subsequent rising CTC levels strongly suggest that the therapeutic regimen with which the patient is being treated is not working. Similar results have been shown with rtPCR assays, although they are not as widely available for routine clinical use. New areas of research are directed toward developing more sensitive means of CTC detection and generating a variety of methods to characterize the

Progress in Molecular Biology
and Translational Science, Vol. 95
DOI: 10.1016/S1877-1173(10)95005-9

molecular and biologic nature of CTCs, such as the status of hormone receptors, epidermal, and other growth factor receptor family members, and indications of stem-cell characteristics.

I. Introduction

The process of metastases of epithelial cancers is complex, involving a variety of phenotypic and morphologic changes by a cancer cell that distinguishes it from its normal counterpart. These include separation from its normal location by loss of adhesion, invasion, and migration into and out of surrounding vascular structures, survival in hostile environments during transit, and the uncontrolled proliferation at other locations.[1] Thus, it is reasonable to hypothesize that detection and monitoring of circulating tumor cells (CTCs) might provide clinically important diagnostic information. The presence of malignant cells in the human vascular space was first reported in 1869 by Ashworth, who described and illustrated obvious evidence of cancer cells within a blood vessel of a patient who had died of cancer.[2] However, despite various attempts, the technology to reliably separate CTCs from the hematopoietic milieu in which they exist prevented clinical application of this observation for the next 130 years. A number of recent technological advances have permitted development of potentially clinical useful assays to detect, enumerate, monitor, and characterize CTCs, and are discussed in this chapter. Because the vast majority of the available data have been generated in regard to breast cancer, this review focuses on that particular disease unless otherwise stated.

II. What Are the Technological Issues?

For the common epithelial malignancies, the number of cancer cells that exist in blood at any given time appears to be quite small. Even in the metastatic setting, CTCs are rare events. Existing technologies suggest that the number of cancer cells per 10 ml of whole blood, if present at all, usually ranges between 1 and 100, compared to billions of red and millions of white corpuscles. Thus, the challenge is to identify these rare events, and then further characterize them to be certain that they are, indeed, malignant and not normal cells.

Several approaches have been taken to accomplish these tasks. Most strategies involve an isolation step, to separate CTCs from red and white blood cells (RBC, WBC), with a second characterization analysis. One broad approach to the isolation step takes advantage of the size and weight difference between cancer and normal blood cells. Both filtration through micromembranes and density gradient separation strategies have been reported. However, in general, these methods have proved too crude to reliably separate the cancer cells with sufficient recovery to be of any clinical meaning. In part, the failure of these approaches is related to the very small number of cells within blood and the loss of cells due to physical manipulation. Furthermore, recent studies have suggested that one feature of the metastatic phenotype may be epithelial–mesenchymal transformation, in which the large and previously inflexible epithelial cell takes on fluid and plastic characteristics of mesenchymal cells.[1] In other words, the cancer cell becomes like a macrophage, giving it the ability to more freely migrate through small passages *in vivo*, and therefore, it might avoid separation by filtration *in vitro*. This drawback may not be insurmountable. Recent reports suggest that highly charged filters may exploit both size and electro-charge differences between epithelial cancer and normal blood cells.[3] Although intriguing, studies with this device are in very early stages and require substantially more investigation.

Biologic strategies to isolate and characterize tumor cells have generally been more successful, to date. These strategies separate or identify the cancer cells from normal hematopoietic constituents by virtue of expression of genes that distinguish the two. The most common approach takes advantage of expression of epithelial-specific genes that are not necessarily related to the malignant phenotype but cannot be detected in normal white or red blood cells. These genes include the cytokeratins (CK),[4] epithelial cellular adhesion molecule (EpCAM),[5] mucin-1 (MUC1),[6] and/or mammoglobin.[7] An alternative approach involves identification of cancer-related or specific molecules. However, the latter has, so far, proved to be insufficiently sensitive (since few cancer-associated molecules are widely expressed across the gamut of any specific type of cancer), or specific (since many of these genes are often expressed in normal leukocytes).[8,9]

Biologic differences between epithelial cancer and normal hematopoietic cells can be identified with a number of techniques. The advent of reverse transcriptase polymerase chain reaction (rtPCR) permits very sensitive detection of minute amounts of transcription of a gene in an environment in which it is not normally produced. rtPCR has been used to detect expression of epithelial-specific genes, such as cytokeratin, as well as cancer-specific genes, such as HER2 in breast cancer.[10] Relative concentrations can be determined using real time rtPCR, and normalized to a standard curve to estimate the number of CTCs per unit volume, such as 1 or 10 ml.

The rtPCR method appears to be quite sensitive.[11] Two potential drawbacks to this method occur because it does not involve any morphologic evaluation of detected events. The first problem is the potential contamination of the preparation, since this technique is very sensitive and can detect even one copy of a transcript, or even illegitimate expression of a putative epithelial-specific gene by leucocytes. For example, expression of the *MUC1* gene is mostly confined to epithelial cells, but has been reported in early leukocyte progenitor cells.[12] Thus, MUC-1 may be detected in circulation in a stressed patient with bone marrow involvement or after chemotherapy or infection. The second is the inability to further characterize the cells for other geno- and pheno-types of interest, since there is no cellular component to these assays.

Another, and recently very commonly applied, technique involves immune-separation of CTCs by conjugating antibodies against a surface antigen of interest, such as EpCAM, with a solid-phase structure, thus positively separating the CTCs from the remaining blood constituents. However, even with this approach WBC continue to contaminate the preparation by both passive and active means. For example, WBC can be seen to phagocytose epithelial cells, and thus partially exposed epithelial membranes are sufficient to result in WBC capture on the solid phase. Thus, a second, characterization step is needed to demonstrate that the captured cells are, indeed, epithelial, and preferably, malignant.

Over the last decade, a highly automated immunoseparation and characterization system, designated CellSearch® (Veridex, LLC; Raritan, NJ),[13] has been developed. The immune-separation component of this system is based on microscopic ferrous particles that remain in suspension and that have been coated with a monoclonal antibody directed against EpCAM (MAb VU-1D9).[13] EpCAM is expressed by 50–100% of cells in approximately 80% of human epithelial malignancies.[14] Following collection of whole blood into a fixative-containing vacutainer tube, the specimen is then incubated with these immunoferrous particles and subjected to a magnetic field that actively removes all magnetic events, eliminating most, but not all, red and white blood cells.

A second analytical step is performed to determine the nature of the separated events. The preparation is incubated with a cocktail of fluorescently labeled monoclonal antibodies directed against CK4–6, 8, 10, 13, and 18 (MAb C11) and CK 19 (MAb A53-B/A2) and against CD 45 (MAb APC), as well as with DAPI. This approach permits the operator to identify epithelial, nucleated events, which are presumably epithelial cells, and to disregard nonnucleated events (DAPI negative), bare nuclei (DAPI positive, CK and APC negative), and/or WBC (APC positive, DAPI positive). This entire process has been completely automated and is approved by the FDA for routine clinical use in patients with breast cancer, as well as colorectal and prostate cancers, to monitor and to help determine the effectiveness of cancer treatment.

This system, of course, also has shortcomings. One potential issue is the relatively poor sensitivity, especially compared to rtPCR. One would expect lower sensitivity with this particular system since it requires that CTCs express both EpCAM for capture and CK for analysis. For example, immunohistochemical studies suggest that up to 20% of breast cancers do not express EpCAM.[15] Indeed, expression of EpCAM may be related to the intrinsic biologic cancer subtype,[16] and, furthermore, cancers may lose EpCAM expression during the metastatic process.[17] Cellular capture assays also require multiple manipulations in which cells are almost certainly mechanically damaged and physically lost.

Moreover, loss of detectable cells occurs due to cellular death during the delay from blood draw to assay. Indeed, a critical component of the CellSearch® system is the development and absolute requirement for use of vacutainers containing fixatives (CellSave™, Veridex, LLC, Raritan, NJ). Without these, delayed analyses were shown to result in much lower CTC counts than immediate evaluation. However, studies have demonstrated that results testing two specimens drawn from the same patient at the same time but analyzed in separate institutions are almost identical when whole blood is collected into CellSave™ tubes.[18] Even with cellular preservative, CTC detection declines after storage for more than 96 h. Thus, this technology also limits use of archived specimens, necessitating prospective studies to enumerate and characterize cells.

Recently, technologies using the same immunoseparation and characterization strategy as CellSearch® but based on different solid-state platforms other than ferrous magnetic separation have been reported.[19,20] In one of these assays, a monoclonal antibody directed against EpCAM has been applied to microposts on a glass chip, and whole blood is applied across the posts by microflow technique. These investigators have reported remarkable sensitivity with this assay. Indeed, their results suggest that nearly 100% of patients with a variety of malignancies have detectable CTCs, often at very high levels of 100–1000 or more cells per 10 ml blood. This assay format also apparently provides the opportunity to further characterize the cells for cancer-specific, and potentially clinically important, molecular features, such as mutations in epithelial growth factor receptor (EGFR).[21] However, at present, this assay is available only in a highly specialized research laboratory; its use appears cumbersome and perhaps operator dependent, and careful studies documenting differential clinical outcomes associated with the presence or absence of CTCs detected by this system have not been reported.

In a second assay, designated "MagSweep®," a neodymium magnetic rod is robotically swept through a well containing the labeled sample.[20] The cells that initially stick to the rod are then washed and released from the rod. In samples of blood spiked with MCF-7 cells, this system was found to have a capture efficiency of 62%, and the captured sample was found to have a purity of 51% MCF-7 cells. The assay has been further evaluated using blood samples from 17 women with metastatic breast cancer and from five healthy volunteers.

CTCs were identified in all the patients with metastatic cancer, and no CTCs were identified in the healthy volunteers. In the patients with metastatic disease, the number of cells isolated from 1 to 100 with an average of 12 cells per 9 ml of whole blood. The prognostic or predictive values of CTCs identified using this system are not yet known.

These technological strategies are not mutually exclusive. Investigators have combined physical separation with molecular characterization, and immunologic separation and characterization with subsequent gene expression analysis. Indeed, the ability to further characterize CTCs above and beyond simple enumeration has been reported for a number of genetic and protein abnormalities, and is discussed below.

III. What Are the Clinical Utilities of CTCs Detection and Enumeration?

Any tumor marker may be valuable in one of several uses, including risk categorization, screening, differential diagnosis, prognosis, prediction of response to specific therapies, and monitoring patients with established cancer.[22] Of these, the most appealing uses for CTCs have been screening for new malignancies, determining prognosis in newly diagnosed cancers, and monitoring to determine disease status.

A. Screening

Development of an effective screening diagnostic tool is very difficult. For screening of a malignancy to be cost-effective, the tool must be sensitive and specific, relatively inexpensive, and easy to apply to a large population, and the disease must be better treated early than late in regard to either improvement in survival or quality of life.[23] In this regard, few diagnostic approaches have gained clinical utility for cancer screening. These include mammography for breast cancer, Pap smear testing for cervical cancer, and either radiographic or colonoscopy screening for colorectal cancers. No serial blood-based assay has been proved adequate for cancer screening. Although prostate-specific antigen (PSA) protein has gained widespread use among clinicians, two prospective randomized clinical trials have suggested that it is either only minimally, or not at all, effective in reducing mortality from prostate cancer.[24,25]

At this time, no assay for CTCs has proved to have sufficient sensitivity and specificity to even remotely approach that is necessary for screening for any epithelial malignancy. There are no reported studies of use of CTCs for this indication.

B. Prognosis

A marker that provides an indication of future behavior of a cancer is of great value in therapeutic planning. Risk of a future event can be determined either before any therapy, or residual risk after some therapy can be determined to assess whether additional treatment might be indicated.

1. Early-Stage Disease

The TNM staging system has been developed to determine prognosis, principally after surgery and or other local therapies in newly diagnosed malignancies. In general, patients who have large tumors and/or positive regional lymph nodes have a higher chance of distant recurrence, and death, than those with small tumors and uninvolved lymph nodes. However, these anatomic staging indications are far from perfect. Therefore, detection of micrometastases in distant sites, such as bone marrow or blood, has been studied intensely. Several separate studies, and an international consortium to pool many studies, have demonstrated that approximately 20% of patients with newly diagnosed early-stage breast cancer have detectable bone marrow micrometastases and that these patients have a significantly worse prognosis than those without marrow involvement.[26] However, it is not clear whether the presence of bone marrow micrometastases adds to already existing prognostic factors, and routine performance of bone marrow evaluation is not recommended for patients with newly diagnosed breast cancer.[27]

Nonetheless, several investigators have evaluated the prognostic effect of CTCs in patients with early-stage malignancy, mostly breast cancer.[28] For example, Ignatiadis *et al.* have reported that approximately 40% of patients with stage I–III breast cancer had evidence of CTCs, as determined by rtPCR for cytokeratin 19.[29] The presence of CTCs was independent of estrogen receptor (ER)or HER2 status. Patients with elevated CTC levels had a higher risk of relapse after approximately 5 years, but only if they had ER and HER2 negative breast cancers. In contrast, investigators using immunomagnetic techniques, specifically CellSearch®, have reported a lower incidence of positivity: approximately 10%.[30,31] These investigators have recently reported that those patients with detectable CTCs either before or after adjuvant chemotherapy have a higher risk of recurrence than those without CTCs.[32]

Currently, the use of CTCs to determine prognosis or "residual risk" after adjuvant chemotherapy for breast, or other malignancies, is investigational[31,33,34] (see above). Although the available data are provocative, no studies have demonstrated that patients with detectable CTCs by any method should be treated differently from those who do not have them.

2. Metastatic Disease

Several studies have suggested that elevated or detectable CTCs in patients with metastatic breast cancer who are starting a new therapy are associated with shorter time to progression and overall survival. Of these, perhaps the most accurate is a prospective study in which more than 175 such patients with < 5CTCs/7.5 ml whole blood, as determined by the CellSearch® system, at baseline were shown to have a substantially longer progression free and overall survival than those with ≥ 5CTCs/7.5 ml whole blood (Fig. 1). These data have been validated by several other studies using this system.[35–37] Other investigators have demonstrated that detection of CTCs using rtPCR is also prognostic for patients with metastatic breast cancer.[7]

Although these data are of biological interest, their clinical utility is unclear. Treatment of patients with metastatic breast cancer is palliative, and therefore the clinician should choose the therapy most likely to work with the fewest side effects. For example, in patients with endocrine-resistant breast cancer, no data have demonstrated that more aggressive therapies, such as combinations or higher-than-standard doses of chemotherapy, are more effective than judicial application of sequential single agents. One exception to this paradigm is for patients with rapidly progressive, visceral disease who have end-organ dysfunction. In this case, one might not have the luxury of empirical use of sequential single agent therapy, and therefore combination chemotherapy is indicated. Although CTC levels at baseline are associated with a significantly worse prognosis, they do not confer the same very poor prognosis as these clinical findings, and there is no evidence that a patient with metastatic breast cancer who has elevated CTC levels should be treated differently from one who does not.[38,39]

Of note, very similar results regarding the prognostic effects of CTCs have been obtained in patients with metastatic colorectal[40] and prostate[41] cancers using the CellSearch® system.

C. Prediction

The presence of a relatively poor prognosis is not, alone, an indication for treatment. Rather, the clinician must also decide how likely the patient may or may not benefit from any, or a specific, therapy. In this regard, predictive factors help to select the proper therapy for patients who will benefit and to avoid exposing others to needless toxic and expensive therapies. Perhaps the best example of a predictive factor in oncology is estrogen receptor content to guide endocrine therapies in patients with breast cancer.[42]

CTCs might serve as predictive factors in one of three ways. First, it is possible that the presence of high levels of CTCs represents something about the biology of that cancer that predicts the relative resistance to systemic

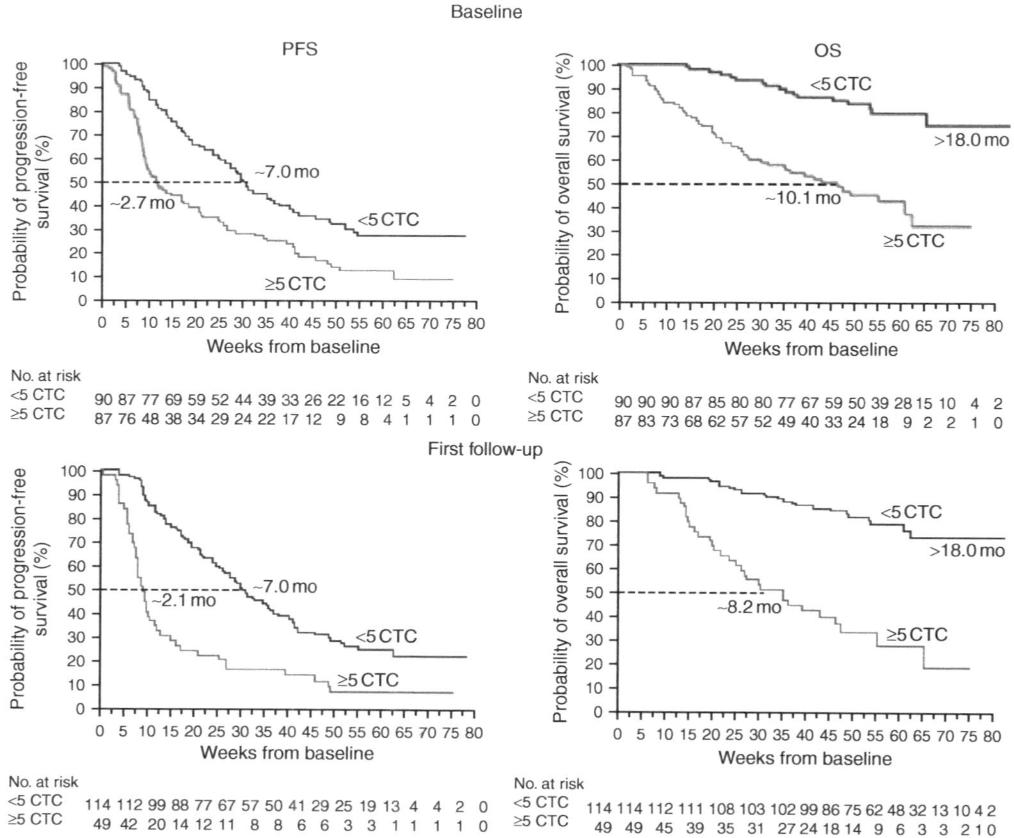

Fig. 1. *Circulating tumor cells predict clinical outcome in patients with metastatic breast cancer.* 177 patients with metastatic breast cancer were tested for circulating tumor cells before initiating and at first follow-up after starting a new systemic therapy. PFS = progression free survival; OS = overall survival. Modified from Ref. 38 with permission.

therapies, in general. However, there are no data to support this theory. Second, and more plausibly, persistently elevated CTC levels after initiating a new type of therapy might represent failure to respond to the selected therapy. Indeed, this phenomenon has been shown to be true in a large, prospective trial of patients with metastatic breast cancer.[38] In this study, patients who had ≥ 5 CTCs/7.5 ml whole blood after one cycle (3–5 weeks) of starting a new systemic treatment had much shorter progression free survival than those who either started with low CTCs or, more importantly, who had ≥ 5 CTCs/7.5 ml at baseline but dropped below this cutoff at the first follow-up (Fig. 1). In this regard, CTC levels at first follow-up could be considered a "retrospective" predictive factor, informing the physician that it is likely that the chosen systemic therapy may be futile. However, it is not clear that such patients (those with persistently elevated CTC levels at first follow-up) benefit from changing therapeutic regimens at such an early time-point, and this question is being addressed in prospective randomized clinical trial being conducted by the Southwest Oncology Group (protocol S0500; Fig. 2).

In addition to the number of CTCs, a third mechanism of prediction of therapeutic effect could be obtained by determination of biologic markers on the cells. This approach may be useful either for monitoring response or to predict sensitivity to specific therapies. For example, it is known that HER2 expression can differ between the cells in the primary tumor and the cells in the metastatic lesions.[43,44] Potentially this phenomenon would predict a tumor that has acquired sensitivity to anti-HER2 therapies such as trastuzumab or lapatinib. HER2 can be detected and quantified on CTCs.[45,46] There are ongoing studies to determine whether the HER2 status of CTCs reflects the HER2 status of the metastatic disease and whether CTC-HER2 status predicts response to anti-HER2 therapy. Additional biologically relevant markers such as Bcl-2 expression and apoptosis can be measured on CTCs.[47] It is hypothesized that the ability to monitor the induction of apoptosis would predict response of a newly started therapy. Similarly, Bcl-2 might predict sensitivity or resistance to different therapeutic classes of drugs.

D. Monitoring

One advantage of assays for CTCs is the relative noninvasiveness of their collection requiring only a simple blood draw. Thus, it is reasonable to assume that monitoring patients with CTCs might be of clinical value. There are three possible clinical situations in which patients might be monitored. The first scenario would be during adjuvant systemic therapy, administered either before or after surgery. Very few studies have addressed this utility, and it is not clear whether changes in CTC levels are sufficient to direct either a change in adjuvant systemic therapy or the proper locoregional therapy, if it is delivered afterwards. Sensitivity is a major obstacle to monitoring CTCs in this setting, at

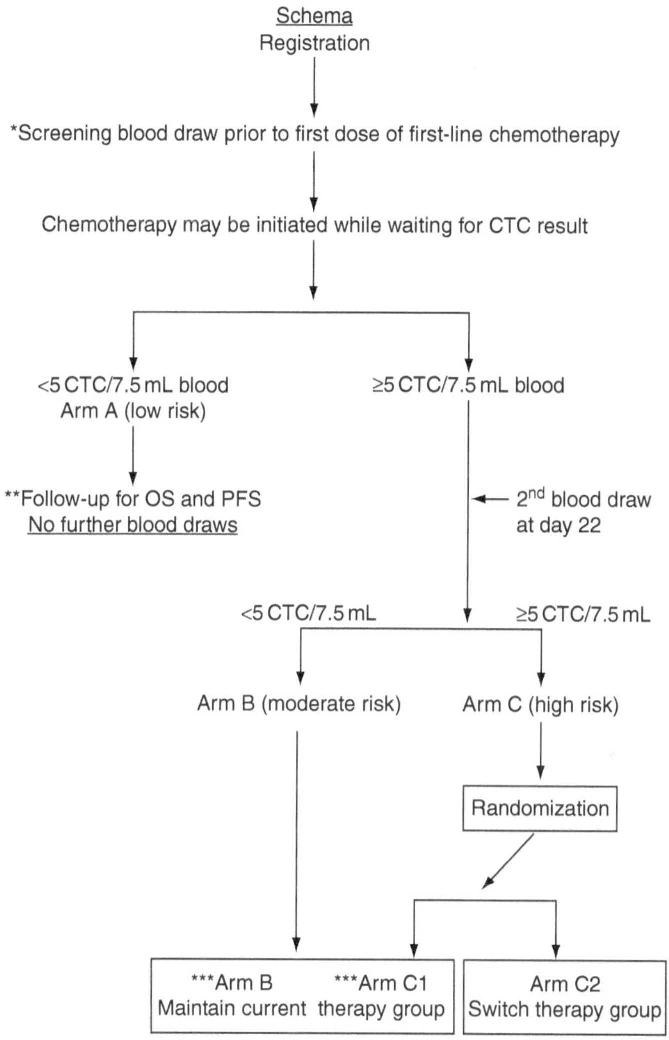

* Patients must be registered prior to initiation of testing (no more than one working day prior to initial CTC submission).

** Patients in the low risk group (arm A) may enroll in other clinical trials while being following for OS and PFS on S0500.

*** Patients in arms B and C1 and their physicians will be blinded to which arm they are in by study design. Protocol requirements are the same for these two arms.

FIG. 2. *Schema for Southwest Oncology Group Protocol S0500.* Patients with hormone refractory metastatic breast cancer who are starting first line chemotherapy are eligible. Those with persistently elevated CTC levels ($\geq$5/7.5 ml whole blood) at first follow-up are randomly assigned to continue that therapy until classic clinical and/or radiographic evidence of progression or to change to a new chemotherapy at that time. Reproduced from Ref. 63 with permission.

least for assays that employ immune-magnetic separation techniques, such as CellSearch®. Only a small proportion of those with early disease have elevated CTC levels, even if they have stage III primary cancers.[31]

A second indication for monitoring CTCs might be to detect impending recurrence in patients who have been rendered free of disease after primary and adjuvant therapy. Very few studies have investigated this strategy. Researchers from Greece have reported that seven of 15 (47%) patients followed on adjuvant tamoxifen who had persistently positive rtPCR-based indications of CTCs developed disease recurrence, compared to six (8.8%) of the 68 persistently negative patients ($P = 0.00026$).[48] Persistence of CK-19 mRNA+ cells was associated with a significantly lower median disease-free interval ($P = 0.0001$) and overall survival ($P = 0.0005$). However, the clinical utility of detection of occult, or impending, relapse in an otherwise asymptomatic patient has not been demonstrated. Indeed, two separate panels convened by the American Society of Clinical Oncology have both concluded that monitoring of such patients with special tests, such as radiographic or scintigraphic imaging or use of circulating serologic tests is not indicated, since it is not clear whether detection of asymptomatic metastases provides an opportunity to improve outcomes.[27,49]

The third, and most clinically relevant at this time, circumstance in which CTCs might be helpful is follow-up of patients with metastatic breast cancer who have been treated for some time with a systemic therapy. In these patients, one might change therapy because of either intolerable toxicity (which can be determined clinically) or evidence of progression. The latter is evaluated by clinical and radiographic means, as well as with circulating serologic markers, such as assays for MUC1 and CEA.[50] As noted, residual elevated CTC levels at first follow-up appear to predict rapid subsequent progression.[38] At least two studies have demonstrated that rising or elevated CTC levels at these later time-points are also strongly associated with progressive metastatic disease.[35,51] Although not recommended by the ASCO Tumor Marker Guidelines Committee, it seems reasonable to use CTCs to complement other means of determination of progression at these time-points, and this utility is cleared by the U.S. Food and Drug Agency.

IV. CTC Characterization

Most of the previous discussions of clinical uses of CTC detection and monitoring has been based on simple enumeration of CTC levels. However, one of the great promises of all of the available techniques involves molecular characterization of the cells. In theory, one could perform genetic, message, or protein analyses that might better refine the prognostic, predictive, or monitoring utilities.

Several methodologic reports have demonstrated the capability of molecular characterization of CTCs. Using immunomagnetic separation, investigators have shown that CTCs can be evaluated for amplification and expression of HER2,[45,46,52,53] IGFR1,[54] urinary plasminogen activator,[52] bcl-2, and M30 antigen (a marker of apoptosis)[47] in breast cancer patients. Using other techniques, CTCs have also been tested for ER,[55] MAGE-A,[56] phosphorylated FAK and PI3-K,[57] HER-2, and mammoglobin.[10,58,59] Other investigators have shown that CTCs from prostate cancer patients can be monitored for ERG, androgen receptor, and PTEN[60] and even global gene expression.[61] A recently reported study demonstrated the ability to determine mutational status of EGFR on CTCs separated from whole blood using a novel fluidic technique through EpCAM-coated microposts.[19,21] Since mutated EGFR has been shown to be predictive for tyrosine kinase inhibitors in nonsmall cell lung cancer, this technique might be useful to select patients for consideration of treatment with these agents.

As exciting as these studies are, none has actually demonstrated that CTC characterization can or should be used to direct therapy. It is possible that CTC genotype and/or phenotype might not be representative of the metastatic tissue from which these cells presumably emanate. Prospective, or well-designed and properly controlled retrospective, clinical trials are needed to demonstrate that patients with marker-positive CTCs are more likely to respond to the targeted agents than those who either have no CTCs or those whose CTCs are negative. Indeed, the heterogeneity of expression of these markers by CTCs is remarkable. As illustrated in Fig. 3, we have shown that HER2 expression on CTCs in a single blood specimen from a single patient can vary widely from highly positive to nearly undetectable levels.[46] These data demonstrate that highly validated indices or profiles will be required before CTC molecular characterization will have clinical utility.[62]

V. Summary

In summary, the field of CTC enumeration and characterization has expanded rapidly in the last decade. There are now several commercially available assays, and at least one of them has FDA clearance for breast, colorectal, and prostate cancer. The main clinical utility of these assays is to help determine progression in patients with metastatic disease, and ongoing studies are investigating other possible uses, for example, at the first follow-up after starting treatment in metastatic patients or as prognostic factors in

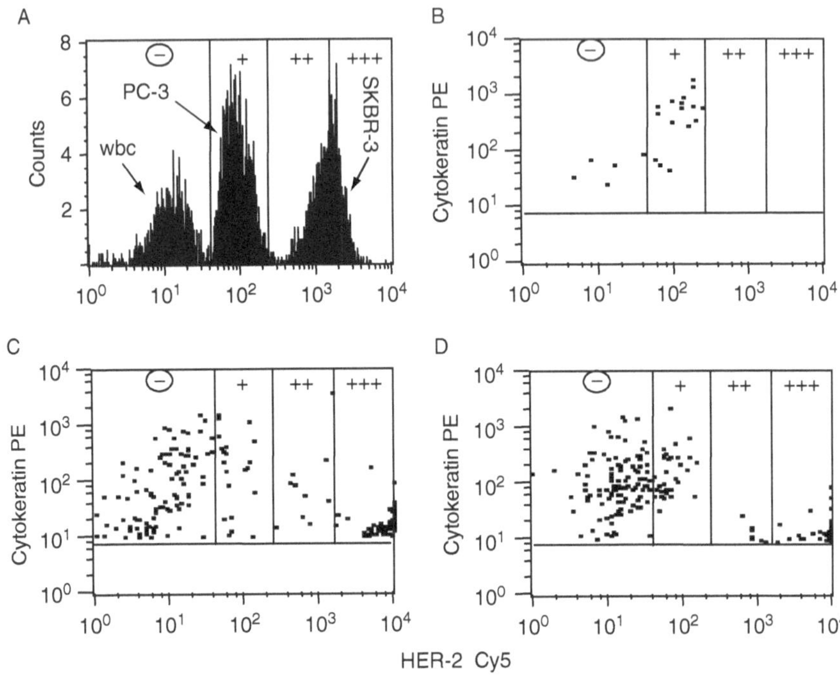

FIG. 3. *HER2 expression on circulating tumor cells.* Circulating tumor cells were separated from whole blood of three patients with metastatic breast cancer and evaluated (B–D) by flow cytometry for cytokeratin (*Y*-axis) and HER2 (*X*-axis) expression. Each square is a nucleated event, and presumably an epithelial cell. In (A), human cells were evaluated by flow cytometry for HER2. WBC = normal white blood cells; no HER2 expression. PC-3 = cultured human prostate cancer cells; HER2 normal (single copy) expression. SKBR-3 = cultured human breast cancer cells; HER2 highly amplified and overexpressed. Reproduced from Ref. 46 with permission.

patients with earlier disease. The ability to further characterize CTCs offers great promise in the future to permit what could be considered a "real-time biopsy" for evaluation of important biomarkers that might reflect the status of the cancer tissue. Well-designed, conducted, and analyzed preclinical and clinical trials are necessary to further extend these exciting early results.

References

1. Yang J, Weinberg RA. Epithelial-mesenchymal transition: at the crossroads of development and tumor metastasis. *Dev Cell* 2008;**14**:818–29.
2. Ashworth TR. A case of cancer in which cells similar to those in the tumours were seen in the blood after death. *Austr Med J* 1869;**14**:146–9.

3. Zheng S, Lin H, Liu JQ, Balic M, Datar R, Cote RJ, et al. Membrane microfilter device for selective capture, electrolysis and genomic analysis of human circulating tumor cells. *J Chromatogr A* 2007;**1162**:154–61.
4. Racila E, Euhus D, Weiss AJ, Rao C, McConnell J, Terstappen LW, et al. Detection and characterization of carcinoma cells in the blood. *Proc Natl Acad Sci USA* 1998;**95**:4589–94.
5. Choesmel V, Pierga JY, Nos C, Vincent-Salomon A, Sigal-Zafrani B, Thiery JP, et al. Enrichment methods to detect bone marrow micrometastases in breast carcinoma patients: clinical relevance. *Breast Cancer Res* 2004;**6**:R556–70.
6. Brugger W, Buhring HJ, Grunebach F, Vogel W, Kaul S, Muller R, et al. Expression of MUC-1 epitopes on normal bone marrow: implications for the detection of micrometastatic tumor cells. *J Clin Oncol* 1999;**17**:1535–44.
7. Ring A, Smith IE, Dowsett M. Circulating tumour cells in breast cancer. *Lancet Oncol* 2004;**5**:79–88.
8. Jung R, Kruger W, Hosch S, Holweg M, Kroger N, Gutensohn K, et al. Specificity of reverse transcriptase polymerase chain reaction assays designed for the detection of circulating cancer cells is influenced by cytokines in vivo and in vitro. *Br J Cancer* 1998;**78**:1194–8.
9. de Graaf H, Maelandsmo GM, Ruud P, Forus A, Oyjord T, Fodstad O, et al. Ectopic expression of target genes may represent an inherent limitation of RT-PCR assays used for micrometastasis detection: studies on the epithelial glycoprotein gene EGP-2. *Int J Cancer* 1997;**72**:191–6.
10. Ignatiadis M, Perraki M, Apostolaki S, Politaki E, Xenidis N, Kafousi M, et al. Molecular detection and prognostic value of circulating cytokeratin-19 messenger RNA-positive and HER2 messenger RNA-positive cells in the peripheral blood of women with early-stage breast cancer. *Clin Breast Cancer* 2007;**7**:883–9.
11. Stathopoulou A, Gizi A, Perraki M, Apostolaki S, Malamos N, Mavroudis D, et al. Real-time quantification of CK-19 mRNA-positive cells in peripheral blood of breast cancer patients using the lightcycler system. *Clin Cancer Res* 2003;**9**:5145–51.
12. Thor A, Viglione MJ, Ohuchi N, Simpson J, Steis R, Cousar J, et al. Comparison of monoclonal antibodies for the detection of occult breast carcinoma metastases in bone marrow. *Breast Cancer Res Treat* 1988;**11**:133–45.
13. Allard WJ, Matera J, Miller MC, Repollet M, Connelly MC, Rao C, et al. Tumor cells circulate in the peripheral blood of all major carcinomas but not in healthy subjects or patients with nonmalignant diseases. *Clin Cancer Res* 2004;**10**:6897–904.
14. Rao CG, Chianese D, Doyle GV, Miller MC, Russell T, Sanders Jr. RA, et al. Expression of epithelial cell adhesion molecule in carcinoma cells present in blood and primary and metastatic tumors. *Int J Oncol* 2005;**27**:49–57.
15. Spizzo G, Went P, Dirnhofer S, Obrist P, Simon R, Spichtin H, et al. High Ep-CAM expression is associated with poor prognosis in node-positive breast cancer. *Breast Cancer Res Treat* 2004;**86**:207–13.
16. Sieuwerts AM, Kraan J, Bolt J, van der Spoel P, Elstrodt F, Schutte M, et al. Anti-epithelial cell adhesion molecule antibodies and the detection of circulating normal-like breast tumor cells. *J Natl Cancer Inst* 2008;**101**:61–6.
17. Braun S, Hepp F, Kentenich CR, Janni W, Pantel K, Riethmuller G, et al. Monoclonal antibody therapy with edrecolomab in breast cancer patients: monitoring of elimination of disseminated cytokeratin-positive tumor cells in bone marrow. *Clin Cancer Res* 1999;**5**:3999–4004.
18. Allard J, Hayes DF, Repollet M, Rao C, Herman ML, Matera J, et al. A cellular preservative improves the specificity and yield of circulating tumor cells in carcinoma patients. *Proc Am Soc Clin Oncol* 2003;**22**:866.
19. Nagrath S, Sequist LV, Maheswaran S, Bell DW, Irimia D, Ulkus L, et al. Isolation of rare circulating tumour cells in cancer patients by microchip technology. *Nature* 2007;**450**:1235–9.

20. Talasaz AH, Powell AA, Huber DE, Berbee JG, Roh KH, Yu W, et al. Isolating highly enriched populations of circulating epithelial cells and other rare cells from blood using a magnetic sweeper device. *Proc Natl Acad Sci USA* 2009;**106**:3970–5.
21. Maheswaran S, Sequist LV, Nagrath S, Ulkus L, Brannigan B, Collura CV, et al. Detection of mutations in EGFR in circulating lung-cancer cells. *N Engl J Med* 2008;**359**:366–77.
22. Henry NL, Hayes DF. Uses and abuses of tumor markers in the diagnosis, monitoring, and treatment of primary and metastatic breast cancer. *Oncologist* 2006;**11**:541–52.
23. Kramer BS. The science of early detection. *Urol Oncol* 2004;**22**:344–7.
24. Schroder FH, Hugosson J, Roobol MJ, Tammela TL, Ciatto S, Nelen V, et al. Screening and prostate-cancer mortality in a randomized European study. *N Engl J Med* 2009;**360**:1320–8.
25. Andriole GL, Crawford ED, Grubb 3rd RL, Buys SS, Chia D, Church TR, et al. Mortality results from a randomized prostate-cancer screening trial. *N Engl J Med* 2009;**360**:1310–9.
26. Braun S, Vogl FD, Naume B, Janni W, Osborne MP, Coombes RC, et al. A pooled analysis of bone marrow micrometastasis in breast cancer. *N Engl J Med* 2005;**353**:793–802.
27. Harris L, Fritsche H, Mennel R, Norton L, Ravdin P, Taube S, et al. American Society of Clinical Oncology 2007 update of recommendations for the use of tumor markers in breast cancer. *J Clin Oncol* 2007;**25**:5287–312.
28. Stathopoulou A, Vlachonikolis I, Mavroudis D, Perraki M, Kouroussis C, Apostolaki S, et al. Molecular detection of cytokeratin-19-positive cells in the peripheral blood of patients with operable breast cancer: evaluation of their prognostic significance. *J Clin Oncol* 2002;**20**:3404–12.
29. Ignatiadis M, Xenidis N, Perraki M, Apostolaki S, Politaki E, Kafousi M, et al. Different prognostic value of cytokeratin-19 mRNA positive circulating tumor cells according to estrogen receptor and HER2 status in early-stage breast cancer. *J Clin Oncol* 2007;**25**:5194–202.
30. Rack K, Schindlbeck C, Hofmann S, Schneeweiss A, Rezai M, Beckmann M, et al. Circulating tumor cells (CTCs) in peripheral blood of primary breast cancer patients. *Proc Am Soc Clin Oncol* 2007;**25S**:10595, Abs 3.
31. Smerage JB, Doyle G, Budd GT, Schott A, Blayney D, Wicha M, et al. Circulating tumor cells in breast cancer patients before and during neoajuvant chemotherapy. *Proc Am Soc Clin Oncol* 2006; **24:**560S, abs 10079.
32. Rack B, Schindlbeck C, Andergassen U, Schneeweiss A, Zwingers T, Lichtenegger W, et al. Use of circulating tumor cells in peripheral blood of breast cancer patients before and after adjuvant chemotehapy to predict increased risk for relapse: the SUCCESS trial. *Proc Am Soc Clin Oncol* 2010;**28**(114s), Abstract #1003.
33. Xenidis N, Vlachonikolis I, Mavroudis D, Perraki M, Stathopoulou A, Malamos N, et al. Peripheral blood circulating cytokeratin-19 mRNA-positive cells after the completion of adjuvant chemotherapy in patients with operable breast cancer. *Ann Oncol* 2003;**14**:849–55.
34. Apostolaki S, Perraki M, Pallis A, Bozionelou V, Agelaki S, Kanellou P, et al. Circulating HER2 mRNA-positive cells in the peripheral blood of patients with stage I and II breast cancer after the administration of adjuvant chemotherapy: evaluation of their clinical relevance. *Ann Oncol* 2007;**18**:851–8.
35. Liu MC, Shields PG, Warren RD, Cohen P, Wilkinson M, Ottaviano YL, et al. Circulating tumor cells: a useful predictor of treatment efficacy in metastatic breast cancer. *J Clin Oncol* 2009;**27**:5153–9.
36. Riethdorf S, Fritsche H, Muller V, Rau T, Schindlbeck C, Rack B, et al. Detection of circulating tumor cells in peripheral blood of patients with metastatic breast cancer: a validation study of the Cell Search system. *Clin Cancer Res* 2007;**13**:920–8.
37. Bardia A, Huang P, Zhang Z, Sokoll L, Ingle J, Carey LA, et al. Circulating tumor cells and CA27.29 as predictors of outcome in patients with metastatic breast cancer in the prospective TBCR 005 biomarker study. *Proc Am Soc Clin Oncol* 2010;**28**(114S), Abstract #1001.

38. Cristofanilli M, Budd GT, Ellis MJ, Stopeck A, Matera J, Miller MC, et al. Circulating tumor cells, disease progression, and survival in metastatic breast cancer. *N Engl J Med* 2004;**351**:781–91.
39. Cristofanilli M, Hayes DF, Budd GT, Ellis MJ, Stopeck A, Reuben JM, et al. Circulating tumor cells: a novel prognostic factor for newly diagnosed metastatic breast cancer. *J Clin Oncol* 2005;**23**:1420–30.
40. Cohen SJ, Punt CJ, Iannotti N, Saidman BH, Sabbath KD, Gabrail NY, et al. Prognostic significance of circulating tumor cells in patients with metastatic colorectal cancer. *Ann Oncol* 2009;**20**:1223–9.
41. de Bono JS, Scher HI, Montgomery RB, Parker C, Miller MC, Tissing H, et al. Circulating tumor cells predict survival benefit from treatment in metastatic castration-resistant prostate cancer. *Clin Cancer Res* 2008;**14**:6302–9.
42. Hammond ME, Hayes DF, Dowsett M, Allred DC, Hagerty KL, Badve S, et al. American Society of Clinical Oncology/College of American Pathologists guideline recommendations for immunohistochemical testing of estrogen and progesterone receptors in breast cancer. *J Clin Oncol* 2010;**28:**2784–95.
43. Luftner D, Henschke P, Kafka A, Anagnostopoulos I, Wiechen K, Geppert R, et al. Discordant results obtained for different methods of HER-2/neu testing in breast cancer—a question of standardization, automation and timing. *Int J Biol Markers* 2004;**19**:1–13.
44. Gutierrez MC, Detre S, Johnston S, Mohsin SK, Shou J, Allred DC, et al. Molecular changes in tamoxifen-resistant breast cancer: relationship between estrogen receptor, HER-2, and p38 mitogen-activated protein kinase. *J Clin Oncol* 2005;**23**:2469–76.
45. Meng S, Tripathy D, Shete S, Ashfaq R, Haley B, Perkins S, et al. HER-2 gene amplification can be acquired as breast cancer progresses. *Proc Natl Acad Sci USA* 2004;**101**:9393–8.
46. Hayes DF, Walker TM, Singh B, Vitetta ES, Uhr JW, Gross S, et al. Monitoring expression of HER-2 on circulating epithelial cells in patients with advanced breast cancer. *Int J Oncol* 2002;**21**:1111–7.
47. Smerage JB, Doyle G, Budd GT, Schott A, Blayney D, Wicha M, et al. Detection of apoptosis and Bcl-2 expression in circulating tumor cells from women being treated for metastatic breast cancer. *Proc Am Assoc Cancer Res* 2006;**47**(187), Abstract # 792.
48. Xenidis N, Markos V, Apostolaki S, Perraki M, Pallis A, Sfakiotaki G, et al. Clinical relevance of circulating CK-19 mRNA-positive cells detected during the adjuvant tamoxifen treatment in patients with early breast cancer. *Ann Oncol* 2007;**18**:1623–31.
49. Khatcheressian JL, Wolff AC, Smith TJ, Grunfeld E, Muss HB, Vogel VG, et al. American Society of Clinical Oncology 2006 update of the breast cancer follow-up and management guidelines in the adjuvant setting. *J Clin Oncol* 2006;**24**:5091–7.
50. Ellis M, Hayes DF, Lippman ME. Treatment of metastatic breast cancer. In: Harris J, Lippman M, Morrow M, Osborne CK, editors. *Diseases of the breast*. Philadelphia: Lippincott Williams & Wilkins; 2004. p. 1101–62.
51. Hayes DF, Cristofanilli M, Budd GT, Ellis MJ, Stopeck A, Miller MC, et al. Circulating tumor cells at each follow-up time point during therapy of metastatic breast cancer patients predict progression-free and overall survival. *Clin Cancer Res* 2006;**12**:4218–24.
52. Meng S, Tripathy D, Shete S, Ashfaq R, Saboorian H, Haley B, et al. uPAR and HER-2 gene status in individual breast cancer cells from blood and tissues. *Proc Natl Acad Sci USA* 2006;**103**:17361–5.
53. Fehm T, Becker S, Duerr-Stoerzer S, Sotlar K, Mueller V, Wallwiener D, et al. Determination of HER2 status using both serum HER2 levels and circulating tumor cells in patients with recurrent breast cancer whose primary tumor was HER2 negative or of unknown HER2 status. *Breast Cancer Res* 2007;**9**:R74.

54. de Bono JS, Attard G, Adjei A, Pollak MN, Fong PC, Haluska P, et al. Potential applications for circulating tumor cells expressing the insulin-like growth factor-I receptor. *Clin Cancer Res* 2007;**13**:3611–6.
55. Neubauer H, Clare SE, Wozny W, Schwall GP, Poznanovic S, Stegmann W, et al. Breast cancer proteomics reveals correlation between estrogen receptor status and differential phosphorylation of PGRMC1. *Breast Cancer Res* 2008;**10**:R85.
56. Kufer P, Zippelius A, Lutterbuse R, Mecklenburg I, Enzmann T, Montag A, et al. Heterogeneous expression of MAGE-A genes in occult disseminated tumor cells: a novel multimarker reverse transcription-polymerase chain reaction for diagnosis of micrometastatic disease. *Cancer Res* 2002;**62**:251–61.
57. Kallergi G, Mavroudis D, Georgoulias V, Stournaras C. Phosphorylation of FAK, PI-3K, and impaired actin organization in CK-positive micrometastatic breast cancer cells. *Mol Med* 2007;**13**:79–88.
58. Ignatiadis M, Kallergi G, Ntoulia M, Perraki M, Apostolaki S, Kafousi M, et al. Prognostic value of the molecular detection of circulating tumor cells using a multimarker reverse transcription-PCR assay for cytokeratin 19, mammaglobin A, and HER2 in early breast cancer. *Clin Cancer Res* 2008;**14**:2593–600.
59. Ntoulia M, Stathopoulou A, Ignatiadis M, Malamos N, Mavroudis D, Georgoulias V, et al. Detection of Mammaglobin A-mRNA-positive circulating tumor cells in peripheral blood of patients with operable breast cancer with nested RT-PCR. *Clin Biochem* 2006;**39**:879–87.
60. Attard G, Swennenhuis JF, Olmos D, Reid AH, Vickers E, A'Hern R, et al. Characterization of ERG, AR and PTEN gene status in circulating tumor cells from patients with castration-resistant prostate cancer. *Cancer Res* 2009;**69**:2912–8.
61. Smirnov DA, Zweitzig DR, Foulk BW, Miller MC, Doyle GV, Pienta KJ, et al. Global gene expression profiling of circulating tumor cells. *Cancer Res* 2005;**65**:4993–7.
62. Simon RM, Paik S, Hayes DF. Use of archived specimens in evaluation of prognostic and predictive biomarkers. *J Natl Cancer Inst* 2009;**101**:1446–52.
63. Hayes DF, Smerage J. Is there a role for circulating tumor cells in the management of breast cancer? *Clin Cancer Res* 2008;**14**:3646–50.

Stem Cells in Normal Development and Cancer

Rosemarie Chirco D'Angelo and Max S. Wicha

Department of Internal Medicine, Division of Hematology and Oncology, University of Michigan Comprehensive Cancer Center, University of Michigan, Ann Arbor, Michigan, USA

In this chapter we provide an overview of stem cells in normal tissues as well as in many different types of cancers. All tissues in the body are derived from organ-specific stem cells that retain the ability to self-renew and differentiate into specific cell types. The cancer stem cell hypothesis suggests that tumors

Progress in Molecular Biology and Translational Science, Vol. 95
DOI: 10.1016/S1877-1173(10)95006-0

arise from cell populations with dysregulated self-renewal. This may be tissue stem cells or more differentiated cells that acquire self-renewal capabilities. In addition, we outline some useful assays for purification and isolation of cancer stem cells including the dye exclusion side population assay, flow cytometry sorting techniques for identification of putative cancer stem cell markers, tumorspheres assay, aldehyde dehydrogenase activity assay, PKH, and other membrane staining used to label the cancer stem cells, as well as *in vivo* xenograft transplantation assays. We also examine some of the cell signaling pathways that regulate stem cell self-renewal including the Notch, Hedgehog, HER2/PI3K/Akt/PTEN, and p53 pathways. We also review information demonstrating the involvement of the microenvironment or stem cell niche and its effects on the growth and maintenance of cancer stem cells. Finally, we highlight the therapeutic implications of targeting stem cells by inhibiting these pathways for the treatment and prevention of cancer.

I. Introduction of Cancer Stem Cells and the Cancer Stem Cell Hypothesis

Cancer is a complex disease and one of the problems for cancer research is that it is very difficult to identify the cell types that are responsible for the initiation and maintenance of the tumor.[1] Several models of tumor initiation have been proposed. In the *classical or stochastic model*, any cell may acquire a mutation which gives it a selective advantage leading to clonal expansion. This clone then acquires further mutations which eventually lead to tumor formation. Therefore, in this model cells within a tumor have equal tumorigenic potential[2] (Fig. 1A). The *hierarchical* or *cancer stem cell model* proposes that these CSCs have gained the capacity to proliferate and may be responsible for generating the bulk of the tumor and tumor progression; suggesting that cells in a tumor are heterogeneous and only a small population of cells are able to initiate tumor growth[1] (Fig. 1B). These models are not mutually exclusive; however, cancer stem cells may be genetically unstable and evolve during tumor progression.

There has been increasing evidence suggesting the existence of stem cells in many different types of cancers, including blood cancers such as leukemia[3] and solid cancers including breast, brain, colon, pancreatic, head and neck cancers, and other types of carcinomas (Table I).[4–8] This has been facilitated through identification of cell surface markers, using fluorescent-activated cell sorting (FACS) techniques and *in vivo* transplantation of potential cancer stem cell populations into immunocompromised mice. Therefore, it has become increasingly important to identify and isolate these types of cells in

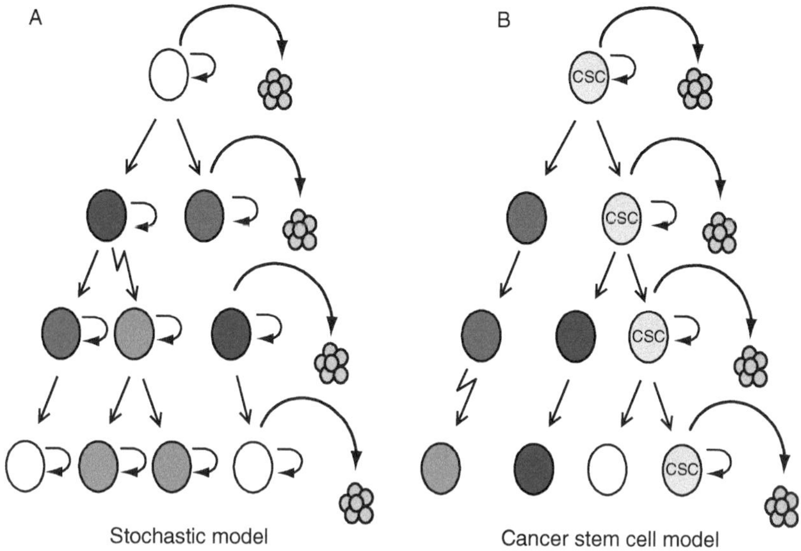

FIG. 1. The stochastic model and the cancer stem cell model. (A) The stochastic model of tumor formation suggests that any cell may acquire a mutation leading to clonal expansion. This clone may then acquire further mutations which eventually lead to tumor formation. (B) The cancer stem cell model suggests that the cancer stem cells have gained the capacity to proliferate and are responsible for generating the bulk of the tumor. These cancer stem cells make up only a small population of the tumor.

tumors and develop strategies to specifically target them in order to more effectively eradicate the tumor. In this chapter, we discuss the similarities of these cancer stem cells to normal stem cells, identifying the assays used to purify and isolate CSCs, investigating the signaling pathways involved in their regulation, as well as discussing potential therapeutic strategies to target CSCs.

II. Comparison of Normal Stem Cells with Cancer Stem Cells

There are three major types of normal stem cells: embryonic, germinal, and somatic.[9] The inner cell mass of the blastocyst generates embryonic stem cells (ESCs). These ESCs display unlimited replication potential and have the capacity of differentiating into every type of daughter cell in the mature organism. ESC lines were first identified in 1998 and their distinct molecular profiles have been determined.[10,11] Some of the genes that were found to be

TABLE I
IDENTIFICATION OF CANCER STEM CELLS IN TUMORS

Tumor type	Cell surface markers
Acute myeloid leukemia	$CD34^{+}CD38^{-}$
Human breast	$CD44^{+}CD24^{-}ESA^{+}$
	$EpCAM^{-}CD49f^{+}$
	High Aldefluor activity
Mouse breast	$CD49f^{hi}CD29^{hi}CD24^{+}Sca1^{-}$
Brain	$CD133^{+}$
Colon	$CD133^{+}$
	$ESA^{+}CD44^{+}CD166^{+}$
	High Aldefluor activity
Pancreatic	$CD24^{+}CD44^{+}ESA^{+}$
	High Aldefluor activity
Ovarian	$CD133^{+}$
Prostate	$CD44^{+}\alpha2\beta1$ $integrin^{high}CD133^{+}$
	High Aldefluor activity
Head and neck	$CD44^{+}$
	High Aldefluor activity
Melanoma	$CD20^{+}$
	$CD133^{+}$
	$ABCB5^{+}$
	$CD271^{+}$
Lung	$CD133^{+}$
	High Aldefluor activity

In recent years, many cancer stem cell-specific markers have been identified in various malignancies including leukemia, breast cancer, brain cancer, colon cancer, pancreatic cancer, ovarian cancer, prostate cancer, head and neck cancer, melanoma, and lung carcinoma. The references for these studies are listed in the text of the chapter.

important in regulating ESCs include the Polycomb genes,[12] Nanog,[13] Oct4,[14] and Sox2.[15] The germinal layer of the embryo generates the germinal cells which have the ability to differentiate into organ-specific cells.[16] The somatic stem cells have the capacity to self-renew and differentiate into many types of cells in an organ or tissue.[9]

The self-renewal of organ-specific adult stem cells allows for the maintenance and homeostasis of a particular organ, just like CSCs may help to maintain the tumor. Two aspects that define adult normal stem cells are that they have the ability to self-renew and to differentiate. During the process of self-renewal, the stem cell can divide symmetrically or asymmetrically to make one or two daughter cells that are exactly like the parental cell, thus allowing for the continuing formation of stem cells in that organ.[17] This is especially important during certain stages of development or after wound healing to increase the stem cell pool of that organ. For example, stem cells within the bone marrow divide rapidly in order to meet the demands of a higher turnover of the different blood cells.[18] When a stem cell differentiates it can divide and give rise to a heterogeneous population of cells that follow tissue-specific lineages that make up the organ.

Symmetric division of stem cells produces two daughter cells that are identical to the mother cell, whereas asymmetric self-renewal produces one identical stem cell and another cell that can generate differentiated progeny. Through asymmetric division, stem cells can divide to give rise to precursors or progenitors committed to a specific differentiation pathway that retain proliferative potential. These proliferating precursor cells are referred to as transit-amplifying (TA) cells. Progenitor cells have limited proliferative capacity and eventually differentiate into tissue-specific mature cells, which eventually undergo apoptosis.[19] Stem cells are quiescent or slowly cycling cells that are maintained in an undifferentiated state until they are needed and these processes of self-renewal and differentiation allow for stem cells to be very tightly regulated and controlled, allowing for the replenishment or repair of damaged tissues.

III. Definition of Cancer Stem Cells and Identification of Cancer Stem Cell Markers

Although the concept of "cancer stem cells" may seem to be relatively new, it arose because cancer cells retain or acquire the same properties of self-renewal and differentiation as characteristic of normal stem cells. The "cancer stem cell hypothesis" suggests that the cancer cells with these properties may be responsible for the formation and maintenance of the tumor. In fact, this is demonstrated by injecting the putative cancer stem cell population into immunocompromised mice such as NOD/SCID (nonobese diabetic/severe combined immune deficiency) mice to access the ability of these cells to generate tumors. This method has been adapted to many types of cancer further described in the sections below.

Cancer cells gain a growth advantage through mutations that either activate oncogenes or inactivate tumor suppressor genes leading to unregulated self-renewal of undifferentiated cells.[20] Over the years, there have been several different theories proposed to explain how a cell becomes a cancer stem cell. The first is that the normal stem cells acquire mutations over time that can get transmitted to their progeny leading to the formation of a tumor. Second, an early progenitor cell may acquire a mutation causing it to gain the capacity to self-renew, a property normally found only in stem cells. Lastly, a late progenitor or fully differentiated cell may gain the ability through dedifferentiation to acquire the properties of stem cells. The work on leukemia stem cells suggests that all three scenarios may occur since the introduction of leukemogenic oncogenes into purified hematopoietic stem cells or progenitors generated murine leukemias.[21–23] Additional work needs to be done to identify and isolate the cancer stem cells and determine their cell of origin.

A. Common Assays to Identify Cancer Stem Cells

In order to study the functional properties of stem cells, it is necessary to identify and then purify them from the bulk cell population making up the whole tumor. This has been proved to be a challenge since the cancer stem cell population constitutes a relatively rare population of cells. Since the defining characteristic properties of a cancer stem cell include self-renewal and differentiation, these properties can be studied utilizing *in vitro* and *in vivo* assays. Some of the most useful techniques for the identification of cancer stem cells are the dye exclusion side population (SP) assay, FACS using cell surface markers, anchorage-independent tumorsphere cultures, aldehyde dehydrogenase (ALDH) activity assay, and stem cell label-retaining assays such as staining with PKH or other cell surface dyes.

1. Dye Exclusion Side Population Assay

One useful assay that may characterize stem cells is the dye exclusion assay. Stem cells have the ability to exclude dyes like Hoechst or rhodamine as a result of the increased expression of ABC membrane transport proteins such as P-glycoproteins or breast cancer resistance proteins (BCRPs) resulting in a SP fraction which can be quantitated by flow cytometry.[24,25] The SP cell fraction has been found to be enriched in cells capable of self-renewal and differentiation with reconstitution of the original cell population suggesting that it may contain the stem cell population.[4,26]

The SP was first identified in normal hematopoietic stem cells where it constituted approximately 0.05% of total bone-marrow cells which were highly enriched in repopulating cells.[27,28] Since this discovery, SPs have been identified in other normal tissues, tumors, and cancer cell lines. Therefore, it has been proposed that the SP fraction can be used to identify CSCs. Several reviews have been written on the subject.[26,29] However, not all of the literature supports the contention that the stem cells are contained within the SP. Therefore, it is necessary to use additional assays to further identify cancer stem cells.

2. FACS of Cell Surface Markers to Isolate the Cancer Stem Cell Populations

Another useful method to identify the cancer stem cell populations is to use cell surface markers to sort cells by flow cytometry. CSCs/progenitors are then assayed by tumor transplantation in immunocompromised mice. The use of flow cytometry and cell surface markers has been very important for the identification and isolation of cancer stem cells in many different cancers. For example, some very useful cancer stem cell markers in the breast are the cell adhesion markers CD44 and CD24. In breast cells, the $CD44^{+}CD24^{-/low}$ population was found to have the ability to self-renew and differentiate and isolation and injection into the mammary fat pad of NOD/SCID mice resulted in increased tumor generation.[4] In pancreatic cancer, the cancer stem cell population is $CD44^{+}CD24^{+}$.[7] CD133 is another cancer stem cell maker in many different types of cancers, including brain, colon, ovarian, head and neck, and lung cancer.[5,6,30–34] However, CD133 seems to be a more controversial cancer stem cell marker because some groups have found that the $CD133^{-}$ population was also tumorigenic.[35] The details about these and other putative cancer stem cell populations in specific cancers are presented in the following section and are listed in Table I.

3. Anchorage-Independent Tumorsphere Cultures

The tumorsphere assay has been utilized to examine stem/progenitor populations *in vitro* for many different cancers including brain and breast stem cells. To generate tumorspheres, cells are plated in low attachment serum free conditions supplemented with growth factors such as basic fibroblast growth factor (bFGF), epidermal growth factor (EGF), and insulin. Most of the cells die but the ones that survive continue to self-renew and divide to make the tumorspheres which are clonally derived colonies of cells derived from a single stem cell.[36] In addition, cells isolated from tumorspheres have a multilineage differentiation potential when given the appropriate signals such as the addition of serum and the extracellular matrix molecule collagen.

In 1992, Reynolds and Weiss were one of the first groups to use this assay to identify undifferentiated neural stem cells in a defined media cultured in the presence of mitogens EGF and bFGF. When these neurospheres were allowed to differentiate by removing mitogens from the medium, many of the cells within the neurospheres differentiated into neurons and astrocytes.[37,38] This assay has since been adopted and used for many different cancers including breast cancer. The mammosphere assay was developed to identify the breast stem cells using either normal breast cells from reduction mammoplasties or various breast cancer cell lines to examine various stem cell regulatory pathways.[39–42] These anchorage-independent growth conditions have also been demonstrated to identify stem cell populations in others cancers such as the prostate, neuroblastoma, ovarian cancer, osteosarcoma, and Ewings sarcoma, pancreatic cancer, hepatocellular carcinoma, and melanoma cells, respectively.[43–49]

4. Aldehyde Dehydrogenase Activity Assay

Another, very useful assay for the identification of cancer stem cells is the ALDH activity assay. ALDH is a detoxifying enzyme responsible for the oxidation of intracellular aldehydes and is thought to play a role in stem cell differentiation through metabolism of retinal to retinoic acid.[50] The ALDH superfamily of enzymes contains more than 17 different genes.[51] The commercially available ALDEFLUOR assay consists of the substrate BODIPY-aminoacetaldehyde (BAAA) labeled with a visible light excitable fluorochrome that is converted into BODIPY-aminoacetate (BAA) by the ALDH family of enzymes which is now a charged molecule and is unable to leave the cell as freely as the unconverted substrate.[52] Therefore, cells expressing high levels of ALDH become brightly fluorescent and can be identified and enumerated using a standard flow cytometer. An inhibitor of the reaction, DEAB is added in a separate reaction to establish the proper gating for the FACS (Fig. 2).

This assay was first used in 1999 to isolate primitive human hematopoietic stem cells using human umbilical cord blood.[52] Since then other studies have shown that murine and human stem/progenitor populations in many different cancers have a high Aldefluor positive population including multiple myeloma, acute myeloid leukemia (AML), breast, hepatocellular carcinomas, lung carcinomas, colon, head and neck, prostate cancer, and osteosarcoma, respectively.[41,42,53–64] In addition to using the Aldefluor assay, it has been demonstrated that ALDH1 immunohistochemistry can also identify the cancer stem cell population.[41] In addition, it is very useful to combine the Aldefluor assay with other cell surface markers or stem cell assays to further purify the cancer stem cell population of interest.

5. Label-Retaining Cell Assay Using PKH Dyes

Cell membrane label-retaining assays using brightly fluorescent dyes such as the PKH 26 or PKH 67 have been used to identify cancer stem cells. These dyes are very useful because they have excellent cell-binding and retention properties and allow the tracking of cell cycle activity *in vitro* and *in vivo*.[65] These vital dyes consist of a fluorophore attached to an aliphatic carbon backbone that irreversibly binds to the lipid bilayer on cell membranes.[66]

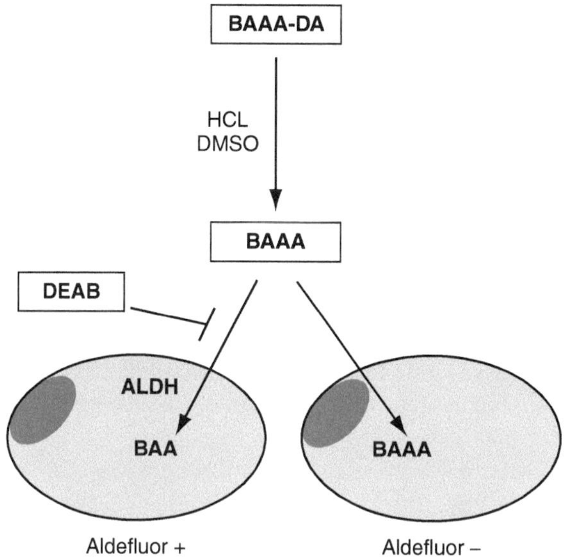

FIG. 2. The Aldefluor assay. The Aldefluor assay consists of the substrate BAAA labeled with a visible light excitable fluorochrome that is converted into BAA by the ALDH family of enzymes which is now a charged molecule and is unable to leave the cell as freely as the unconverted substrate. Therefore, cells expressing high levels of ALDH activity become brightly fluorescent and can be identified and enumerated using a standard flow cytometer. The inhibitor of the reaction DEAB is also added in a separate reaction to establish the proper gating for the FACS.

When the cell actively divides, it distributes the dye among its daughter cells, and therefore the intensity of the stain decreases in an exponential fashion with each division of the TA progenitors.[67] Although stem cells retain a capacity for self-renewal, throughout most of their lifetime they exist in a relatively quiescent state. During asymmetric self-renewal the cell that retains stem cell properties enters a quiescent state while the other daughter cell, termed a transient amplifying cell undergoes rapid proliferation followed by differentiation. Thus, although the stem cell retains the POTENTIAL to proliferate, it retains PKH dye due to its relative quiescence. Boyd was the first to suggest that these membrane dyes label growth inhibited cells.[68]

This method was first used by Hendrix and colleagues in 1996, to fluorescently label murine hematopoietic stem cells.[69] Since then it has been very useful to track the homing of short- and long-term repopulating cells to different hematopoietic organs following transplantation because the cells can be sorted by flow cytometry and then reimplanted.[67,70] It has been suggested that long-term repopulating cells remain quiescent in the bone marrow shortly after engraftment, whereas short-term cells divide more rapidly.[71] The PKH dye has also been used to study asymmetric self-renewal of a hematopoietic progenitor cell line using an automated time-lapse fluorescent microscope system to determine changes in cell size and fluorescence intensity during culture.[72] In addition, PKH dye has been able to be detected *in vivo* by fluorescence microscopy.[73]

More recently, PKH dyes have been used to identify specific cancer stem cell populations by staining the cells with the dye and growing either *in vitro* tumorsphere cultures or injecting the cells *in vivo* to examine PKH high, low, and negative populations (Fig. 3). Using ovarian cancer cells, the Bapat group identified that label-retaining PKH high cells exhibited the highest self-renewal potential and expression of the stem cell markers. Furthermore, treatment with

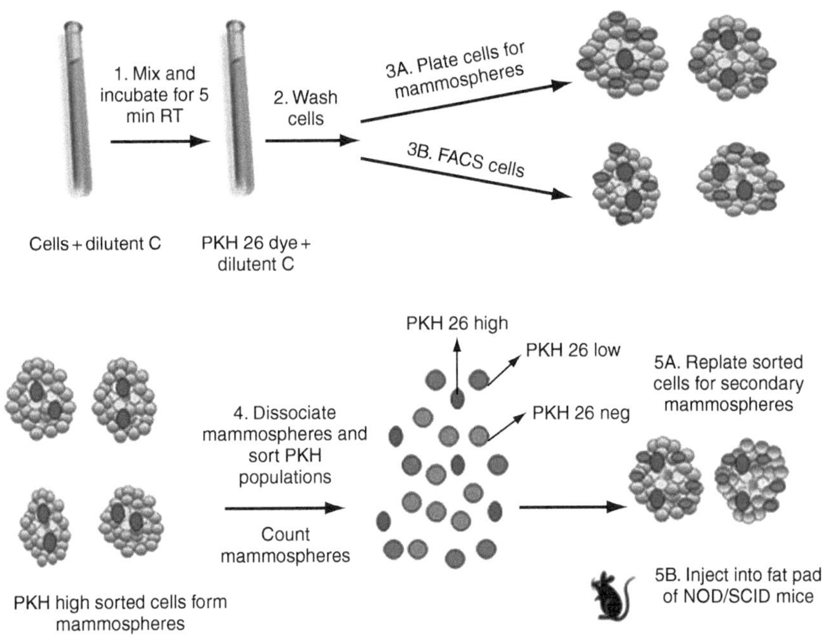

FIG. 3. The PKH 26 staining method and formation of mammospheres. (1) Resuspend cells in the Dilutent C buffer and mix dye with Dilutent C in another tube. Mix the contents of the tubes together and incubate for 5 min. (2) Wash cells and then (3A) plate cells for mammospheres or (3B) FACS for PKH positive cells only. (4) After a week, count and dissociate mammospheres and FACS for the three different populations: PKH negative, PKH low, or PKH high. (5) These populations can either be replated for secondary mammospheres or injected into NOD/SCID mice to look at tumorigenicity.

chemotherapy enriched the PKH high cells; while, the PKH low subset consisted of proliferation-arrested aneuploid cells and euploid progenitor cells.[66] Though this method is very useful for labeling cells of interest, it was recently shown dyes may be transferred to surrounding cells resulting in generation of artifacts.[74] Therefore, it is more useful if this method is combined with other stem cell assays to further validate the stem cell phenotype.

The PKH label-retaining studies have also been used in conjunction with mammosphere assays in tumor models and in human primary breast cells. The method for labeling cells with PKH dye, formation of mammospheres, and injection of PKH populations into the mammary fat pad of NOD/SCID mice is demonstrated in Fig. 3. Recently, the Pelicci group showed that only the PKH high cells in wild-type mice were able to generate mammospheres, that these cells divided asymmetrically, and that as little as one PKH high cell could reconstitute the cleared mammary fat pad.[75] To distinguish the stem cell population from the TA progenitors of the normal primary human breast, Pece *et al.*, used the mammosphere assay to show that during the growth of a mammosphere a small number of mammary stem cells retain the PKH high population (under 1%), while the bulk population is PKH negative and composed of daughter cells.[76] In addition, the PKH high population could also reconstitute the cleared mammary fat pad with as little as one cell.

IV. Identification of Cancer Stem Cells

Stem cells are defined by their ability to self-renew as well as to differentiate into cells forming the bulk of the tumor. The first stem cells to be identified in normal tissue were those of the hematopoietic system. McCulloch and Till identified that three major lineages of the blood system are derived from one common precursor stem cell.[77] The process of cancer cells undergoing differentiation was first demonstrated in the 1960s with embryonic carcinomas or teratocarcinomas by Pierce and his group. They showed that these embryonic cancers can give rise to multiple lineages and that these somatic tissues were benign.[78,79] Later, it was demonstrated that these teratocarcinomas could also give rise to normal chimeric mice.[80,81]

A. Cancer Stem Cells in Hematological Malignancies

The first demonstration of existence of cancer stem cells was in hematologic cancers. Furth and Kahn demonstrated that a single murine leukemia cell can give rise to a tumor in mice.[82] However, it was the work of John Dick and colleagues that first demonstrated the existence of cancer stem cells in AML. They were able to purify the rare cancer stem cell population by labeling human AML cells with monoclonal antibodies against the cell surface markers CD34 and CD38, markers that had previously been used to isolate normal hematopoietic stem cells. They FACS the cells and transplanted the different populations into NOD/SCID immunocompromised mice and demonstrated that $CD34^{+}CD38^{-}$ fraction generated leukemic grafts that closely resembled the disease found in the original patient.[3] Further seminal studies by this group have demonstrated these AML stem cells are hierarchically organized, able to be serially passaged in NOD/SCID mice, and retain differentiation capacity.[83,84]

B. Cancer Stem Cells in the Normal Breast and Breast Cancer

Following the identification of cancer stem cells in leukemia, scientists questioned if similar cells could be identified in solid tumors. The breast consists of a very complicated network of ducts with different cells types with regenerative properties during puberty and pregnancy, suggesting existence of a stem cell.[85] Breast tissue is made up of a network of ducts ending in smaller ductal structures called terminal ductal lobular units (TDLU). A breast stem cell can differentiate into two types of cells found in the TDLU, a luminal epithelial cell or a myoepithelial cell. The luminal lineage is further divided into ductal and alveolar cells that constitute the alveolar units that expand during pregnancy. Myoepithelial cells are basal and surround the ducts contracting to help secrete milk through the network of ducts during the process of lactation.

The development of the *in vivo* mammary reconstitution assay, tissue dissociation techniques, and mammary fat pad transplant assay lead to the identification of many important cancer stem/progenitor markers in the human and mouse breast. The fat pad transplantation assay created by DeOme *et al.*, consists of removal of the epithelial component of a young mouse fat pad and then the breast cells are injected into the stromal part and the ductal outgrowths are counted.[86] Using this assay outgrowth could be generated using either explants or cell suspension of mammary cells and it was discovered that these outgrowths could be serially transplanted and contained precursor cells.[87–91] Kordon and Smith, utilized retroviral integration sites to demonstrate that a single stem cell was capable of reconstituting the entire mammary gland.[92] Over the years, this assay has been improved by Kupperwaser's group by "humanizing" the fat pad by mixing irradiated fibroblasts with nonirradiated fibroblasts providing growth factors necessary resulting in a more human-like stroma in immunocompromised NOD/SCID mice.[93] In 2008, Eirew *et al.* developed another *in vivo* stem cell breast assay, whereas dissociated cells are suspended with fibroblasts embedded in collagen gels which are then implanted into the renal capsule of immunodeficient mice.[94]

Al-Hajj and coworkers first reported in 2003 that a small population of cells could give rise to breast tumors. They isolated the breast cancer stem cells based on the differential cell surface expression of adhesion markers CD44 and CD24 using flow cytometry. Using primary tumor samples or pleural effusions that eliminated all of the contaminated hematopoietic cells or endothelial cells, often referred to as lineage negative (Lin^-) cells, they found that the cancer stem cell population is characterized as $Lin^-CD44^+CD24^{-/low}$.[4] In addition, $Lin^-CD44^+CD24^{-/low}$ cells that expressed epithelial surface antigen (ESA), a marker found in epithelial cancers were further enriched in their ability to form tumors. As few as 200 cells could be serially transplanted, each time regenerating a heterogeneous population that recapitulated the original tumor. These CSCs demonstrated self-renewal and differentiation, the two important hallmarks of a cancer stem cell.

More recently, human and mouse breast cells that have a high expression of CD49f (alpha 6 integrin) and negative or very low expression of the epithelial molecule EpCAM also have mammary regenerative capacity *in vivo* using the mammary transplantation methods described above.[94,95] In addition, the $EpCAM^+CD49f^+$ population has been demonstrated to be the luminal progenitor population, while $EpCAM^+CD49f^-$ are the committed luminal progenitors. Using, the humanized mouse model and the ALDH activity assay, Ginestier found that $Aldefluor^+$ cells contain the stem/progenitor activity.[41]

A number of groups have studied normal and malignant stem cells in the mouse mammary gland. The work of the Eaves and Visvader groups has been instrumental in this area. They have found that mouse breast stem cells are enriched in the $CD49f^{hi}CD29^{hi}CD24^+Sca1^-$ fraction.[96–98] Shackleton *et al.* demonstrated that a single genetically tagged mammary stem cell could regenerate an entire mammary epithelial tree, thus demonstrating that an epithelial organ could be reconstituted from a single stem cell.[96]

C. Cancer Stem Cells in the Brain

Brain tumors were the next type of cancer in which a putative cancer stem cell was identified. Anatomically, the brain can be divided broadly into three parts: the cerebrum, brain stem, and cerebellum. Structurally it is made up of many different types of cells including the nerve cells or neurons and the glial cells. There are three main types of glial cells: astrocytes, oligodendrocytes, and microglial cells. Neural stem cells are defined as multipotential progenitor cells with the ability to undergo continuous self-renewal and to differentiate into a complete range of neural lineages.[99] In 1992, Reynolds and Weiss first isolated neuronal stem cells from the subventricular zone (SVZ) of the mouse brain and cultured them as neurospheres.[37] Uchida *et al.* demonstrated that the cell surface marker CD133 (Prominin-1), originally shown to be a hematopoietic stem cell marker, may be a potential neural stem cell marker. $CD133^+$ sorted cells initiated neurosphere cultures, and the progeny of clonogenic cells could differentiate into both neurons and glial cells.[38] Later, it was demonstrated that only the $CD133^+$ cells were capable of tumor initiation in an *in vivo* NOD/SCID mouse brain.[5] Another report, by Hemmati *et al.*, substantiated this finding of cancer stem cells in pediatric brain tumors.[100] In addition, it was demonstrated glioblastoma cell lines contain neural stem cells that can form neurospheres, self-renew, and differentiate into multiple lineages *in vitro*, as well as generate tumors *in vivo*.[101,102] Some other stem cell markers have been reported to be expressed in different types of brain cancers including the intermediate filament protein Nestin and Musashi-1.[103–105] In addition, the stem cell marker CD44, is overexpressed in brain tumors[106] and it mediates glioma cell adhesion and invasion.[107] Lastly, the cell surface protein neuron-glial 2 (NG2), a 300-kD membrane spanning chondroitin sulfate proteoglycan has been shown to be a potential stem cell marker, $NG2^+$ cells are multipotent *in vivo*, differentiating into neurons and astrocytes[108,109] and *in vitro* differentiating in all CNS lineages[110] and pericytes.[111]

D. Cancer Stem Cells in the Gastrointestinal Tract

1. Colon Cancer

Existence of stem cells in colon cancer was reported in 2007 by several research groups.[6,30,112] $CD133^+$ cells were found to be tumorigenic in immunocompromised mice and the tumors could be serially transplanted for several generations. However, CD133 has recently been challenged as a marker for colon cancer cells because $CD133^+$ and $CD133^-$ cells were both able to initiate tumors.[35] A separate report suggested that the phenotype of colon cancer stem cells is $ESA^+CD44^+CD166^+$.[112] Another study reported that colon cancer stem cells expressed ALDH1 immunostaining and Aldefluor activity. Huang *et al.* discovered that during colon cancer progression there was an increase in $ALDH1^+$ cells and implantation of $Aldefluor^+$ cells generated xenograft tumors, while $Aldefluor^-$ cells did not form tumors in NOD/SCID mice.[62]

2. Pancreatic Cancer

Pancreatic cancer is a highly lethal disease, which is usually diagnosed at an advanced state for which there is no effective therapies. In 2007, The Simeone group used the same markers that were used to isolate breast stem cells. However, they found that cells expressing all three surface markers $CD44^+CD24^+ESA^+$, comprising only 0.2–0.8% of the human pancreatic cancer cells had the highest tumorigenic initiating potential in NOD/SCID mice. These cells also exhibited the stem cell properties of self-renewal and the ability to produce differentiated progeny.[7] Recently, it was reported that in pancreatic tumors there is a distinct population of $CD133^+$ cancer stem cells that was associated with the metastatic phenotype of the tumor.[113]

E. Cancer Stem Cells in Ovarian Cancer

Ovarian cancer is another very aggressive tumor that is associated with a lack of early symptoms, rapid metastases, and poor prognosis. Bapat and colleagues were the first to identify ovarian cancer stem cells, by isolating specific clones among a mixed population of cells derived from the ascites of a patient with advanced ovarian cancer that had stem cell-like characteristics, grew in an anchorage-independent manner *in vitro* as spheroids, and maintained the capacity to differentiate. In addition, tumors established from these clones resembled human disease.[114] An SP has also been identified in mouse ovarian cancer cell lines.[115] Zhang *et al.* found the association of $CD44^+CD117^+$ (c-kit) cells had cancer stem cell characteristics. As few as 100 dissociated spheroid cells formed tumors in mice which recapitulated the original tumor.[45] Very recently, CD133 has also been reported to be an important marker for ovarian cancer stem cells, $CD133^+$ cells derived from ovarian tumors were capable of self-renewal, and they were associated with increased tumor aggression in xenografts.[31,116] Lastly, it has been reported that $CD44^+MyD88^+$ ovarian cancer cells are able to form spheroids in suspension and recapitulate the original tumor *in vivo* and are relatively resistant to chemotherapy.[117]

F. Cancer Stem Cells in Prostate Cancer

Androgen deprivation therapy for prostate cancer kills most of the cells within the tumor, however most patients go on to develop androgen-independent disease that is resistant to treatment. This suggests that there may be a cancer stem cell population in prostate cancer. In 2005, prostate cancer stem cells were identified as having a $CD44^{+}/\alpha 2\beta 1$ integrin$^{high}/CD133^+$ phenotype.[118] In addition, similar tumorigenic populations have been isolated from prostate cell lines and xenografts.[119,120] In addition, the Aldefluor assay has also been used to identify mouse and human prostate cancer stem cells. Two studies found that high Aldefluor activity enhanced the proliferative potential and tumorigenic capacity *in vitro* and *in vivo*, respectively.[63,121] Several ALDH isoforms were found to be expressed in human prostate cancer cells and clinical specimens of primary prostate tumors.[121]

G. Cancer Stem Cells in Head and Neck Cancer

Like many epithelial tumors, head and neck squamous cell carcinoma (HNSCC) contains a heterogeneous population of cancer cells with a cancer stem cell population. $CD44^+$ cells comprising less than 10% of the cell population were found to have a significant tumorigenic potential, while $CD44^-$ cells did not, these tumors were found to express the basal marker CK 5/14.[8] Other work has been done with cancer cell lines to show that CD133 could isolate a subpopulation of cells that have cancer stem cell properties and can form tumors,[32,33] and only a very small fraction of highly expressing CD133 cells showed high coexpression of CD29 or CD44.[122] More recently, ALDH expressing cancer stem cells have been reported in primary human head and neck cancers. Clay *et al.* found that cells with ALDH high activity compromise a small subpopulation that was tumorigenic in NOD/SCID mice at very low numbers.[123]

H. Cancer Stem Cells in Melanoma

Melanoma is a disease that is made up of a heterogeneous population of cells within the tumor; this has led researchers to question if it follows a cancer stem cell model. In 2005, Fang *et al.* reported that the B cell marker CD20 isolated cells that have self-renewal properties.[124] Other markers such as CD133[125] or ABCB5[126] have been used to isolate cancer stem cell populations in melanomas. However, a more recent publication by the Morrison group has discovered that despite this data, melanoma cells may not follow a stem cell model. They discovered this by modifying the NOD/SCID xenotransplantation assay, including the use of more highly immunocompromised NOD/SCID interleukin-2 receptor gamma chain null (IL-2rg−/−) mice. Using these mice, they saw a dramatic increase in the frequency of tumor-initiating cells up to 25% compared with 1 in a million using the traditional NOD/SCID model.[127] This suggested the need to consider the assays used to access CSC properties.

Very recently in the literature there were two studies that further added to the controversy of the existence of a cancer stem cell marker in melanoma. The first study is consistent with the proposition that melanoma does not follow a stem cell model. Using the H3K4 demethylase JARID1B (KDM5B/PLU-1/RBP2-H1) as a biomarker, Roesch *et al.* found a small subpopulation of slow cycling melanoma cells.[128] When the $JARID1B^+$ cells were isolated they gave rise to a highly proliferative progeny, knockdown of this population results in a decrease in tumor growth, suggesting this subpopulation is essential for continuous tumor growth. They found that expression of JARID1B does not follow a stem cell model because JARID1B negative cells can become positive and all melanoma cells are tumorigenic even with a single cell injection. However, this study has been criticized since it utilized established melanoma cell lines rather than primary tumors. Another study using isolated melanoma cells obtained from primary melanomas by the Weissman group demonstrated that expression of the neural crest stem cell marker CD271 enriched for cells able to initiate tumors in highly immunocompromised NOD/SCID Rag-$2^{(-/-)}\gamma_{ac}{}^{(-/-)}$ mice. Furthermore, $CD271^+$ cells were tumor-initiating in 90% of the human melanomas tested and they were able to give rise to $CD271^+$ and $CD271^-$ cells.[129] In addition, $CD271^+$ melanoma cells transplanted into engrafted human skin or bone of Rag-$2^{(-/-)}\gamma_{ac}^{(-/-)}$ mice generated melanomas and were capable of metastasis, while the isolated $CD271^-$ population did not form melanomas. Together, these studies suggest that melanoma may follow a stem cell model. However, the percent of tumor-initiating cells may increase as the cancer evolves.

I. Cancer Stem Cells in Lung Cancer

Very recently, researchers have been exploring the existence of cancer stem cells in lung cancer. It was reported that the $CD133^+$ cells are the lung cancer tumorigenic population, while the $CD133^-$ cells did not have tumor-initiating activity. $CD133^+$ cells were able to generate differentiated lung cancer cells under the appropriate culture conditions.[34] In addition, cancer stem cells were isolated with relatively high ALDH1 activity *in vitro*, and these cells have the stem cell capabilities of self-renewal, differentiation, resistance to chemotherapy, and they express the CSC surface marker CD133.[130]

V. Activation of Signaling Pathways and Targeted Therapies for Cancer Stem Cells

Advances in stem cell research have enabled the elucidation of important signaling pathways that are activated in cancer stem cells in many different types of cancers. These pathways are candidates for targeted treatment against cancer stem cell populations. In order to effectively target CSCs, it is necessary to understand how these pathways interact and crosstalk to regulate certain cancer stem cell processes including growth, self-renewal and differentiation, and cell death. Some of the pivotal pathways that have been shown to regulate normal and malignant stem cells include: Notch, Hedgehog, HER-2/PI3K/Akt/PTEN, Wnt, and p53. In this chapter, we discuss how these pathways are activated and how they regulate normal and cancer stem cells. We then discuss potential therapies that target these pathways.

A. Notch Signaling Pathway

The Notch pathway is an evolutionally conserved signaling pathway that plays an important role in many stem cell processes such as cell fate and determination, differentiation, proliferation, and apoptosis or cell death.[131] In humans, the Notch family is made up of four different members Notch[1–4] and five different ligands including (Jagged 1-2, Delta-like 1, 3, and 4). The interaction between these transmembrane receptors and the ligands at the cell surface of two adjacent cells is needed for activation of the pathway. The Notch receptors consist of three domains each with a different cellular function: including the extracellular, transmembrane, and intracellular domain. The Notch receptors are precursor proteins that need to be cleaved in order to be activated. The unprocessed precursors are cleaved in the trans-Golgi network by a furin-like convertase and then it gets reassembled on the cell surface.[132]

The interaction of the ligand and receptor induces a conformation change in the Notch receptors resulting in two proteolytic cleavages. First, TACE (TNFalpha-converting enzyme), a member of the ADAM (a disintegrin and metalloprotease domain) family of metalloproteases cleaves the Notch receptors at the extracellular domain (ECD; S2 cleavage).[133,134] The second cleavage is at the transmembrane domain (S3) and it is mediated by the γ-secretase complex. The γ-secretase complex consists of presenilin 1 and 2, nicastrin, Pen-2, and Aph1.[135,136] After these two cleavage steps, the Notch intracellular domain (NICD) gets released to the cytoplasm and translocates to the nucleus to activate the transcription of Notch target genes. In the nucleus Notch binds to a transcriptional repressor CSL/CBF1 or RBP-jk, and upon Notch binding to the repressor, it is switched into an activated state and there is an additional recruitment of coactivators, such as p300, PCAF, and Mastermind-like 1 (MAML1), that aid in the activation of Notch target genes.[137,138] Some of the primary Notch target genes are the members of the Hairy enhancer of split (*Hes*) or Hairy related (*Hey* or *Hrt*) families of helix-loop-helix transcriptional factors. Other Notch target genes include c-Myc, cyclin D1, *NF-κB*, and p21/Waf1. Within the cell, Notch signaling is modified by the cytoplasmic negative regulatory protein Numb.[139] Numb has been found to be asymmetrically segregated into one of two daughter cells and inhibits Notch by promoting the ubiquitination of membrane-bound Notch-1 receptor.[140,141] The activation of the Notch receptor, translocation to the nucleus, and transcription of Notch target genes is depicted in Fig. 4.

It has been reported that the Notch signaling network is frequently deregulated in many human cancers.[142] Deletions of Notch-1 were discovered in T-cell acute lymphoblastic leukemia (T-ALL) to be an activating mutation in 50% of all T-ALL.[143] In addition, a high level of expression of Notch-1 and the ligand Jagged-1 is associated with poor prognosis in breast cancer[144] and metastasis in prostate cancer.[145] Upregulated expression of Notch receptors and their ligands in solid tumors have also been seen in many other types of cancers and excellent reviews have been written.[16,142] In contrast to its role in tumor promotion, Notch-1 has also been found to act as a tumor suppressor in some cancers including skin cancers.[146,147]

Notch signaling is required for the self-renewal of ESCs and has been implicated in the maintenance of normal stem cells and CSC populations. Notch-1 was found to be essential for generation of hematopoietic stem cells

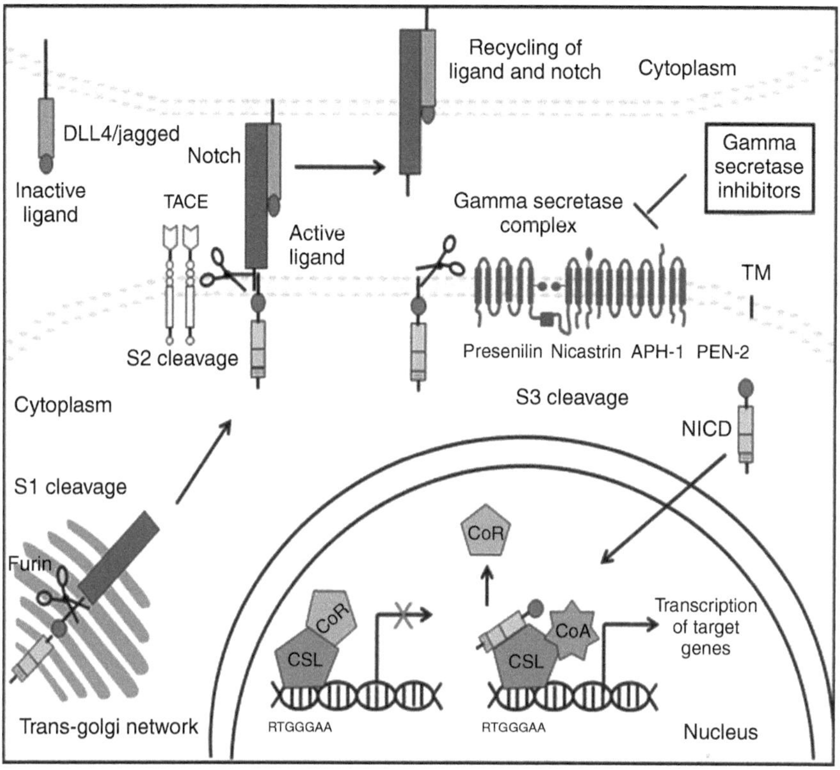

FIG. 4. Notch pathway. Unprocessed Notch precursors are cleaved in the trans-Golgi network by a furin-like convertase (S1 cleavage) which are reassembled at the cell surface. The interaction of the ligand and receptor induces a conformation change in the Notch receptors resulting in two proteolytic cleavages. First, TACE (TNFalpha-converting enzyme), a member of the ADAM (a disintegrin and metalloprotease domain) family of metalloproteases cleaves the Notch receptors at the ECD (S2 cleavage). The second cleavage is at the transmembrane domain (S3 cleavage) and it is mediated by the γ-secretase complex. After these two cleavage steps, the Notch intracellular domain (NICD) gets released to the cytoplasm and translocates to the nucleus to activate the transcription of Notch target genes. In the nucleus Notch binds to a transcriptional repressor CSL/CBF1 or RBP-jk, upon Notch binding to the repressor it is switched into an activated state and there is an additional recruitment of coactivators that aid in the activation of Notch target genes. Gamma-secretase inhibitors inhibit the activation of NICD and the translocation of Notch into the nucleus.

and maintenance of neural stem cells.[148,149] In 2004, the Wicha group showed that activation of Notch signaling by addition of the Delta ligand promoted stem cell self-renewal and differentiation of human mammary progenitor cells.[150] Clarke and colleagues demonstrated that Musashi-1 and Notch-1 regulate human breast stem cells enriched using the cell surface makers

$CD44^+CD24^-$.[151,152] Recently, this group showed that Notch-4 activity was increased in breast CSC, and that inhibition of Notch-4 signaling reduced breast CSCs and completely inhibited tumor initiation.[153] In glioma cancer, there was an increase in expression of Notch ligands and receptors and Notch signaling was found to promote the formation of cancer stem cell-like cells resulting in the generation of increased neurospheres with expression of nestin. These cells could be induced to differentiate into neurons, astrocytes, or oligodendrocytes.[154] Fan *et al.* found that Notch blockade reduced $CD133^+$ medulloblastoma cells and then frequency of apoptotic cells was increased 10-fold following Notch blockade in $nestin^+$ cells compared to $nestin^-$ cells.[155] The Notch pathway is known to interact and crosstalk with multiple oncogenic pathways that are also cancer stem cell regulators including NF-kB, Akt, Sonic hedgehog (SHh), mammalian target of rapamycin (mTOR), Ras, Wnt, epidermal growth factor receptor (EGFR) and platelet-derived growth factor (PDGF) signaling; some of these pathways are discussed below.[156–162]

A number of genetic and pharmacologic strategies to inhibit the Notch pathway have been developed. These include downregulating the Notch receptors with antisense[163] or RNA interference, monoclonal antibodies[164] treatment with soluble receptor decoys that act by sequestering Notch ligands[156] and pharmacological inhibitors of γ-secretase which prevent the second ligand-induced proteolytic cleavage of Notch receptors (Fig. 4). All of these strategies are discussed in more detail in several reviews on the inhibition of the Notch signaling pathway.[165]

The γ-secretase drugs were initially developed for treatment of Alzheimer's disease. These drugs inhibit all of the four Notch receptors and are associated with toxicities including the gastrointestinal system resulting in intestinal goblet cell hyperplasias.[166] Therefore, it may be more useful to develop drugs that inhibit specific Notch receptors for a particular type of cancer. In addition, it also may be more useful to combine these drugs with standard chemotherapeutic agents or other cancer stem cell selective pathway inhibitors. γ-Secretase inhibitors have been reported to have antitumor effects in many different types of cancers including melanoma and sarcoma, medulloblastoma, T-ALL, and intestinal adenomas.[167–170] Furthermore, a phase I clinical trial for the Notch inhibitor, MK0752 (developed by Merck, Whitehouse Station, NJ) has been launched for relapsed or refractory T-ALL patients and advanced breast cancers.

B. Hedgehog Signaling Pathway

The proteins of the Hedgehog (Hh) family are mediators of fundamental processes in ESCs regulating their self-renewal and differentiation. Three hedgehog genes have been identified in mammals: Desert hedgehog (DHh), Indian hedgehog (IHh), and Sonic hedgehog (SHh), all of which are secreted glycoproteins and act as ligands for receptor complexes on adjacent cells.[171]

The Hedgehog receptor complex is comprised of at least three proteins, Patched 1 (PTCH1), Smoothened (SMOH), and Hedgehog interacting protein (HIP).[172,173]

In the absence of ligands, the 12 transmembrane protein PTH1 binds to the 7 transmembrane SMOH protein and represses its function by preventing its localization to the cell surface.[174] When Hh ligands are present, the Hh signaling cascade is initiated by Hh binding to the PTH1 protein on the target cell, allowing for SMOH to activate the transcription factors GLI1, GLI2, and GLI3.[175] The GLI proteins translocate to the nucleus and control the transcription of Hedgehog target genes including GLI1 and PTCH1. GLI1 and GLI2 function mostly as activators and GLI3 as a repressor.[176] The Suppressor of fused (SUFU) protein inhibits the activation of all of the GLI proteins.[177] The activation of the Hedgehog signaling pathway in the presence of the Hg ligand is depicted in Fig. 5.

The first description of the activation of the Hedgehog signaling pathway in cancer was the Gorlin syndrome which was found to be caused by a mutation in the PTCH1 protein.[178] This disease is characterized by the presence of multiple basal cell carcinomas (BCCs) of the skin which predisposes patients to develop other types of cancer including medulloblastoma, a tumor of cerebellar granule neuron progenitor cells, and rhabdomyosarcoma, a muscle tumor. It was later discovered that a large majority of BCCs and medulloblastomas resulted from hyperactivated Hedgehog signaling with increased levels of GLI1 and PTCH1 in tumor cells, or PTCH1 or SUFU mutations, respectively.[179–181]

In other types of tumors, the aberrant activation of the hedgehog signaling pathway is not as well understood. However, Hh has been found to be overexpressed in small-cell lung cancer (SCLC). Watkins *et al.* reported that 25% of human SCLC samples had relatively high expression of both SHh and GLI1 and that growth of SCLC cell lines could be blocked *in vitro* using the Smoothened inhibitor cyclopamine or a monoclonal anti-Hh antibody. Furthermore, cyclopamine prevented growth of SCLC in immunocompromised mice.[182,183] The Hh pathway has also been show to be important in pancreatic tumors. In 2003, Thayer *et al.* demonstrated abnormal expression of SHh, PTCH1, and SMOH in human pancreatic tumors, as well as PTCH1 and SMOH positive cancer cell lines that were growth inhibited by cyclopamine *in vitro* and had delayed tumor growth in xenographs of nude mice *in vivo*.[184]

The Hh signaling is also important in the maintenance of ESC pools including the epidermis and mouse neocortical cells.[185,186] In addition, Hh signaling has also been demonstrated to play a role in cancer stem cells. The pancreatic cancer stem cells population $CD44^+CD24^+ESA^+$ has enhanced expression of the Sonic Hedgehog protein.[7] In glioblastoma multiforme, Hh regulates the ESC genes, OCT4, SOX2, and BMI1, and the self-renewal of $CD133^+$ glioma cancer stem cells.[187] For human normal mammary stem cells, the Hedgehog

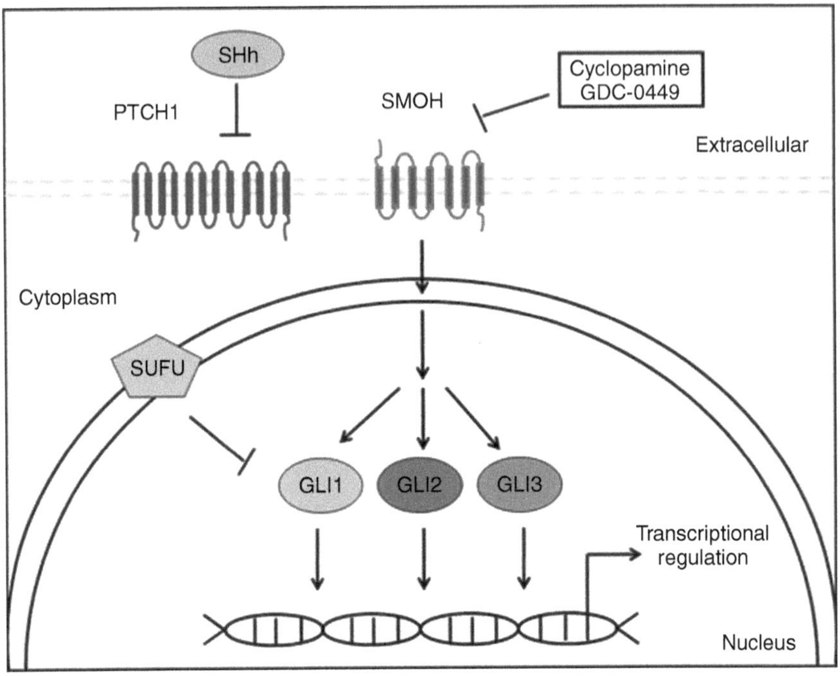

FIG. 5. Hedgehog pathway. The Hedgehog receptor complex is comprised of at least three proteins, Patched 1 (PTCH1), Smoothened (SMOH), and Hedgehog interacting protein (HIP). In the absence of ligands, the 12 transmembrane protein PTH1 binds to 7 transmembrane SMOH protein and represses its function by preventing its localization to the cell surface. When Hh ligands are present, the Hh signaling cascade is initiated by Hh binding to the PTH1 protein on the target cell, now allowing for SMOH to activate the transcription factors GLI1, GLI2, and GLI3. The GLI proteins translocate to the nucleus and control Hedgehog target gene transcription. The Suppressor of fused (SUFU) protein is the inhibitor for the activation of all of the GLI proteins. Cyclopamine and GDC-0449 are inhibitors of the Hedgehog signaling pathway.

pathway has been shown to be important for the self-renewal and multilineage differentiation of the lineage negative $CD44^{+}CD24^{-/low}$ progenitor/stem cell population. In 2006, Liu *et al.* demonstrated that the Hedgehog proteins PTH1, GLI1, and GLI2 are expressed in the mammary stem/progenitor cells cultured as mammospheres and that these genes are downregulated once the cells are put under differentiation conditions.[40] In addition, it was discovered that these effects were mediated by the polycomb gene BMI-1.

Hedgehog signaling may also act on the stromal cells in tumors regulating tumor growth through paracrine mechanisms. Yauch *et al.* reported that Hh ligands fail to activate signaling in tumor cells and found that there is a ligand-dependent activation of the Hh pathway in the stromal microenvironment. In addition they found that specific inhibition of Hh signaling using small molecule inhibitors, a neutralizing anti-Hh antibody or genetic deletion of Smoothened in the mouse stroma resulted in growth inhibition in xenograft tumor models.[188]

To antagonize the activation of the Hedgehog signaling cascade, inhibitors have been used that inhibit or act downstream of SMOH. The most widely used and commercially available inhibitor of the Hh pathway is cyclopamine, a natural-product SMOH inhibitor derived from corn lilies (Fig. 5). Phase I clinical trials of advanced BCC and medulloblastoma with the orally bioavailable small molecule Hh pathway inhibitor GDC-0449 have been initiated (Fig. 5). GDC-0449 induced rapid regression of the tumor and suppression of the hedgehog pathway in the *Ptc1*$^{+/-}$ murine medulloblastoma model.[189] In patients, it has been well tolerated with limited side effects and has shown some promise in the regression of these types of cancers.[189–191] Phase II studies utilizing these compounds in combination with cytotoxic agents are in development.

C. HER2/PI3K/Akt/PTEN and Wnt Pathway Signaling Cascade

The HER2/Neu growth factor receptor pathway is known to play a role in many different types of cancers. More recently, it has been discovered that this protein also plays a role in the signaling of cancer stem cells. HER2 is a member of the HER, ErbB, or EGFR family of receptor tyrosine kinases (TKs). The EGFR family members all have the same structure: an N-terminal ECD, a single transmembrane helix, a TK domain, and an intracellular regulatory domain.[192] Activation of the HER receptors is the initiating event in signal transduction. Upon ligand binding at the ECD, there is receptor dimerization which triggers the intrinsic TK activity resulting in phosphorylation of the TK residues in the intracellular domain. These phosphorylation sites act as docking sites for other signaling molecules and result in activation of intracellular signaling pathways. This family of receptors has the ability to heterodimerize with other members or form homodimers, in fact, HER2 is the preferred coreceptor for the other members of the family.[193] There are more than 10 different known ligands including EGF, neuregulin, and betacellulin.[194] However, none of these ligands binds to the ECD of HER2 with high affinity.[195]

In cancer, HER2 overexpression results in ligand-independent signaling, with an increase in cell proliferation and inhibition of apoptosis through activation of the phosphatidylinositol 3-kinase (PI3K, p85 subunit)/Akt (protein kinase B) and mitogen activated protein kinase (MAPK) survival signaling pathways. Akt is a central regulator of the Wnt and PI3K signaling pathways, and the tumor suppressor PTEN is an Akt pathway inhibitor and it is frequently mutated in human cancer.[196] Activation of the Akt pathway results in cell cycle progression due to downregulation of the cell cycle inhibitor p27 and inhibition of apoptosis thorough a reduction in the proapoptotic molecules BAD and caspase-9.[197] The HER-2 activation and signaling cascade is shown in Fig. 6.

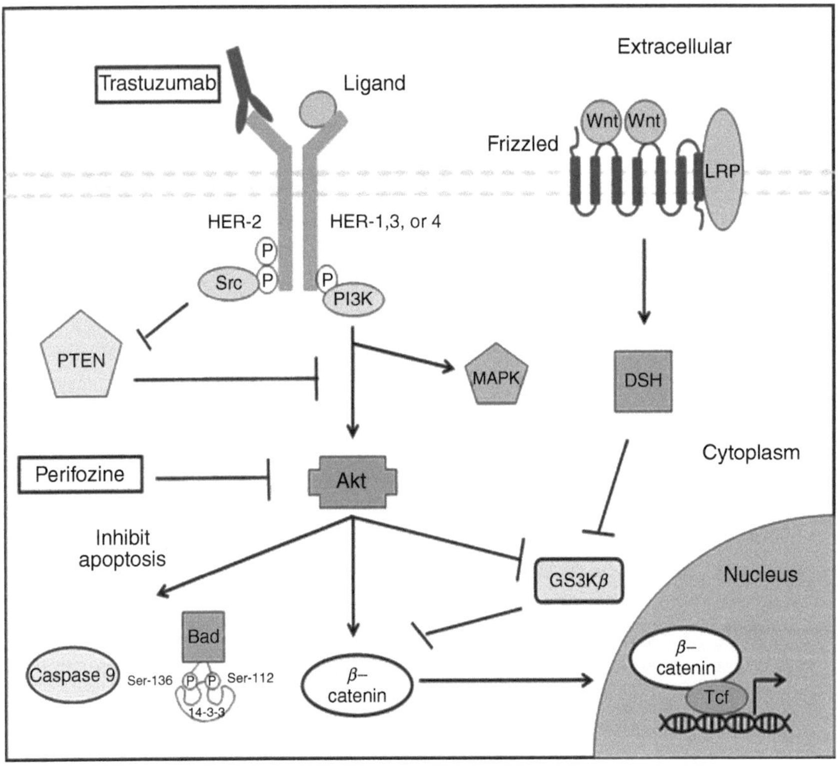

FIG. 6. HER2/PI3K/Akt/PTEN and Wnt pathway signaling cascade. Activation of the HER receptors is the initiating event in signal transduction. Upon ligand binding at the ECD, there is receptor dimerization which triggers the intrinsic TK activity resulting in phosphorylation of the TK residues in the intracellular domain. These phosphorylation sites act as docking sites for other signaling molecules such as phosphatidylinositol 3-kinase (PI3K, p85 subunit)/Akt (protein kinase B) and mitogen activated protein kinase (MAPK) survival signaling pathways. The tumor suppressor PTEN is an Akt pathway inhibitor and it is frequently mutated in human cancer. Trastuzumab is an anti-HER2 monoclonal antibody therapy used in cancers that express high levels of HER2, it has also been demonstrated to reduce the breast cancer stem cell population in Her2 overexpressing cells. Activation of the Akt pathway results in cell cycle progression due to downregulation of the cell cycle inhibitor p27 and inhibition of apoptosis thorough a reduction in the proapoptotic molecules BAD and caspase-9. In breast cancer cells, a loss in PTEN results in activation of the downstream Wnt signaling pathway through phosphorylation and inactivation of GSK/3β as well as direct phosphorylation at serine 552 and nuclear translocation of β-catenin. The Akt inhibitor perifosine is able to target the tumorigenic cell population in breast tumor xenografts.

Overexpression of the HER2/Neu protein is seen in approximately 25–30% of breast cancers and is primarily due to amplification and not an activating mutation of the HER2 gene and this is associated with a poor prognosis.[198,199] In addition, there have been several reviews describing the overexpression of HER2 in other carcinomas, including ovarian, gastric, bladder, and others.[200,201] The development of HER2-targeted therapies to treat HER2 overexpressing breast cancers, including the monoclonal antibody Herceptin or trastuzumab and the small molecule TK inhibitor lapatinib have been very important advancements in breast cancer therapy (Fig. 6). When used in combination with conventional chemotherapy, these drugs have improved progression-free survival and overall survival of patients with advanced disease and reduced the recurrence rate by $\sim$50% in women with a HER2 amplification.[202] Due to the clinical benefit of these breast cancer patients, these HER2-targeted drugs are now being tested in various other HER2 overexpressing cancers. However, despite this significant clinical advancement, only 1/3 of HER2 tumors respond to this treatment and resistance may develop with chronic exposure. Therefore, more work needs to be done to study the resistance to this type of treatment and combinations of drugs that target multiple pathways need to be tested. Increasing evidence suggests that resistance may be due to the loss of the tumor suppressor PTEN, resulting in aberrant activation of the PI3K/Akt survival signaling pathway.[203]

Recent studies have suggested that HER2-targeted therapies may be beneficial because they are targeting the cancer stem cell populations. Korkaya *et al.* demonstrated that overexpression of HER2 in breast carcinoma cell lines increased the Aldefluor positive cancer stem cell population resulting in increased invasion and metastasis that was reduced with Herceptin treatment. This was demonstrated using the *in vitro* mammosphere assay, Aldefluor assay, and injection of cells into the mammary fat pad of NOD/SCID mice.[204] Magnifico and colleagues showed that HER2 expression was higher in the stem cell population compared with the bulk population and the CSCs were sensitive to Herceptin treatment.[205] Interestingly, they also demonstrated that HER2 is a downstream target gene of the Notch signaling pathway, as shown by reduction of the HER2 expression with treatment with a gamma-secretase inhibitor.

Recent evidence suggests that PI3K/Akt/PTEN pathway is very important in the regulation of breast cancer stem cells. The Wicha group demonstrated that knockdown of PTEN leads to activation of the Akt pathway resulting in enrichment of normal and malignant human mammary stem/progenitor cells *in vitro* and *in vivo.*[206] Akt activation in turn results in activation of the downstream Wnt signaling pathway through phosphorylation and inactivation of GSK/3β as well as direct phosphorylation on serine 552 followed by nuclear translocation of β catenin. The Akt inhibitor perifosine was able to target the tumorigenic cell population in breast tumor xenografts. Therefore, it has been suggested that PI3K or Akt selective inhibitors may prove to be useful in the targeting of breast cancer stem cells. The interaction of the HER2 and Wnt pathway in breast cancer stem cells is depicted in Fig. 6.

D. The p53 Pathway

The tumor suppressor p53 is known as the guardian of genome. It acts in response to a cell undergoing DNA damage induced by a variety of stress signals and primes cells to either undergo apoptosis or cell cycle arrest in many different types of cancers. More recently, it has been found that the p53 pathway also plays a vital role in stem cell self-renewal of induced pluripotent stem (iPS) cells generated from somatic cells. In fact, reprogramming somatic cells to iPS cells has been accomplished by expressing pluripotency factors and oncogenes, including the three transcription factors Oct4 (also known as Pou5f1), Klf4, and Sox2.[207] In 2009, five independent laboratories have identified that p53 is an important checkpoint during the reprogramming of differentiated adult somatic cells to iPS cells.[208–212] The presence of p53 acts as a barrier to the iPS cell generation, when p53 was absent or deleted there was an increase in the number of iPS cells.

Recently, the role of p53 in the regulation of cancer stem cell self-renewal via symmetric or asymmetric cell division has been investigated. Cicalese and colleagues used the PKH 26 cell surface dye labeling assay described above to track the mouse breast stem cells *in vitro* and *in vivo*. Initially, they labeled the cells and grew them as mammosphere cultures. After 1 week, only 1% of mammosphere cells remained PKH high, these cells were able to form secondary mammospheres and retained mammary outgrowth potential.[75] Real-time imaging of these cells as they divide revealed that 80% of the divisions were asymmetric. In contrast, the mammary cells of $p53^{-/-}$ mice contained a higher proportion of mammary stem cells, and *in vitro* these cells underwent a higher ($\sim$75%) proportion of symmetric cell divisions. Additionally, restoration of p53 function with Nutlin3, an inhibitor of MDM-2 dependent p53 degradation, reversed cell divisions back to asymmetric and reduced mammosphere and tumor formation.[75]

E. The Tumor Microenvironment/CSC Niche and Cytokine Signaling Loops

The microenvironment plays an important role in the regulation of the tumor cells by secreting factors that promote the growth and progression of cancer.[213] Recently, the surrounding tumor microenvironment or cancer stem cell niche has been found to play a role in the regulation of the self-renewal and proliferation of cancer stem cells. Paracrine interactions between the tumor stem cells, their differentiated progeny, and the microenvironment regulate tumorigenesis and cancer stem cell functions. The tumor niche is made up of many different types of cells including: fibroblasts, endothelial, hematopoietic, adipocytes, and extracellular matrix components. All of these cell types and molecules communicate together through a complex network of growth factors, cytokines such as the interleukin (IL) family, and receptors.[214] An example of this is seen in the hematopoietic system where myeloproliferative disease can arise as a result of mutations that only affect the bone-marrow microenvironment and not the hyperproliferative hematopoietic cells themselves, revealing the capability of the microenvironment to be the sole cause of hematopoietic disorders.[215] In addition, normal cells can also regulate metastasis: the addition of bone-marrow-derived human mesenchymal stem cells was found to stimulate secretion of the chemokine CCL5 (RANTES) which acts on the cancer cells in a paracrine fashion leading to enhancement of their motility, invasion, and metastasis.[216]

There is also evidence that the cancer stem cell population and signaling pathways described above may be affected by the microenvironment. For example, endothelial cells secrete factors that promote the self-renewal, inhibit their differentiation, and enhance the self-renewal of normal neural stem cells in culture through activation of Notch and Hes1.[217] In addition, dividing neural progenitors sometimes reside close to blood vessels in the brain providing a vascular niche.[218]

Cytokine signaling networks are an essential component of the microenvironment and they play an important role in tumorigenesis and the behavior of cancer stem cells. In colon cancer, $CD133^+$ CSCs produced and utilized IL-4 to protect themselves from apoptosis. Therefore, treatment with an IL-4 antagonist or IL-4 neutralizing antibody sensitized the $CD133^+$ cells to standard chemotherapy and enhanced the antitumor efficacy.[219] Interleukin-8 expression regulated angiogenesis, tumorigenicity, and metastases in androgen-independent prostate cancer by induction of the metalloproteinase MMP-9.[220] In breast cancer, an increase in serum IL-6 levels correlates with a poor disease outcome and reduced prognosis.[221] *In vitro*, IL-6 gene expression was upregulated in mammospheres obtained from aggressive ductal breast carcinomas and that treatment with IL-6 triggered the Notch-3-dependent signaling pathway promoting self-renewal, hypoxia survival, and the invasive potentials of normal and tumor mammospheres.[221] The Wicha group demonstrated that in many different breast cancer stem cells Aldefluor positive cells had an overexpression of CXCR1, a receptor for the cytokine IL-8 or addition of exogenous IL-8 increased the CSC population.[42] In addition, when they block CXCR1 with a specific blocking antibody or repertaxin, a small molecule CXCR1 inhibitor, there was a depletion of the CSC population in breast cancer cell lines *in vitro*. *In vivo*, treatment with repertaxin resulted in reduction of tumor volume, metastasis, and secondary tumor formation in human breast cancer xenografts with addition of the chemotherapy drug docetaxel.[222] There was massive apoptosis in the bulk tumor via FAS ligand/FAS receptor mediated extrinsic apoptotic signaling mediated by the FAK/Akt/FOX03A pathway. This suggests that repertaxin sensitized CSCs to be killed by FAS mediated apoptosis and that CXCR1 inhibition may be a very useful way to target the breast cancer stem cells.

These studies suggest that it may be beneficial clinically to develop specific cancer stem cell therapies that target these cytokine loops, thereby targeting the CSC populations and the tumor microenvironment. The CXCR1 inhibitor Repertaxin was developed to prevent graft rejection and was shown to be relatively nontoxic in phase I clinical trials. In addition, monoclonal antibody therapy against IL-6 or its receptor is being evaluated in clinical trials for multiple myeloma.[223] These targeted therapies can be given alone or in combination with standard chemotherapy. The combination approach may be more useful because it can help to sensitize the cancer stem cells to bystander induced apoptosis mediated by the chemotherapy.

F. The Role of MicroRNAs and Regulation of Cancer Stem Cell Signaling

Emerging evidence suggests that microRNAs (miRNAs) play an essential role in regulating the genes that control self-renewal, differentiation, and division of normal stem, cancer, and CSCs. MiRNAs are a conserved class of naturally occurring 21–23 nucleotide noncoding RNAs that are processed from larger hairpin structures that regulate gene expression by binding to mRNAs, resulting in mRNA degradation or translational inhibition.[224] The biogenesis or processing of miRNAs is a very complex process. First, miRNAs are transcribed by the RNA polymerase II enzyme producing a long primary-miRNA (pri-miRNA) with a cap structure at the 5′ end along with polyadenylation at the 3′ end, very similar to mRNA.[225,226] The pri-miRNAs contain hairpin-shaped structures of ~ 70 nucleotides that get cleaved by a nuclear microprocessor complex to become the hairpin intermediates pre-miRNA. These pre-miRNAs then get transported from the nucleus to the cytoplasm, where further cleavage takes place by the RNase III endonuclease Dicer-1 and its essential transactivating response RNA binding protein (TRBP) producing a short imperfect double-stranded miRNA duplex. The intermediate miRNA duplex is unwound into mature miRNA by helicase. Next, the mature miRNA are loaded into the Argonaute 2 protein of the RNA-induced silencing complex (RISC) to target mRNA molecules with complementary sequences to direct posttranscriptional gene silencing (PTGS) and/or mRNA destabilization.[227,228] The miRNA forms 6–8 complementary base pair interactions with the 3′ untranslated regions (UTR) of their mRNAs.[229] At sites with extensive pairing complementarity, miRNAs can direct Argonaute-catalyzed mRNA cleavage.[230–232]

The biogenesis of miRNAs has been shown to be essential for development and function of normal stem cells. For example, studies using mouse Dicer (dcr-1) mutants resulted in animal death early in development with depletion of normal stem cells, suggesting Dicer is essential for the maintenance of stem cell pools.[233] In ESCs, mutation of dcr-1 resulted in severe defects in ESC differentiation *in vivo* and *in vitro*; however, reintroduction of Dicer-1 reversed these defects.[234] In fact, 36 miRNAs have been identified to be expressed in human ESCs relative to differentiated cells.[235]

Increased levels of certain miRNAs can lead to aberrant expression of oncogenes or a loss of a tumor suppressor gene leading to increased tumorigenesis. The miRNA miR-21 has been found to be overexpressed and function as an oncogene in breast tumor tissues through regulation of the antiapoptotic gene Bcl-2 and Programmed Cell Death 4 (PCD4) resulting in increased tumor growth and decreased cell death.[236–238] Another miRNA, the miR-17-92 cluster, which is comprised of seven miRNAs, can accelerate c-Myc induced B cell lymphoma development and is highly expressed in many tumors, including lung tumors.[239,240] Some of the potential targets of miR-17-92 include E2F1 (which promotes cell proliferation) and the tumor suppressor genes PTEN (which promotes apoptosis) and RB2.[241] In medulloblastomas, miR-17-92 has been shown to interact with the Hedgehog stem cell pathway.[242] In addition, miR-135a and miR-135b were found to be greatly upregulated in colorectal adenomas and carcinomas, resulting in downregulation of the APC gene leading to an accumulation of β-catenin and upregulation of the Wnt signaling pathway.[243]

On the other hand, other miRNAs are decreased in cancerous cells, these miRNAs are referred to as tumor suppressor miRNAs (TSmiRNAs). In cancer, the expression of TSmiRNAs is downregulated resulting in increased progression of the disease.[244] Recently, it has been discovered that TSmiRNAs play an important role in the self-renewal and differentiation of cancer stem cells through regulation of the cancer stem cell pathways mentioned above. One example of a TSmiRNA is let-7, relative to normal lung samples, let-7 was reduced in various cancer cell lines and tumors.[245,246] Let-7 was not expressed in breast cancer stem cells and its expression increased with differentiation, however infecting these cells with let-7-lentivirus reduced proliferation, mammosphere formation, and tumor formation and metastasis in NOD/SCID mice.[247] It was found that let-7 negatively regulates ras and is very important in the self-renewal of cancer stem cells. Other miRNAs have also been found to be reduced in breast cancer stem cells. Shimono and colleagues identified that three clusters of miRNAs including miR-200c-141, miR-200b-200a-429, and miRNA-183-96-182 were downregulated in human breast CSCs, normal human and murine mammary stem/progenitor cells, and embryonal carcinoma cells. They also discovered that expression of BMI1, a known regulator of stem cell self-renewal, was a target of miR-200c.[248]

In leukemia, miR-15a and miR-16-1 are expressed at low levels with an increased expression in the antiapoptotic molecule bcl-2.[249] In prostate cancer cells, downregulation of these miRNAs resulted in upregulation of Wnt signaling.[250] In high-grade gliomas, miR-128 levels were significantly reduced and it was discovered that miR-128 caused a downregulation of Bmi-1 signaling.[251] In medulloblastoma, miR-199b-5p was found to regulate the Notch signaling pathway by downregulating the expression of HES1. Overexpressing miR-199b-5p blocked Notch signaling and decreased the cancer stem cell $CD133^+$ population.[252] In medulloblastoma, MiR-125b, miR-326, and miR-324-5p negatively regulate SMOH, a protein that is an activator of the Hedgehog signaling pathway, therefore, reduced expression of these miRNAs result in increased Hedgehog activity.[253] MiR-34 is reduced in pancreatic and gastric cancers, and when restored it can inhibit tumorsphere growth *in vitro* and tumor initiation *in vivo* through regulation of Bcl-2, Notch, and HMGA2.[254,255]

The targeting of miRNAs for anticancer therapy is a new area of study and has potential for reducing the cancer stem cell population. If a certain miRNA is overexpressed and plays an oncogenic role then a potential therapy would be to downregulate the miRNA using an antagonist. For example, miR-21 is increased in breast cancer, expression of an anti-miR-21 oligonucleotide suppressed cell growth *in vitro* and tumorigenesis *in vivo* in MCF-7 breast cancer cells.[237] On the other hand, if the miRNA is a TSmiRNA, the best strategy would be to use miRNA mimics or lentiviruses to restore the tumor suppressor potential.[244] A lentivirus that expressed miR-34a was found to inhibit cancer cell growth and tumorspheres and it was able to restore the tumor suppressor effect in pancreatic cancer stem cells.[254,255] However, the challenge of this type of therapy is the delivery of these anti-miRNA oligonucleotides or miRNA mimetics to human patients because they would have to be packaged in an effective gene delivery vehicle. MiRNA therapy is discussed in more detail in several reviews and with time it may provide a useful approach for the targeting of cancer stem cells.[244]

G. Natural Compounds for Treatment and Prevention of Cancer Stem Cells

Targeting of the cancer stem cell population is a new therapeutic strategy that may be useful for the prevention and treatment of cancer. Many laboratory groups are discovering natural compounds that can modulate the cancer stem cell number by targeting stem cell pathways, antiapoptotic mechanisms, and induction of differentiation.[256] Phytochemicals, compounds made from fruits, vegetables, and grains, possess these anticancer properties and may be very useful for the prevention of many different cancers. For example, curcumin from turmeric was shown downregulate the Notch and Wnt cancer stem cell signaling pathways.[257,258] A more recent study by the Wicha group found that curcumin and piperine had a direct effect on the self-renewal of stem cells, treatment with both dietary polyphenols inhibited mammosphere formation, serial passaging, reduced the percent of $ALDH^+$ cells, and inhibited the Wnt signaling.[259] A similar effect was found for sulforaphane, a natural compound derived from broccoli/broccoli sprouts.[260] In addition, other dietary compounds have also been found to regulate molecules in the Wnt stem cell signaling pathway including apple-derived quercetin and epigallocatechin-3-gallate (EGCG) a molecule from green tea.[261] Vitamin D_3 has been shown to regulate stem cell differentiation and may be useful as a cancer preventive treatment.[262] A review by Kawasaki *et al.* further describes the targeting of cancer stem cells with many different dietary polyphenols.[256]

VI. Therapeutic Implications for Targeting Cancer Stem Cells

The existence of cancer stem cells has very important implications for the development of cancer therapeutics. The cancer stem cell hypothesis suggests that by targeting the CSC population there will be a reduction in cells that are responsible for the growth of the tumor. In addition to being responsible for the development of the tumor, cancer stem cells have also been found to be more aggressive, invasive, and metastatic than cells forming the tumor bulk. For example, Aldefluor positive populations of mammary carcinoma cell lines display increased invasive characteristics as well as increased ability to metastasize when injected into the left ventricle of NOD/SCID mice.[58] In addition, Balic *et al.* found an increase in the breast cancer stem cell population $CD44^+CD24^-$ in metastatic bone-marrow sites in patients with breast carcinoma.[263] Therefore, treatment of the cancer stem cell populations may lead to a reduction in clinical metastasis.

Recent evidence suggests that another survival advantage that cancer stem cells have gained is the resistance to both radiation and standard chemotherapy. There are several hypotheses as to how stem cells have acquired this resistance. Since stem cells are quiescent or slowly dividing, they are more resistant to cell cycle active chemotherapeutic drugs. In addition, since cancer stem cells have a higher level of drug transporters such as the adenosine triphosophate-binding cassette proteins or the BCRP, more of the drug is effluxed out of the cell, so the chemotherapy would be less effective. Also, the enzyme ALDH is highly expressed in cancer stem cells and it has been found to metabolize certain chemotherapeutic drugs, such as cyclophosphamide.[264] Therefore, it would be beneficial clinically to use drugs that target cancer stem cells either alone or in combination with radiation or standard chemotherapy to gain the ability to kill more tumors cells resulting in a better clinical outcome and survival of cancer patients.

Since cancer stem cells may make up only a small population of the tumor, changes need to be made to measure the success and outcome of a potential therapy targeting cancer. Currently, clinical trials use the Response Evaluation Criteria in Solid Tumors (RECIST) to measure a reduction or shrinkage in tumor size. However, this tumor regression does not always correlate with increased survival.[265] With cancer stem cell treatments, this reduction may not be evident, which suggests that there are limitations to the present therapies and trial designs and their ability to target cancer stem cells. Therefore, clinical trials need to be restructured to have the ability to directly measure the effect of these interventions on the cancer stem cell population. One possible design would be to conduct a neoadjuvant trial or to give the drug before the patient's surgery and then follow the patient outcome overtime by examining the different cancer stem markers in a patient's biopsy. In addition, the cells from the biopsy can be used to conduct other stem cell assays such as the formation of tumorspheres as a read out of the drugs ability to target cancer stem cells. This was done in a recent neoadjuvant study that demonstrated an increase in the proportion of $CD44^{+}CD24^{-}$ breast cancer stem cells after chemotherapy. However, after anti-HER2 treatment with lapatinib, there was a decrease in the cancer stem cell population suggesting that anti-HER2 therapy may be a beneficial cancer stem cell treatment.[266] Therefore, currently in the clinic and the laboratory the use of cancer stem cell treatments is causing physicians and researchers to rethink how to treat patients and this approach may require development of novel clinical trial design.

VII. Conclusions

In this chapter we have compared cancer stem cells with normal stem cells. The most important properties of both are their ability to self-renew and differentiate into the multiple cells that make up a given organ. The cancer stem cell hypothesis suggests that it is the small cancer-initiating population that is responsible for perpetuating the tumor as well as generating the cells that form the bulk of a tumor. In order to study the cancer stem cells in any type of cancer, many different assays have been developed including the SP assay, FACS using cancer stem cell markers, Aldefluor assay or ALDH immunostaining, and PKH dye labeling assays. All of these assays can be used alone or in combinations to further elucidate the cancer stem cell population of interest. These assays have been used to isolate cancer stem cell populations in many different cancers discussed in this chapter and listed in Table I. Now that these cancer stem cell populations have been identified, the next logical step would be to devise strategies to target and kill them. One potential strategy would be to target the cancer stem cell pathways discussed here such as Notch, Hedgehog, HER2, or Wnt either alone or in combinations using multiple drugs. Another useful strategy would be to use standard chemotherapy with the cancer stem cell targeting therapies to further reduce tumor size and the cancer stem cell population. Lastly, to see the clinical benefit for these types of treatments it is necessary to redesign future clinical trials to look at different clinical endpoints for these types of drugs. This chapter has demonstrated that the treatment of cancer stem cells is a very exciting and promising therapy and will hopefully lead to an improved outcome and survival of patients with cancer.

References

1. Wang JC, Dick JE. Cancer stem cells: lessons from leukemia. *Trends Cell Biol* 2005;**15**:494–501.
2. Nowell PC. The clonal evolution of tumor cell populations. *Science* 1976;**194**:23–8.
3. Lapidot T, Sirard C, Vormoor J, Murdoch B, Hoang T, Caceres-Cortes J, et al. A cell initiating human acute myeloid leukaemia after transplantation into SCID mice. *Nature* 1994;**367**:645–8.
4. Al-Hajj M, Wicha MS, Benito-Hernandez A, Morrison SJ, Clarke MF. Prospective identification of tumorigenic breast cancer cells. *Proc Natl Acad Sci USA* 2003;**100**:3983–8.
5. Singh SK, Hawkins C, Clarke ID, Squire JA, Bayani J, Hide T, et al. Identification of human brain tumour initiating cells. *Nature* 2004;**432**:396–401.
6. O'Brien CA, Pollett A, Gallinger S, Dick JE. A human colon cancer cell capable of initiating tumour growth in immunodeficient mice. *Nature* 2007;**445**:106–10.
7. Li C, Heidt DG, Dalerba P, Burant CF, Zhang L, Adsay V, et al. Identification of pancreatic cancer stem cells. *Cancer Res* 2007;**67**:1030–7.
8. Prince ME, Sivanandan R, Kaczorowski A, Wolf GT, Kaplan MJ, Dalerba P, et al. Identification of a subpopulation of cells with cancer stem cell properties in head and neck squamous cell carcinoma. *Proc Natl Acad Sci USA* 2007;**104**:973–8.
9. Kakarala M, Wicha MS. Cancer stem cells: implications for cancer treatment and prevention. *Cancer J* 2007;**13**:271–5.
10. Thomson JA, Itskovitz-Eldor J, Shapiro SS, Waknitz MA, Swiergiel JJ, Marshall VS, et al. Embryonic stem cell lines derived from human blastocysts. *Science* 1998;**282**:1145–7.
11. Assou S, Le Carrour T, Tondeur S, Strom S, Gabelle A, Marty S, et al. A meta-analysis of human embryonic stem cells transcriptome integrated into a web-based expression atlas. *Stem Cells* 2007;**25**:961–73.
12. O'Carroll D, Erhardt S, Pagani M, Barton SC, Surani MA, Jenuwein T. The polycomb-group gene Ezh2 is required for early mouse development. *Mol Cell Biol* 2001;**21**:4330–6.
13. Chambers I, Colby D, Robertson M, Nichols J, Lee S, Tweedie S, et al. Functional expression cloning of Nanog, a pluripotency sustaining factor in embryonic stem cells. *Cell* 2003;**113**:643–55.
14. Niwa H, Miyazaki J, Smith AG. Quantitative expression of Oct-3/4 defines differentiation, dedifferentiation or self-renewal of ES cells. *Nat Genet* 2000;**24**:372–6.

15. Graham V, Khudyakov J, Ellis P, Pevny L. SOX2 functions to maintain neural progenitor identity. *Neuron* 2003;**39**:749–65.
16. Wang Z, Li Y, Banerjee S, Sarkar FH. Emerging role of Notch in stem cells and cancer. *Cancer Lett* 2009;**279**:8–12.
17. He S, Nakada D, Morrison SJ. Mechanisms of stem cell self-renewal. *Annu Rev Cell Dev Biol* 2009;**25**:377–406.
18. Bapat S. Cancer stem cells: identification and targets. Wiley, Hoboken, N.J. 2009.
19. Bapat SA. Evolution of cancer stem cells. *Semin Cancer Biol* 2007;**17**:204–13.
20. Pardal R, Molofsky AV, He S, Morrison SJ. Stem cell self-renewal and cancer cell proliferation are regulated by common networks that balance the activation of proto-oncogenes and tumor suppressors. *Cold Spring Harb Symp Quant Biol* 2005;**70**:177–85.
21. Cozzio A, Passegue E, Ayton PM, Karsunky H, Cleary ML, Weissman IL. Similar MLL-associated leukemias arising from self-renewing stem cells and short-lived myeloid progenitors. *Genes Dev* 2003;**17**:3029–35.
22. Krivtsov AV, Twomey D, Feng Z, Stubbs MC, Wang Y, Faber J, et al. Transformation from committed progenitor to leukaemia stem cell initiated by MLL-AF9. *Nature* 2006;**442**:818–22.
23. Somervaille TC, Cleary ML. Identification and characterization of leukemia stem cells in murine MLL-AF9 acute myeloid leukemia. *Cancer Cell* 2006;**10**:257–68.
24. Zhou FC, Kelley MR, Chiang YH, Young P. Three to four-year-old nonpassaged EGF-responsive neural progenitor cells: proliferation, apoptosis, and DNA repair. *Exp Neurol* 2000;**164**:200–8.
25. Bunting KD. ABC transporters as phenotypic markers and functional regulators of stem cells. *Stem Cells* 2002;**20**:11–20.
26. Hadnagy A, Gaboury L, Beaulieu R, Balicki D. SP analysis may be used to identify cancer stem cell populations. *Exp Cell Res* 2006;**312**:3701–10.
27. Goodell MA, Brose K, Paradis G, Conner AS, Mulligan RC. Isolation and functional properties of murine hematopoietic stem cells that are replicating in vivo. *J Exp Med* 1996;**183**:1797–806.
28. Goodell MA, Rosenzweig M, Kim H, Marks DF, DeMaria M, Paradis G, et al. Dye efflux studies suggest that hematopoietic stem cells expressing low or undetectable levels of CD34 antigen exist in multiple species. *Nat Med* 1997;**3**:1337–45.
29. Smalley MJ, Clarke RB. The mammary gland "side population": a putative stem/progenitor cell marker? *J Mammary Gland Biol Neoplasia* 2005;**10**:37–47.
30. Ricci-Vitiani L, Lombardi DG, Pilozzi E, Biffoni M, Todaro M, Peschle C, et al. Identification and expansion of human colon-cancer-initiating cells. *Nature* 2007;**445**:111–5.
31. Curley MD, Therrien VA, Cummings CL, Sergent PA, Koulouris CR, Friel AM, et al. CD133 expression defines a tumor initiating cell population in primary human ovarian cancer. *Stem Cells* 2009;**27**:2875–83.
32. Zhou L, Wei X, Cheng L, Tian J, Jiang JJ. CD133, one of the markers of cancer stem cells in Hep-2 cell line. *Laryngoscope* 2007;**117**:455–60.
33. Wei XD, Zhou L, Cheng L, Tian J, Jiang JJ, Maccallum J. In vivo investigation of CD133 as a putative marker of cancer stem cells in Hep-2 cell line. *Head Neck* 2009;**31**:94–101.
34. Eramo A, Lotti F, Sette G, Pilozzi E, Biffoni M, Di Virgilio A, et al. Identification and expansion of the tumorigenic lung cancer stem cell population. *Cell Death Differ* 2008;**15**:504–14.
35. Shmelkov SV, Butler JM, Hooper AT, Hormigo A, Kushner J, Milde T, et al. CD133 expression is not restricted to stem cells, and both CD133+ and CD133− metastatic colon cancer cells initiate tumors. *J Clin Invest* 2008;**118**:2111–20.

36. Ghods AJ, Irvin D, Liu G, Yuan X, Abdulkadir IR, Tunici P, et al. Spheres isolated from 9L gliosarcoma rat cell line possess chemoresistant and aggressive cancer stem-like cells. *Stem Cells* 2007;**25**:1645–53.
37. Reynolds BA, Tetzlaff W, Weiss S. A multipotent EGF-responsive striatal embryonic progenitor cell produces neurons and astrocytes. *J Neurosci* 1992;**12**:4565–74.
38. Reynolds BA, Weiss S. Generation of neurons and astrocytes from isolated cells of the adult mammalian central nervous system. *Science* 1992;**255**:1707–10.
39. Dontu G, Abdallah WM, Foley JM, Jackson KW, Clarke MF, Kawamura MJ, et al. In vitro propagation and transcriptional profiling of human mammary stem/progenitor cells. *Genes Dev* 2003;**17**:1253–70.
40. Liu S, Dontu G, Mantle ID, Patel S, Ahn NS, Jackson KW, et al. Hedgehog signaling and Bmi-1 regulate self-renewal of normal and malignant human mammary stem cells. *Cancer Res* 2006;**66**:6063–71.
41. Ginestier C, Hur MH, Charafe-Jauffret E, Monville F, Dutcher J, Brown M, et al. ALDH1 is a marker of normal and malignant human mammary stem cells and a predictor of poor clinical outcome. *Cell Stem Cell* 2007;**1**:555–67.
42. Charafe-Jauffret E, Ginestier C, Iovino F, Wicinski J, Cervera N, Finetti P, et al. Breast cancer cell lines contain functional cancer stem cells with metastatic capacity and a distinct molecular signature. *Cancer Res* 2009;**69**:1302–13.
43. Dubrovska A, Kim S, Salamone RJ, Walker JR, Maira SM, Garcia-Echeverria C, et al. The role of PTEN/Akt/PI3K signaling in the maintenance and viability of prostate cancer stem-like cell populations. *Proc Natl Acad Sci USA* 2009;**106**:268–73.
44. Hansford LM, McKee AE, Zhang L, George RE, Gerstle JT, Thorner PS, et al. Neuroblastoma cells isolated from bone marrow metastases contain a naturally enriched tumor-initiating cell. *Cancer Res* 2007;**67**:11234–43.
45. Zhang S, Balch C, Chan MW, Lai HC, Matei D, Schilder JM, et al. Identification and characterization of ovarian cancer-initiating cells from primary human tumors. *Cancer Res* 2008;**68**:4311–20.
46. Fujii H, Honoki K, Tsujiuchi T, Kido A, Yoshitani K, Takakura Y. Sphere-forming stem-like cell populations with drug resistance in human sarcoma cell lines. *Int J Oncol* 2009;**34**:1381–6.
47. Hong SP, Wen J, Bang S, Park S, Song SY. CD44-positive cells are responsible for gemcitabine resistance in pancreatic cancer cells. *Int J Cancer* 2009;**125**:2323–31.
48. Tan S, Chen JS, Sun LJ, Yao HR. Selective enrichment of hepatocellular cancer stem cells by chemotherapy. *J Int Med Res* 2009;**37**:1046–56.
49. Na YR, Seok SH, Kim DJ, Han JH, Kim TH, Jung H, et al. Isolation and characterization of spheroid cells from human malignant melanoma cell line WM-266-4. *Tumour Biol* 2009;**30**:300–9.
50. Chute JP, Muramoto GG, Whitesides J, Colvin M, Safi R, Chao NJ, et al. Inhibition of aldehyde dehydrogenase and retinoid signaling induces the expansion of human hematopoietic stem cells. *Proc Natl Acad Sci USA* 2006;**103**:11707–12.
51. Yoshida A, Rzhetsky A, Hsu LC, Chang C. Human aldehyde dehydrogenase gene family. *Eur J Biochem* 1998;**251**:549–57.
52. Storms RW, Trujillo AP, Springer JB, Shah L, Colvin OM, Ludeman SM, et al. Isolation of primitive human hematopoietic progenitors on the basis of aldehyde dehydrogenase activity. *Proc Natl Acad Sci USA* 1999;**96**:9118–23.
53. Armstrong L, Stojkovic M, Dimmick I, Ahmad S, Stojkovic P, Hole N, et al. Phenotypic characterization of murine primitive hematopoietic progenitor cells isolated on basis of aldehyde dehydrogenase activity. *Stem Cells* 2004;**22**:1142–51.

54. Hess DA, Meyerrose TE, Wirthlin L, Craft TP, Herrbrich PE, Creer MH, et al. Functional characterization of highly purified human hematopoietic repopulating cells isolated according to aldehyde dehydrogenase activity. *Blood* 2004;**104**:1648–55.
55. Hess DA, Wirthlin L, Craft TP, Herrbrich PE, Hohm SA, Lahey R, et al. Selection based on CD133 and high aldehyde dehydrogenase activity isolates long-term reconstituting human hematopoietic stem cells. *Blood* 2006;**107**:2162–9.
56. Matsui W, Huff CA, Wang Q, Malehorn MT, Barber J, Tanhehco Y, et al. Characterization of clonogenic multiple myeloma cells. *Blood* 2004;**103**:2332–6.
57. Pearce DJ, Taussig D, Simpson C, Allen K, Rohatiner AZ, Lister TA, et al. Characterization of cells with a high aldehyde dehydrogenase activity from cord blood and acute myeloid leukemia samples. *Stem Cells* 2005;**23**:752–60.
58. Charafe-Jauffret E, Ginestier C, Iovino F, Tarpin C, Diebel M, Esterni B, et al. Aldehyde dehydrogenase 1-positive cancer stem cells mediate metastasis and poor clinical outcome in inflammatory breast cancer. *Clin Cancer Res* 2010;**16**:45–55.
59. Ma S, Chan KW, Lee TK, Tang KH, Wo JY, Zheng BJ, et al. Aldehyde dehydrogenase discriminates the CD133 liver cancer stem cell populations. *Mol Cancer Res* 2008;**6**:1146–53.
60. Ucar D, Cogle CR, Zucali JR, Ostmark B, Scott EW, Zori R, et al. Aldehyde dehydrogenase activity as a functional marker for lung cancer. *Chem Biol Interact* 2009;**178**:48–55.
61. Chen YC, Chen YW, Hsu HS, Tseng LM, Huang PI, Lu KH, et al. Aldehyde dehydrogenase 1 is a putative marker for cancer stem cells in head and neck squamous cancer. *Biochem Biophys Res Commun* 2009;**385**:307–13.
62. Huang EH, Hynes MJ, Zhang T, Ginestier C, Dontu G, Appelman H, et al. Aldehyde dehydrogenase 1 is a marker for normal and malignant human colonic stem cells (SC) and tracks SC overpopulation during colon tumorigenesis. *Cancer Res* 2009;**69**:3382–9.
63. Burger PE, Gupta R, Xiong X, Ontiveros CS, Salm SN, Moscatelli D, et al. High aldehyde dehydrogenase activity: a novel functional marker of murine prostate stem/progenitor cells. *Stem Cells* 2009;**27**:2220–8.
64. Wang L, Park P, Zhang H, La Marca F, Lin CY. Prospective identification of tumorigenic osteosarcoma cancer stem cells in OS99-1 cells based on high aldehyde dehydrogenase activity. *Int J Cancer*, doi: 10.1002/ijc. 25331.
65. Oh DJ, Lee GM, Francis K, Palsson BO. Phototoxicity of the fluorescent membrane dyes PKH2 and PKH26 on the human hematopoietic KG1a progenitor cell line. *Cytometry* 1999;**36**:312–8.
66. Kusumbe AP, Bapat SA. Cancer stem cells and aneuploid populations within developing tumors are the major determinants of tumor dormancy. *Cancer Res* 2009;**69**:9245–53.
67. Lanzkron SM, Collector MI, Sharkis SJ. Hematopoietic stem cell tracking in vivo: a comparison of short-term and long-term repopulating cells. *Blood* 1999;**93**:1916–21.
68. Boyd FT. Identification of growth inhibited cells by retention of a lipophilic fluorescent dye. *Cell Growth Differ* 1993;**4**:777–84.
69. Hendrikx PJ, Martens CM, Hagenbeek A, Keij JF, Visser JW. Homing of fluorescently labeled murine hematopoietic stem cells. *Exp Hematol* 1996;**24**:129–40.
70. Szilvassy SJ, Bass MJ, Van Zant G, Grimes B. Organ-selective homing defines engraftment kinetics of murine hematopoietic stem cells and is compromised by Ex vivo expansion. *Blood* 1999;**93**:1557–66.
71. Lanzkron SM, Collector MI, Sharkis SJ. Homing of long-term and short-term engrafting cells in vivo. *Ann NY Acad Sci* 1999;**872**:48–54, discussion 54–6.
72. Lee GM, Fong SS, Oh DJ, Francis K, Palsson BO. Characterization and efficacy of PKH26 as a probe to study the replication history of the human hematopoietic KG1a progenitor cell line. *In Vitro Cell Dev Biol Anim* 2002;**38**:90–6.

73. Askenasy N, Farkas DL. Optical imaging of PKH-labeled hematopoietic cells in recipient bone marrow in vivo. *Stem Cells* 2002;**20**:501–13.
74. Lassailly F, Griessinger E, Bonnet D. Microenvironmental contaminations induced by fluorescent lipophilic dyes used for non-invasive in vitro and in vivo cell tracking. *Blood* 2010;**115**:5347–54.
75. Cicalese A, Bonizzi G, Pasi CE, Faretta M, Ronzoni S, Giulini B, et al. The tumor suppressor p53 regulates polarity of self-renewing divisions in mammary stem cells. *Cell* 2009;**138**:1083–95.
76. Pece S, Tosoni D, Confalonieri S, Mazzarol G, Vecchi M, Ronzoni S, et al. Biological and molecular heterogeneity of breast cancers correlates with their cancer stem cell content. *Cell* 2010;**140**:62–73.
77. McCulloch EA, Till JE. The radiation sensitivity of normal mouse bone marrow cells, determined by quantitative marrow transplantation into irradiated mice. *Radiat Res* 1960;**13**:115–25.
78. Pierce GB, Dixon Jr. FJ. Testicular teratomas. I. Demonstration of teratogenesis by metamorphosis of multipotential cells. *Cancer* 1959;**12**:573–83.
79. Pierce Jr. GB, Dixon Jr. FJ, Verney EL. Teratocarcinogenic and tissue-forming potentials of the cell types comprising neoplastic embryoid bodies. *Lab Invest* 1960;**9**:583–602.
80. Mintz B, Illmensee K. Normal genetically mosaic mice produced from malignant teratocarcinoma cells. *Proc Natl Acad Sci USA* 1975;**72**:3585–9.
81. Brinster RL. Participation of teratocarcinoma cells in mouse embryo development. *Cancer Res* 1976;**36**:3412–4.
82. Furth J, Kahn M. The transmission of leukemia of mice with a single cell. *Am J Cancer* 1937;**31**:276–82.
83. Bonnet D, Dick JE. Human acute myeloid leukemia is organized as a hierarchy that originates from a primitive hematopoietic cell. *Nat Med* 1997;**3**:730–7.
84. Hope KJ, Jin L, Dick JE. Acute myeloid leukemia originates from a hierarchy of leukemic stem cell classes that differ in self-renewal capacity. *Nat Immunol* 2004;**5**:738–43.
85. Visvader JE. Keeping abreast of the mammary epithelial hierarchy and breast tumorigenesis. *Genes Dev* 2009;**23**:2563–77.
86. Deome KB, Faulkin Jr. LJ, Bern HA, Blair PB. Development of mammary tumors from hyperplastic alveolar nodules transplanted into gland-free mammary fat pads of female C3H mice. *Cancer Res* 1959;**19**:515–20.
87. Hoshino K, Gardner WU. Transplantability and life span of mammary gland during serial transplantation in mice. *Nature* 1967;**213**:193–4.
88. Daniel CW, De Ome KB, Young JT, Blair PB, Faulkin Jr. LJ. The in vivo life span of normal and preneoplastic mouse mammary glands: a serial transplantation study. *Proc Natl Acad Sci USA* 1968;**61**:53–60.
89. Daniel CW, Young LJ. Influence of cell division on an aging process. Life span of mouse mammary epithelium during serial propagation in vivo. *Exp Cell Res* 1971;**65**:27–32.
90. Young LJ, Medina D, DeOme KB, Daniel CW. The influence of host and tissue age on life span and growth rate of serially transplanted mouse mammary gland. *Exp Gerontol* 1971;**6**:49–56.
91. Smith GH, Medina D. A morphologically distinct candidate for an epithelial stem cell in mouse mammary gland. *J Cell Sci* 1988;**90**(Pt 1):173–83.
92. Kordon EC, Smith GH. An entire functional mammary gland may comprise the progeny from a single cell. *Development* 1998;**125**:1921–30.
93. Kuperwasser C, Chavarria T, Wu M, Magrane G, Gray JW, Carey L, et al. Reconstruction of functionally normal and malignant human breast tissues in mice. *Proc Natl Acad Sci USA* 2004;**101**:4966–71.

94. Eirew P, Stingl J, Raouf A, Turashvili G, Aparicio S, Emerman JT, et al. A method for quantifying normal human mammary epithelial stem cells with in vivo regenerative ability. *Nat Med* 2008;**14**:1384–9.
95. Lim E, Vaillant F, Wu D, Forrest NC, Pal B, Hart AH, et al. Aberrant luminal progenitors as the candidate target population for basal tumor development in BRCA1 mutation carriers. *Nat Med* 2009;**15**:907–13.
96. Shackleton M, Vaillant F, Simpson KJ, Stingl J, Smyth GK, Asselin-Labat ML, et al. Generation of a functional mammary gland from a single stem cell. *Nature* 2006;**439**:84–8.
97. Stingl J, Eirew P, Ricketson I, Shackleton M, Vaillant F, Choi D, et al. Purification and unique properties of mammary epithelial stem cells. *Nature* 2006;**439**:993–7.
98. Sleeman KE, Kendrick H, Ashworth A, Isacke CM, Smalley MJ. CD24 staining of mouse mammary gland cells defines luminal epithelial, myoepithelial/basal and non-epithelial cells. *Breast Cancer Res* 2006;**8**:R7.
99. Hulspas R, Quesenberry PJ. Characterization of neurosphere cell phenotypes by flow cytometry. *Cytometry* 2000;**40**:245–50.
100. Hemmati HD, Nakano I, Lazareff JA, Masterman-Smith M, Geschwind DH, Bronner-Fraser M, et al. Cancerous stem cells can arise from pediatric brain tumors. *Proc Natl Acad Sci USA* 2003;**100**:15178–83.
101. Galli R, Binda E, Orfanelli U, Cipelletti B, Gritti A, De Vitis S, et al. Isolation and characterization of tumorigenic, stem-like neural precursors from human glioblastoma. *Cancer Res* 2004;**64**:7011–21.
102. Yuan X, Curtin J, Xiong Y, Liu G, Waschsmann-Hogiu S, Farkas DL, et al. Isolation of cancer stem cells from adult glioblastoma multiforme. *Oncogene* 2004;**23**:9392–400.
103. Jang T, Litofsky NS, Smith TW, Ross AH, Recht LD. Aberrant nestin expression during ethylnitrosourea-(ENU)-induced neurocarcinogenesis. *Neurobiol Dis* 2004;**15**:544–52.
104. Krupkova Jr. O, Loja T, Zambo I, Veselska R. Nestin expression in human tumors and tumor cell lines. *Neoplasma* 2010;**57**:291–8.
105. Uchida K, Mukai M, Okano H, Kawase T. Possible oncogenicity of subventricular zone neural stem cells: case report. *Neurosurgery* 2004;**55**:977–8.
106. Pilkington GJ, Akinwunmi J, Ognjenovic N, Rogers JP. Differential binding of anti-CD44 on human gliomas in vitro. *NeuroReport* 1993;**4**:259–62.
107. Merzak A, Koocheckpour S, Pilkington GJ. CD44 mediates human glioma cell adhesion and invasion in vitro. *Cancer Res* 1994;**54**:3988–92.
108. Dayer AG, Cleaver KM, Abouantoun T, Cameron HA. New GABAergic interneurons in the adult neocortex and striatum are generated from different precursors. *J Cell Biol* 2005;**168**:415–27.
109. Belachew S, Chittajallu R, Aguirre AA, Yuan X, Kirby M, Anderson S, et al. Postnatal NG2 proteoglycan-expressing progenitor cells are intrinsically multipotent and generate functional neurons. *J Cell Biol* 2003;**161**:169–86.
110. Sellers DL, Horner PJ. Instructive niches: environmental instructions that confound NG2 proteoglycan expression and the fate-restriction of CNS progenitors. *J Anat* 2005;**207**:727–34.
111. Dore-Duffy P, Katychev A, Wang X, Van Buren E. CNS microvascular pericytes exhibit multipotential stem cell activity. *J Cereb Blood Flow Metab* 2006;**26**:613–24.
112. Dalerba P, Dylla SJ, Park IK, Liu R, Wang X, Cho RW, et al. Phenotypic characterization of human colorectal cancer stem cells. *Proc Natl Acad Sci USA* 2007;**104**:10158–63.
113. Hermann PC, Huber SL, Herrler T, Aicher A, Ellwart JW, Guba M, et al. Distinct populations of cancer stem cells determine tumor growth and metastatic activity in human pancreatic cancer. *Cell Stem Cell* 2007;**1**:313–23.
114. Bapat SA, Mali AM, Koppikar CB, Kurrey NK. Stem and progenitor-like cells contribute to the aggressive behavior of human epithelial ovarian cancer. *Cancer Res* 2005;**65**:3025–9.

115. Szotek PP, Pieretti-Vanmarcke R, Masiakos PT, Dinulescu DM, Connolly D, Foster R, et al. Ovarian cancer side population defines cells with stem cell-like characteristics and Mullerian Inhibiting Substance responsiveness. *Proc Natl Acad Sci USA* 2006;**103**:11154–9.
116. Baba T, Convery PA, Matsumura N, Whitaker RS, Kondoh E, Perry T, et al. Epigenetic regulation of CD133 and tumorigenicity of CD133+ ovarian cancer cells. *Oncogene* 2009;**28**:209–18.
117. Alvero AB, Chen R, Fu HH, Montagna M, Schwartz PE, Rutherford T, et al. Molecular phenotyping of human ovarian cancer stem cells unravels the mechanisms for repair and chemoresistance. *Cell Cycle* 2009;**8**:158–66.
118. Collins AT, Berry PA, Hyde C, Stower MJ, Maitland NJ. Prospective identification of tumorigenic prostate cancer stem cells. *Cancer Res* 2005;**65**:10946–51.
119. Patrawala L, Calhoun T, Schneider-Broussard R, Li H, Bhatia B, Tang S, et al. Highly purified CD44+ prostate cancer cells from xenograft human tumors are enriched in tumorigenic and metastatic progenitor cells. *Oncogene* 2006;**25**:1696–708.
120. Patrawala L, Calhoun-Davis T, Schneider-Broussard R, Tang DG. Hierarchical organization of prostate cancer cells in xenograft tumors: the CD44+alpha2beta1+ cell population is enriched in tumor-initiating cells. *Cancer Res* 2007;**67**:6796–805.
121. van den Hoogen C, van der Horst G, Cheung H, Buijs JT, Lippitt JM, Guzman-Ramirez N, et al. High aldehyde dehydrogenase activity identifies tumor-initiating and metastasis-initiating cells in human prostate cancer. *Cancer Res* 2010;**70**:5163–73.
122. Harper LJ, Piper K, Common J, Fortune F, Mackenzie IC. Stem cell patterns in cell lines derived from head and neck squamous cell carcinoma. *J Oral Pathol Med* 2007;**36**:594–603.
123. Clay MR, Tabor M, Owen JH, Carey TE, Bradford CR, Wolf GT, et al. Single-marker identification of head and neck squamous cell carcinoma cancer stem cells with aldehyde dehydrogenase. *Head Neck* 2010;**32**:1195–201.
124. Fang D, Nguyen TK, Leishear K, Finko R, Kulp AN, Hotz S, et al. A tumorigenic subpopulation with stem cell properties in melanomas. *Cancer Res* 2005;**65**:9328–37.
125. Monzani E, Facchetti F, Galmozzi E, Corsini E, Benetti A, Cavazzin C, et al. Melanoma contains CD133 and ABCG2 positive cells with enhanced tumourigenic potential. *Eur J Cancer* 2007;**43**:935–46.
126. Schatton T, Frank MH. Cancer stem cells and human malignant melanoma. *Pigment Cell Melanoma Res* 2008;**21**:39–55.
127. Quintana E, Shackleton M, Sabel MS, Fullen DR, Johnson TM, Morrison SJ. Efficient tumour formation by single human melanoma cells. *Nature* 2008;**456**:593–8.
128. Roesch A, Fukunaga-Kalabis M, Schmidt EC, Zabierowski SE, Brafford PA, Vultur A, et al. A temporarily distinct subpopulation of slow-cycling melanoma cells is required for continuous tumor growth. *Cell* 2010;**141**:583–94.
129. Boiko AD, Razorenova OV, van de Rijn M, Swetter SM, Johnson DL, Ly DP, et al. Human melanoma-initiating cells express neural crest nerve growth factor receptor CD271. *Nature* 2010;**466**:133–7.
130. Jiang F, Qiu Q, Khanna A, Todd NW, Deepak J, Xing L, et al. Aldehyde dehydrogenase 1 is a tumor stem cell-associated marker in lung cancer. *Mol Cancer Res* 2009;**7**:330–8.
131. Yin L, Velazquez OC, Liu ZJ. Notch signaling: emerging molecular targets for cancer therapy. *Biochem Pharmacol* 2010;**80**:690–701.
132. Blaumueller CM, Qi H, Zagouras P, Artavanis-Tsakonas S. Intracellular cleavage of Notch leads to a heterodimeric receptor on the plasma membrane. *Cell* 1997;**90**:281–91.
133. Brou C, Logeat F, Gupta N, Bessia C, LeBail O, Doedens JR, et al. A novel proteolytic cleavage involved in Notch signaling: the role of the disintegrin-metalloprotease TACE. *Mol Cell* 2000;**5**:207–16.

134. Mumm JS, Schroeter EH, Saxena MT, Griesemer A, Tian X, Pan DJ, et al. A ligand-induced extracellular cleavage regulates gamma-secretase-like proteolytic activation of Notch1. *Mol Cell* 2000;**5**:197–206.
135. De Strooper B, Annaert W, Cupers P, Saftig P, Craessaerts K, Mumm JS, et al. A presenilin-1-dependent gamma-secretase-like protease mediates release of Notch intracellular domain. *Nature* 1999;**398**:518–22.
136. Schroeter EH, Kisslinger JA, Kopan R. Notch-1 signalling requires ligand-induced proteolytic release of intracellular domain. *Nature* 1998;**393**:382–6.
137. Wu L, Griffin JD. Modulation of Notch signaling by mastermind-like (MAML) transcriptional co-activators and their involvement in tumorigenesis. *Semin Cancer Biol* 2004;**14**:348–56.
138. Wallberg AE, Pedersen K, Lendahl U, Roeder RG. p300 and PCAF act cooperatively to mediate transcriptional activation from chromatin templates by notch intracellular domains in vitro. *Mol Cell Biol* 2002;**22**:7812–9.
139. Lu B, Rothenberg M, Jan LY, Jan YN. Partner of Numb colocalizes with Numb during mitosis and directs Numb asymmetric localization in Drosophila neural and muscle progenitors. *Cell* 1998;**95**:225–35.
140. McGill MA, McGlade CJ. Mammalian numb proteins promote Notch1 receptor ubiquitination and degradation of the Notch1 intracellular domain. *J Biol Chem* 2003;**278**:23196–203.
141. Bray SJ. Notch signalling: a simple pathway becomes complex. *Nat Rev Mol Cell Biol* 2006;**7**:678–89.
142. Miele L, Golde T, Osborne B. Notch signaling in cancer. *Curr Mol Med* 2006;**6**:905–18.
143. Weng AP, Ferrando AA, Lee W, Morris 4th JP, Silverman LB, Sanchez-Irizarry C, et al. Activating mutations of NOTCH1 in human T cell acute lymphoblastic leukemia. *Science* 2004;**306**:269–71.
144. Reedijk M, Odorcic S, Chang L, Zhang H, Miller N, McCready DR, et al. High-level coexpression of JAG1 and NOTCH1 is observed in human breast cancer and is associated with poor overall survival. *Cancer Res* 2005;**65**:8530–7.
145. Santagata S, Demichelis F, Riva A, Varambally S, Hofer MD, Kutok JL, et al. JAGGED1 expression is associated with prostate cancer metastasis and recurrence. *Cancer Res* 2004;**64**:6854–7.
146. Allenspach EJ, Maillard I, Aster JC, Pear WS. Notch signaling in cancer. *Cancer Biol Ther* 2002;**1**:466–76.
147. Dotto GP. Notch tumor suppressor function. *Oncogene* 2008;**27**:5115–23.
148. Kumano K, Chiba S, Kunisato A, Sata M, Saito T, Nakagami-Yamaguchi E, et al. Notch1 but not Notch2 is essential for generating hematopoietic stem cells from endothelial cells. *Immunity* 2003;**18**:699–711.
149. Hitoshi S, Alexson T, Tropepe V, Donoviel D, Elia AJ, Nye JS, et al. Notch pathway molecules are essential for the maintenance, but not the generation, of mammalian neural stem cells. *Genes Dev* 2002;**16**:846–58.
150. Dontu G, Jackson KW, McNicholas E, Kawamura MJ, Abdallah WM, Wicha MS. Role of Notch signaling in cell-fate determination of human mammary stem/progenitor cells. *Breast Cancer Res* 2004;**6**:R605–15.
151. Clarke RB, Anderson E, Howell A, Potten CS. Regulation of human breast epithelial stem cells. *Cell Prolif* 2003;**36**(Suppl 1):45–58.
152. Clarke RB, Spence K, Anderson E, Howell A, Okano H, Potten CS. A putative human breast stem cell population is enriched for steroid receptor-positive cells. *Dev Biol* 2005;**277**:443–56.
153. Harrison H, Farnie G, Howell SJ, Rock RE, Stylianou S, Brennan KR, et al. Regulation of breast cancer stem cell activity by signaling through the Notch4 receptor. *Cancer Res* 2010;**70**:709–18.

154. Zhang XP, Zheng G, Zou L, Liu HL, Hou LH, Zhou P, et al. Notch activation promotes cell proliferation and the formation of neural stem cell-like colonies in human glioma cells. *Mol Cell Biochem* 2008;**307**:101–8.
155. Fan X, Matsui W, Khaki L, Stearns D, Chun J, Li YM, et al. Notch pathway inhibition depletes stem-like cells and blocks engraftment in embryonal brain tumors. *Cancer Res* 2006;**66**:7445–52.
156. Nickoloff BJ, Qin JZ, Chaturvedi V, Denning MF, Bonish B, Miele L. Jagged-1 mediated activation of notch signaling induces complete maturation of human keratinocytes through NF-kappaB and PPARgamma. *Cell Death Differ* 2002;**9**:842–55.
157. Osipo C, Golde TE, Osborne BA, Miele LA. Off the beaten pathway: the complex cross talk between Notch and NF-kappaB. *Lab Invest* 2008;**88**:11–7.
158. Rangarajan A, Syal R, Selvarajah S, Chakrabarti O, Sarin A, Krishna S. Activated Notch1 signaling cooperates with papillomavirus oncogenes in transformation and generates resistance to apoptosis on matrix withdrawal through PKB/Akt. *Virology* 2001;**286**:23–30.
159. Nair P, Somasundaram K, Krishna S. Activated Notch1 inhibits p53-induced apoptosis and sustains transformation by human papillomavirus type 16 E6 and E7 oncogenes through a PI3K-PKB/Akt-dependent pathway. *J Virol* 2003;**77**:7106–12.
160. Mungamuri SK, Yang X, Thor AD, Somasundaram K. Survival signaling by Notch1: mammalian target of rapamycin (mTOR)-dependent inhibition of p53. *Cancer Res* 2006;**66**:4715–24.
161. Nakamura T, Tsuchiya K, Watanabe M. Crosstalk between Wnt and Notch signaling in intestinal epithelial cell fate decision. *J Gastroenterol* 2007;**42**:705–10.
162. Wang Z, Banerjee S, Li Y, Rahman KM, Zhang Y, Sarkar FH. Down-regulation of notch-1 inhibits invasion by inactivation of nuclear factor-kappaB, vascular endothelial growth factor, and matrix metalloproteinase-9 in pancreatic cancer cells. *Cancer Res* 2006;**66**:2778–84.
163. Garces C, Ruiz-Hidalgo MJ, Font de Mora J, Park C, Miele L, Goldstein J, et al. Notch-1 controls the expression of fatty acid-activated transcription factors and is required for adipogenesis. *J Biol Chem* 1997;**272**:29729–34.
164. Yasutomo K, Doyle C, Miele L, Fuchs C, Germain RN. The duration of antigen receptor signalling determines CD4+ versus CD8+ T-cell lineage fate. *Nature* 2000;**404**:506–10.
165. Nickoloff BJ, Osborne BA, Miele L. Notch signaling as a therapeutic target in cancer: a new approach to the development of cell fate modifying agents. *Oncogene* 2003;**22**:6598–608.
166. Evin G, Sernee MF, Masters CL. Inhibition of gamma-secretase as a therapeutic intervention for Alzheimer's disease: prospects, limitations and strategies. *CNS Drugs* 2006;**20**:351–72.
167. Curry CL, Reed LL, Golde TE, Miele L, Nickoloff BJ, Foreman KE. Gamma secretase inhibitor blocks Notch activation and induces apoptosis in Kaposi's sarcoma tumor cells. *Oncogene* 2005;**24**:6333–44.
168. Hallahan AR, Pritchard JI, Hansen S, Benson M, Stoeck J, Hatton BA, et al. The SmoA1 mouse model reveals that notch signaling is critical for the growth and survival of sonic hedgehog-induced medulloblastomas. *Cancer Res* 2004;**64**:7794–800.
169. O'Neil J, Calvo J, McKenna K, Krishnamoorthy V, Aster JC, Bassing CH, et al. Activating Notch1 mutations in mouse models of T-ALL. *Blood* 2006;**107**:781–5.
170. van Es JH, van Gijn ME, Riccio O, van den Born M, Vooijs M, Begthel H, et al. Notch/gamma-secretase inhibition turns proliferative cells in intestinal crypts and adenomas into goblet cells. *Nature* 2005;**435**:959–63.
171. Echelard Y, Epstein DJ, St-Jacques B, Shen L, Mohler J, McMahon JA, et al. Sonic hedgehog, a member of a family of putative signaling molecules, is implicated in the regulation of CNS polarity. *Cell* 1993;**75**:1417–30.
172. Hatsell S, Frost AR. Hedgehog signaling in mammary gland development and breast cancer. *J Mammary Gland Biol Neoplasia* 2007;**12**:163–73.

173. Bak M, Hansen C, Friis Henriksen K, Tommerup N. The human hedgehog-interacting protein gene: structure and chromosome mapping to 4q31.21->q31.3. *Cytogenet Cell Genet* 2001;**92**:300–3.
174. Denef N, Neubuser D, Perez L, Cohen SM. Hedgehog induces opposite changes in turnover and subcellular localization of patched and smoothened. *Cell* 2000;**102**:521–31.
175. Rubin LL, de Sauvage FJ. Targeting the Hedgehog pathway in cancer. *Nat Rev Drug Discov* 2006;**5**:1026–33.
176. Ruiz i Altaba A. Catching a Gli-mpse of Hedgehog. *Cell* 1997;**90**:193–6.
177. Huangfu D, Anderson KV. Signaling from Smo to Ci/Gli: conservation and divergence of Hedgehog pathways from Drosophila to vertebrates. *Development* 2006;**133**:3–14.
178. Johnson RL, Rothman AL, Xie J, Goodrich LV, Bare JW, Bonifas JM, et al. Human homolog of patched, a candidate gene for the basal cell nevus syndrome. *Science* 1996;**272**:1668–71.
179. Dahmane N, Lee J, Robins P, Heller P, Ruiz i Altaba A. Activation of the transcription factor Gli1 and the Sonic hedgehog signalling pathway in skin tumours. *Nature* 1997;**389**:876–81.
180. Unden AB, Zaphiropoulos PG, Bruce K, Toftgard R, Stahle-Backdahl M. Human patched (PTCH) mRNA is overexpressed consistently in tumor cells of both familial and sporadic basal cell carcinoma. *Cancer Res* 1997;**57**:2336–40.
181. Taylor MD, Liu L, Raffel C, Hui CC, Mainprize TG, Zhang X, et al. Mutations in SUFU predispose to medulloblastoma. *Nat Genet* 2002;**31**:306–10.
182. Watkins DN, Berman DM, Baylin SB. Hedgehog signaling: progenitor phenotype in small-cell lung cancer. *Cell Cycle* 2003;**2**:196–8.
183. Watkins DN, Berman DM, Burkholder SG, Wang B, Beachy PA, Baylin SB. Hedgehog signalling within airway epithelial progenitors and in small-cell lung cancer. *Nature* 2003;**422**:313–7.
184. Thayer SP, di Magliano MP, Heiser PW, Nielsen CM, Roberts DJ, Lauwers GY, et al. Hedgehog is an early and late mediator of pancreatic cancer tumorigenesis. *Nature* 2003;**425**:851–6.
185. Zhou JX, Jia LW, Liu WM, Miao CL, Liu S, Cao YJ, et al. Role of sonic hedgehog in maintaining a pool of proliferating stem cells in the human fetal epidermis. *Hum Reprod* 2006;**21**:1698–704.
186. Palma V, Ruiz i Altaba A. Hedgehog-GLI signaling regulates the behavior of cells with stem cell properties in the developing neocortex. *Development* 2004;**131**:337–45.
187. Clement V, Sanchez P, de Tribolet N, Radovanovic I, Ruiz i Altaba A. HEDGEHOG-GLI1 signaling regulates human glioma growth, cancer stem cell self-renewal, and tumorigenicity. *Curr Biol* 2007;**17**:165–72.
188. Yauch RL, Gould SE, Scales SJ, Tang T, Tian H, Ahn CP, et al. A paracrine requirement for hedgehog signalling in cancer. *Nature* 2008;**455**:406–10.
189. Rudin CM, Hann CL, Laterra J, Yauch RL, Callahan CA, Fu L, et al. Treatment of medulloblastoma with hedgehog pathway inhibitor GDC-0449. *N Engl J Med* 2009;**361**:1173–8.
190. Yauch RL, Dijkgraaf GJ, Alicke B, Januario T, Ahn CP, Holcomb T, et al. Smoothened mutation confers resistance to a Hedgehog pathway inhibitor in medulloblastoma. *Science* 2009;**326**:572–4.
191. Von Hoff DD, LoRusso PM, Rudin CM, Reddy JC, Yauch RL, Tibes R, et al. Inhibition of the hedgehog pathway in advanced basal-cell carcinoma. *N Engl J Med* 2009;**361**:1164–72.
192. Park JW, Neve RM, Szollosi J, Benz CC. Unraveling the biologic and clinical complexities of HER2. *Clin Breast Cancer* 2008;**8**:392–401.
193. Graus-Porta D, Beerli RR, Daly JM, Hynes NE. ErbB-2, the preferred heterodimerization partner of all ErbB receptors, is a mediator of lateral signaling. *EMBO J* 1997;**16**:1647–55.
194. Olayioye MA, Neve RM, Lane HA, Hynes NE. The ErbB signaling network: receptor heterodimerization in development and cancer. *EMBO J* 2000;**19**:3159–67.

195. Alroy I, Yarden Y. Biochemistry of HER2 oncogenesis in breast cancer. *Breast Dis* 2000;**11**:31–48.
196. Keniry M, Parsons R. The role of PTEN signaling perturbations in cancer and in targeted therapy. *Oncogene* 2008;**27**:5477–85.
197. Stiles BL. PI-3-K and AKT: onto the mitochondria. *Adv Drug Deliv Rev* 2009;**61**:1276–82.
198. Slamon DJ, Clark GM, Wong SG, Levin WJ, Ullrich A, McGuire WL. Human breast cancer: correlation of relapse and survival with amplification of the HER-2/neu oncogene. *Science* 1987;**235**:177–82.
199. Sjogren S, Inganas M, Lindgren A, Holmberg L, Bergh J. Prognostic and predictive value of c-erbB-2 overexpression in primary breast cancer, alone and in combination with other prognostic markers. *J Clin Oncol* 1998;**16**:462–9.
200. Holbro T, Hynes NE. ErbB receptors: directing key signaling networks throughout life. *Annu Rev Pharmacol Toxicol* 2004;**44**:195–217.
201. Hynes NE, Lane HA. ERBB receptors and cancer: the complexity of targeted inhibitors. *Nat Rev Cancer* 2005;**5**:341–54.
202. Korkaya H, Wicha MS. HER-2, notch, and breast cancer stem cells: targeting an axis of evil. *Clin Cancer Res* 2009;**15**:1845–7.
203. Nagata Y, Lan KH, Zhou X, Tan M, Esteva FJ, Sahin AA, et al. PTEN activation contributes to tumor inhibition by trastuzumab, and loss of PTEN predicts trastuzumab resistance in patients. *Cancer Cell* 2004;**6**:117–27.
204. Korkaya H, Paulson A, Iovino F, Wicha MS. HER2 regulates the mammary stem/progenitor cell population driving tumorigenesis and invasion. *Oncogene* 2008;**27**:6120–30.
205. Magnifico A, Albano L, Campaner S, Delia D, Castiglioni F, Gasparini P, et al. Tumor-initiating cells of HER2-positive carcinoma cell lines express the highest oncoprotein levels and are sensitive to trastuzumab. *Clin Cancer Res* 2009;**15**:2010–21.
206. Korkaya H, Paulson A, Charafe-Jauffret E, Ginestier C, Brown M, Dutcher J, et al. Regulation of mammary stem/progenitor cells by PTEN/Akt/beta-catenin signaling. *PLoS Biol* 2009;**7**: e1000121.
207. Takahashi K, Yamanaka S. Induction of pluripotent stem cells from mouse embryonic and adult fibroblast cultures by defined factors. *Cell* 2006;**126**:663–76.
208. Hong H, Takahashi K, Ichisaka T, Aoi T, Kanagawa O, Nakagawa M, et al. Suppression of induced pluripotent stem cell generation by the p53-p21 pathway. *Nature* 2009;**460**:1132–5.
209. Kawamura T, Suzuki J, Wang YV, Menendez S, Morera LB, Raya A, et al. Linking the p53 tumour suppressor pathway to somatic cell reprogramming. *Nature* 2009;**460**:1140–4.
210. Li H, Collado M, Villasante A, Strati K, Ortega S, Canamero M, et al. The Ink4/Arf locus is a barrier for iPS cell reprogramming. *Nature* 2009;**460**:1136–9.
211. Marion RM, Strati K, Li H, Murga M, Blanco R, Ortega S, et al. A p53-mediated DNA damage response limits reprogramming to ensure iPS cell genomic integrity. *Nature* 2009;**460**:1149–53.
212. Utikal J, Polo JM, Stadtfeld M, Maherali N, Kulalert W, Walsh RM, et al. Immortalization eliminates a roadblock during cellular reprogramming into iPS cells. *Nature* 2009;**460**:1145–8.
213. Tlsty TD, Coussens LM. Tumor stroma and regulation of cancer development. *Annu Rev Pathol* 2006;**1**:119–50.
214. Brisken C, Duss S. Stem cells and the stem cell niche in the breast: an integrated hormonal and developmental perspective. *Stem Cell Rev* 2007;**3**:147–56.
215. Walkley CR, Olsen GH, Dworkin S, Fabb SA, Swann J, McArthur GA, et al. A microenvironment-induced myeloproliferative syndrome caused by retinoic acid receptor gamma deficiency. *Cell* 2007;**129**:1097–110.

216. Karnoub AE, Dash AB, Vo AP, Sullivan A, Brooks MW, Bell GW, et al. Mesenchymal stem cells within tumour stroma promote breast cancer metastasis. *Nature* 2007;**449**:557–63.
217. Shen Q, Goderie SK, Jin L, Karanth N, Sun Y, Abramova N, et al. Endothelial cells stimulate self-renewal and expand neurogenesis of neural stem cells. *Science* 2004;**304**:1338–40.
218. Palmer TD, Willhoite AR, Gage FH. Vascular niche for adult hippocampal neurogenesis. *J Comp Neurol* 2000;**425**:479–94.
219. Todaro M, Alea MP, Di Stefano AB, Cammareri P, Vermeulen L, Iovino F, et al. Colon cancer stem cells dictate tumor growth and resist cell death by production of interleukin-4. *Cell Stem Cell* 2007;**1**:389–402.
220. Inoue K, Slaton JW, Eve BY, Kim SJ, Perrotte P, Balbay MD, et al. Interleukin 8 expression regulates tumorigenicity and metastases in androgen-independent prostate cancer. *Clin Cancer Res* 2000;**6**:2104–19.
221. Knupfer H, Preiss R. Significance of interleukin-6 (IL-6) in breast cancer (review). *Breast Cancer Res Treat* 2007;**102**:129–35.
222. Ginestier C, Liu S, Diebel ME, Korkaya H, Luo M, Brown M, et al. CXCR1 blockade selectively targets human breast cancer stem cells in vitro and in xenografts. *J Clin Invest* 2010;**120**:485–97.
223. Kastritis E, Charidimou A, Varkaris A, Dimopoulos MA. Targeted therapies in multiple myeloma. *Target Oncol* 2009;**4**:23–36.
224. Hatfield S, Ruohola-Baker H. microRNA and stem cell function. *Cell Tissue Res* 2008;**331**:57–66.
225. Lee Y, Kim M, Han J, Yeom KH, Lee S, Baek SH, et al. MicroRNA genes are transcribed by RNA polymerase II. *EMBO J* 2004;**23**:4051–60.
226. Cai X, Hagedorn CH, Cullen BR. Human microRNAs are processed from capped, polyadenylated transcripts that can also function as mRNAs. *RNA* 2004;**10**:1957–66.
227. Hammond SM, Bernstein E, Beach D, Hannon GJ. An RNA-directed nuclease mediates post-transcriptional gene silencing in Drosophila cells. *Nature* 2000;**404**:293–6.
228. Bartel DP. MicroRNAs: target recognition and regulatory functions. *Cell* 2009;**136**:215–33.
229. Elbashir SM, Lendeckel W, Tuschl T. RNA interference is mediated by 21- and 22-nucleotide RNAs. *Genes Dev* 2001;**15**:188–200.
230. Hutvagner G, Zamore PD. A microRNA in a multiple-turnover RNAi enzyme complex. *Science* 2002;**297**:2056–60.
231. Song JJ, Smith SK, Hannon GJ, Joshua-Tor L. Crystal structure of Argonaute and its implications for RISC slicer activity. *Science* 2004;**305**:1434–7.
232. Yekta S, Shih IH, Bartel DP. MicroRNA-directed cleavage of HOXB8 mRNA. *Science* 2004;**304**:594–6.
233. Bernstein E, Kim SY, Carmell MA, Murchison EP, Alcorn H, Li MZ, et al. Dicer is essential for mouse development. *Nat Genet* 2003;**35**:215–7.
234. Kanellopoulou C, Muljo SA, Kung AL, Ganesan S, Drapkin R, Jenuwein T, et al. Dicer-deficient mouse embryonic stem cells are defective in differentiation and centromeric silencing. *Genes Dev* 2005;**19**:489–501.
235. Suh MR, Lee Y, Kim JY, Kim SK, Moon SH, Lee JY, et al. Human embryonic stem cells express a unique set of microRNAs. *Dev Biol* 2004;**270**:488–98.
236. Frankel LB, Christoffersen NR, Jacobsen A, Lindow M, Krogh A, Lund AH. Programmed cell death 4 (PDCD4) is an important functional target of the microRNA miR-21 in breast cancer cells. *J Biol Chem* 2008;**283**:1026–33.
237. Si ML, Zhu S, Wu H, Lu Z, Wu F, Mo YY. miR-21-mediated tumor growth. *Oncogene* 2007;**26**:2799–803.

238. Yan LX, Huang XF, Shao Q, Huang MY, Deng L, Wu QL, et al. MicroRNA miR-21 overexpression in human breast cancer is associated with advanced clinical stage, lymph node metastasis and patient poor prognosis. *RNA* 2008;**14**:2348–60.
239. He L, Thomson JM, Hemann MT, Hernando-Monge E, Mu D, Goodson S, et al. A microRNA polycistron as a potential human oncogene. *Nature* 2005;**435**:828–33.
240. Hayashita Y, Osada H, Tatematsu Y, Yamada H, Yanagisawa K, Tomida S, et al. A polycistronic microRNA cluster, miR-17-92, is overexpressed in human lung cancers and enhances cell proliferation. *Cancer Res* 2005;**65**:9628–32.
241. Lu Y, Thomson JM, Wong HY, Hammond SM, Hogan BL. Transgenic over-expression of the microRNA miR-17-92 cluster promotes proliferation and inhibits differentiation of lung epithelial progenitor cells. *Dev Biol* 2007;**310**:442–53.
242. Uziel T, Karginov FV, Xie S, Parker JS, Wang YD, Gajjar A, et al. The miR-17 92 cluster collaborates with the Sonic Hedgehog pathway in medulloblastoma. *Proc Natl Acad Sci USA* 2009;**106**:2812–7.
243. Nagel R, le Sage C, Diosdado B, van der Waal M, Oude Vrielink JA, Bolijn A, et al. Regulation of the adenomatous polyposis coli gene by the miR-135 family in colorectal cancer. *Cancer Res* 2008;**68**:5795–802.
244. DeSano JT, Xu L. MicroRNA regulation of cancer stem cells and therapeutic implications. *AAPS J* 2009;**11**:682–92.
245. Johnson CD, Esquela-Kerscher A, Stefani G, Byrom M, Kelnar K, Ovcharenko D, et al. The let-7 microRNA represses cell proliferation pathways in human cells. *Cancer Res* 2007;**67**:7713–22.
246. Esquela-Kerscher A, Trang P, Wiggins JF, Patrawala L, Cheng A, Ford L, et al. The let-7 microRNA reduces tumor growth in mouse models of lung cancer. *Cell Cycle* 2008;**7**:759–64.
247. Yu F, Yao H, Zhu P, Zhang X, Pan Q, Gong C, et al. let-7 regulates self renewal and tumorigenicity of breast cancer cells. *Cell* 2007;**131**:1109–23.
248. Shimono Y, Zabala M, Cho RW, Lobo N, Dalerba P, Qian D, et al. Downregulation of miRNA-200c links breast cancer stem cells with normal stem cells. *Cell* 2009;**138**:592–603.
249. Cimmino A, Calin GA, Fabbri M, Iorio MV, Ferracin M, Shimizu M, et al. miR-15 and miR-16 induce apoptosis by targeting BCL2. *Proc Natl Acad Sci USA* 2005;**102**:13944–9.
250. Bonci D, Coppola V, Musumeci M, Addario A, Giuffrida R, Memeo L, et al. The miR-15a-miR-16-1 cluster controls prostate cancer by targeting multiple oncogenic activities. *Nat Med* 2008;**14**:1271–7.
251. Godlewski J, Nowicki MO, Bronisz A, Williams S, Otsuki A, Nuovo G, et al. Targeting of the Bmi-1 oncogene/stem cell renewal factor by microRNA-128 inhibits glioma proliferation and self-renewal. *Cancer Res* 2008;**68**:9125–30.
252. Garzia L, Andolfo I, Cusanelli E, Marino N, Petrosino G, De Martino D, et al. MicroRNA-199b-5p impairs cancer stem cells through negative regulation of HES1 in medulloblastoma. *PLoS ONE* 2009;**4**:e4998.
253. Ferretti E, De Smaele E, Miele E, Laneve P, Po A, Pelloni M, et al. Concerted microRNA control of Hedgehog signalling in cerebellar neuronal progenitor and tumour cells. *EMBO J* 2008;**27**:2616–27.
254. Ji Q, Hao X, Zhang M, Tang W, Yang M, Li L, et al. MicroRNA miR-34 inhibits human pancreatic cancer tumor-initiating cells. *PLoS ONE* 2009;**4**:e6816.
255. Ji Q, Hao X, Meng Y, Zhang M, Desano J, Fan D, et al. Restoration of tumor suppressor miR-34 inhibits human p53-mutant gastric cancer tumorspheres. *BMC Cancer* 2008;**8**:266.
256. Kawasaki BT, Hurt EM, Mistree T, Farrar WL. Targeting cancer stem cells with phytochemicals. *Mol Interv* 2008;**8**:174–84.

257. Wang Z, Zhang Y, Banerjee S, Li Y, Sarkar FH. Notch-1 down-regulation by curcumin is associated with the inhibition of cell growth and the induction of apoptosis in pancreatic cancer cells. *Cancer* 2006;**106**:2503–13.
258. Jaiswal AS, Marlow BP, Gupta N, Narayan S. Beta-catenin-mediated transactivation and cell-cell adhesion pathways are important in curcumin (diferuylmethane)-induced growth arrest and apoptosis in colon cancer cells. *Oncogene* 2002;**21**:8414–27.
259. Kakarala M, Brenner DE, Korkaya H, Cheng C, Tazi K, Ginestier C, et al. Targeting breast stem cells with the cancer preventive compounds curcumin and piperine. *Breast Cancer Res Treat* 2009;**122**:777–85.
260. Li Y, Zhang T, Korkaya H, Liu S, Lee HF, Newman B, et al. Sulforaphane, a dietary component of broccoli/broccoli sprouts, inhibits breast cancer stem cells. *Clin Cancer Res* 2010;**16**:2580–90.
261. Pahlke G, Ngiewih Y, Kern M, Jakobs S, Marko D, Eisenbrand G. Impact of quercetin and EGCG on key elements of the Wnt pathway in human colon carcinoma cells. *J Agric Food Chem* 2006;**54**:7075–82.
262. Nagler A, Riklis I, Kletter Y, Tatarsky I, Fabian I. Effect of 1, 25 dihydroxyvitamin D3 and retinoic acid on normal human pluripotent (CFU-mix), erythroid (BFU-E), and myeloid (CFU-C) progenitor cell growth and differentiation patterns. *Exp Hematol* 1986;**14**:60–5.
263. Balic M, Lin H, Young L, Hawes D, Giuliano A, McNamara G, et al. Most early disseminated cancer cells detected in bone marrow of breast cancer patients have a putative breast cancer stem cell phenotype. *Clin Cancer Res* 2006;**12**:5615–21.
264. Sladek NE. Aldehyde dehydrogenase-mediated cellular relative insensitivity to the oxazaphosphorines. *Curr Pharm Des* 1999;**5**:607–25.
265. Brekelmans CT, Tilanus-Linthorst MM, Seynaeve C, vd Ouweland A, Menke-Pluymers MB, Bartels CC, et al. Tumour characteristics, survival and prognostic factors of hereditary breast cancer from BRCA2-, BRCA1- and non-BRCA1/2 families as compared to sporadic breast cancer cases. *Eur J Cancer* 2007;**43**:867–76.
266. Li X, Lewis MT, Huang J, Gutierrez C, Osborne CK, Wu MF, et al. Intrinsic resistance of tumorigenic breast cancer cells to chemotherapy. *J Natl Cancer Inst* 2008;**100**:672–9.

Bioinformatics and Systems Biology of Cancers

Gilbert S. Omenn

Department of Internal Medicine, Department of Human Genetics, School of Public Health, Center for Computational Medicine and Bioinformatics, University of Michigan, Ann Arbor, Michigan, USA

Molecular databases and bioinformatics methods and tools are essential for modern cancer research. Multilevel analyses of all the protein-coding genes, thousands of proteins, and hundreds of metabolites require integration in terms of signaling and metabolic pathways and networks. This chapter provides background and examples of genomic, gene expression, epigenomic, proteomic, and metabolomic investigations of cancer progression and emergence of invasive and metastatic properties of cancers.

Progress in Molecular Biology
and Translational Science, Vol. 95
DOI: 10.1016/S1877-1173(10)95007-2

1877-1173/10 $35.00

I. Introduction

Bioinformatics is a rapidly growing scientific discipline, embracing the development and application of algorithms and methods to turn biological data into knowledge of biological systems, often stimulating further experimentation and/or clinical research. It comprises (a) theory and methods, with algorithms, statistical methods, machine learning, ontologies, and visualization, and with measures of scope, accuracy, completeness, and replicability; (b) organized structures for streams of data, ranging from base reads from gene expression microarrays, DNA and RNA sequencing, mass spectrometry proteomics experiments, comparative analysis of structures and alignment for protein sequences and genomic sequences across species to physiological and clinical information; (c) applied methods and data mining for sequence analysis, whole genome assembly, protein structure prediction from amino acid sequences, characterization of alternative splicing, and deduction and modeling of mechanisms for gene regulation and dysregulation; and (d) multiscalar, multilevel analyses from individual molecules to large complexes, organelles, cells, complete organisms, and populations. The overall aim is to understand the molecular and functional consequences of changes in health and disease and to predict the molecular mediation of effects from environmental perturbations, development and aging, and therapeutic and preventive interventions.

Biology is an information science requiring bioinformatics for the capture, storage, analysis, mining, and modeling of data about genome sequences, gene expression, and regulation of gene expression through histones, DNA methylation, miRNAs, and transcription factors; protein structures, protein expression, and posttranslational modifications of proteins, including proteolysis and protein degradation; metabolomics; and multilevel links to phenotypes. Figure 1 shows that informatics analyses can begin from clinical or biological phenotypes and link those phenotypes to experimental perturbations that produce changes in the multiple levels of gene and protein expression. The explosion of data from genome sequencing, SNP and haplotype analysis, next-generation DNA and RNA sequencing, structural proteomics, and protein expression has required rapidly increasing efficiency of computers, data storage, and algorithms for efficient large-scale analysis and modeling.

Many databases represent important collaborative endeavors of particular scientific communities. For example, the Protein Data Bank (PDB) was established at Brookhaven National Laboratory in 1971 as an archive for crystal structures; in the early years of the field, it seemed that each solved structure was a Nobel Prize event. There were just seven protein structures in the beginning.[1] By the early 1990s, the leading journals required a PDBid for

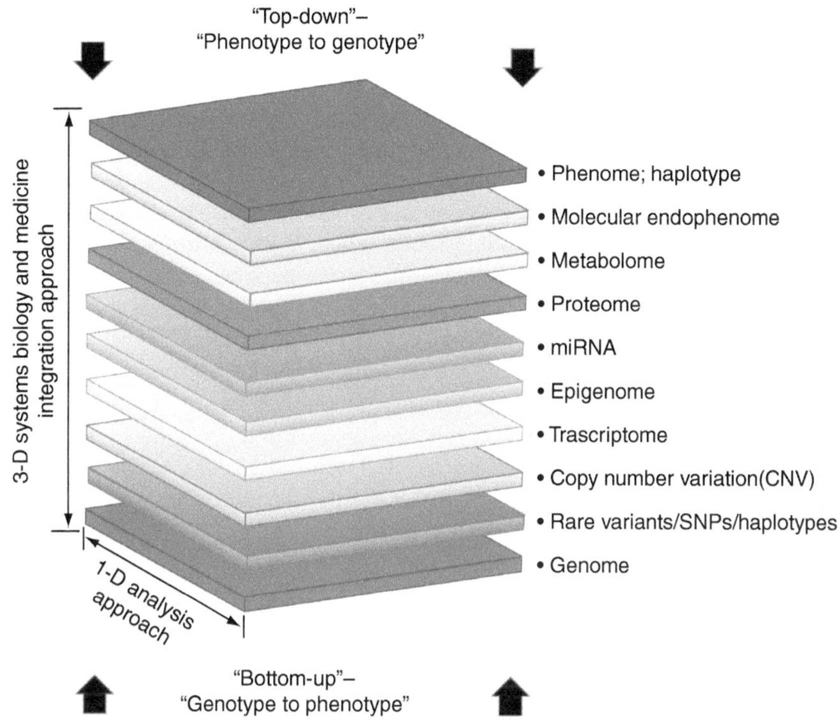

FIG. 1. Bottom-up and top-down multiscalar' omics analyses to link genotype and phenotype. From Athey, Omenn *et al.*, National Center for Integrative Biomedical Informatics Annual Research Conference, April 20, 2010, Ann Arbor, Michigan.

publication and the National Institute for General Medical Sciences (NIH/NIGMS) had adopted the guidelines of the International Union of Crystallography. By 2002 there were 17,000 entries. Since 1998, PDB has been managed by the Research Collaboratory for Structural Bioinformatics (RSCB). In 1991 a companion Nucleic Acid Database (NDB) was created.

Under the leadership of the National Library of Medicine of the NIH and its National Center for Biological Information, numerous other databases have been created and expanded with rapid development of molecular studies, such as Gene Expression Omnibus. In the proteomics domain, there is a coordinated effort called ProteomeXchange organized by the Human Proteome Organization[2,3] to have major datasets contributed to the distributed file-sharing system Tranche at www.proteomecommons.org, then automatically sent to the Protein Identification Database PRIDE at the European Bioinformatics Institute (www.ebi.ac.uk/pride), where the primary data and original analysis can be accessed, and then to PeptideAtlas at the Institute for Systems Biology, where related datasets are combined and all datasets are reanalyzed with uniform criteria through the TransProteomicPipeline (www.peptideatlas.org). Many academics have started comprehensive databases, including Oncomine, a compendium of cancer microarray gene expression datasets with embedded statistical, informatics, and graphing tools.[4] Oncomine was utilized to test for heterogeneity of gene expression patterns among patients with prostate cancers, leading to the remarkable discovery of fusion genes in prostate cancers and other solid tumors.[5,6]

There are multiple protein–protein interactions resources, including MiMI, Michigan Molecular Interactions, which is a synthesis of six such databases.[7] Finally, there are multiple open source and proprietary pathways and networks databases (Table I for NCBI resources and http://ncibi.org for tools and tutorials from our National Center for Integrative Biomedical Informatics, NCIBI).

II. The Cancer Biomedical Informatics Grid (caBIG)

The NIH National Cancer Institute Center for Biomedical Informatics and Information Technology established the Cancer Biomedical Informatics Grid (caBIG) to share data, information, and knowledge, and enhance collaboration. The aims are to accelerate research progress, reduce the burden of cancer at the population level, and facilitate development of more personalized healthcare. NCI-designated cancer centers, research-oriented hospitals, and many additional research groups have invested a lot of effort to implement caBIG tools and infrastructure and to link with genomics databases, tissue repositories, and clinical databases. As of May 2009, the NCI claimed that 1500 individuals and 450 organizations were connected through caBIG and caBIG-compatible.[8] A full list of tools available for adoption or adaptation is available at https://cabig.nci.nih.gov/tools/toolsearch_view.

The principles underlying caBIG are open access, open development, open source, and federation, with interoperable software, data standards, and computing infrastructure. The Life Sciences and Translational Research area includes caArray, caTissue Suite, geWorkbench, cancer genome-wide association studies (caGWAS), cancer Bench to Bedside (caB2B), caIntegrator2, and caGrid. The tools in the Clinical Trials and Imaging area include cancer adverse event reporting system (caAERS), cancer central clinical participant registry (C3PR), cancer data exchange (caXchange), patient study calendar (PSC), clinical trials data management system (CDMS) integration, national

TABLE I

RESOURCES AND DATABASES AT NATIONAL CENTER FOR BIOLOGICAL INFORMATION AT THE NATIONAL LIBRARY OF MEDICINE/NATIONAL INSTITUTES OF HEALTH (NLM/NIH)

Literature Databases:
Bookshelf: A collection of searchable biomedical books linked to PubMed.
PubMed: Allows searching by author names, journal titles, and a new Preview/Index option. PubMed database provides access to over 18 million MEDLINE citations back to the mid-1950s. It includes History and Clipboard options which may enhance your search session. NCBI provides a simple PubMed tutorial.
PubMed Central: The U.S. National Library of Medicine digital archive of life science journal literature. Access is completely free and unrestricted.
OMIM: Online Mendelian Inheritance in Man is a database of human genes and genetic disorders authored and edited by Dr. Victor A. McKusick and colleagues at Johns Hopkins and elsewhere. NCBI provides a short tutorial for searching the OMIM database.
OMIA: Online Mendelian Inheritance in Animals is a database of genes, inherited disorders and traits in animal species other than human and mouse.
Journals: Search the journals database for links to journals in the Entrez system, including the genetic database.
A *Selection of Molecular Databases*:
Nucleotide Database: The nucleotide database contains sequence data from GenBank, EMBL, and DDBJ, the members of the tripartite, international collaboration of sequence databases. Nucleotide allows the user to retrieve nucleotide sequences in both GenBank and FASTA formats.
Protein Database: The protein database contains sequence data from the translated coding regions from DNA sequences in GenBank, EMBL, and DDBJ as well as protein sequences submitted to UniProt (collaboration between PIR, EBI, and SIB) and the Protein Data Bank (PDB) (sequences from solved structures).
Structure Database: The structure database or Molecular Modeling Database (MMDB) contains experimental data from crystallographic and NMR structure determinations. The data for MMDB are obtained from the PDB. The NCBI has cross-linked structural data to bibliographic information, to the sequence databases, and to the NCBI taxonomy. NCBI provides a structure database tutorial (requires installation of Cn3D).
PopSet Database: A PopSet is a set of DNA sequences that have been collected to analyze the evolutionary relatedness of a population. The population could originate from different members of the same species, or from organisms from different species. The PopSet database contains aligned sequences submitted as a set resulting from a population genetic, phylogenetic, or mutation study describing evolutionary events and population variation. The PopSet database contains both nucleotide and protein sequence data.
Taxonomy Database: The taxonomy database contains the names of all organisms that are represented in the genetic databases with at least one nucleotide or protein sequence. You can search for nucleotide, protein, and structure data from specific taxonomic groupings, from the domain level (archaea, bacteria, eukaryota) down to the species level.
Gene Expression Database: The gene expression database is a gene expression/molecular abundance repository and a curated, online resource for gene expression data browsing, query, and retrieval.
A *Selection of Genome Databases*:

(*Continues*)

TABLE I (*Continued*)

Genomes: The Genomes database provides views for a variety of genomes, complete chromosomes, contiged sequence maps, and integrated genetic and physical maps. The whole genomes of over 4500 organisms can be found here. All three main domains of life—bacteria, archaea, and eukaryota—are represented, as well as many viruses and organelles.

Clusters of Orthologous Groups (COG) Database: Phylogenetic classification of proteins encoded in completed genomes. COGs were identified by comparison of protein sequences from 43 complete genomes, representing 30 major phylogenetic lineages. Each COG consists of individual orthologous proteins or orthologous sets of paralogs from at least three different lineages. Assuming that orthologs have similar functions, the COG grouping allows transfer of functional information from one member to the entire COG.

Homologene: A database of calculated and curated orthologs. The calculated homologs are the result of nucleotide sequence comparisons between organisms.

Entrez Gene: A single query interface to curated sequence information. Includes information on official nomenclature, aliases, sequence accessions, phenotypes, homology, map locations and related web sites.

TaxPlot: This feature enables the user to compare the similarity of a query genome to different species.

SKY/M-FISH & CGH Database: This Spectral Karyotyping (SKY), Multiplex Fluorescence *In Situ* Hybridization (M-FISH) and Comparative Genomic Hybridization (CGH) provides a public platform for investigators to share and compare molecular cytogenetic data.

A Selection of Tools:

Entrez: Entrez is a retrieval system designed for searching several linked databases, including Nucleotide, Protein, Genome, Structure, and PopSet. Entrez categories can be searched using subject, author, or unique identifiers such as accession numbers, phrases, truncated terms, and combined sets. There is also a simple Entrez tutorial.

BLAST: BLAST (Basic Local Alignment Search Tool) is a set of similarity search programs designed to explore all of the available sequence databases regardless of whether the query is protein or DNA. For a better understanding of BLAST you can refer to the BLAST Course which explains the basics of the BLAST algorithm, or to the NCBI BLAST tutorial.

Map Viewer: Integrated views of chromosome maps for 78 organisms. Useful for the identification and localization of genes.

ORF Finder: Graphical tool which finds all ORF (Open Reading Frames) based on a set of criteria. Can be used with standard and alternative genetic codes.

VecScreen: A tool for identifying segments of a nucleic acid sequence that may be of vector, linker, or adapter origin prior to sequence analysis or submission. VecScreen was developed to combat the problem of vector contamination in public sequence databases.

Spidey: Aligns an mRNA sequence to a genomic sequence. Can determine the intron/exon structure, returning one or more models of genomic structure.

Other resources are available to assist with downloading large amounts of specific data in batches (to create so-called "batch-files") from sequence databases like GenBank and TIGR (J. Craig Venter Institute). These resources include scripting tools such as BioPerl and Biopython.

NCBI (National Center for Biotechnology Information) is a resource for molecular biology information. NCBI creates and maintains public databases, conducts research in computational biology, develops software tools for analyzing genome data, and disseminates biomedical information. The NCBI site is constantly being updated and some of the changes include new databases and tools for data mining.

NCBI offers several searchable literature, molecular and genomic databases, and many bioinformatic tools. An up-to-date list of databases and tools can be found on the NCBI Sitemap and Entrez Data Model. Location: www.ncbi.nlm.nih.gov.

biomedical imaging archive (NBIA), annotation imaging markup (AIM), and eXtensible imaging platform (XIP). The Enterprise Support Network has six knowledge centers and support service providers. In practice there are many challenges ensuring interoperability between existing institutional and vendor-provided systems and these caBIG developments.

III. TCGA: The Cancer Genome Anatomy Project

With the near-completion of the Human Genome Sequence and dramatic stepwise enhancements of high-throughput genomic technology platforms and databases, the National Cancer Program and the scientific community have undertaken a comprehensive survey of human cancer genomes.[9] Through integrated multidimensional analyses, The Cancer Genome Atlas aims to discover and catalog major cancer-causing genome alterations in large cohorts of patients with various human tumors. Such multiscalar analyses generate an avalanche of data requiring bioinformatics methods and large-scale computational capacity for study design, database development, data storage and retrieval, data mining, and pathway-, network-, and systems-level disease modeling.

All TCGA data are deposited at the Data Coordinating Center for public access (http://cancergenome.nih.gov/dataportal/), classified by data type (mutations, gene expression, clinical), and data level for structured access with patient privacy protections. See TCGA Data Primer (http://tcga-data.nci.nih.gov/docs/TCGA_Data_Primer.pdf). Biospecimen collection and handling is controlled by detailed protocols. For example, several platforms were utilized for copy number (CN) assays (Agilent 244K, Affymetrix SNP6.0, Illumina 550K) and several algorithms (GISTIC, GTS, RAE) were used to identify variation. mRNA and microRNA expression profiles were produced with Affymetrix U133A, Affymetrix Exon 1.0ST, custom Agilent 244K, and Agilent miRNA platforms.

A. Glioblastoma Multiforme—Genomic Analyses

The first cancer selected for study by TCGA was glioblastoma (grade IV), the most common primary tumor of the brain in adults. These tumors arise *de novo* without history of low-grade disease; those tumors that arise by progression from low-grade tumors are called secondary glioblastomas. Patients with newly diagnosed GBM have a median survival of about 12 months, with poor responses to all therapies. At the start, there was cumulative knowledge of dysregulation of growth factor signaling due to mutational activation or amplification of receptor tyrosine kinase (RTK) genes, activation of the phosphatidylinositol-3-OH kinase (PI3K) pathway, and inactivation of the TP53 and retinoblastoma (Rb) tumor suppressor genes.[10] Genome-wide association studies had shown extensive genomic heterogeneity and molecular subclasses.

The TCGA Research Network[10] GBM pilot project determined DNA CN, gene expression, and DNA methylation aberrations in specimens from 206 GBM patients, with nucleotide sequence aberrations in 91 of the 206. The analysis generated new insights into the roles of mutations of ERBB2, NF1, and TP53; uncovered frequent mutations and in-frame deletions of the PIK3R1 regulatory subunit gene; found previously unreported homozygous deletions of NF1 and PARK2 and amplifications of AKT3; and proposed a network of pathways altered in GBM. SNP analyses identified copy-neutral loss of heterozygosity, most commonly involving a region of chromosome 17p containing TP53. In samples from 21 treated patients, GBMs had an association between MGMT promoter methylation and a hypermutator phenotype due to mismatch repair deficiency of MLH1, MSH2, MSH6, or PMS2. All mutations and CN variants identified were compared with the information available in the Human Gene Mutation Database to determine which were already known and which were novel.

There was extensive heterogeneity of molecular abnormalities across the GBMs. To construct an integrated view of common genetic alterations, TCGA investigators mapped validated somatic nucleotide substitutions, homozygous deletions, and focal amplifications onto major pathways implicated in GBM, obtaining a highly interconnected network involving three major pathways: RTK signaling and TP53 and RB tumor suppressor pathways. By CN data, 66%, 70%, and 59% of the 206 samples had somatic alterations in core components of RB, TP53, and RTK/RAS/PI3K; in the subset of 91 samples with sequencing data, those percentages rose to 87%, 78%, and 88%. Consistent with the likelihood that more than one pathway must be altered to generate malignant phenotypes, 74% of the specimens had aberrations in all the three pathways. There was some evidence for coactivation of RTKs (EGFR, ERBB2, PDGFRA, MET). In any single pathway, there are multiple potential sites for dysregulation; for example, with TP53, there were ARF deletions, MDM2 and MDM4 amplifications, and mutations of TP53 itself. Some specific mutations may be amenable to therapy with available agents; for example, patients with CDK4/CKD6 amplifications or CDKN2C/CDKN2A deletions or inactivating mutations could benefit from CDK inhibitor drugs which would not be effective in patients with RB1 mutations. Other drugs might target other specific mutated or amplified gene products.

Methylation status of MGMT predicts sensitivity to the drug temozolomide, an alkylating agent that seems effective in some patients with newly diagnosed GBM. These unbiased, systematic genome analyses of large sample cohorts led to the proposal of how to reduce the risk of early resistance to effective therapy with the alkylating agent.

Freire *et al.*[11] applied additional techniques to identify and quantitate CN alterations in GBM, starting with the data in The Cancer Genome Atlas. They built a catalog of aberrant regions with aCGH data from 167 patients. Using an information theory approach, they produced new software and a Cancer Genome Browser for visualization and data processing (http://code.google.com/p/cancergenome/). They calculated Shannon's entropy as a measure of deviation (amplitude and prevalence) from the common state of CN 2 at recurrent regions of aberration across patients. After filtering, they found 31 regions, of which 10 involved genes known to modulate cell proliferation in GMB: EGFR, MDM2, MDM4, CDK4, PTEN, PDFGRA, CDKN2A, CDKN2C, NF1, and CHD5. The 21 others were novel findings. They compared their findings with those from the TCGA Network[10] that utilized GISTIC and GTS, with arbitrary thresholds. The methods were reinforcing, except that the entropy method was less able to detect events with prevalence $< 5\%$. They intend to incorporate all the three tools in the Browser.

A common research strategy, especially for cancers, is to produce animal models convenient for experimental manipulations. Hodgson *et al.*[12] profiled DNA CN and mRNA expression in 21 independent GBM tumor lines maintained as subcutaneous xenografts in mice (GBMX). The Cancer Genome Atlas was used to compare results in the animal model with previously reported results from patients. The predominant CN signature in both groups was chromosome-7 gain/chromosome-10 loss, a signature for poor prognosis. Amplification and overexpression of known GBM oncogenes EGFR, MDM2, CDk6, and MYCN were confirmed; several novel genes were noted.

Cerami *et al.*[13] utilized an automated network analysis to distinguish "driver" mutations from incidental "passenger" mutations in the TCGA GBM dataset. Their hypothesis is that cellular networks, comprising protein–protein interactions and signaling pathways in their Human Interaction Network, contain functional modules and that tumors target specific modules that are critical to tumor growth. They organized the analysis around the "hallmarks of cancers"[14]: self-sufficiency in growth, evasion of apoptosis, sustained angiogenesis, tissue invasion, and metastasis. They expected universal alterations at the module level to emerge from wide diversity at the genetic level. Despite considerable patient-to-patient variation among the TCGA GBM cases for which both CN and sequence mutation data were available (and after seven hypermutable cases were excluded), two large modules were identified involving signaling via TP53, Rb, PI3K, and RTKs. New candidate altered modules included AGAP2/CENTG1, an activator of the PI3K pathway, and DCTN2, with four genes whose protein products are located in the centrosome and microtubule organizing center, critical to correct segregation of chromosomes during cell division. Their software tool NetBox (http://cbio.mskcc.org/netbox) was made publicly available. It is preloaded with the Human Interaction Database (based on HPRD, Reactome, NCI/Nature Pathway Interaction Database, and MSKCC Cancer Cell Map), with a command line interface for connecting genes into a network, identifying linker genes, partitioning the network into modules, and testing random background models; results can be uploaded into Cytoscape for visualization.

Of the eight genes described as significantly mutated in the frequency analysis by TCGA Network,[10] 7 (all but NF1) were confirmed in this study in the two largest modules. Many of the targets of high-level focal amplification or homozygous deletions were also identified in these networks. The TP53 and RB pathways are captured by the RB1 module in the network analysis, and the RTK-associated genes and pathways are captured by the PIK3R1 module.

Deletions and amplifications are CN variations that are important mutational phenomena. Gaire *et al.*[15] introduced a statistical methodology, MIRAGAA, for MicroRNA and Genome Aberration Analyzer, to assess coordinated changes of genome CNs and miRNA expression, applied to TCGA GBM datasets. Their hypothesis was that, since the functional effects of a change in gene function can be achieved by more than one mechanism, we should not expect to find the same genome aberration or miRNA expression change across a large number of samples.

Further development of the TCGA GBM project will markedly increase the number of specimens and the sequencing of genes and miRNA targets, utilize additional curated protein–protein interaction and miR target prediction programs, integrate epigenetic methylation and histone data, explore GBM subtyping, and compare findings with the next TCGA tumor types (ovarian, lung). 'Omics analyses of CD133+ GBM stem cells, compared with CD133− GBM cells, show meaningful differential expression. At the Institute for Systems Biology, multifaceted GBM studies will include high-throughput genomic and RNA sequencing of large numbers of individual cells, beginning to address the important cellular heterogeneity of individual cancers.

The search for reliable subclassification of GBM is ongoing. Noushmehr *et al.*[16] profiled promoter DNA methylation alterations in 272 GBM tumors as part of TCGA. They found a glioma-CpG island methylator phenotype (G-CIMP) and concerted hypermethylation at a large number of loci. These findings were validated in a fresh set of GBM samples. G-CIMP tumors are highly associated with isocitrate dehydrogenase 1 (IDH1) somatic mutations at position R132 and lower grade gliomas, though they may progress to become secondary GBM. The patients are younger and have longer survival than patients with higher grade tumors.

B. Glioblastoma Multiforme—Proteomic Analyses

Proteomics has also been applied to stratification of GBM, with quite interesting findings. Brennan *et al.*[17] performed targeted proteomic analysis of 27 surgical specimens of glioma from Memorial Sloan-Kettering Cancer Center, using 55 antibodies for Western blot experiments that could distinguish activated from total protein. The targets were PDGFB, PDGFR-alpha, PDGFR-beta, and phospho-PDGFR-beta; EGFR and phospho-EGFR; downstream pathways Ras (BRAF, total and phospho-MEK, ERK), Akt (PTEN, total AKT, phospho-AKT, RHEB), Notch (DLL1, Jagged, full length Notch 1 and 2, cleaved Notch 1 and 2, and downstream target Hes1), Wnt (beta-catenin), SHH, and NF1. Band intensity was quantitated by comparison with actin and tubulin. The aim was to identify patterns of coordinate activation among relevant signal transduction pathways and compare these proteome-based findings with the integrated analysis of genomic and gene expression data from 243 GBM cases available in the TCGA. Based on principal components analysis and unsupervised clustering, three subclasses of GBM emerged, associated with predominance of EGFR activation (15 proteins), PDGFR activation (17 proteins), or loss of the RAS regulator NF1 (8 proteins). The EGFR signaling subgroup had elevated expression of Notch ligands, cleaved Notch receptor, and downstream target Hes1. The PDGF subgroup had high levels of PDGFB ligand and phosphorylation of PDGFR-beta and NFkB. Finally, loss of NF-1 was associated with lower MAPK and PI3K activation and overexpression of YKL40, a mesenchymal marker. It is quite striking that these molecular changes represent posttranslational modifications that would not be detectable at the mRNA level, including Notch cleavage, kinase phosphorylation, and stabilization of beta-catenin.

These three signaling classes match distinct, but heterogeneous, transcriptional subclasses of primary GBM available from the TCGA public portal. Unsupervised clustering reflected confounding influence of necrosis, inflammatory cells, brain parenchyma, and other variables. Brennan managed to reduce these confounders and enhance the comparison of proteomic and transcriptomic subclassification. The proportions of the three groups were relatively equal. They also examined mutation and CN genomic aberrations in these datasets and found supporting evidence for the classification, though no single CN aberration distinguished any class. The intrinsic cellular heterogeneity of gliomas is masked by the use of homogenized tissue in these TCGA genomic and proteomic analyses.

Clinically, the two subtypes of GBM are distinguished by histologic grade at diagnosis, with primary GBM grade 4 and secondary GBM lower grade. It turns out that these two subtypes draw from similar sets of molecular events: amplification and activating mutations in EGFR, overexpression of PDGF and its receptors, and loss of tumor supressors INK4a/ARF, p53, and PTEN. The standard first-line chemotherapy for GBM is radiation combined with the alkylating agent Temozolomide. In general, single-agent therapy directed at the two most commonly altered receptors in unselected populations has had poor results. Hopefully, the new pathway-related stratification of gliomas will lead to specific, more effective chemotherapy.

C. Molecular Systems Biology of the Brain: Modules Affected in Prion Disease

The search for organ-specific molecular signatures is a major focus at the Institute for Systems Biology, starting with brain and liver. A very interesting example of combining molecular analyses with modules and pathways and clinical pathophysiology is the Hwang *et al.*[18] study of prion disease in the mouse. This chapter is likely to become a landmark in systems biology, both for its design and specific methods and for its novel findings.[19]

Since the emergence of 'omics technologies, global analyses of gene expression (mRNA) and proteins have yielded increasingly long lists of disease-associated molecules. Distinguishing true-positive from false-positive signals and organizing the findings into pathways, networks, and modules related to histopathological and clinical phenotypes in temporal and spatial dimensions is a general set of challenges. The task is magnified by the complexity of the brain, the peculiarities of the transmissible protein agents of prion diseases, and the variability in both prion properties and genetic makeup of infected organisms, with analogies to cancer biology.

Remarkably, these problems were turned into levers to enhance the studies by Hwang *et al.*[18] With two prion strains, characterized by different incubation times, and mice from six different genetic backgrounds, including strains with altered prion protein (PrP) expression levels, they performed a subtractive analysis that drastically reduced biological and experimental noise and focused on sets of genes reflecting the disease process in common across the host genotypes and infectious agent strains. The pathological/clinical endpoint was defined as "disease incubation time" from inoculation at age 5 weeks to advanced clinical impairment, ranging from 56 to 392 days. Genome-wide gene expression in whole brain homogenates was analyzed over 8–10 time points, with 1- to 4-week time intervals adjusted to the wide range of incubation times.

From the massive amount of data, the authors extracted a core of just 333 genes which were differentially expressed in all five of the combinations involving mice with normal levels of prion protein (compared with 7400 genes differentially expressed in at least one of those five backgrounds). These 333 genes were considered central to prion disease; 161 were mapped onto functional pathways using protein–protein interaction, metabolic, and signaling pathway information from public databases. Visualization of changes in gene expression in critical biological modules functioning in cellular and subcellular compartments over the months of disease progression provided a dynamic scheme for the processes which characterize the molecular conversion of benign prion protein (PrP^{C}) to disease-causing PrP^{Sc} isoforms accumulating in lipid rafts, followed by the three stages of neuropathology: synaptic degeneration, activation of microglia and astrocytes, and neuronal cell death. Notably, 178 genes not previously associated with prion disease were identified among the 333 differentially expressed, highly associated genes, including sets encoding functional modules for androgen, iron, and arachidonate/prostaglandin metabolism (http://prion.systemsbiology.net).

There are many implications of this study. The same principle of interaction of host and infectious agent variation can be applied to eco-genetic systems analysis of other infections (tuberculosis, malaria, HIV, influenza, *Escherichia coli*, etc.), as well as environmental and behavioral variables that act on genetic variation to modify risk and manifestations of disease. From the methodological point of view, the subtractive design strategy adopted by Hwang *et al.*[18] reduces biological and experimental/technical noise in large-scale datasets. Finally, the kinds of neuropathological responses appear to be limited, so other degenerative disorders, including forms of Alzheimer disease, and even neoplastic diseases, may activate the same molecular and cellular processes and express similar molecular signatures. For example, there were clues from altered cholesterol, sphingolipid, and glycosaminoglycan homeostasis that might justify proposing statins and other drugs for prevention of both prion disease and Alzheimer disease.

Functional validation of the roles of specific genes and of identified modules in the definable stages of disease progression must proceed beyond selective RT-PCR. At the system-level, it will be interesting to investigate the functional and pathophysiological consequences of the dynamic changes in network architecture observed by the authors' network. This study examined only the transcriptome. Epigenomics and miRNA analysis will inform gene regulation, and proteomics and metabolomics will confirm and reveal new downstream effector pathways and molecular targets for therapeutic and preventive interventions. Relevant regions of the brain could be compared, especially the thalamus where prion replication seems to start. Validation of the mouse model also must overcome a large experience that animal models are often quite different from the human disease.

An area for future research is the creation of both approximate and rigorous mathematical models to describe the process and predict the dynamic behavior of genes, mRNAs, miRNAs, proteins, and metabolites in the disease process. Global sensitivity analysis, switching from mathematical to numerical analysis, should be more effective than qualitative applied methods with nonlinear differential equations.[20] Determination of the kinetic parameters governing each step in prion activation and progression of disease would promote modeling of temporal and spatial dynamics,[21] while biochemical pathways can be reconstructed using mass action time-series data from perturbed systems.[22] The resulting networks can be queried for alignment, integration, and evolution.[23] Prion disease research will generate molecular explanations at the level of 3D structures, chemical modifications, and patterns of misfolding for the distinct strains of infectious prions with differences in sites of infection in the brain, duration of incubation, and other properties governing interactions with the host. At the practical level, brain-specific plasma markers for the core processes discovered here could become assays for testing asymptomatic cattle and people for prion infections. Brain-specific proteins have already been detected in the peripheral blood in this mouse model (unpublished).

IV. Alternative Splicing: Discovery of a New Class of Protein Cancer Biomarker Candidates

Alternative splicing increases protein diversity without significantly increasing genome size. It is now recognized to be very common throughout the human genome. Cancer-specific splicing events have been reported at the mRNA level in colon, bladder, and prostate tissues, with diagnostic and prognostic implications.[24] Splice events that affect the protein coding region of the mRNA give rise to proteins differing in sequence and activities; splicing within the noncoding regions can result in changes in regulatory elements, such as translation enhancers or RNA stability domains, which may dramatically influence protein expression.[25] Several databases with alternatively spliced transcripts are available.[26] We have utilized the ECgene database,[27] which is based on evidence collected from clustering of ESTs, mRNA sequences, and gene model predictions. We are characterizing human tumor specimens, plasma from patients with tumors, and tumor specimens and plasma from genetically defined mouse strains that are models of human pancreatic and human breast cancers. The National Cancer Institute has made such mouse models of human cancers a major thrust (http://mouse.ncifcrf.gov) in the search for biological understanding of mechanisms of cancer initiation and cancer progression and the companion search for biomarkers for diagnosis, prognosis, and response to therapy.[28]

A. The Modified ECGene Database of Potential Translation Products

ECgene combines genome-based EST clustering and a transcript assembly procedure to construct gene models that encompass all alternative splicing events. The reliability of each isoform was assessed from the nature of cluster members and from the minimum number of clones required to reconstruct all exons in the transcript.[29] We combined Ensembl (version 40) with ECgene database (mm8, build 1); the transcript sequences were translated in three reading frames; within each dataset, the first instance of each protein sequence longer than 14 amino acids was recorded. The resulting proteins from both database translations were combined and filtered for redundancy, with preference given to Ensembl. We added a collection of common protein contaminant sequences, and then generated and added a set of reversed sequences as an internal control for false identifications. The total for the mouse was 10.4 million protein sequence entries.

A comparable process generated a human modified ECgene database with 14.2 million entries, which we are using for the studies of human specimens. The mzXML files containing the mass spectral information are searched against the modified ECgene database using X!Tandem software.[30] Peptides are integrated to a list of proteins using TransProteomic Pipeline and/or the Michigan Peptide to Protein Integration workflow, and further analyzed.[31–33] Peptides that were identified by X!Tandem search with X!Tandem expect value < 0.001 or with three or more spectra with expect value < 0.01 had a false discovery rate (FDR) $< 1\%$, based on peptides identified from reverse sequences compared to total peptides identified after applying the threshold. To characterize alternatively spliced peptides and proteins, we used InterProScan and Motif Scan,[34] Gene Ontology, and FuncAssociate, and displayed protein–protein interactions with the Cytoscape plug-in for MiMI.[7]

B. Identification of Splice Variant Proteins in Plasma of Mice with Pancreatic Cancer

Pancreatic ductal adenocarcinoma (PDAC) is the most lethal of human cancers, due to absence of methods for early diagnosis and chemoresistance of advanced disease. Five-year survival is $< 5\%$ of patients, with 31,000 deaths per year in the United States.[35] The KRAS G12D activation and p16/Ink4a and p19/Arf-p53 deletions mouse model of PDAC was genetically engineered by DePinho and Bardeesy to match the molecular lesions of human PDAC; it recapitulates the histopathologic progression and clinical effects of the human disease in a highly reproducible and synchronous fashion. The tumors express pancreatic ductal markers (CK-19) and apical mucins (Muc1, Muc5AC), show activation of Hedgehog, Notch, and EGFR developmental signaling pathways, harbor genomic alterations syntenic to human PDAC, and exhibit proliferative stroma.[36–38] We exploited this model to test our hypothesis that cancer-specific ASVs could be identified in mass spectrometric analyses of plasma proteins from mice carrying these molecular lesions, compared with wild-type mice.[31]

Plasma samples were processed into 163 fractions after immunodepletion of the three most abundant proteins, (albumin, immunoglobulins, and transferrin), which account for 90% of protein mass, and acrylamide labeling, then digested with trypsin and analyzed with a ThermoFinnigan LTQ-FT mass spectrometer.

As outlined in Fig. 2, our integrated analysis revealed 420 distinct splice isoforms, of which 92 were novel, not matching any previously annotated mouse protein sequence. For seven of those novel variants, we prepared primers and validated the predicted sequences in the mRNA with qRT-PCR for all seven. Isotopic labeling of cysteine-containing peptides with D3 versus D0 acrylamide for the tumor-bearing mice and wild-type controls, respectively, permitted relative quantitation of 28 of the 92 novel proteins (those whose ASV peptides contained cysteine). Differential expression was demonstrated for peptides from novel variants of muscle-type pyruvate kinase, malate dehydrogenase 1, glyceraldehyde-3-phosphate dehydrogenase (G3PD), proteoglycan 4, minichromosome maintenance complex component 9, high mobility group box 2, and hepatocyte growth factor activator. Upon annotation, we presented[31] literature evidence that many of the ASVs may well be involved in pancreatic cancer, including alpha-fetoprotein, apolipoprotein E, ceruloplasmin, fibronectin, glyceraldehyde-3-phosphate dehydrogenase, hemopexin, peptidyl-prolylisomerase, and tubulin alpha among the novel ASVs, and acyl coA acetyl-transferase, chromograinin b, granulin, insulin-like growth factor binding protein 2, and regenerating islet-derived 3alpha among the known ASVs that also had significant differential expression (upregulation).

From a systems biology point of view, one of the most interesting proteins is pyruvate kinase (PK), the critical enzyme in the metabolic switch to aerobic glycolysis in cancers, known since 1929 as "the Warburg Effect."[39,40] Aerobic glycolysis refers to persistence of high lactate production in the presence of oxygen. PK catalyzes transfer of a phosphoryl group from phosphoenolpyruvate to ADP, generating ATP. Most adult tissues express the M1 isoform, whereas tumors (and embryonic tissues), rapidly growing tissues, express the M2 isoform. Our analyses showed positively correlated upregulation of peptide and mRNA expression of the novel variant of pkm2 in plasma from tumor-bearing mice, compared with the wild type. Using the University of California Santa Cruz Genome Browser and UCSC Blast, we mapped the peptide sequences to the genomic structure of the mouse muscle-type pyruvate kinase and identified a 42 bp region that corresponds to the 14-amino acid-peptide CLAAALIVTESGR (aa 482 to 495). This peptide was identified from five different spectra. It had never before been reported in the mouse, but it aligns perfectly with the homologous portion of the human, rat, and chicken muscle PK M1 isoform.

Tumor type pkm2 has been reported as a metabolic marker specifically for pancreatic cancers.[41] We are now exploring computerized structural prediction algorithms to characterize the effects of alternative splicing and of phosphorylation. Warburg noted that cancer cells take up glucose at higher rates than normal tissue, but use a smaller fraction of the glucose for oxidative phosphorylation, even when oxygen is not limited. Such aerobic glycolysis is due to reprogramming of metabolic genes to permit a greater fraction of glucose

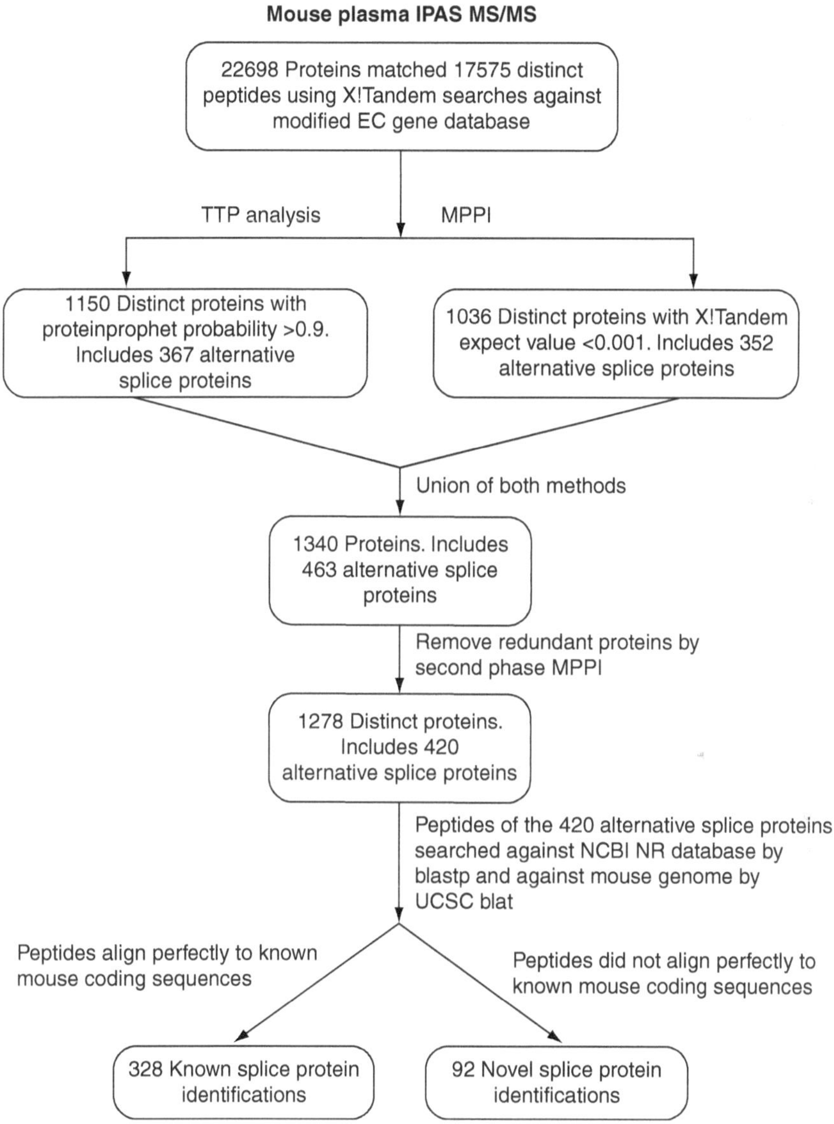

FIG. 2. Multi-step workflow for discovery and identification of alternative splice variant proteins associated with pancreatic cancer from tandem mass spectrometry of peptides. Workflow shows multi-step analysis from X!Tandem search of Intact Protein Analysis System MS/MS, through TransProteomicPipeline and Michigan Peptide-to-Protein Integration algorithms, leading to 420 alternatively spliced proteins, 328 previously reported and 92 novel. [From reference [31], with permission of *Cancer Research*].

metabolites to be incorporated into macromolecule synthesis rather than burned to carbon dioxide. Oncogenic forms of fibroblast growth factor receptor type 1 inhibit the pyruvate kinase M2 (PKM2) isoform by direct phosphorylation of PKM2 tyrosine residue 105 (Y105)[42] This site-specific phosphorylation inhibits the formation of active, tetrameric PKM2 from less active dimers by disrupting binding of the PKM2 cofactor fructose-1,6-bisphosphate. A PKM2 mutant with phenylalanine substituted for this tyrosine (Y105F) in cancer cells leads to decreased cell proliferation under hypoxic conditions, increased oxidative phosphorylation with reduced lactate production, and reduced tumor growth in xenografts in nude mice.

Growth factor signaling pathways activate protein tyrosine kinases and decrease the specific activity of pyruvate kinase (measured without regard to isoforms). Christofk *et al.*[43] used a proteomic screen with an immobilized phosphotyrosine (pTyr)-peptide library affinity matrix and SILAC labeling to identify novel pTyr-binding proteins from HeLa cell lysates. Binding of pTyr-peptides to PKM2 releases the allosteric activator fructose-1,6-bisphosphate, leading to inhibition of PKM2 enzyme activity. pTyr signaling stimulated by several growth factors diverts glucose metabolites from energy production to anabolic processes to support rapid growth of cancer cells, including nucleic acid and fatty acid biosynthesis. The M2 isoform is the only PK isoform that binds pTyr peptides (M1, liver, and red blood cell isoforms do not). PKM1 and PKM2 are identical in sequence except for a 56-amino acid stretch encoded by an alternatively spliced region involving exon 10 in PKM2, which forms an allosteric pocket unique to PKM2 in which the FBP activator can bind. Mutation of lys-433 at the lip of the pocket to glutamate interferes with the binding. Boxer *et al.* have now generated substituted *N,N′*-diarylsulfonamide molecules that activate the PKM2 and alter the Warburg effect, which could become a new antiproliferation therapeutic strategy.[44]

We also searched our peptide findings for variants of proteins chosen as possible pancreatic cancer biomarkers, from among 1442 proteins identified, in a parallel study of this same mouse model that was the source of the data for this analysis.[38] We found variants of three of their nine proteins assayed by ELISA in humans in our list of 420 splice variant proteins: lipocalin 2 (*lcn2*), regenerating islet-derived 3 (*reg3a*), and tumor necrosis factor receptor superfamily member 1A (*tnfrsf1a*); according to our quantitative expression analysis, the *tnfrsf1a* showed > 2-fold increase in expression in plasma from the tumor-bearing mice compared with wild-type mouse plasma.

Meanwhile, Harsha *et al.*[45] created a pancreatic cancer data compendium for biomarker candidates from published microarray and proteomic datasets from both exocrine and endocrine neoplasms of the pancreas. Data were pulled from GEO, ArrayExpress, and Oncomine.[4,46,47] They annotated the lists with evidence of these molecules in pancreatic juice, plasma, or serum and on plasma membranes of cells. They also compared results for pancreatitis. The manual curation of the literature consumed 7000 person-hours. A total of 1868 genes were reported as overexpressed only in mRNA analyses, with 441 over-expressed in both mRNA and protein studies; 207 molecules were overexpressed only at the protein level, and 648 proteins altogether. Two proteins, CECAM1 and MUC1 seemed to outperform the clinically used marker CA 19-9.[48,49]

C. Identification of Splice Variant Proteins in Tumor Tissue of Mice with Her2/Neu Breast Cancer

We analyzed LC–MS/MS datasets from tumor and normal mammary tissue from a mouse model of HER2/neu-driven breast cancer.[32] Whiteaker *et al.*[50] had identified 6758 peptides, representing > 700 proteins; we downloaded the mzXML dataset from their submission to PeptideAtlas.[51] The lysates from individual tissue specimens were pooled from 5 tumor-bearing mice and from 5 normal mice and analyzed with LC–MS/MS. In our workflow (Fig. 2), MPPI was sufficient without TPP (and its Q3Ratio and XPRESS features) since the lysates were not labeled.

We found a total of 608 distinct alternative splice variants, 540 known and 68 novel. Of the 68 novel splice variant proteins, 54 were from the tumor and 23 from the control sample, with 9 in common. Of the 15 biomarker candidates Whiteaker *et al.*[50] had confirmed as overexpressed in tumor lysates with quantitative MRM-MS, we found that 10 had splice variants in our analysis. However, we do not know the activities of different isoforms of these or any other proteins from proteomics analyses.

Among these 68 novel proteins we demonstrated variants resulting from new translation start sites, new splice sites, extension or shortening of exons, deletion or switch of exons, intron retention, and translation in an alternative reading frame. To validate the protein findings, we were able to design optimal primers for qRT-PCR analysis for 32 of the 45 novel peptides found only in the tumor sample. Each was amplified successfully; 31 of the 32 were validated, and 29 of the 31 showed increased mRNA expression.[32]

In our annotations, 16 of the novel peptides found only in the tumor sample and with increased mRNA expression by PCR were highlighted because of functional motifs potentially significant in cancers. There were two variants with interesting annotations for BRCA. First, the peptide sequence "FSRAEAEGPGQACPPRPFPC" is in the second intronic region of leucine-zipper-containing LZF (rogdi) gene (Fig. 3). Using Splice Site Prediction by Neural Network from the Berkeley Drosophila Genome Project (http://www.fruitfly.org/seq_tools/splice.html), we found a predicted donor splice site "gact-gaggtgaggtg" where the novel peptide was identified as coding sequence with a

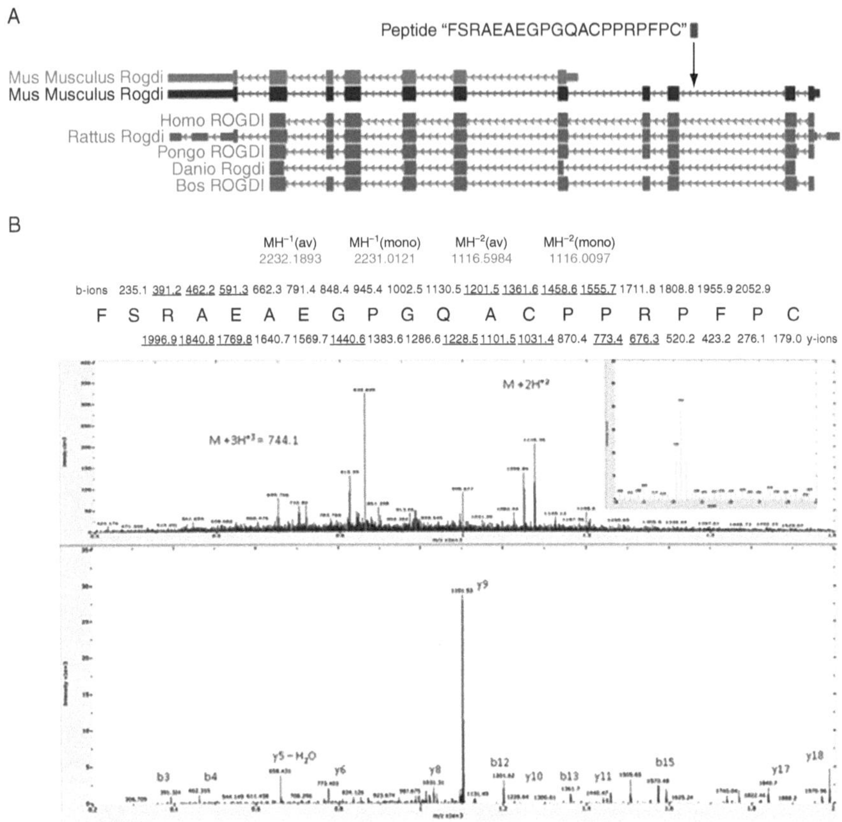

FIG. 3. Genomic structure of alternative splice variant for Rodgi leucine zipper protein from studies of Her2/neu breast cancer model. (A) Variant identified by a unique peptide in the tumor lysate sample, aligned to the second intronic region of the mouse Rogdi gene by UCSC blat (chr16: 5,012,685–5,012,744). The wide blocks are the exons; the narrow are the UTR regions; the line with arrows denotes the noncoding intronic regions. (B) MS1 (top) and MS/MS (bottom) spectral images of the novel peptide identified from the Rogdi gene. The inset of the MS1 spectrum shows a clearly defined precursor ion isotope envelope. (From Ref. 33 with permission of *Disease Markers*.)

splice site prediction score of 0.93. Functional motifs identified in this section of intronic sequence include LIG_BRCT_BRCA1_1, a phosphopeptide motif which interacts directly with the carboxy-terminal domain of BRCA1. Second, the peptide "GSGLVPTLGRGAETPVSGAGATRGLSR" aligned to the first intronic region of transcription factor sox7; the same LIG_BRCT_BRCA1_1 motif was found in this intronic region. We intend to model interactions with BRCA1 for both of these protein domains.

The peptide "IYYSFGALKLGCFNFPLLKFL" aligns perfectly to a region in mouse chromosome 7 with sequence conservation in five other species, including human; two functional motifs link this unnamed protein to a tyrosine-based sorting signal TRG_ENDOCYTIC_2 responsible for interaction with the mu subunit of Adaptor Protein (AP) complex and to a MAP kinase docking function via LIG_MAPK_2. This signal motif is particularly interesting due to tyrosine-based internationalization of the *neu* proto-oncogene product.[52] Then there are 12 variants with casein kinase II (CK2) phosphorylation, protein kinase phosphorylation (PKC), or N-myristoylation sites.[32] One of these is a new variant of pyruvate kinase muscle type (pkm2) identified by the peptide "GHPGPEVWGGAGCGHGVCIFPAAVGAVEASFK"; the first 20 amino acids are from the middle section of exon 6 and the remaining 12 amino acids are from the middle section of exon 9. Two N-myristoylation sites and one PKC phosphorylation site were found in this peptide sequence.

Employing spectral counting, we found 53 known splice variants differentially expressed. Using MotifScan with prosite patterns and prosite profiles as search parameters, we focused on the top 5 frequently occurring prosite patterns; CK phosphorylation, PKC phosphorylation, and n-myristoylation sites were found 1.5 times more frequently in these 53 variants than in 53 randomly selected normal proteins. We refer to these 53 known alternative splice variants and the 45 novel proteins found only in tumor sample as "tumor-associated splice variants."

Finally, we utilized GeneGo Metacore™ software to characterize significant biological process networks. Cytoskeletal rearrangement, integrin-mediated cell adhesion, and translation initiation were found in common among the top-ranking networks from both all tumor-associated variants and the variants identified in the tumor sample. Figure 4 shows the direct protein interactions displayed by Cytoscape with MiMI plug-in; 177 of 460 input gene symbols are interacting. The gene names in bold denote differentially expressed alternative splice variants, including many of those we had already annotated as candidates for a role in the systems biology of breast cancer. Cdc42, arhgdia, and rdx are among these proteins implicated in breast cancer mechanisms.[32] Proline-rich motifs in specific splice variants of rdx and other proteins involved in extracellular matrix and cell motility can be contrasted with other known isoforms of these genes, which do not contain the proline-rich region. These motifs in these proteins participate in delivering actin monomers to cellular locations where ruffles, filopodia, and microspikes—actin-rich membrane protrusions—are formed.

We are in the earliest stages of identifying and evaluating alternative splice variants for their potential roles in systems biology of cancers, especially critical features like initiation, progression, cell motility, invasiveness, and metastasis. These results illustrate the use of algorithms and databases for annotation of

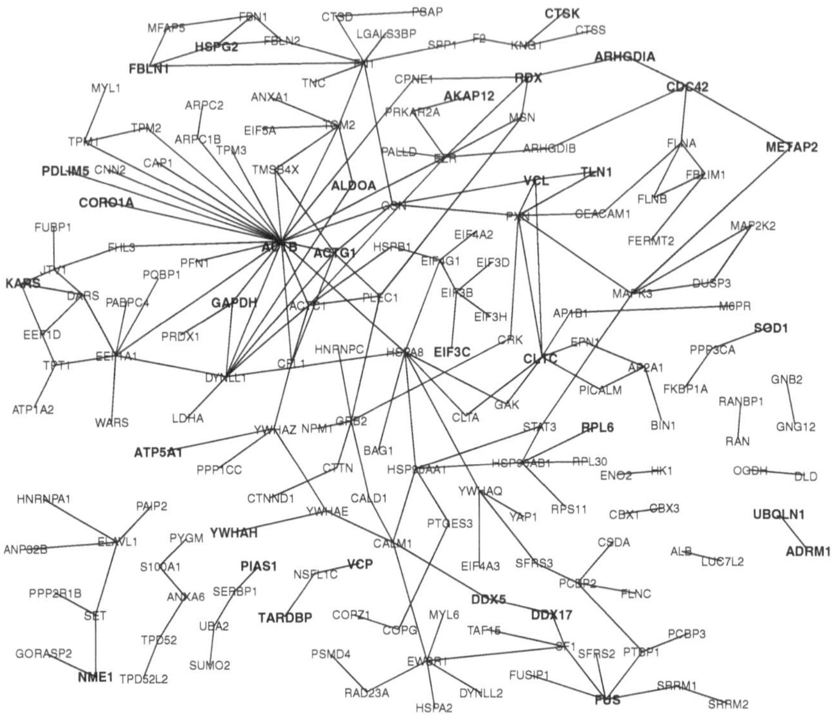

FIG. 4. Cytoscape visualization with MiMI-plug-in of protein–protein interactions involving alternative splice variants in Her2/neu breast cancer model. The input gene list contained the alternative splice variants detected only in the tumor sample. Only direct interactions between input genes are shown. Gene symbols in bold are those annotated in the original publication. See text. (From Ref. 33 with permission of *Disease Markers*.)

biological significance. Further research is needed to delineate major subtypes of common cancers, model the likely changes in structure and function of splice variants compared with prevalent isoforms, and propose and validate uses of these proteins as targets for therapy and biomarkers for diagnosis, prognosis, and response to treatments. In addition, the coding functions of the reverse transcriptome[53] and the regulatory functions of both strands warrant investigation, accelerated by next-generation sequencing methods.

D. Assembling a Genome-Wide Splicing Code

Barash *et al.*[54] introduced a bioinformatics scheme that uses combinations of hundreds of RNA features to predict tissue-dependent changes in alternative splicing for thousands of exons. They used data profiling 3665 cassette-type alternative exons across 27 mouse tissues, including whole embryo and a variety of adult tissues. The splicing code identifies distinct regulatory programs in different tissues, mutation-verified regulatory sequences, and new classes of splicing patterns. Regulatory strategies include large combinations of features, low exon inclusion levels that are overcome by features in specific tissues, features appearing deep into introns, and modulation of splice variant levels by transcript structure characteristics. Tissue-dependent splicing is regulated by trans-acting factors, cis-acting RNA sequence motifs, and other RNA features like exon length and secondary structure. The code should facilitate reliable predictions of the regulatory properties of previously uncharacterized exons and the effects of mutations within regulatory elements.

V. Concepts Tools for Systems Biology Analysis

Our group at the National Center for Integrative Biomedical Informatics (NCIBI), led by Maureen Sartor, has created a very useful web-based gene set functional enrichment and relation mapping tool we call ConceptGen.[55] This resource (see http://portal.ncibi.org/gateway) offers 20,000 concepts from 14 different types of biological knowledge, including data, visualization, and interactivity not similarly available in earlier gene set enrichment or gene set relation mapping tools. Those tools, including DAVID/EASE,[56] GSEA/MSigDB,[57] GeneCards and GeneDecks,[58] and Oncomine,[59] have been used extensively in annotation and interpretation of gene expression, proteomic, metabolomic, and transcription factor binding data. These multifaceted resources commonly use differential gene expression profiles; Gene Ontology biological processes, molecular functions, and cellular compartments, and KEGG, Panther, and/or Biocarta pathways. ConceptGen has such additional information as Online Mendelian Inheritance in Man (OMIM), MiMI (a synthesis of several different protein–protein interaction databases, with provenance), medical literature-derived concepts (MeSH), drug targets (Drug Bank), Tranfac transcription factor targets, Pfam protein families, MiRBase for microRNA targets, Metabolite for metabolic interactions. Gene set relation mapping goes beyond enrichment testing through visualization that enhances exploratory analysis and hypothesis generation. Two concepts are related when they have significantly more genes in common than expected by chance; these relationships can form a network. Concepts and data may cluster into distinct groups that have previously unsuspected relationships between concepts from diverse types of concepts, as listed above. Visualization of related concepts can be presented graphically or in heatmaps.

An interesting overview application was the question "Which genes are most often and least often differentially expressed in gene expression datasets?" The range was from zero to 108 (53%) of the 203 experiments analyzed. Genes most often and most significantly differentially expressed were enriched for cell cycle, cell proliferation, apoptosis, transcription factors, and immediate-early proteins. Least often differentially expressed were G-protein coupled receptors, sensory receptors, ion channels, and neurotransmitter binding. Obviously, these results reflect biases in reported studies, with a heavy emphasis on cancer studies.

We highlighted the usefulness of ConceptGen with two applications: (1) the time-course of epithelial–mesenchymal transition as an *in vitro* model of invasion, motility, and metastasis induced in A549 human lung adenocarcinoma cells by TGF-beta[60,61]; and (2) the discovery of sarcosine as a metabolite biomarker and biomediator of poor prognosis in prostate cancer.[62] ConceptGen Network graphs and heatmaps showed extracellular matrix, cell adhesion, cell movement and metallothioneins enriched 8 h after TGF-beta induction of EMT; at 1 h, JUN and TGFBR1 were enriched hubs. ConceptGen can also begin with metabolic compounds and genes encoding the metabolic enzymes that synthesize or catabolize those compounds, themselves mapped with our Metscape tool and MetOntology. Six metabolites were shown to change significantly in LnCaP human prostate cancer cells undergoing progression from benign to localized to metastatic prostate cancer.[62,63] Gene Ontology categories amino acid metabolism, nitrogen breakdown, and amino methyltransferase activity were enriched, all clustering around the compound sarcosine (methylglycine) in the heatmap view. The previously unconnected concepts amino acid metabolism and methylation were linked through methylglycine, based on the bioinformatics. Then extensive wet-lab experiments showed that adding sarcosine or the precursor glycine to the medium of nonaggressive prostate cancer cells made them highly invasive and motile. Transfecting those cells with the gene for glycine methyltransferase or blocking degradation with siRNA treatment of sarcosine dehydrogenase had the same effect. Starting from a highly invasive cell line and performing complementary manipulations to reduce sarcosine levels made those cells noninvasive. Thus, sarcosine appears to be not just a biomarker, but a mediator, of metastatic prostate cancer.[62] Additional processes enriched were aminoacyl-tRNA biosynthesis, independently identified as dysregulated by androgen in prostate cancer[64] and procollagen-proline dioxygenase activity.[55]

ConceptGen can also be used to annotate and potentially characterize genes of unknown function. Sartor *et al.* demonstrated this application with the gene Chac1, cation transport regulatory homolog 1, from *E. coli*. The gene was identified as the most significantly differentially expressed gene in an unrelated RNA-Seq study; it is not annotated to any GO or pathway term; and little is known of its protein. The ConceptGen query yielded several related genes known to be involved in the unfolded protein response and apoptosis. When the 100 top ranking genes were uploaded into a private ConceptGen account for analysis of enriched concepts, apoptosis, amino acid transport, tRNA synthetase activity, and Cebpb, ATF4, and Chop/Ddit3 gene activities were most prominent. The connection between CHAC1 and the ATF4/ATF3/CHOP cascade and apoptosis has now been identified experimentally.[65] ConceptGen and the complementary gene enrichment/gene relation mapping tools should be helpful in predicting pathways and processes of other unannotated genes.

VI. Determining the Activity of All 21,000 Protein-Coding Genes in the Human Genome

Global molecular analyses are intended to facilitate a comprehensive functional view of the cell and the organism. However, it seems that our minds are more comfortable with a focus on the top 10 or top 20 genes or proteins in lengthy lists of differential expression, or with selective annotation of a small number with particularly interesting biological implications, as with our splice variant proteins. Performing annotations on a large scale was demonstrated above with the brain prion disease story. Performing genome-wide experimental analysis can be illustrated with the following story.

Neumann *et al.*[66] conducted a remarkable application of widely available methods to reduce protein expression with siRNA and measure resulting defects in cell anatomy or structure with time-lapse microscopy and temporal event-order maps of mitosis in HeLa cells. They used computational tools in an automated pipeline to systematically screen the entire ENSEMBL human genome. More than 500 genes were implicated in mitosis, most for the first time, and then validated experimentally using recombineering technology. In all, they had 572 validated mitotic genes found with at least two different siRNAs in the first-pass screen, plus another 677 scored with only one siRNA. They tested whether the mouse equivalent of a small subset of 21 of their candidate genes could prevent the phenotypic defects caused by reducing the expression of the human gene; most of the cases were confirmed. Species sequence differences prevented the siRNAs from silencing the mouse equivalent genes in these cases.

Phenotypic signatures can be compared at http://www.mitocheck.org for potential results from perturbations caused by developmental processes, disease states, pharmaceuticals, and other culture treatment conditions. The findings could be enhanced with profiles of transcription, pull-downs of protein complexes, and analyses of signal transduction and gene regulation by proteins and miRs and epigenomic modifiers with the HeLa cells and other cell lines. As illustrated by their imaging of the spindle microtubules, the primary screen can be complemented by visualization of other elements of the mitotic machinery, such as centrosomes, spindle microtubules, and kinetochores. This approach will be particularly fruitful for cancer biology, since so many cancer therapeutics perturb the cell cycle.

The siRNA scheme utilized 2 or 3 independent sequences for each gene and reduced mRNA levels to below 30% of original levels for 97% of the 1000 genes quantitatively tested. The experiments were performed in triplicate for 2 days, with monitoring of fluorescent chromosomes after tagging of histone 2B with green fluorescent protein. A computational chromosomal morphology recognition pipeline with Python, the C++ image processing library VIGRA, the classification library libSVM, and R for plotting and statistical analysis was essential to manage the data from 190,000 time-lapse movies of 19 million cell divisions. About 200 quantitative features were extracted from each nucleus, leading to classification into 16 morphological classes and positional tracking of the nucleus. Live imaging not only picked up the rare, transient mitotic events, but also revealed primary defects and secondary consequences. For example, polylobed, grape, and binuclear cells arise from premature nuclear assembly, chromosome segregation errors, or cytokinesis failures. The most common consequence of a mitotic delay was polylobed or grape-shaped nuclei indicative of mitotic exit with aberrant chromosome segregation that did not cause cell death; thus, these effects persist. The event order map for binuclear cells indicative of cytokinesis defects and failure of cell body separation seems generally well tolerated by the cell. Only 22% (124/572) of the validated mitotic hits exhibited cell death phenotypes; thus, cell survival is not a very reliable indicator of mitotic defects, even in HeLa cells, which have negligible p53 DNA damage response and may present a sensitized background for apoptosis. Cell migration was influenced by 360 genes, with no correlation with effects on mitosis.

Other comprehensive efforts to characterize the functional genome are being explored. Under the leadership of the international Human Proteome Organization, there is an effort underway to identify the protein products of all the approximately 21,000 protein-coding genes, their tissue distribution, and their protein–protein interactions.[2] The Swedish Human Protein Atlas (www.proteinatlas.org) already has immunohistochemical results for each of about 8000 proteins in 48 normal human tissues and a dozen each of 20 tumor types. It is feasible to estimate the intensity of IHC staining and the intracellular localization of the staining, as well as the proportion of cells stained.

VII. Bioinformatics and Systems Biology of Metastasis: The Case of Lung Cancers

Lung cancers are the leading cancer cause of death in the United States, with more deaths than from breast, prostate, colorectal, and pancreatic cancers combined. Since 1986, deaths from lung cancers have outnumbered deaths from breast cancers in women. The primary cause—tobacco smoking and exposure to second-hand cigarette smoke—is indisputable, offering important directions for prevention. There are some additional environmental exposures known to cause lung cancers (asbestosis, radon, chromium, nickel, bis-chloromethylether, generally synergistic with cigarette smoking). The 5-year survival rate is still only 15% of patients, about the same as when I graduated medical school 45 years ago. Even among stage I patients, as judged by radiologists, surgeons, and pathologists, about half of the patients' tumors show molecular signatures more similar to stage III patients. Thus, understanding the basis for metastasis and predicting the prognosis for patients with lung cancers (and other cancers) should have highest priority. Keshamouni *et al.*[64] have provided a comprehensive treatise on this subject.

Metastasis is an active process employing genetic and epigenetic mechanisms to form a cell capable of responding to certain chemotactic signals that direct motility, avoiding immune responses, being refractory to growth inhibitors, proliferating independently of growth factors for sustained cell division, interacting with other cells to be cotranslocated, and implanting in distant locations, most often through the lymphatic system and less often directly into blood vessels. Rather than a late process in cancer progression, metastatic potential is now viewed, based on gene expression and proteomic profiling, to be present in localized tumor cells capable of early spread. Alternatively, this metastatic potential may depend upon a small subpopulation of tumor cells, the cancer stem cells. The phenomenon of epithelial–mesenchymal transition is well recognized *in vivo* and, after TGF-beta induction, *in vitro*, facilitating molecular studies such as[60] and many others cited in Keshamouni *et al.*[68] Lung cancer is following glioblastoma and ovarian cancers in the progressive expansion of the TCGA program.

VIII. Special Resources for Pharmacogenomics of Cancer Therapies

There is often striking individual variation in the metabolism of drugs and in response to the actions of the drugs on receptors, enzymes, and other molecular targets. The finding that EGFR-inhibiting monoclonal antibodies (Cetuximab, Panitumumab) used in treatment of colorectal cancers (and other cancers) are ineffective in patients whose tumors have activating mutations of KRAS was predicted and served to explain that such an activated KRAS pathway provided a bypass of the inhibition of the EGFR pathway. In the case of Herceptin, the drug is effective only in patients whose breast cancers have highly amplified Her2/neu (ERBB2) receptors on the tumor cells.

An organized bioinformatics effort in relation to pharmacogenetics and pharmacogenomics, led over the past decade by Russell Altman at Stanford, has created PharmGKB, a knowledge base that captures the relationships among drugs, diseases/phenotypes, and genes involved in pharmacokinetics (PK) and pharmacodynamics (PD). The manually curated information includes literature annotations, primary data sets, PK and PD pathways, and expert-generated summaries of PK/PD relationships between drugs, diseases/phenotypes, and genes.[69] PharmGKB currently has literature annotations documenting the relationships of over 500 drugs, 450 diseases, and 600 variant genes. In addition to gene–drug relationships, the PharmGKB contains data on gene variation, gene expression, gene–disease relationships, drug action, and pathways. To serve whole genome studies, PharmGKB has added new functionalities, including browsing the variant display by chromosome and cytogenetic locations, allowing the user to view SNP variants not located within a gene, and providing tools that facilitate quality control and compare data sources, such as dbSNP, JSNP, and HapMap. PharmGKB is accessible at www.pharmgkb.org.

PharmGKB is organized to serve four main user groups: gene-oriented users, drug-oriented users, bioinformaticians, and clinical/disease-oriented investigators. PharmGKB organizes knowledge pertinent to each group into user-based views and resources, with annotated information about 200 well-documented Very Important Pharmacogenes. VIP gene pages highlight the key variants, haplotypes, splice variants, drugs, diseases, and phenotypes associated. Curated phenotype data sets are reviewed and include meaningful phenotypic annotations related to pharmacogenomic research, from clinical-metabolite data to protein constructs. PharmGKB is providing support for consortia of investigators interested in pooling pharmacogenomic data sets in order to improve population coverage and statistical power. Historically, pharmacogenetic studies focused on single genes with Mendelian inheritance patterns. Now there is growing interest in how pathways of interacting genes can affect both drug metabolism and drug response and how additional genes can modify primary drug-related traits and organ and cellular responses. Hansen *et al.*[70] have developed a genome-wide candidate gene list and a scheme that ranks 12,460 genes for their potential relevance to a specific query gene and its clinical indications or adverse effects. They started with 2488 genetic interactions from PharmGKB and 33600 physical drug target interactions from DrugBank, mapped to the InWeb interactome that encodes 313,524 physical interactions among 12,460 human gene products. Using gene–drug interactions, networks of gene–gene interactions, protein–protein interactions, and available measures of drug–drug similarity, Hansen *et al.*

ranked the genes for likelihood of interaction with the drug, illustrating the approach with gefitinib, carboplatin, gemcitabine, and warfarin. They confirmed some known gene–drug interactions and revealed novel interactions for experimental assessment.

In June 2010, Cancer Research UK announced a pharmacogenomics pilot project presented as a step toward personalizing cancer care through genetic testing. A partnership among the research community, the National Health Service, drug and diagnostic companies, and the government is planned, together with engagement of patients willing to undergo genetic testing on their tumors for research linking treatment responses to genotypes.

IX. Conclusion

We remain locked into a "geographic classification" of cancers, based on the organ site of origin and the histological grade and clinical stage. The more we learn about molecular and cellular heterogeneity, the more it becomes clear that there are numerous differences among cancers of the same organ origin, histology, and stage. As illustrated by Her2/neu-amplified, ER+, and triple-negative subtypes of adenocarcinoma of the breast, we will progressively learn enough about the molecular signatures and pathogenesis of cancers that we can treat the subtype or even the individual patient much more specifically. Among the molecular types of lung adenocarcinomas, at least two cases have now been noted (described at a Symposium at Harvard Medical School 28 May, 2010) with Her2/neu amplification; as might be predicted, these patients responded well to specific therapy with Herceptin.

Our goal should be to learn enough about cancer biology to recognize the heterogeneity of mechanisms regardless of the site of origin, to choose or design therapies targeted to the causal mechanism, and to gain sufficient evidence of efficacy and effectiveness to treat based on biomarkers before the cancer has grown sufficiently to be "seen" by palpation, imaging, or physical complications. Such an ambitious goal will also require that our most specific, most effective new drugs be utilized for early-stage patients, especially when those patients have molecular signatures of poor prognosis/high metastatic potential, rather than being reserved to patients with far advanced disease and multiple aberrant pathways and networks. The systems biology approaches and bioinformatics tools illustrated in this Chapter point the way to such clinical advances.

Acknowledgment

I appreciate the assistance of Denise Taylor-Moon and Rajasree Menon, PhD, in the preparation of this chapter.

References

1. Bourne PE, Weissig H, editors. *Structural bioinformatics*. Hoboken, NJ: Wiley-Liss; 2003.
2. HUPO. A gene-centric human proteome project. *Mol Cell Proteomics* 2010;**9**:427–9.
3. Omenn GS, Aebersold R, Paik YK. 7(th) HUPO World Congress of Proteomics: launching the second phase of the HUPOPlasma Proteome Project (PPP-2) 16–20 August 2008, Amsterdam, The Netherlands. *Proteomics* 2009;**9**:4–6.
4. Rhodes DR, Yu J, Shanker K, Deshpande N, Varambally R, Ghosh D, et al. ONCOMINE: a cancer microarray database and integrated data-mining platform. *Neoplasia* 2004;**6**:1–6.
5. Chinnaiyan AM. Gene fusions in solid tumors. In: Ruddon R, editor. *Introduction to the molecular biology of cancer*. Chennai: Elsevier; 2010.
6. Tomlins SA, Rhodes DR, Perner S, Dhanasekaran SM, Mehra R, Sun XW, et al. Recurrent fusion of TMPRSS2 and ETS transcription factor genes in prostate cancer. *Science* 2005;**310**:644–8.
7. Gao J, Ade AS, Tarcea VG, Weymouth TE, Mirel BR, Jagadish HV, et al. Integrating and annotating the interactome using the MiMI plugin for Cytoscape. *Bioinformatics* 2009;**25**:137–8.
8. caBIG Overview (2009). NIH publication 09-7438.
9. TCGA . TCGA Network; comprehensive genomic characterization defines human glioblastoma genes and core pathways. *Nature* 2008;**455**:1061–8.
10. Network T. Comprehensive genomic characterization defines human glioblastoma genes and core pathways. *Nature* 2008;**455**:1061–8.
11. Freire P, Vilela M, Deus H, Kim Y-W, Koul D, Colman H, et al. Exploratory analysis of the copy number alterations in glioblastoma multiforme. *PLoS ONE* 2008;**3**:e4076.
12. Hodgson JG, Yeh R-F, Ray A, Wang NJ, Smirnov I, Yu M, et al. Comparative analyses of gene copy number and mRNA expression in glioblastoma multiforme tumors and xenografts. *Neuro Oncol* 2009;**11**:477–87.
13. Cerami E, Demir E, Schultz N, Taylor BS, Sander C. Automated network analysis identifies core pathways in glioblastoma. *PLoS ONE* 2010;**5**:e8918.
14. Hanahan D, Weinberg RA. The hallmarks of cancer. *Cell* 2000;**100**:57–70.
15. Gaire RK, Bailey J, Bearfoot J, Campbell IG, Stuckey PJ, Haviv I. MIRAGAA—a methodology for finding coordinated effects of microRNA expression changes and genome aberrations in cancer. *Bioinformatics* 2010;**26**:161–7.
16. Noushmehr H, Weisenberger DJ, Diefes K, Phillips HS, Pujara K, Berman BP, et al. Identification of a CpG island methylator phenotype that defines a distinct subgroup of glioma. *Cancer Cell* 2010;**17**:510–22.
17. Brennan C, Momota H, Hambardzumyan D, Ozawa T, Tandon A, Pedraza A, et al. Glioblastoma subclasses can be defined by activity among signal transduction pathways and associated genomic alterations. *PLoS ONE* 2009;**4**:e7752.
18. Hwang D, Lee IY, Yoo H, Gehlenborg N, Cho J-H, Petritis B, et al. A systems approach to prion disease. *Mol Syst Biol* 2009;**5**:252.

19. Omenn GS. A landmark systems analysis of prion disease of the brain. *Mol Syst Biol* 2009;**5**:254.
20. Chen WW, Schoeberl B, Jasper PJ, Niepel M, Nielsen UB, Lauffenburger DA, et al. Input–output behavior of ErbB signaling pathways as revealed by a mass action model trained against dynamic data. *Mol Syst Biol* 2009;**5**:239.
21. Kholodenko BN. Cell-signalling dynamics in time and space. *Nat Rev Mol Cell Biol* 2006;**7**:165–76.
22. Srividhya J, Crampin EJ, McSharry PE, Schnell S. Reconstructing biochemical pathways from time course data. *Proteomics* 2007;**7**:828–38.
23. Sharan R, Ideker T. Modeling cellular machinery through biological network comparison. *Nat Biotechnol* 2006;**24**:427–33.
24. Thorsen K, Sorensen KD, Brems-Eskildsen AS, Modin C, Gaustadnes M, Hein A-MK, et al. Alternative splicing in colon, bladder, and prostate cancer identified by exon array analysis. *Mol Cell Proteomics* 2008;**7**:1214–24.
25. Bracco L, Kearsey J. The relevance of alternative RNA splicing to pharmacogenomics. *Trends Biotechnol* 2003;**21**:346–53.
26. Larsson TP, Murray CG, Hill T, Fredriksson R, Schiöth HB. Comparison of the current RefSeq, Ensembl and EST databases for counting genes and gene discovery. *FEBS Lett* 2005;**579**:690–8.
27. Kim P, Kim N, Lee Y, Kim B, Shin Y, Lee S. ECgene: genome annotation for alternative splicing. *Nucleic Acids Res* 2005;**33**:D75–9.
28. Marks C. Mouse models of human cancers consortium (MMHCC) from the NCI. *Dis Model Mech* 2009;**2**:111.
29. Kim N, Shin S, Lee S. ECgene: genome-based EST clustering and gene modeling for alternative splicing. *Genome Res* 2005;**15**:566–76.
30. Craig R, Beavis RC. TANDEM: matching proteins with tandem mass spectra. *Bioinformatics* 2004;**20**:1466–7.
31. Menon R, Zhang Q, Zhang Y, Fermin D, Bardeesy N, DePinho RA, et al. Identification of novel alternative splice isoforms of circulating proteins in a mouse model of human pancreatic cancer. *Cancer Res* 2009;**69**:300–9.
32. Menon R, Omenn GS. Proteomic characterization of novel alternative splice variant proteins in HER2/neu-induced breast cancers. *Cancer Res* 2010;**70**:3440–9.
33. Omenn GS, Yocum AK, Menon R. Alternative splice variants, a new class of protein cancer biomarker candidates: findings in pancreatic cancer and breast cancer with systems biology implications. *Dis Markers* 2010;**28**:1–11.
34. Quevillon E, Silventoinen V, Pillai S, Harte N, Mulder N, Apweiler R, et al. InterProScan: protein domains identifier. *Nucleic Acids Res* 2005;**33**:W116–20.
35. Jemal A, Siegel R, Ward E, Hao Y, Xu J, Thun MJ. Cancer statistics, 2009. *CA Cancer J Clin* 2009;**59**:225–49.
36. Bardeesy N, Aguirre AJ, Chu GC, Cheng K-h, Lopez LV, Hezel AF, et al. Both p16Ink4a and the p19Arf-p53 pathway constrain progression of pancreatic adenocarcinoma in the mouse. *Proc Natl Acad Sci USA* 2006;**103**:5947–52.
37. Aguirre AJ, Bardeesy N, Sinha M, Lopez L, Tuveson DA, Horner J, et al. Activated Kras and Ink4a/Arf deficiency cooperate to produce metastatic pancreatic ductal adenocarcinoma. *Genes Dev* 2003;**17**:3112–26.
38. Faca VM, Song KS, Wang H, Zhang Q, Krasnoselsky AL, Newcomb LF, et al. A mouse to human search for plasma proteome changes associated with pancreatic tumor development. *PLoS Med* 2008;**5**:e123.
39. Warburg O. On the origin of cancer cells. *Science* 1956;**123**:309–14.

40. Christofk HR, Vander Heiden MG, Harris MH, Ramanathan A, Gerszten RE, Wei R, et al. The M2 splice isoform of pyruvate kinase is important for cancer metabolism and tumour growth. *Nature* 2008;**452**:230–3.
41. Kumar Y, Gurusamy K, Pamecha V, Davidson BR. Tumor M2-pyruvate kinase as tumor marker in exocrine pancreatic cancer: a meta-analysis. *Pancreas* 2007;**35**:114–9.
42. Hitosugi T, Kang S, Vander Heiden MG, Chung T-W, Elf S, Lythgoe K, et al. Tyrosine phosphorylation inhibits PKM2 to promote the Warburg effect and tumor growth. *Sci Signal* 2009;**2**:ra73.
43. Christofk HR, Vander Heiden MG, Wu N, Asara JM, Cantley LC. Pyruvate kinase M2 is a phosphotyrosine-binding protein. *Nature* 2008;**452**:181–6.
44. Boxer MB, Jiang J-k, Vander Heiden MG, Shen M, Skoumbourdis AP, Southall N, et al. Evaluation of substituted *N,N′*-diarylsulfonamides as activators of the tumor cell specific M2 isoform of pyruvate kinase. *J Med Chem* 2009;**53**:1048–55.
45. Harsha HC, Kandasamy K, Ranganathan P, Rani S, Ramabadran S, Gollapudi S, et al. A compendium of potential biomarkers of pancreatic cancer. *PLoS Med* 2009;**6**:e1000046.
46. Barrett T, Troup DB, Wilhite SE, Ledoux P, Rudnev D, Evangelista C, et al. NCBI GEO: mining tens of millions of expression profiles—database and tools update. *Nucleic Acids Res* 2007;**35**:D760–5.
47. Parkinson H, Kapushesky M, Shojatalab M, Abeygunawardena N, Coulson R, Farne A, et al. ArrayExpress—a public database of microarray experiments and gene expression profiles. *Nucleic Acids Res* 2007;**35**:D747–50.
48. Argani P, Rosty C, Reiter RE, Wilentz RE, Murugesan SR, Leach SD, et al. Discovery of new markers of cancer through serial analysis of gene expression: prostate stem cell antigen is overexpressed in pancreatic adenocarcinoma. *Cancer Res* 2001;**61**:4320–4.
49. Simeone DM, Ji B, Banerjee M, Arumugam T, Li D, Anderson MA, et al. CEACAM1, a novel serum biomarker for pancreatic cancer. *Pancreas* 2007;**34**:436–43.
50. Whiteaker JR, Zhang H, Zhao L, Wang P, Kelly-Spratt KS, Ivey RG, et al. Integrated pipeline for mass spectrometry-based discovery and confirmation of biomarkers demonstrated in a mouse model of breast cancer. *J Proteome Res* 2007;**6**:3962–75.
51. Deutsch EW. The peptideatlas project. In: Hubbard S, Jones AR, editors. *Proteome bioinformatics*. Totowa, NJ: Humana Press; 2010. p. 285–96.
52. Gilboa L, Ben-Levy R, Yarden Y, Henis YI. Roles for a cytoplasmic tyrosine and tyrosine kinase activity in the interactions of neu receptors with coated pits. *J Biol Chem* 1995;**270**:7061–7.
53. He Y, Vogelstein B, Velculescu VE, Papadopoulos N, Kinzler KW. The antisense transcriptomes of human cells. *Science* 2008;**322**:1855–7.
54. Barash Y, Calarco JA, Gao W, Pan Q, Wang X, Shai O, et al. Deciphering the splicing code. *Nature* 2010;**465**:53–9.
55. Sartor MA, Mahavisno V, Keshamouni VG, Cavalcoli J, Wright Z, Karnovsky A, et al. ConceptGen: a gene set enrichment and gene set relation mapping tool. *Bioinformatics* 2010;**26**:456–63.
56. Dennis G, Sherman B, Hosack D, Yang J, Gao W, Lane HC, et al. DAVID: database for annotation, visualization, and integrated discovery. *Genome Biol* 2003;**4**:P3.
57. Subramanian A, Tamayo P, Mootha VK, Mukherjee S, Ebert BL, Gillette MA, et al. Gene set enrichment analysis: a knowledge-based approach for interpreting genome-wide expression profiles. *Proc Natl Acad Sci USA* 2005;**102**:15545–50.
58. Stelzer G, Inger A, Olender T, Iny-Stein T, Dalah I, Harel A, et al. GeneDecks: paralog hunting and gene-set distillation with GeneCards annotation. *OMICS* 2009;**13**:477–87.
59. Rhodes DR, Kalyana-Sundaram S, Mahavisno V, Varambally R, Yu J, Briggs BB, et al. Oncomine 3.0: genes, pathways, and networks in a collection of 18,000 cancer gene expression profiles. *Neoplasia* 2007;**9**:166–80.

60. Keshamouni VG, Michailidis G, Grasso CS, Anthwal S, Strahler JR, Walker A, et al. Differential protein expression profiling by iTRAQ-2DLC–MS/MS of lung cancer cells undergoing epithelial–mesenchymal transition reveals a migratory/invasive phenotype. *J Proteome Res* 2006;**5**:1143–54.
61. Keshamouni VG, Jagtap P, Michailidis G, Strahler JR, Kuick R, Reka AK, et al. Temporal quantitative proteomics by iTRAQ 2D-LC–MS/MS and corresponding mRNA expression analysis identify post-transcriptional modulation of actin-cytoskeleton regulators during TGF-beta-Induced epithelial–mesenchymal transition. *J Proteome Res* 2009;**8**:35–47.
62. Sreekumar A, Poisson LM, Rajendiran TM, Khan AP, Cao Q, Yu J, et al. Metabolomic profiles delineate potential role for sarcosine in prostate cancer progression. *Nature* 2009;**457**:910–4.
63. Beecher C. Cancer metabolomics. In: Ruddon R, editor. *Introduction to the molecular biology of cancer.* Chennai: Elsevier; 2010.
64. Vellaichamy A, Sreekumar A, Strahler JR, Rajendiran T, Yu J, Varambally S, et al. Proteomic interrogation of androgen action in prostate cancer cells reveals roles of aminoacyl tRNA synthetases. *PLoS ONE* 2009;**4**:e7075.
65. Mungrue IN, Pagnon J, Kohannim O, Gargalovic PS, Lusis AJ. CHAC1/MGC4504 is a novel proapoptotic component of the unfolded protein response, downstream of the ATF4-ATF3-CHOP cascade. *J Immunol* 2009;**182**:466–76.
66. Neumann B, Walter T, Heriche J-K, Bulkescher J, Erfle H, Conrad C, et al. Phenotypic profiling of the human genome by time-lapse microscopy reveals cell division genes. *Nature* 2010;**464**:721–7.
67. Beer DG, Kardia SL, Huang CC, Giordano TJ, Levin AM, Misek DE, et al. Gene expression profiles project survival of patients with lung adenocarcinoma. *Nat Med* 2002;**8**:816–24.
68. Keshamouni V, Arenberg D, Kalenkerian G. *Lung cancer metalasis: Novel Biological Mechanisms and Impact on Clinical Practice.* Philadelphia: Springer; 2010.
69. Hernandez-Boussard T, Whirl-Carrillo M, Hebert JM, Gong L, Owen R, Gong M, Gor W, Liu F, Truong C, Whaley R, Woon M, Zhou T, Altman RB, Klein TE. The pharmacogenetics and pharmacogenomics knowledge base: accentuating the knowledge. *Nucl. Acids Res.* 2008;**36**:D913–918.
70. Hansen NT, Brunak S, Altman RB. Generating genome-scale candidate gene lists for pharmacogenomics. *Clin Pharmacol Ther* 2009;**86**:183–9.

Progress in Cancer Nanotechnology

ISTVAN J. MAJOROS,* BRENT B. WARD,† KYUNG-HOON LEE,‡ SEOK KI CHOI,* BAOHUA HUANG,* ANDRZEJ MYC,* AND JAMES R. BAKER*

*Michigan Nanotechnology Institute for Medicine and Biological Sciences, University of Michigan, Ann Arbor, Michigan, USA

†Oral and Maxillofacial Surgery, University of Michigan Hospitals, Ann Arbor, Michigan, USA

‡Michigan Nanotechnology Institute for Medicine and Biological Sciences and Department of Chemistry, University of Michigan, Ann Arbor, Michigan, USA

Progress in Molecular Biology
and Translational Science, Vol. 95
DOI: 10.1016/S1877-1173(10)95008-4

I. Introduction and Historical Perspective

The now famous statement "there's plenty of room at the bottom" spoken by Richard Feynman in 1959 at the American Physical Society provided the basis for the concept of nanotechnology. Feynman suggested fanciful ideas such as mechanical factories of tiny robots that would lead to the creation of billions of tiny factories to make nanometer-scale materials. He gave the example of a reduction in print font size that might permit the Encyclopedia Britannica to fit on the head of a pin. While the concept of "nanobots" remains science fiction, great advances in the engineering of small materials have even allowed for "printing" of Feynman's "pin-head" encyclopedia. More importantly, it inspired a field of endeavor now called "nanotechnology," a term coined by Professor Noro Taniguchi of the Tokyo University of Science in 1974.

A strict definition of nanotechnology refers to molecular devices smaller than 100 nm, which are therefore on the "nano" scale. One nanometer (nm) is one billionth or 10^{-9} of a meter. For size reference, the hemoglobin molecule is 5 nm, most viruses are 10–100 nm, and a human cell 10,000–20,000 nm (Fig. 1). Soon after mechanical and electrical approaches to nanotechnology development became feasible, biologists began to explore opportunities for advancement in living systems. "Biological Approaches and Novel Applications for Molecular Nanotechnology" was the first scientific conference on this topic and was held in 1996.

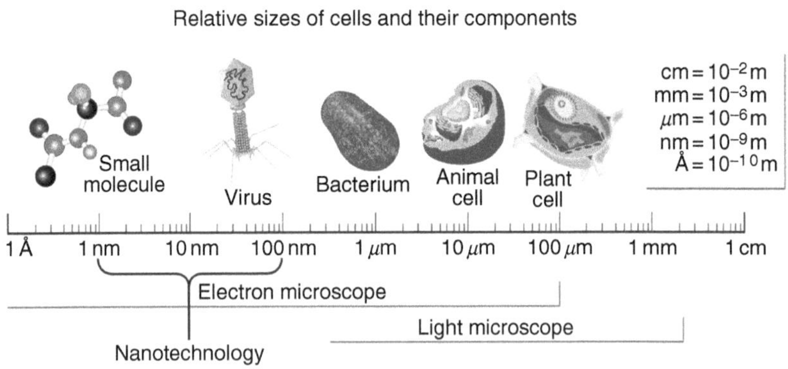

FIG. 1. Scale of nanotechnology in comparison to biological structures such as cells and their components.

Nanomedicine has now emerged as a specific focus in nanotechnology, and refers to medical interventions at the molecular scale for curing disease or repairing damaged tissues, such as bone, muscle, or nerve.[1] This field combines the expertise of individuals in medicine, mathematics, biology, engineering, chemistry, and computer science for the creation of devices for human application.

Proof of the intense interest in this technology is the creation of major funding at the national level, including eight centers of nanotechnology excellence and 12 cancer nanotechnology platform partnerships through the NCI Alliance for Nanotechnology in Cancer Program. Likewise, other institutes, including both public and private, are providing opportunities for research funding to advance our current knowledge in this field.

While many fields are now employing nanomedicine, the major focus remains in oncology. Potential applications involve the ability to specific targeted treatment of cancer, and also enhanced ability for the prevention, early diagnosis, and monitoring of oncologic disease progression through superior imaging techniques. According to the National Cancer Institute, "nanotechnology will serve as multifunctional tools that will not only be used with any number of diagnostic and therapeutic agents, but will change the very foundations of cancer diagnosis, treatment, and prevention."[1] The envisioned multifunctional "smart" nanodevice would contain the ability to (a) detect cancer at an early stage, (b) enhance imaging to pinpointing tumor location, (c) deliver drugs effective against the tumor, (d) and allow for monitoring of cancer cell eradication.

The National Cancer Institute's Cancer Nanotechnology Plan includes the following six major challenge areas thought to focus and maximize efforts in a variety of fields associated with these technological advances:

1. Prevention and Control of Disease
2. Early Detection and Proteomics
3. Imaging Diagnostics
4. Multifunctional Therapeutics
5. Quality of Life Enhancement in Cancer Care
6. Interdisciplinary Training

II. Targeted Therapy

The clearest example of the power nanotechnology provides for cancer therapy is embodied in the concept of targeted therapy. In this process, agents can be selectively delivered to cells as a "silver bullet" to kill only the diseased cells. The term "silver bullet" originates from folklore in which a silver bullet was the only type of bullet for firearms effective against a number of mystical foes.

In modern use for cancer therapy, it refers to an effort to create devices which specifically target disease cells while leaving the host otherwise untouched. This avoids the toxicity of traditional chemotherapy and also improves its efficacy.

A Medline search for "targeted therapy" or "targeted drug delivery" reveals the explosion of interest and discovery in this area. Two articles from 1902–1978 are available with the first mention attributed to the use of propanolol for essential hypertension by Lauro *et al.* In the 1980s, 53 articles were published with the concept of targeted approaches which blossomed in the 1990s to 261. Since that time, significant expansion in this field has occurred with 1429 articles for targeted approaches from 2000 to 2006. This increase in information for targeting is attributable to a number of factors, including our expanding knowledge of the cancer process, and technological advances which allow us to interact with cancer cells in a selective fashion.

Advances in our understanding of cancer have resulted in significant knowledge of the cellular and molecular changes leading to malignancy. Folds and colleagues elaborated the concept of tumor progression in the 1950s,[2] which was shortly followed by cytogenetic evidence of chromosomal changes associated with cancer in the 1960s. With the development of molecular techniques, the theory that tumorogenesis could result from a single altered cell evolved.[2] Ultimately, a multistep model, with acquisition of various cellular abnormalities at each stage, was proposed.[2] While not all of the steps involved have been completely understood, our understanding of the cancer cell and of the role of the surrounding network of tissues is increasing.[2] Markers of cell transformation and cancer have been identified in some cases, and approaches are being made to utilize these markers as targets in treatment strategies. In addition, other normal markers that are overexpressed in cancer cells offer another option for targeting for selective improvements in cancer treatment.

For effective targeting, some selective process must occur that enhances treatment of cancer cells while minimizing toxicity to normal tissues. While not all advances in targeted therapy are considered nanomedicine, there are many exciting prospects for future clinical application that employ nanomaterials to target cancer. In this chapter, our focus will be on current advances in nanotechnology for cancer diagnosis and therapy with an emphasis on those agents that provide opportunities for targeted cancer cell destruction.

III. Computer Simulations as an Approach to Develop Nanotechnology in Cancer

Nanoplatform approaches to cancer include drug delivery nanovectors for the administration of targeted therapeutic and imaging agents.[3] Nanotechnology has enhanced the development of a wide range of new drug delivery systems such as nanovectors, liposomal drug carriers, polymers, or polymer micelle-based delivery vehicles.[3–5] Nanovectors significantly improve the pharmacokinetics and biodistribution of the free drugs and reduce their side effects.[4,6,7] Nanovectors can be generated for novel and personalized therapeutic agents by combining them with preferred therapeutic and biological targeting moieties.[3,8,9]

Molecular modeling can be an important methodology to design and improve nanovectors, to understand the physicochemical properties of the nanodevices, and to provide predictive capabilities for the rational design of targeted drug delivery carriers and devices. It also provides nanoscale images at atomic resolution, predicts the nanoscale interactions of nanodevices with their biological environments, and evaluates strategies for redesigning biopolymers for bionanotechnological uses.[10] Owing to rapid development of the computational algorithms and methodology, and the availability of inexpensive yet powerful computational resources, biomolecular simulations are widely used in characterizing and designing nanoparticles (NPs) with useful functionalities, and this has markedly contributed to our understanding of these technologies.[11–13] Molecular dynamics (MD) simulations of cancer nanotechnology are one of the essential approaches to design efficient drug delivery nanodevices, to understand their physical properties and interactions with their biological environments.[13,14]

Herein, we will discuss several important drug delivery vehicles from the standpoint of the computational approach to designing cancer nanotechnology. Those nanodevices include liposomes, peptide-based nanotubes,[8] and dendron polymers.[15] To design nanovectors that provide optimal and effective functionality, one must understand their structure, dynamics, and interactions with biological environments. In this respect, computational methods are very helpful to design new and efficient drug and gene delivery systems predicting their required composition, geometry, and dimensions.

Multiscale simulations have shown consistency with experimental measurements and provided atomic-scale insights into structure and dynamical behavior of dendrimers and their interactions with other molecules such as biological membranes and polyelectrolytes.[16–18] MD simulations on dendrimers were carried out to understand the morphological and topological structures of these polymers, as a function of generation, by calculating moment of inertia.[19] This study demonstrated changes in morphology of the dendrimer structures with generation number.[19] The dendrimer generations 1–3 are asymmetric, while generations 5–7 are almost spherical. Simulations on the dendrimers were also performed to understand the effect of the pH,[20–23] salt,[22] charges,[24] and solvent[23,25] on their structural properties. Recent MD simulations on G4 PAMAM (poly(amidoamine)) dendrimers at various pH levels indicate how the PAMAM dendrimer could be used as drug delivery vehicles based upon pH-induced conformational changes from a "dense core" at high pH to a "dense shell" at low pH (Fig. 2).[21]

MD simulation studies were used to evaluate structural properties and biological function of dendrimer conjugated with folic acid (FA) by attaching different capping groups to increase selective binding with the receptor.[26] The primary amines on the surface of the dendrimer caused charged-based,

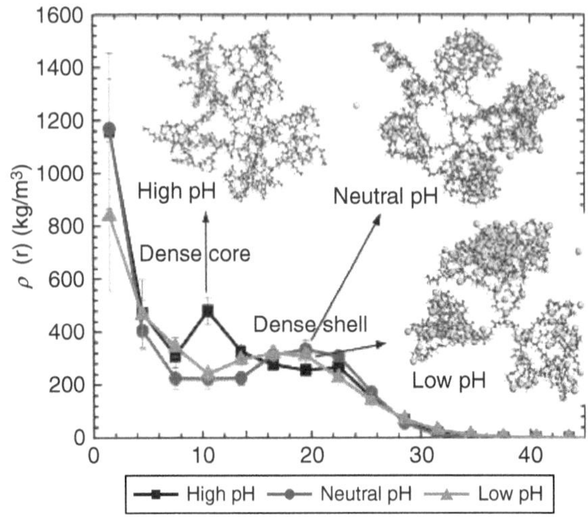

FIG. 2. Radial density distribution (from the center of mass) of G4-NH_2 PAMAM dendrimers at different pH levels (averaged over 200 ps). Snapshots are developed from MD simulations are shown in the inset.[24]

nonspecific interactions that diminished specific targeting.[26] Carboxy groups on the surface of the dendrimers produced local branch aggregation, whereas amide groups on the surface tended to result in an overall relaxation of the molecular structures.[26] MD simulations were also used to investigate the size and morphology changes of dendrimers and dendrimer-entrapped gold and silver PAMAM dendrimers, and to show that metal and inorganic NPs entrapped within a dendrimer interact with cells in a similar manner to the dendrimer without conjugating metals (Fig. 3).[27–29]

Interactions of dendrimers with the drugs, lipid bilayers, and proteins are important to understand for designing optimal targeted drug delivery molecules based on dendrimers.[30–32] Coarse-grained MD simulations were used to investigate dendrimer–lipid interactions, demonstrating that charged G7 PAMAM dendrimers form large holes in -α-dimyristoylphosphatidylcholine (DMPC) bilayers, while G5 dendrimers are capable of expanding smaller holes.[33] The interactions were also investigated by varying surface charge density and size of the dendrimer, showing that uncharged G3 and G5 dendrimers were not inserted into the bilayers, while the charged and partially charged G5 dendrimers were fully inserted into the bilayer.[34] The results of these simulations agree with qualitative experimental results and offer an insight into the bilayer-penetrating mechanism of dendrimers.[34] All-atom simulations may be more helpful to design and engineer newer classes of dendrimers optimized for biological functionality (Fig. 4).

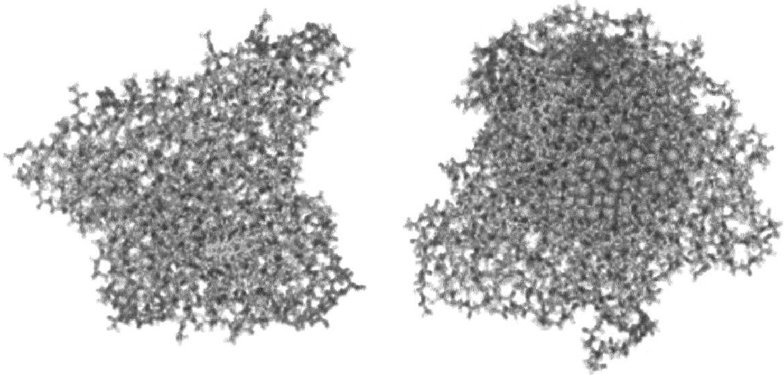

FIG. 3. Equilibrated configurations of G5.NHAc-FI-FA dendrimers (surface acetylated fluorescein thioisocyanate (FITC) and folic acid (FA) G5 PAMAM dendrimer) (left) and [$(Au_0)_{51.2}$-G5 NHAc-FI-FA] DENP (Dendrimer-Entrapped gold NanoParticle with similar modifications) (right) after 100 ps MD simulations. The yellow and pink moieties on both configurations represent FITC and FA molecules, respectively.[32] (For interpretation of the references to color in this figure legend, the reader is referred to the Web version of this chapter.)

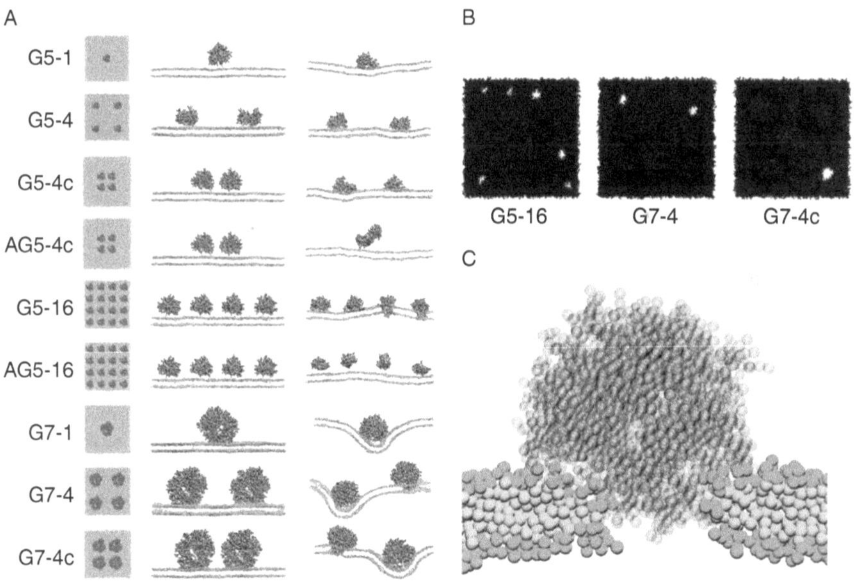

FIG. 4. (A) Top (left image) and side views (middle image) of the beginning (left panel) and end (right panel) of coarse-grained simulations of dendrimers with DMPC bilayers. Gray and green dots represent dendrimers and headgroups of the DMPC bilayer, respectively. The explicit water molecules, DMPC tails and ions are omitted for clarity. Note that side views show only one cross-section of the system and cannot capture all dendrimers. (B) Top view of the DMPC bilayer at the end of simulations G5-16, G7-4, and G7-4c. (G5-16 means generation 5 and separated 16 dendrimers and G7-4c means generation 7 and clustered four dendrimers.) (C) An image of the dendrimer-induced pore in a DMPC bilayer at the end of the simulation G7-4c. Transparent gray dots represent a G7 dendrimer. Green and yellow dots represent head and tail groups of DMPC, respectively.[34] (For interpretation of the references to color in this figure legend, the reader is referred to the Web version of this chapter.)

Computational methods have been applied to design and optimize liposomal delivery vehicles and to explore various thermodynamic and kinetic processes related to membrane binding and transport.[35] Computational simulations that predict lipid bilayer permeability can be applied to design liposomal delivery vehicles, since drug release rates are controlled by membrane transport.[35] MD simulations have predicted permeability coefficients of solutes across lipid bilayers and investigated a variety of permeation properties that are not easily obtained from experiments.[36] MD simulation studies have also demonstrated structural and dynamic aspects of gel phase lipid bilayers in membranes.[37–39]

Carbon nanotubes are also used as drug delivery vehicles and have be modeled with chemical receptors on the surface to target a particle site of the cell and release the contents.[40,41] Molecular modeling was employed to study the interaction of carbon nanotubes with cisplatin, a platinum-based chemotherapy anticancer drug.[42] Recent coarse-grained MD simulations indicate that the interaction between carbon nanotube and cell is dominated by van der Waals interactions and hydrophobic forces and that the interaction mechanism is affected by the size of the carbon nanotubes.[43] Thin carbon nanotubes directly pierce through cell membrane, whereas larger tubes enter the cell through a wrapping mechanism.[43] The simulation results were qualitatively consistent with the experimental observation that the carbon nanotube can enter animal cells.[41] Based upon the fact that living cells are covered with glycocalyx, there is a size limit of carbon nanotubes, which reach the surface of the player.[43] One should consider the optimal size window of the nanotubes designed as drug delivery carriers.[43]

Computer simulation has been well established to understand NP properties such as size, composition, morphology, and distribution of the atom density of the NPs. The information obtained from the simulations can help and improve the rational design of NP size, shape, or surface properties for biomedical nanotechnology applications in cancer. The computational approach employed for these nanovectors may be extended to other materials used as drug delivery vehicles.

IV. Nanomolecular Carriers for Drugs and Imaging Agents

A. Dendrimers

Dendritic macromolecules, also known as dendrimers, are uniform spherical nanostructures ranging from 10 to 200 Å in diameter. Dendrimers have been used as a backbone for the attachment of several types of biological materials in order to construct nanomolecules for both diagnostic and therapeutic applications (Fig. 5). Functional attachments have included iron oxide for targeted imaging, a phiphilux G1D2 sensor for monitoring apoptosis, drugs including methotrexate (MTX) and Taxol chemotherapeutics, and FA, RGD

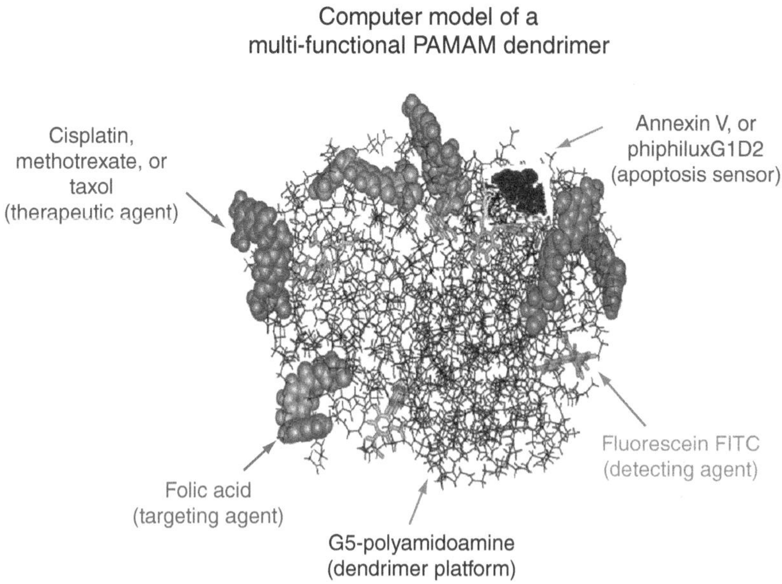

FIG. 5. Computer model of a multifunctional PAMAM dendrimer.

peptides and antibody fragments as targeting ligands. The exact structure, conjugation, and numbers of targeting molecules and drugs are crucial to the functioning of these molecules; thus the need for the precise synthesis and analysis tools of nanotechnology. Using a cell line that overexpresses the alpha folate receptor (FR), folate-targeted dendrimer chemotherapy has shown a 10–50-fold increase in therapeutic efficacy, whether analyzed *in vitro* or *in vivo* xenograft mouse tumor models.

An additional advantage of targeting *in vivo* has been a significant decrease in systemic toxicity compared to free drug.[26,44,45] Expanding on this technology, it will be possible to create a multitude of different drug conjugates whereby both the targeted agent and the chemotherapeutic agent utilized could be specifically modified, based on tumor receptor characteristics and pathophysiological properties. Additional dendrimers have been fabricated to target the epidermal growth factor receptor (EGFR), prostate-specific membrane antigen (PSMA), and $\alpha_v\beta_3$ integrin (RGD peptides). In use, a patient's tumor could be screened for cell surface receptor type in order to make decisions regarding which targeting agent would have the greatest potential for success.

The dendritic macromolecules are able to travel through the body due to their small, nanoscale sizes (3–10 nM), allowing delivery of attached drugs or other designed attachments. Dendrimers are compatible with the biological system given that their sizes, shapes, functionalities, and solubility are appropriate and biocompatible. Most widely used dendrimers are PAMAM ammonia or ethylene diamine core dendrimers with generations 3, 4, and 5, and closely resemble in size of antibodies, enzymes, or globular proteins, such as match in size and shape insulin (3.0 nM), cytochrome C (4.0 nM), and hemoglobin (5.5 nM), respectively.[46]

Each dendrimer carrier has an initiator core and arms made of the repeated monomer subunits that act as branching points. Generations or layers are synthesized in multiple reaction steps to add layers of monomers until the overall structure reaches the desired size. Dendrimers that are synthesized in this way from monomers also create a structure with multiple surface functionalities for conjugation of functional groups or drugs. Dendrimers have generated interest for drug delivery because of their uniformity and solubility, especially structures including PAMAM, modified poly(propyleneimine) (POPAM), and aromatic ether-type dendrimers.

As the number of layers of the dendrimer increases, the repetitive synthetic steps produce a highly branched macromolecule with a three-dimensional structure characterized by interior cavities capable of encasing therapeutic drugs or imaging units, and reactive terminal groups capable of attaching drugs, dyes, and/or a variety of sensors. The synthetic routes utilize molecular engineering techniques to produce mono- or multifunctional dendritic drug carriers with a strikingly precise structure; a necessity for attachment of various molecules for visualization, cancer cell detection/targeting, and malignant cell destruction.

The mathematical descriptions of PAMAM, POPAM, and POMAM (hybrid) dendrimers[47] can be used theoretically to determine the molecular weights and the number of terminal and tertiary amine groups for each of these types of dendrimers dependent on the generation.

General notation for the formula of the molecular weight for each generation is given by

$$\mathrm{MW} = \mathrm{MW}_{\mathrm{core}} + (\text{Molecular weight of monomers})F_{\mathrm{c}}\left(\frac{F_{\mathrm{r}}^{g+1} - 1}{F_{\mathrm{r}} - 1}\right) \qquad (1)$$

where F_{c} is the core functionality (EDA core functionality = 4, NH_3 core functionality = 3), F_r is the multiplicity of the repeating unit (MA + EDA's multiplicity is 2), and g is the generation number.

The number of terminal groups for a dendrimer can be determined through the use of the formula

$$\text{Number of terminal groups}, Z = F_c F_r^g \qquad (2)$$

The number of tertiary amines for a dendrimer can also be calculated utilizing the following equation:

$$\text{Number of tertiary amines} = T_c + F_c\left(\frac{2^g - 1}{2 - 1}\right) \quad (3)$$

where T_c is the number of tertiary amines of the core (EDA core has two tertiary amines).

Chemotherapeutic drugs, targeting ligands, that attach to overexpressed receptors present on the cancer cell surfaces, and fluorophores, for visualization of the location of the dendrimer within a system, are usually attached to the reactive terminal groups.

Certain terminal groups having the best combinations of the least steric hindrance, low hydrophobic interaction and/or hydrogen bonding, and other effects that facilitate the conjugation of chemical entities such as drugs, targeting unit, and other molecules.

B. Other Polymeric Carriers

1. Magnetic Polymeric Composites

Another NP approach to drug delivery involves polymer matrices combined with powdered drugs and magnetic beads. This potentially could deliver drugs by differing rates of diffusion through the polymer matrix in the presence of a magnetic field. Research by Langer and coworkers[46] has shown that drug diffusion is slow through matrix pores without the presence of a magnetic field, yet upon introduction of an external oscillating magnetic field, drug release is increased. The sizes of the holes of the matrix in this setting change in such a manner that larger amounts of drug are released within a shorter period of time. The study concluded that rates of drug release were approximately 30 times higher in the presence of a magnetic field, and that alterations in strength and frequency of the magnetic field allowed for precise control and modification of release rate.[46] Patient compliance might be an issue with carriers of this type, but biocompatibility[46] was apparently excellent for these devices.

2. Hydrophobic Polymers

Hydrophobic polymers like poly(ethylene–vinyl acetate) or poly(lactic–glycolic acid) copolymers are also capable polymer matrices for drug delivery devices. When they are dissolved in solvents and mixed with proteins or drugs ranging in size, the system exhibits slow release over a controlled number of days. The principle behind the idea of using hydrophobic polymers originates from the fact that hydrophobic–hydrophilic interactions with small molecules affect mobility of molecules; and copolymers create pores/channels with appropriate sizes and hydrophilic interaction strength to retain the drugs. By controlling implant geometry, the release can be altered to maintain a constant rate.[46] In order for a carrier to erode heterogeneously, it is best for it to be hydrophobic, but it also must possess water-sensitive linkages.[48] Poly(anhydrides) fall into this category, as their anhydride bonds are sensitive to water and cleavage.

Alteration in pore size is a key to achieving different release rates. By controlling the molecular weight, composition, size, and concentration of particles added measuring the diffusion by the polymer provides the ability to determine pore size and to control rate of release. The determination of channel size is important as this will control the rate of diffusion of the protein or drug in and out of the system.[46] Interestingly, in the absence of drugs, several poly(ethylene–vinyl acetate) polymer matrices had no pores with size suitable for drug release. However, when introducing a drug or protein into the matrix, pores were formed due to phase separation or incompatibility between the matrix and drug/protein.

Besides a poly(ethylene–vinyl acetate) matrix, the approaches using this concept involve poly(anhydride)-type polymers as delivery agents for proteins and low molecular weight chemotherapeutic drugs. A variety of synthetic routes exist to prepare poly(anhydride)-type matrices such as melt-condensation, ring-opening polymerization, interfacial condensation, dechlorynation, and the use of dehydrative coupling agents.[49]

A new concept, "polymer degradation," is introduced by the use of poly(anhydride) matrices. During this controlled degradation, the drug or protein is in the interior of the matrix. As the anhydride bond is sensitive to the presence of water, it reacts when placed in a biological situation and starts to dissolve or decompose, thus allowing for the release of the drug or the protein of the interior compartments.

3. Homopolymers, Cross-Linked Polymers

A number of investigators have been focusing on using a variety of homo- and cross-linked polymers for drug delivery. Most notable and interesting are the studies that deal with cross-linked hydrogels, polypyrrole (PPy), poly-*N*-(2-hydroxypropyl) methacrylamide (HPMA), poly(ethylene glycol) (PEG) conjugates, and synthetic cyclodextrin-based drug delivery systems. In other work, poly(lactic acid) (PLA) and poly(glycolic acid) (PGA) carriers have also been conjugated with Taxol for use as an anticancer therapeutic.[50] PLA and PGA have been coupled with different types of drugs, adenosine triphosphate (ATP), lysozyme, and growth factors. These systems are biocompatible, water soluble, stable at pH 7.4, and capable of releasing the carried substrates. Thus, there are a number of options with polymers for drug delivery.

C. Liposomes

Liposomes are spherical vesicles with membrane composed of a phospholipid and cholesterol bilayer usually with an aqueous solution at its core. They represent one of the first "nanotechnologies" to enter clinical medicine, although some forms are larger than the technical limits of nanotechnology. Described in 1961 by Sir Alec Bangham, they have been utilized for drug delivery due to their ability to sequester DNA or drugs that would normally not enter the intercellular compartment. However, when these materials are encased in a liposome, they can be delivered to cells through diffusion as well as receptor-mediated events. Liposomes have therefore been used to deliver a wide variety of therapeutics and imaging agents, including small molecule drugs, gene therapies, and antisense oligonucleotides.[51,52]

An additional current strategy using liposomes has been developed whereby targeted cationic liposomes deliver functioning tumor suppressor gene, p53, to a variety of tumors, including head and neck cancer, melanoma, and breast cancer.[53] A planned clinical trial using liposomes employing transferrin to target patients with Head and Neck cancer selectively delivers p53 gene to sensitize cancer cells, making them more susceptible to standard therapeutic regimens. Liposomes are being utilized in a large number of later stage phase II and III clinical trials for chemotherapy. The list of trials includes applications for solid tumors in nearly every site of the body as well as hematological malignancy.

D. Buckyballs

Named after Dr. R. Buckminster Fuller, buckyballs mimic the structure of his geodesic dome and were developed in 1985. They are molecules composed entirely of carbon–carbon bonds that are brought together chemically in order to form tubes, spheres, or ellipsoid shapes for a variety of applications. Water-soluble C_{60} derivatives have been demonstrated to cross cell membranes.[54] As a drug delivery agent for cancer, conjugations of buckyballs to antibodies and chemotherapy have been investigated. Multiple buckyballs may be attached to a single antibody, which could allow for targeted delivery of a number of different chemotherapeutic agents on a single antibody specific to tumor (Fig. 6).

Ashcroft *et al.* reported their ability to covalently attach ZME-018 proteins (murine anti-gp240 melanoma antibody) to C_{60}-paxitaxel conjugate in a targeted therapy model.[55] An alternative strategy recently reported, uses fullerene–peptide conjugates to form "Bucky amino acid" as delivery for intracellular drug delivery. Using cell lines, targeting could be specific to either the cytoplasm or the peri-nuclear space if peptides with nuclear localization sequences were coupled to the material.[56]

Using similar chemistry, single-walled carbon nanotubes (SWNT) have been used for delivery of small interfering RNA (siRNA) into tumor cells both *in vitro* and in mouse tumor models. In these models, siRNA was successfully delivered and served to silence the targeted genes producing tumor growth arrest in culture and suppressing growth in nude mice injected tumors.[57]

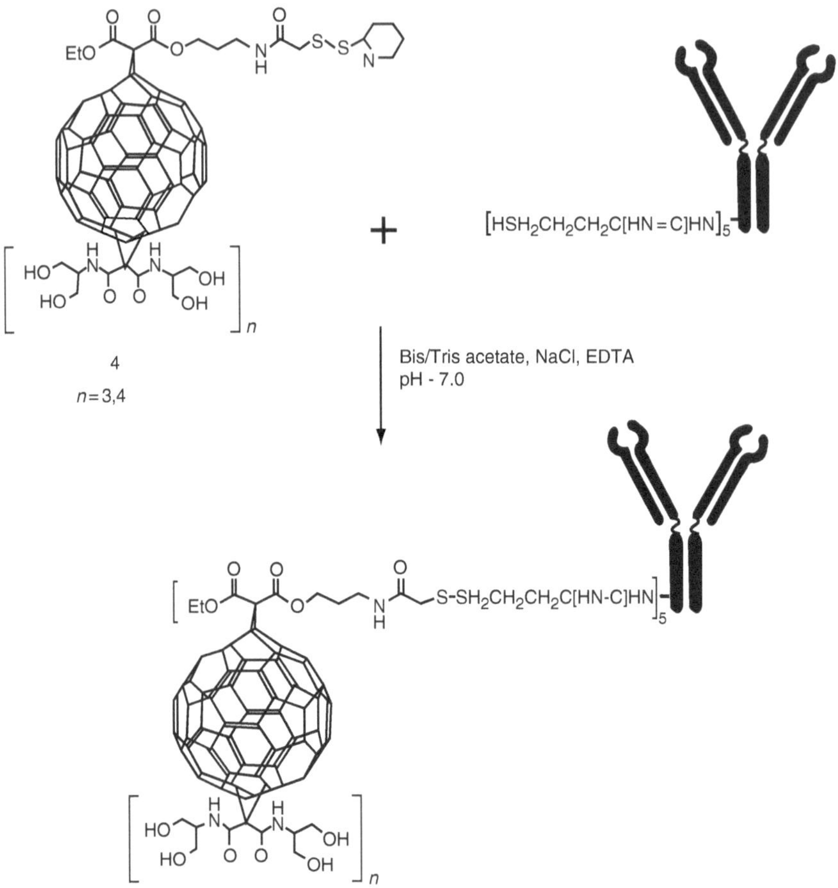

FIG. 6. Schema of buckyball conjugation to a single antibody.[59]

E. Quantum Dots

Quantum dots are nanosized crystals that emit variable wavelengths of visible light when exposed to ultraviolet light. Latex beads, filled with quantum dots, can be created to bind to specific sequences of DNA. The quantum dots can then be utilized to screen cells for specific DNA sequences, providing a spectral bar code for DNA identification. The use of DNA sequences specific for cancer mutations allows the beads only to bind in cells with the desired disease.[58] Quantum dots have also been conjugated to a variety of targeting agents, including transferring,[59] peptides,[60] and antibodies.[61] Recently, the ability to track *in vivo, single* quantum dots conjugated to monoclonal anti-her2 antibody in mice has been reported.[62] With continued development, potential clinical applications for quantum dot conjugates include imaging for primary disease as well as detection of sentinel or regional lymph node involvement, and distant metastatic disease.

F. Nanoshells

Nanoshells and nanorods are nanosized beads and rods with gold of varied thickness which can absorb specific wavelengths of light converting this energy to heat. These devices can be used for imaging and for selective cell destruction when activated with light from the near-infrared region. Nanoshells can be linked to antibody for specific delivery into cells with desired characteristics and have been shown in cell line models to offer selective tumor death while leaving normal cells unaffected. El-Sayed *et al.* demonstrated the ability to couple anti-epidermal growth factor receptor (anti-EGFR) monoclonal antibody to nanoshells for targeting to HOC 313 and HSC3 oral epithelial cell lines. Conjugated nanorods specifically bind to the overexpressing EGFR cells, resulting in strongly scattered red light on dark field microscopy which could distinguish these cells from nonmalignant counterparts. In addition, after exposure to 800 nm continuous red laser, malignant cells required only half the laser energy for photothermal destruction compared to controls.[63]

In preclinical models for tumor treatment in mice, 20 billion gold nanoshells have been injected IV and demonstrated preferential migration to tumors. Following photothermal treatment, 55% of mice receiving nanoshell therapy/ laser treatment survived to the end of the 35-day study with complete tumor regression and no regrowth after 90 days. Conversely, no mice in the control groups survived beyond 35 days.[64] This technology combines the possibility for imaging as well as the ability to offer treatment once tumor is bound.

V. Nanotechnology in Cancer-Targeted Delivery of Therapeutic Agents

A. Introduction

One of the primary goals in cancer-targeted drug delivery is to enhance the therapeutic index of an anticancer drug by facilitating the drug uptake by a target cell.[65–68] While many classes of anticancer chemotherapeutic drugs are available for the treatment of various cancers, almost all of the drugs show relatively low therapeutic indices ($TI = LD_{50}/ED_{50}$) primarily due to their nonselective uptake into cancer cells as well as normal healthy cells, and as a result, undesired adverse effects. Even with those drugs termed "molecularly targeted anticancer agents," including small molecule inhibitors that target selectively specific members of target receptor tyrosine kinases (e.g., imatinib, erlotinib),[69] each of these drugs is not able to selectively target malignant cells only. Compared to the molecularly targeted actions involved in small molecule-based anticancer therapies, the nanotechnology-based anticancer strategy aims to enhance the therapeutic benefits of a drug by enabling its selective delivery to a targeted cancer cell. Typically, this delivery strategy is achieved by a NP carrier functionalized with targeting ligands that are chosen to bind certain protein receptors on the cancer cell surface and to initiate the cellular uptake of the nanoconjugate.[66,67,70–72] This strategy has been successfully applied for the delivery of several anticancer therapeutics, including MTX,[71,73] paclitaxel,[65,70] doxorubicin,[74] and cisplatin.[67] The effectiveness of the cancer-targeted drug delivery can be determined by several factors pertinent to the design of such therapeutic NPs, including the shape and size of the NP carrier, the amounts of ligands presented, and the type of the linker used to attach drug molecules to the NP that determines the extent of drug release.[45,75–77] This review chapter aims to highlight key aspects involved in the NP design and the modes of action in cancer-targeting nanotechnology.

B. Targeting Ligands

Over more than a decade, the biomedical applications of drug-carrying NPs for anticancer delivery have been demonstrated by the use of targeting ligands specific for cell surface receptors overexpressed in cancer cells (Fig. 7). Those cancer-implicated receptors include folic acid receptor (FAR)[78,79] that recognizes FA, $\alpha_v\beta_3$ integrin[80–82] that recognizes RGD (and a cyclic form) as a binding motif, PSMA,[83] and transferrin receptor.[84] In addition, epidermal growth factor (EGF)-binding receptors are also targeted that include Her2,[85] and EGFR.[86,87] The design for the cellular uptake of such NPs requires the covalent attachment of specific targeting ligand molecules to the NP surface in order to achieve the selective adhesion of a NP to the targeted cell surface. In this design, each of the ligands is attached in multiple copies because such a multivalent format for ligand presentation is highly required during the receptor-mediated endocytosis in order to achieve tight NP-cell association.[88–91] Therefore, in a typical design for targeted NPs, each NP is covalently conjugated with multiple copies of a targeting ligand on its periphery, and further functionalized to carry therapeutic or imaging molecules as the payloads for cellular delivery.[44,45,92,93]

C. Anticancer Drugs and Therapeutic Nanoconjugates

Practical applicability of the cancer-targeting nanotechnology has been demonstrated by using a number of important anticancer chemotherapeutics, such as MTX,[71,73] cisplatin,[67] doxorubicin,[74] and paclitaxel,[65,70] where the anticancer efficacy of each of these drugs is compromised by its adverse effects (Fig. 8).

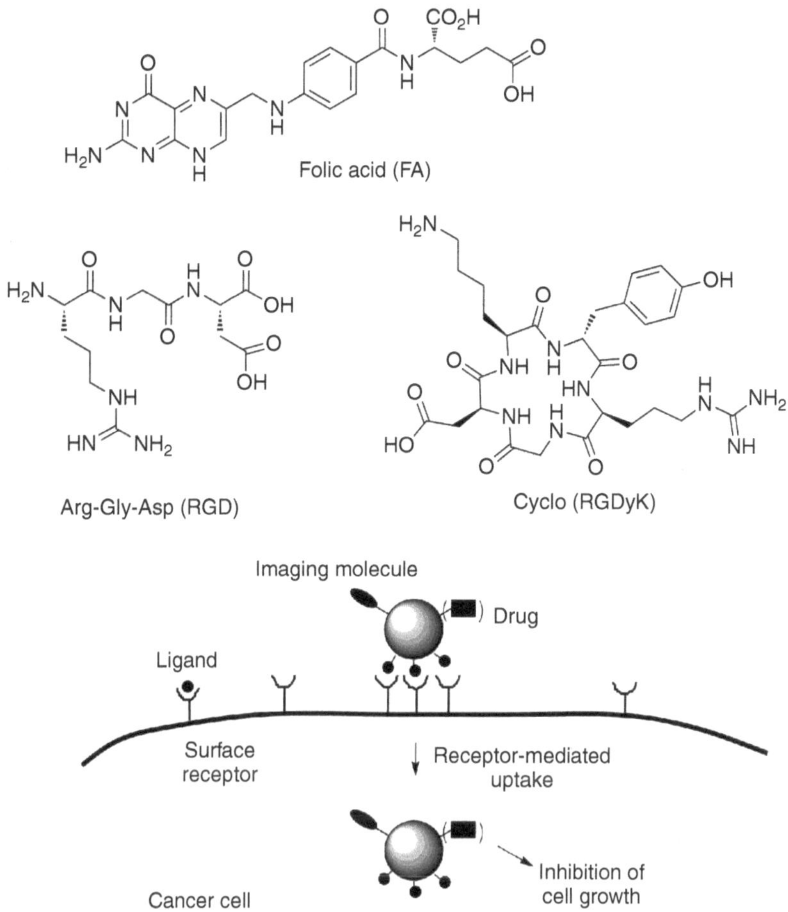

FIG. 7. Structures of selected targeting ligands, and a schematic for the mechanism of action by a cancer-targeting nanoconjugate.

MTX belongs to the class of antifolate cytotoxic drugs and acts by inhibiting dihydrofolate reductase (DHFR) ($K_i = 0.0034$ nM),[94,95] a cytosolic enzyme that catalyzes the reduction of dihydrofolate to tetrahydrofolate in the *de novo* biosynthesis of thymidine for DNA synthesis. Despite its potent activity against cancer cells, MTX has the dose-limiting systemic toxicity,[96] and therefore has been actively investigated for targeted delivery aiming to improve its therapeutic index. In this approach, MTX is delivered as a conjugated form after covalent attachment to a nanocarrier, such as dendrimers,[44,71,73,97–99] dextran,[100] oligopeptide,[101] albumin protein,[102] and iron oxide NPs,[103] through an amide[98–100,102–104] or a less stable

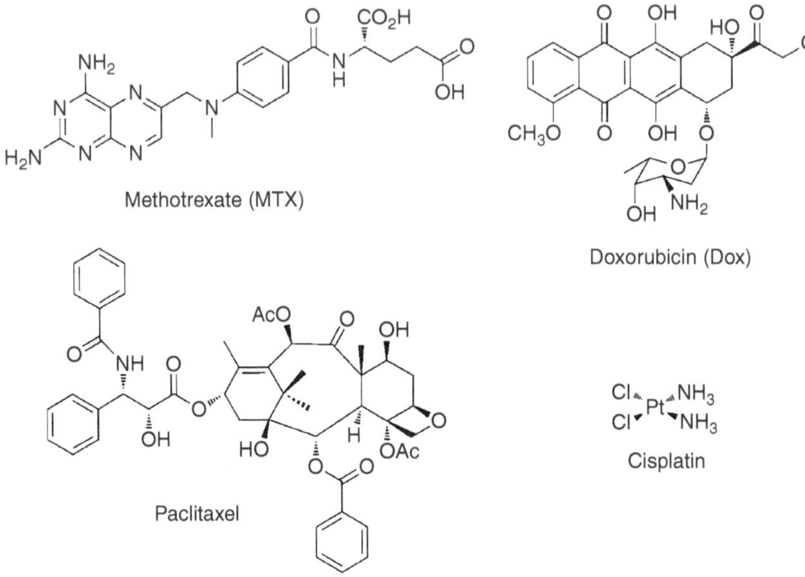

FIG. 8. Structures of selected anticancer therapeutic molecules.

ester bond.[44,71,73] In our recent study, we synthesized a nanoconjugate **1** (G5-FA-MTX; Fig. 9), which comprises FA, a high affinity ligand for the FAR ($K_d = 0.4$ nM),[105] and MTX, both attached to the surface of a generation 5 poly(amidoamine) (G5 PAMAM) dendrimer.[106] This conjugate **1** was selectively uptaken into a FR-upregulated KB cell through the mechanism that involves the receptor-mediated endocytosis and was able to subsequently inhibit cell growth by the activity of the drug payload.[71] Many other MTX-nanoconjugates are also reported to display potent cytotoxicity *in vitro*, of which certain of those tested in animal models also displayed selective tumor targeting,[44,98,99] prolonged systemic circulation,[44,98,100] and ultimately enhanced therapeutic index.[44] Paclitaxel belongs to the taxol class of anticancer drugs which works by inducing apoptosis.[107] The biological activity is attributed to its binding to microtubules within a dividing cell during mitosis, and as a result, the microtubule stabilization that leads to preventing cell division.[108] Though taxol displays potent cytotoxicity by the molecular mechanism at the microtubule site, this activity is not specific for cancer cells but also observed in normal cells. In a proof-of-principle study for targeted taxol delivery, we synthesized a multifunctional PAMAM dendrimer conjugated with paclitaxel, and FA as a FR-targeting ligand (**2**, G5-FA-Tx; Fig. 9).[70] In the cell studies performed *in vitro*, this targeted taxol conjugate showed potent and selective cytotoxicity to the KB

1 (G5-FA-MTX)

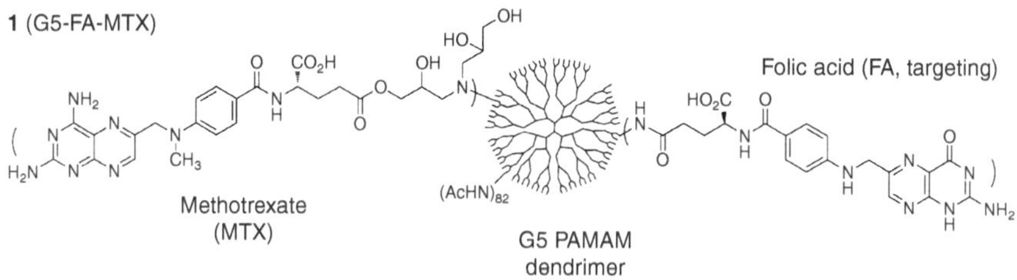

2 (G5-FA-Tx)

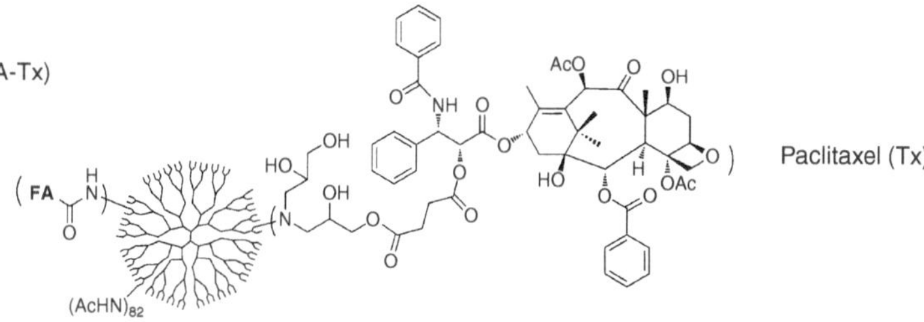

3 (G5-FA-Dox)

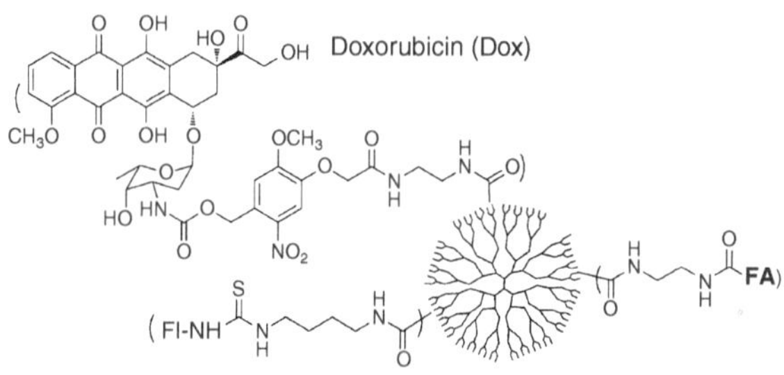

4 (SWCN-FA-Pt)

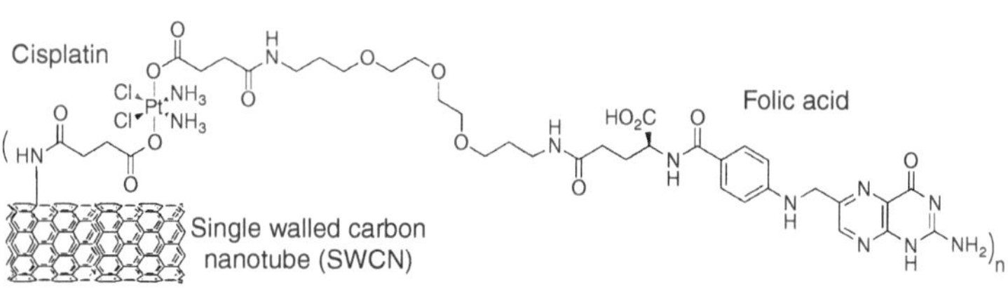

FIG. 9. Structures of selected therapeutic nanoconjugates.

cells with upregulated FR. In a similar principle, paclitaxel is delivered to cancer cells after conjugation to a nanocarrier that targets certain cancer surface markers which are aberrantly and abundantly expressed on many cancer cells. Such examples reported by other laboratories include those carrier-ligand pairs based on heparin-FA[109] glycan polymer-glucose ligand,[110] poly(lactide)-FA,[111] and poly(glutamic acid)/poly(lactide)-galactoside.[112]

Doxorubicin belongs to the class of anthracycline anticancer drugs that are closely related with daunomycin.[113] It has a potent anticancer activity which is caused by its inhibition of DNA replication after tight intercalation into dsDNA. However due to its lack of the cancer-specific activity and acute systemic toxicity, doxorubicin has been investigated as one of the therapeutic candidates for cancer-targeted delivery. We prepared a FR-targeting PAMAM dendrimer nanoconjugate that releases doxorubicin via a photochemical mechanism (**3**, G5-FX-Dox; Fig. 9).[114] This photocaged doxorubicin conjugate was able to selectively bind FAR overexpressing KB cells and to display high cytotoxicity following exposure to UV light which triggers the drug release. A number of other studies that describe cancer-targeted delivery of doxorubicin are reported from other laboratories. Selected examples of those studies are based on the use of nanoscale carrier-ligand systems, such as micellar NPs (from chitosan copolymer)–FA,[115] liposome-peptide ligand for neural cell adhesion molecule (NCAM),[116] liposome-RGD,[117] nanoscale graphene oxide-FA,[118] micelle-RGD,[119] poly(leucine–aspartic acid–valine), as both a targeting ligand and a drug carrier.[120]

Platinum-based anticancer agents have been also considered for cancer-targeted delivery by nanotechnology. Therapeutic issues pertinent to these low molecular weight platinum drugs include not only the lack of their cancer cell specificity but also unfavorable pharmacokinetics such as short blood circulation times that limit their ability for sustained tumor uptake and sufficient level of intracellular DNA cross-linking as the mechanism of drug action. A representative example from other laboratories is shown in Fig. 9 where cisplatin is attached to the SWNT as a longboat-shaped delivery system and where FA is also linked together as a targeting ligand.[121] This delivery system demonstrated its ability to target tumor cells that overexpress the FR on the surface and to deliver the Pt cargos into the cancer cells by endocytosis. The report noted that once inside the cell, cisplatin was formed by reductive release from the delivery system and caused DNA damage after diffusion into the nucleus. Another targeted delivery system applied for cisplatin was also demonstrated by PSMA-targeting NP carriers based on poly(D,L-lactic-*co*-glycolic acid)-PEG polymers.[122] In this system, cisplatin was encapsulated as the Pt(IV) species into a NP conjugated with PSMA-binding aptamers on the surface and delivered specifically to prostate cancer cells.

D. Mechanisms of Targeted Drug Action

Drug-carrying targeted nanoconjugates are designed such that as soon as the therapeutic agents are released, they become active. The release only happens under particular conditions. The release mechanisms currently being developed are based primarily on the cleavage reactions of the drug linker domain catalyzed by endogenous physiological factors such as reduction,[65] low pH,[67] and hydrolytic enzymes.[68] After cellular uptake, the therapeutic nanoconjugates are proposed to enter into subcellular acidic compartments such as endosomes (pH $\approx$ 5–6.5) where the drug molecules are released chemically or enzymatically by linker hydrolysis from its conjugates.[123,124]

Such drug release may also occur later in lysosomes which are more acidic and contain digestive peptidases.[125] Generally, in targeted drug delivery, free drug molecules are expected to account for the primary mechanism of drug action provided that a cellular condition is satisfied where the drug is released extensively.

However dependent on drug classes, certain of the drugs delivered are still active as the form tethered to the NP carrier, and such tethered drugs can make a significant contribution toward the cytotoxicity. In the study intended to understand its mechanism(s) of action of drug after cellular entry, we studied the chemical stability of the MTX ester linkage of **1** under the conditions that mimic the acidic environment of FR-containing endosomes.[126] Interestingly, this MTX conjugate was resistant to the hydrolytic release of the drug under an acidic or esterase-mediated cell-free condition. In a separate cell-free enzyme assay, the MTX conjugate was able to inhibit purified DHFR nearly as potent as free MTX.[126] These observations serve the evidence that the active species responsible for cell growth inhibition might comprise MTX either released as well as still conjugated.

In other drug classes such as doxorubicin, and platinum agents that target dsDNA in the nucleus, each of the drugs must be released from the carrier system to be active. Figure 10 shows selected examples of drug release mechanisms where the linker is cleaved by acidic conditions, reduction, or esterases in cytosol or plasma. Examples for the acidic labile linkers are based on the hydrazone linkage to doxorubicin, or the ester bond to pacliatxel.[70,127] Cisplatin is derivatized to Pt(IV) species that allows to incorporate a linker structure, and released as the active Pt(II) species upon reductive cleavage following the delivery of the NP. Beyond the passive control mechanism of drug release catalyzed by endogenous physiological factors as described, recently we reported a photochemical-based active control mechanism for drug release after targeted cellular uptake. In this scenario, the folate-targeted doxorubicin conjugate **3** (Figs. 9 and 10) is specifically taken up by the FR-abundant tumor cell and following the irradiation by UV light, the drug is specifically released inside the cancer cell (Table I).

E. Summary

Nanotechnology is uniquely suited to provide multifunctional platforms for the cancer-targeted delivery of therapeutic agents and imaging molecules. Applications of such multifunctional NPs have been well demonstrated by targeting

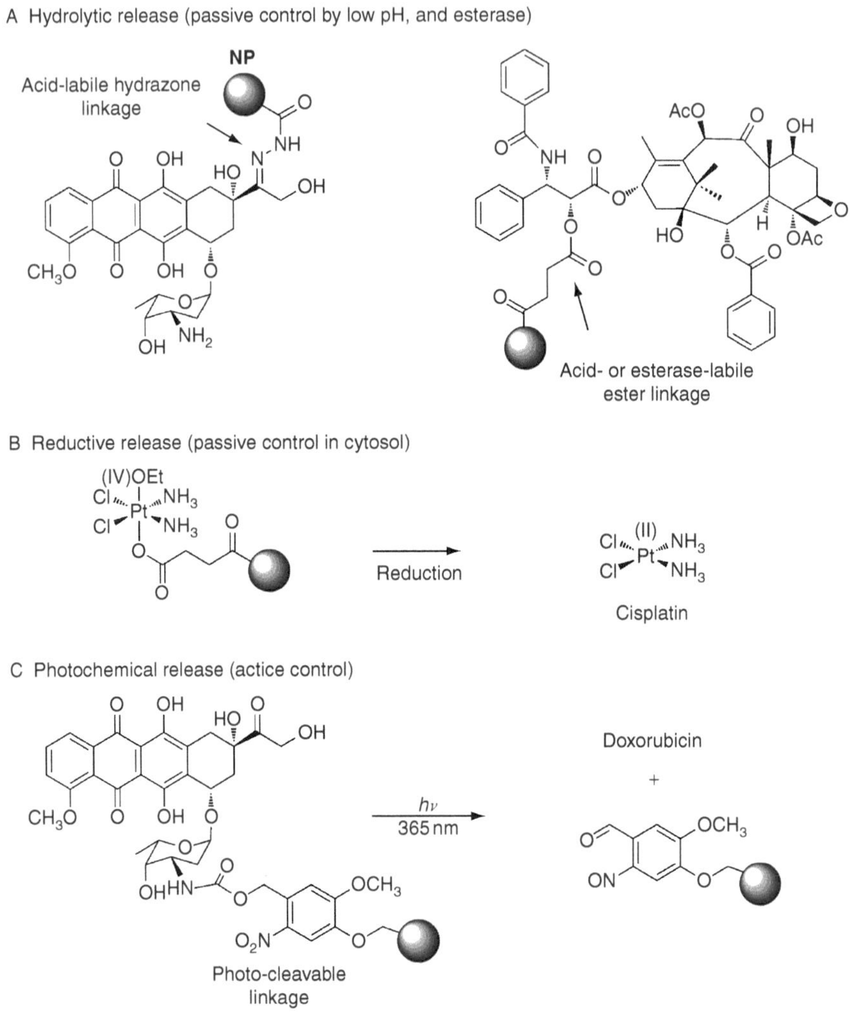

FIG. 10. Mechanisms of drug release in therapeutic drug delivery.

cancer-implicated surface molecules such as FAR, $\alpha_v\beta_3$ integrin, and PSMA. The effectiveness of the targeted approach depends on the specific design features that relate to nanometer-sized carriers, the targeting ligand specific for a certain receptor molecule on the cancer cell, the potency of the payloads like small molecule drug molecules, and the mechanisms for drug release.

TABLE I

NANOCONJUGATES FOR TARGETED DELIVERY OF ANTICANCER THERAPEUTIC MOLECULES

Nanocarrier	Targeting molecule (ligand)	Therapeutic molecule (drug)	Therapeutic target or action	Drug release mechanism
Dendrimer	Folic acid	Methotrexate	Dihydrofolate reductase	Encapsulation[117]; linker hydrolysis; conjugated[44,71,73,118,128]
Dendrimer	Anti-Her2 antibody	Methotrexate		Conjugated[86,94,99]
Dendrimer	Folic acid	Paclitaxel	β-Tubulin (microtubule)	Linker hydrolysis[70]
Heparin	Folic acid	Paclitaxel		Linker hydrolysis[103]
Poly(lactide)	Folic acid	Paclitaxel		Linker hydrolysis[104]
Glycan polymer	Glucose	Paclitaxel		Linker hydrolysis[102]
Dendrimer	Folic acid	Doxorubicin	DNA intercalation; topoisomerase II	Linker hydrolysis[119]; linker reduction; photochemical,[120,129,130] encapsulation[121]
Chitosan copolymer	Folic acid	Doxorubicin		Linker hydrolysis[71]
Liposome	Arg-Gly-Asp	Doxorubicin		Encapsulation[108]
Liposome	Neural cell adhesion molecule-targeting ligand	Doxorubicin		Encapsulation[107]
HPMA copolymer	Immunoglobulin	Doxorubicin		Linker hydrolysis[73]
HMPA copolymer	Hyaluronan	Doxorubicin		Enzymatic linker cleavage[122]
PEG Micelle		Doxorubicin		Micelle reduction[123]
Liposome	Folic acid	Doxorubicin		Encapsultion[79,125,131,132]
SWCN	Folic acid	Cisplatin		Encapsulation[111]
PLGA-PEG	Aptamer	Cisplatin	DNA crosslink	Linker hydrolysis[112]

VI. Targeted Imaging

Early detection and treatment of cancer will lead to more efficient treatment and improved survival. Noninvasive *in vivo* imaging tools have been rapidly increasing and these are important to cancer detection as technology progresses. These tools can be used to identify and monitor tumor growth, as well as identify cancer-related physiological events in tumors. Targeted cancer imaging has seen intense interest recently because the specificity of these methods could aid in distinguishing cancer from surrounding normal tissue, leading to the better detection and delineation of tumor. Clinical applications of positron emission tomograph (PET) and nuclear scintillation scanning and radiographic imaging rely on labeled probes, while the majority of other imaging techniques, including ultrasound and magnetic resonance imaging (MRI), rely on energy interacting with tissues in the body.[133] All of these techniques, however, have clinical uses that can be improved by NP-delivered probes.

Tumor-specific imaging techniques have used metal NPs or NP nanocomposites as contrast agents. Targeted NPs for imaging, in a manner similar to drug targeting, are used to enhance the efficacy of imaging and, in general, rely on two kinds of targeting mechanisms: passive and active targeting. Passive targeting often involves long-circulating NPs that exploit structural abnormalities in the vasculature of tumors and inflammatory and infectious sites. This phenomenon, known as the enhance permeability and retention (EPR) effect,[134,135] is based on the concept that these tissues possess "leaky" vasculature that allows macromolecules and NPs to extravasate and accumulate more readily. While this approach has been used extensively in research, its clinical utility appears limited. The more promising approach toward increasing the local accumulation of NPs in diseased tissue is active or specific targeting. Again, like drug targeting, this involves conjugation of targeting molecules that possess a high affinity for unique molecular signatures found on malignant cells.[136] Often augmented by the EPR effect, these receptor–ligand or antigen–antibody interactions provide an effective strategy to improve their residence time in malignant tissues such as tumors. Targeting ligands, such as proteins,[137] peptides,[138] aptamers,[139–141] and small molecules,[142] have been investigated to increase the site-specific accumulation of NPs. In some cases, specific binding can also facilitate internalization of the NP by receptor-mediated endocytosis. Because of the many surface functional groups, many NPs have been used for this approach, but dendrimers are very suitable to serve as multifunctional platforms for combined targeted drug delivery and imaging. Antibodies, angiogenesis markers, short peptides, and small molecules are used as targeting groups for targeted imaging of cancer. Table II shows ligands that are commonly used for targeted imaging of cancer when the imaging is based on the use of dendrimer.

TABLE II
TARGETED IMAGING OF CANCER USING DENDRIMER PLATFORM

Ligand	Targeting	Application	Reference
Folic acid	Folate receptor	MRI, optical, *in vivo* and *in vitro*	136–139
RGD peptide	$\alpha_v\beta_3$ Integrin	MRI, optical, *in vitro*	140,141
EGF	EGFR	Optical, *in vitro*	142
Cell-penetrating peptide	MMP-2 and -9 protease	MR and optical, *in vivo*	143
LHRH peptide	LHRH receptors	N/A	144
J591 antibody	PSMA	Optical, *in vitro*	145
MORF oligomer		Radioactivity *in vitro*	146
ClPhIQ acid	TSPO (PBR)	Optical, *in vitro*	147
Aptamer	Tenascin-C	Optical, *in vitro*	148

The FR is one of the most studied targets for imaging as it is overexpressed in several human tumors. Conjugates of FA linked via its gamma-carboxyl have the ability to target cancer cells through folate-receptor-mediated endocytosis.[143] Because of this, FA is the most common targeting ligand for the purposes of cancer imaging. Generally, FA is conjugated with the carrier surface through covalent bonds, then a chelator (for MRI) or a dye (for optical imaging) is conjugated to the same carrier. Compared with targeted small imaging molecules, dendrimer conjugates can enhance the system circulation time up to 24–72 h. The work of Phillip Low and his colleagues pioneered the use of folate targeting for nuclear medicine imaging of tumors.

One of the first examples of NP FA-targeted, gadolinium-loaded G5 PAMAM dendrimer NPs for tumor-specific MR contrast enhancement showed a prolonged postinjection imaging of tumors.[144] G5 PAMAM dendrimer was conjugated with 4.5 FA molecules. DOTA-NCS was then conjugated to the primary amine functional groups, followed by the loading of $GdCl_3$. A control compound without FA was also synthesized. The results demonstrated that the Gd(III)-DOTA-G5-FA contrast NPs specifically bind to xenograft tumors established with human epithelial cancer cells (KB) which overexpress FRs and that the NPs generate a statistically significant signal enhancement in tumors when compared to the signal enhancement generated by the nontargeted Gd(III)-DOTA-G5 contrast agent. *In vivo* 3D MR data also showed that the targeted NP enhanced the signal intensity in tumors in a statistically significant manner when compared with nontargeted NPs.

Targeted gadolinium compounds for MRI imaging have relied on the development of nanoplatforms that can carry a high payload of gadolinium and can enhance the longitudinal relaxivities (R_1) per gadolinium. As the R_1 value for chelated gadolinium is typically only between 5 and 30 mM^{-1} s^{-1} when attached to these nanoparticulate carriers, these contrast agents clearly benefited most from their ability to carry a high gadolinium payload. Rather than loading chelators on the surface of the NPs, Chen *et al.* demonstrated that T_1 relaxation times were enhanced in clusters of PAMAM dendrimer loaded with Gd-DTPA throughout the cluster structure targeted with FA (Scheme 1).[145] The overall R_1 of the whole nanocluster is estimated to be 3.6×10^6 mM^{-1} s^{-1} because of the extremely high payload of Gd in the structure.[145] *In vivo* study with mice with KB cell xenografts showed that these dendrimer nanoclusters exhibited a statistically significant improvement in image contrast compared with targeted dendrimers. Nanoclusters were fabricated by cross-linking PAMAM fifth-generation dendrimers, using a bi-functional amine-reactive crosslinker.[1] Following DNC formation, paramagnetic Gd3+ ions were conjugated to DNCs by DTPA.[2,146] The resulting paramagnetic DNCs were further functionalized with the tumor-targeting ligand FA[147] and the fluorescence dye FITC.[72]

Other types of molecules have also been shown to be functional for targeting imaging using dendrimers. A protease-activated, cell-penetrating peptide was used as the targeting moiety to target Gd-DOTA-loaded PAMAM dendrimer for *in vivo* fluorescence and MR imaging.[148] A range of 15–30 Gd-DOTA molecules and Cy5 dye were loaded onto one G5 PAMAM dendrimer which was also conjugated with targeting peptides. *In vivo* imaging results showed the visualization of protease by T_1-weighted MRI on animals bearing HT-1080 xenografts for 48 h. The dual-labeled, targeted dendrimer model was also used to enhance the edge of the tumors.[148] In addition, RGD peptide,[81,149] EGF,[87]

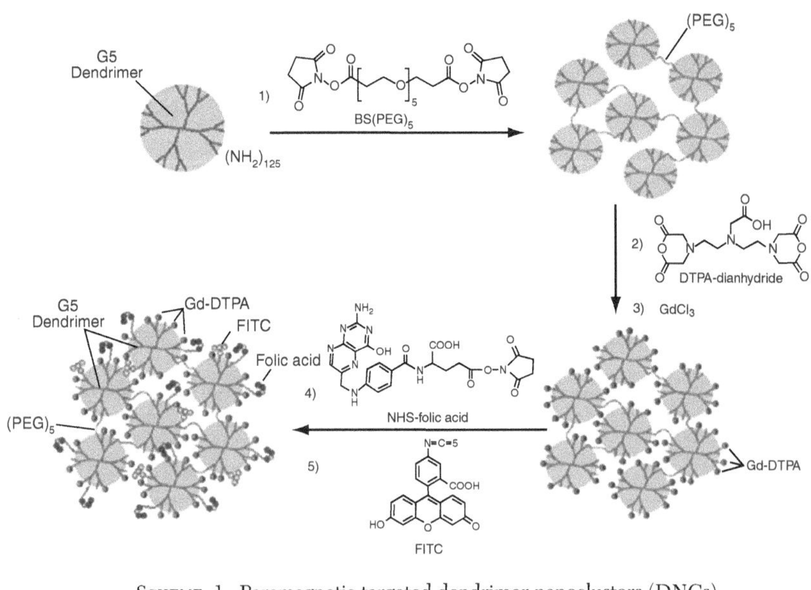

SCHEME. 1. Paramagnetic targeted dendrimer nanoclusters (DNCs).

and J591 antibody[150] have also been conjugated to PAMAM dendrimer with fluorescent labels. *In vitro* confocal microscopy results all showed high uptake by cells overexpressing the corresponding receptors.

In summary, active targeting of imaging agents with NPs offers an approach to improve the identification and delineation of early-stage tumors.

VII. Apoptosis Sensors

A. Introduction

Kerr *et al.*[151] originally described two forms of cell death, necrosis and apoptosis, which may occur in the absence of pathological manifestations. The term apoptosis is used to describe a process in which a cell actively participates in its own destruction. Apoptosis, or programmed cell death (PCD), is a normal physiological process which occurs during embryonic development as well as in maintenance of tissue homeostasis. Additional events of apoptosis include the condensation of the cytoplasm and nucleus due to lysosomal rupture and cell shrinkage, degradation of cellular proteins, membrane blebbing, condensation of nuclear chromatin, and internucleosomal cleavage of DNA. Cells that have undergone shrinkage show an increased presence of caspase-3-like activity and have fragmented DNA.[152]

The overabundance or deficiency of apoptosis in cells can lead to recognizable disease. Many illnesses can be commonly associated with a defect in apoptosis. Alzheimer's disease and stroke can be attributed to an excess of apoptosis. Cancer and autoimmune diseases, however, can be attributed to a lack of apoptosis.[153]

While the process of apoptosis is potentially reversible in its early stages, once caspase activity has begun, the process becomes irreversible. Caspases, a structurally related group of cysteine aspartate-specific proteases, cleave peptide bonds following specific recognition sequences. They play a central role in apoptosis of vertebrate cells. Early apoptotic events consist of changes in the plasma membrane, including increased permeability, a loss of membrane symmetry, and construction of membrane-bound apoptotic bodies. Externalization of phosphatidyl serine from the inner to the outer plasma membrane leaflet also occurs (via caspase activation) and can be used as an indicator of the initialization of apoptosis.[152]

In the past several years, great emphasis has been placed on the mitochondria as important components to the apoptotic process.[153] There are at least three general mechanisms occurring in the apoptotic cascade which involve cell mitochondria: (i) disruption of the electron transport chain, which results in loss of cell metabolism and ATP production; (ii) release of caspase-activating proteins to the cytosol; and (iii) alteration of the redox potential.[154]

An alteration of cell metabolism, largely due to a decrease in mitochondrial respiration, is a key step toward cell death.[155] Disruptions of the mitochondria begin with a decrease in membrane potential. Research by Heiskanen *et al.*[156] supports the belief that a decrease in membrane potential does not occur partially in all mitochondria within a cell, but occurs fully (full depolarization) within a finite number of mitochondria within a cell. Depolarization leads to the mitochondrial permeability transition pore (PT pore) and allows for the passage of small molecules and ions, equilibrating the ion concentrations between the matrix and the intermembrane space of the mitochondria. This equilibrium causes destruction of the respiratory chain and rupturing of the mitochondrial outer membrane. The folded cristae of the inner membrane have a much larger surface area than the outer membrane and therefore, the expansion of the inner membrane causes the outer membrane to rupture, resulting in the release of cytochrome c and other caspase-activating proteins from the mitochondria into the cytosol.[154,156] The release of cytochrome c into the cytosol activates caspase-9, which cleaves procaspase-3 into its active form, caspase-3, which is largely responsible for the biochemical and morphological changes by which we so commonly characterize the apoptotic pathway. The final stage of apoptosis is characterized by cell fragmentation into "apoptotic bodies" which are rapidly eliminated by phagocytic cells without eliciting significant inflammatory damage to surrounding cells.

Apoptosis is important to the treatment of cancer, as most therapies induce cancer cell death by a variety of biological, chemical, and physical events, such as growth factor withdrawal, signal transduction after engagement of cell surface receptors ("death receptors"—Fas/Cd95/Apo-1, etc.), chemotherapeutic drugs, and radiation (UV or γ). Thus, one of the best methods for monitoring the response of tumors to therapy is by monitoring apoptosis. A variety of methods are used to detect whether or not the process of apoptosis is occurring. Single-dye detection using rhodamine derivatives as well as double-dye fluorescence resonance energy transfer (FRET) detection are both commonly used as apoptosis detectors.

B. Apoptosis Detection

Table III lists apoptotic events separated by location in the cell which are detectable or made visible by imaging methods currently in use. These events are most commonly detected by surface-staining dyes, flow cytometry, or by the other methods listed in italics. Apoptosis detection is often performed using single- or double-dye detection methods. They can be used as free dyes or may be coupled to a dendrimer or another delivery device and targeted to specific cells in order to determine whether apoptosis has occurred.

TABLE III
THE OCCURRENCE OF APOPTOTIC EVENTS MONITORED BY USING DIFFERENT DETECTION METHODS

DNA cleavage and nuclear events	Biochemical events	Mitochondrial events	Changes in surface morphology and composition
Segmentation in chromatin, nuclei	Caspase activity, FRET detection, rhodamine derivatives, various fluorophores	Permeability transition, detection by vital dyes	Time lapse characterization of surface morphology
Chromatin condensation	Detection of caspase cleavage products	Mitochondrial antigens	Phospholipid externalization, Annexin V binding
DNA cleavage *in situ* by TUNEL method, detection of DNA fragmentation, strand breaks	Transglutaminase activity	Cytochrome *c* release and alterations	Changes in membrane permeability, DAPI, Hoechst
Anti-single-stranded DNA antibody	PARP activity	Metabolic activity	[a]Dyes and methods used for visualization of these events are italicized
Hairpin oligos to detected double-stranded breaks	Death antigens		

C. Single-Dye Apoptosis Sensor

Single-dye detection methods are often the less complex choice for apoptosis sensing. There are a wide variety of techniques in practice that utilize single-dye detection. These include detection by coumarin-based dyes; rhodamine derivative dye; terminal deoxynucleotidyl transferase (TdT)-mediated end-labeling of DNA strand breaks; (TUNEL method); use of Annexin V for the detection of phosphatidyl serine on cell membranes, usually in combination with propidium iodide (PI); and utilization of DNA-binding dyes such as the Hoechst dye and DAPI. The staining is analyzed by using microscope or flow cytometry.[157]

Detection of caspase-3 activity is used to determine whether the apoptotic process is irreversible. Coumarin-based fluorogenic substrates such as Ac-DEVD-AFC and Z-DEVD-AMC are often used to detect caspase-3 activity in cellular lysates; however, more effort has been placed on utilization of rhodamine-based derivatives for caspase-3 detection, as coumarin-based fluorogenic substrates have low extinction coefficients. It has been reported by Liu *et al.*[158] that the $(Z\text{-}DEVD)_2$-Rh 110 exhibits a much higher turnover rate than the coumarin-based Ac-DEVD-AFC substrate and is at least 10-fold more sensitive than Z-DEVD-AMC under their assay conditions.[158] The products released from coumarin-based substrates also have short excitation and emission wavelengths, which limit assay sensitivity, and are cell impermeable; therefore, they only work with cellular extracts.[159] Residual fluorescence of the substrates as well as poor wavelength separation of their products due to cellular autofluorescence give way to high background interference in microscopic assays.[16,158]

Rhodamine derivatives, including rhodamine 110 and rhodamine 123, are often used to detect caspase activity as well as changes in mitochondrial membrane potential. There are many advantages to using rhodamine-based substrates over other dyes. They have longer excitation and emission wavelengths, which reduces interference from testing compounds; the uncleaved substrate does not fluoresce and therefore has a very low background signal; and lastly, rhodamine and its derivatives are also cell-permeable, which allows for flow cytometry to measure the amount of fluorescence emitted, and therefore clearly verify the occurrence of the apoptotic process.[160]

While it is reportedly difficult to couple Rhodamine 110 to the tetrapeptide (a chain of four amino acids, D-E-V-D, which is cleaved by caspase-3 after the second D) in order to form the fluorogenic substrates, sequential coupling of the amino acids to Rhodamine 110 has been the most suitable way of preparing the substrate.[16,158] Use of two DEVD blocking groups to prevent fluorescence of the substrate leads to the necessity of two hydrolysis reactions for cleavage of both the DEVD sites in $(Z\text{-}DEVD)_2$-Rh 110. This limits the linear dynamic range of the substrate, and the creation of a Rhodamine 110-based substrate with only one hydrolysable amide group (one DEVD site) would allow for more efficiency in the use of the Rhodamine 110 derivative as a fluorogenic substrate.

Another method used for apoptosis detection is known as TdT-mediated digoxigenin nick end labeling, that is, the "TUNEL" method. DNA fragmented by activated endonucleases into segments 180–200 bps in length are labeled by dUTP, which is later detected by light or fluorescence microscopy. This method is performed *in situ* for observation of apoptosis at the single cell level, where it is often difficult to distinguish between the occurrence of apoptosis or necrosis in a cell. While considered the standard for detection of DNA fragmentation, a hallmark of the apoptotic process, the TUNEL method often fails to distinguish between apoptosis and necrosis, as cells in late phases of necrosis undergo DNA damage, which would stain as a false positive for apoptosis. Another drawback of this method is that it cannot be used in live cells.[161]

An early marker of the apoptotic process is indicated by the externalization of phosphatidyl serine from the inner to the outer plasma membrane leaflet. The Ca^{2+}-dependent protein Annexin V is used to detect the presence of phosphatidyl serine on cell membranes. The externalization of phosphatidyl serine occurs via caspase activation; therefore, its presence, indicated by the binding of Annexin V to the phosphatidyl serine, acts as a good marker of the initialization of apoptosis.[157] Due to the morphological differences that characterize apoptosis and necrosis, the use of Annexin V in combination with PI staining makes it possible to differentiate between apoptotic and necrotic cells.[161] An apoptotic cell membrane is impermeable until the final stages of the death process, while necrotic cell membrane is permeable. In later stages of apoptosis, however, apoptotic cells may stain both annexin V and PI positive due to loss of membrane integrity; therefore, it is impossible to differentiate between apoptotic and necrotic cells during the late stage of apoptosis.[162]

Another method to visualize apoptosis can be achieved by the use of dyes that bind internally to DNA fragments produced during apoptosis. Hoechst dyes (33342, 33258) are particularly useful, as they are fluorescent dyes that permanently stain DNA for visualization of chromatin condensation, conformation changes, and nuclear fragmentation, which provides a simple method for detection of late events in the apoptotic process *in vivo*.[163]

Chromatin condensation, fragmentation, and caspase activity can also be analyzed by double staining with DAPI and PhiPhiLux-G_6D_2. PhiPhiLux-G_6D_2 is a fluorogenic substrate that is cleaved in a DEVD-dependent manner to produce rhodamine molecules, which fluoresce red, whereas DAPI stains nuclei (apoptotic or viable) blue.[164] PhiPhiLux, which is used to measure caspase activation, poorly penetrates the cell membrane and needs a longer incubation time.[159] Using this technique, it is impossible to measure apoptotic events shorter than the time of staining, which is approximately 1 h.

Green fluorescent protein (GFP) is a versatile protein that has the ability to be used in a variety of circumstances to evaluate apoptosis. GFP localization within cells provides the ability to differentiate between necrotic cells and cells undergoing apoptosis. GFP also allows monitoring of the morphological changes occurring within cells undergoing apoptosis and the "real time" monitoring of the apoptotic or necrotic processes without damage or alteration to the cell. In this situation, lower GFP fluorescence indicates a loss of cell viability.[165] GFP stains the same cells that PI and Annexin V staining can stain; however, it has been documented that by Strebel *et al.*[157] that the use of GFP requires fewer steps to achieve similar results.

Flow cytometry analysis can quantify apoptosis both in a single cell and in a population by detecting the loss and extent of DNA fragmentation, morphological changes in the cytoplasm, as well as changes in membrane permeability.[162] Apoptotic cells give lower forward and higher side scatter values than viable cells due to their smaller size and higher cytoplasm condensation. Various fluorogenic substrates as described earlier can be used in combination with this detection method in order to quantitatively determine whether apoptosis is taking place.

Thus, single-dye stains using targeted NPs can be used to detect apoptosis in cancer cells.

D. Double-Dye FRET Reagent-Based Apoptosis Sensor

FRET detection is used to examine structural and dynamic characteristics of biological molecules in aqueous solution.[166] Macromolecules to undergo examination are labeled with covalently linked donor and acceptor fluorophores. Under continuous laser illumination, the donor fluorophore transfers energy to the acceptor fluorophore, where it is reemitted as fluorescence.

Generally, there are two ways to measure the FRET effect with steady-state microscopy. The first method of detection of the FRET effect measures the decrease of donor fluorophore emission and the increase in acceptor fluorophore emission in a two-dye system. The second method of detection of the FRET effect is called acceptor photobleaching. In this method, the donor is excited, the acceptor is photobleached, and the donor emission is measured before and after acceptor photobleaching.[167] However, this method cannot be used *in vivo*.

In order for the FRET effect to occur, it is necessary that the fluorescence emission wavelength of the donor fluorophore molecule overlap the excitation wavelength of the acceptor molecule, and these two molecules are spatially separated by no more than 20–80 Å (Fig. 11).[168] It is assumed that due to the peptide conformation in aqueous solution, the donor and acceptor molecules are in close proximity to one another (10–100 Å), which allows for the energy transfer from the donor fluorophore to the acceptor fluorophore. The transfer of energy due to the FRET effect can be detected by the reduction of fluorescence from the donor fluorophore and an increase in the intensity of the fluorescent emissions from the acceptor fluorophore.[168] Donor fluorophore colors are restricted to blue or cyan, which have limited fluorescent capabilities and also undergo auto fluorescence, and these dyes are paired with green- or yellow-colored acceptor fluorophores. FRET has great importance in the field of apoptosis detection and has been widely implemented in the study of PCD. The detection of apoptosis in cells is made possible by the activation of caspase-3. The property of caspase-3, which cleaves certain cellular substrates during apoptosis, can be utilized to cleave the donor–receptor system, and FRET detection can be used to monitor apoptosis.

If the linker between the donor and acceptor fluorophores contains a cleavable substrate (i.e., amino acid sequence D-E-V-D), then caspase-3, which actively cleaves cellular substrates possessing the sequence D-E-V-D, can cleave the linker between the donor and acceptor fluorophores in the molecular probe.[169] The cleavage of the peptide by caspase-3 between valine and aspartic acid in the recognition sequence D-E-V-D results in the elimination of the FRET effect, as the donor and acceptor fluorophores are no longer joined. Flow cytometry is used to quantify the amount of fluorescence present. By observing the intensity shift between the emissions of the donor and acceptor fluorophores, it is possible to determine the change in the FRET effect as a function of the cleavage of the linker by the caspase-3.[169]

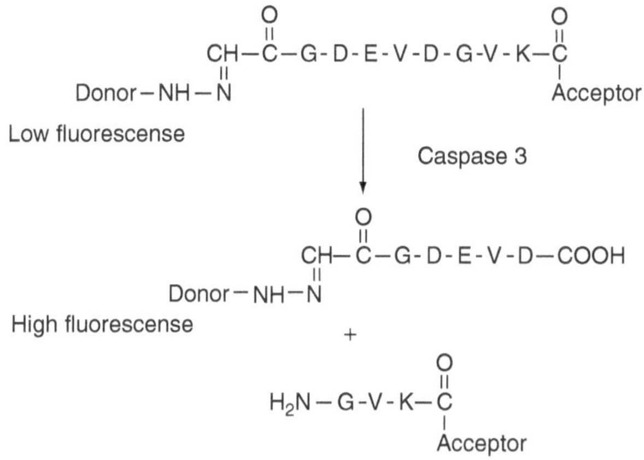

FIG. 11. Image of overlapping acceptor/donor spectra: J(λ) = spectral overlap integral of the absorption spectrum of the acceptor and fluorescence, emission spectrumof the donor $=\int e_A(l)F_D(l)\,l^4 dl$, $cm^3M^{-1}$where ε_A = extinction coefficient of acceptor; FD = fluorescence emission intensity of donor as a fraction of the total integrated intensity.

There is, however, one problem with the reliance on caspase-3 to cleave the D-E-V-D substrate within the linker between the donor and acceptor fluorophores. It is that the linker on the dendrimer or other carrier may not be fully accessible for cleavage by the enzyme for structural reasons, which can hinder the detection abilities of the FRET effect on apoptosis. The linker, for example, may not be cleaved due to orientation of the donor and acceptor fluorophores to one another as well as due to the distance between the two fluorophores.[169] This method, nonetheless, has shown to be able to detect apoptosis exceptionally well in living, intact cells.

Below we briefly describe the design and synthesis of a NP-based FRET-apoptosis device. Evaluation of the FRET effect necessitates the calculation of the transfer efficiency of the photons between the donor and acceptor. The number of photons emitted from the donor and acceptor are counted within a certain time frame, and the ratio of these numbers is called the energy-transfer efficiency. This ratio is dependent on the distance between the donor and acceptor fluorophores. The distance between the two fluorophores must be calculated in order to determine the dynamic and conformational changes of the macromolecule under consideration.[170] The efficiency of the energy transfer can be calculated using the following formula, $E(r)=(1+(r/R_0)^6)^{-1}$, where $E(r)$ is the efficiency of the Förster energy transfer between the fluorophore donor and acceptor, which is separated by a distance r, and where R_0 is the Förster radius, (i.e., the distance at which the efficiency is ½). The larger the efficiency $E(r)$ of the energy transfer, the shorter the distance (r) between the fluorophores. This formula is valid due to the angular averaging of the dipole–dipole interaction, which occurs due to the fast movement of the fluorophores. It must be noted, however, that the value r fluctuates due to the dynamics of the macromolecule to which it is attached, as the molecule under observation will change shape, undergo folding and unfolding of its structure, and various other conformational changes.[170] If the time window of observation is such that enough photons have been emitted for calculation and the distance between the donor and acceptor fluorophores does not change, then the efficiency can be turned into a useful calculation of distance r. It must also be noted that at a constant concentration of free and associated FRET pairs, the emission of the FRET donor is inversely proportional to the mole fraction of associated molecules.[168]

While use of double-dye (fluorophore) FRET is the most commonly used FRET detection technique, however the use of triple- and possible quadruple-dye detection has also been under investigation. Research by Liu and Lu[166] has shown that it is possible to label biomolecules that possess many arms with three or even four dyes for maximal detection of conformational changes as a result of fluorescent resonance energy transfer. A dual FRET sensors system excited with a single light has been also designed, and it has been proved that the system works. Niino *et al.*[171] constructed FRET sensors with Sapphire/RFP to combine with CFP/YFP and accomplished simultaneous imaging of cAMP and cGMP in a single cell. The requirements for the multiple dyes and the distances involved make NPs excellent platforms in which to develop these detectors. For example, the independent labeling of dendrimer arms using fluorophores of different colors will allow for determination of the spatiotemporal dynamics of various intracellular signals, including the apoptotic process. However, one problem encountered with using four dyes for detection is the possible nonoverlap of spectra between donor and acceptor fluorophores; this destroys the FRET effect, thus preventing the energy transfer between the fluorophores.[166]

E. Conclusion

Nanotechnology has emerged on the scientific forefront just within the past two decades, but within this short time, researchers utilizing this technology have already pioneered numerous advancements in a variety of fields as never before. Molecular engineering on the nanoscale has allowed scientists the capability of synthesizing and modifying both biological and synthetic molecules on the molecular level for use in animal as well as human systems to detect apoptosis. The advancement of technology is, however, always occurring at a rapid-fire place, so what has been reviewed and discussed here represents a mere fraction of what is currently going on in the scientific world. The material presented in this section must be assessed with this in mind. The application of molecular engineering has allowed for the creation of multifunctional dendritic nanobiomolecules, capable of delivering targeted therapeutic drugs while also having the ability to fluoresce and detect cell death. Synthesis of these uniform, nanoscale molecules provides for "biodevices" enable scientists to perform precise measurements and delivery drugs and other molecules to specific cells, especially cancer cells. Continued research and advancements within the fields of molecular engineering and cancer research will bring even more complex and exciting results in the near future.

VIII. Future Direction in Research and Technology

Advances in nanotechnology and targeted drug delivery will change treatment approaches in the near and long-term future. As our ability to understand the complex processes involved in the diseased cell increases, new opportunities for application of nanotechnology will emerge through the delineation of appropriate therapeutic targets. Development of new strategies and the expansion of currently technologies hold great promise in cancer and many other diseases. Nanotechnology offers the unique opportunity of developing

TABLE IV
Some Nanodevices in Cancer Diagnosis and Therapy

Device	Development stage	Cancer application	Current state
Dendrimer	*In vivo* (mouse)	ImagingDrug Delivery	1. 10–50-fold increase in chemotherapeutic efficacy with decreased toxicity.2. Diagnostic MRI enhancement3. Apoptosis sensing
Liposomes	Human clinical trials	Imaging Drug delivery	Clinical trials of liposomal encapsulated chemotherapeutics presently under study in a wide variety of malignancies.
Buckyballs/ Nanotubes	*In vivo* (mouse)	Diagnostic and therapeutic	Conjugation to antibodies and chemotherapy. Delivery of siRNA
Quantum dots	*In vivo* (mouse)	Imaging	Capable of imaging tumors with defined DNA sequences as well as lymphatic mapping in animal models.
Nanoshells	*In vivo* (mouse)	ImagingPhotothermal ablationDrug delivery	Preferential migration to tumor with imaging and treatment capability via photothermal ablation. Targeted drug delivery via antibody and chemotherapeutic conjugation

personalized therapeutics based on targeting and treating specific receptors and abnormalities of a patient's tumor. The limited number of strategies reviewed here has provided only a small number of examples in an ever-growing field of nanotechnology and targeted therapeutics for drug delivery. As predicted by the NCI, it is likely that nanotechnology will enhance all current aspects of cancer prevention, detection, and treatment (Table IV).

References

1. NIH Roadmap for Medical research.
2. NIH/NCI, U. S. D. o. h. a. h. s. Cancer nanotechnology: going small for big advances (2004).
3. Ferrari M. Cancer nanotechnology: opportunities and challenges. *Nature* 2005;**5**:161–71.
4. Pangburn TO, Petersen MA, Waybrant B, Adil MM, Kokkoli E. Peptide- and aptamer-functionalized nanovectors for targeted delivery of therapeutics. *J Biomed Eng* 2009;**131**:1–20.
5. Medina OP, Zhu Y, Kairemo K. Targeted liposomal drug delivery in cancer. *Curr Pharm Des* 2004;**10**:2981–9.
6. Orive G, Hernandez RM, Gascon AR, Pedrax JL. Micro and nano drug delivery systems in cancer therapy. *Cancer Ther* 2005;**3**:131–8.
7. Suri SS, Fenniri H, Singh B. Nanotechnology-based drug delivery systems. *J Occup Med Toxicol* 2007;**2**:1–6.
8. Hughes GA. Nanostructure-mediated drug delivery. *Nanomedicine* 2004;**1**:22–30.
9. Tian P. Molecular dynamics simulations of nanoparticles. *Annual Rep Prog Chem Sect C* 2008;**104**:142–64.
10. Lu D, Aksimentiev A, Shih AY, Cruz-Chu E, Freddolino PL, Arkhipov A, et al. The role of molecular modeling in bionanotechnology. *Phys Biol* 2006;**3**:S40–53.
11. Tian P. Molecular dynamics simulations of nanoparticles. *Annu Rep Prog Chem Sect C* 2008;**104**:142–64.
12. Haddish-Berhane N, Rickus JL, Haghighi K. The role of multiscale computational approaches for rational design of conventional and nanoparticle oral drug delivery systems. *Int J Nanomedicine* 2007;**2**:315–31.
13. Discher DE, Ortiz V, Srinivas G, Klein ML, Kim Y, Christian D, et al. Emerging applications of polymersomes in delivery: from molecular dynamics to shrinkage of tumors. *Prog Polym Sci* 2007;**32**:838–57.
14. Hanahan D, Weinberg RA. The hallmarks of cancer. *Cell* 2000;**100**:57–70.
15. Kandasamy SK, Lee H, Larson RG. *Computational simulations of dendrimers.* Chicago: Pan Stanford; 2009.
16. Giupponi G, Buzza MA, Adolf DB. Are polyelectrolyte dendrimers stimuli responsive? *Macromolecules* 2007;**40**:5959–65.
17. Nisato G, Ivkov R, Amis EJ. Size invariance of polyelectrolyte dendrimers. *Macromolecules* 2000;**33**:4172–6.
18. Ballauff M, Likos CN. Dendrimers in solution: Insight from theory and simulation. *Angew Chem Int Ed* 2004;**43**:2998–3020.
19. Naylor AM, Goddard III WA. Starburst dendrimers. 5. Molecular shape control. *J Am Chem Soc* 1989;**111**:2339–41.

20. Lee I, Athey BD, Wetzel AW, Meizner W, Baker Jr. JR. Structural molecular dynamics studies on polyamidoamine dendrimers for a therapeutic application: effects of pH and generation. *Macromolecules* 2002;**35**:4510–20.
21. Liu Y, Bryantsev VS, Diallo MS, Goddard III WA. PAMAM dendrimers undergo pH responsive conformational changes without swelling. *J Am Chem Soc* 2009;**131**:2798–9.
22. Welch P, Muthukumar M. Tuning the density profile of dendrite polyelectrolytes. *Macromolecules* 1998;**31**:5892–7.
23. Maiti PK, Cagin T, Lin S-T, Goddard III WA. Effect of solvent and pH on the structure of PAMAM dendrimers. *Macromolecules* 2005;**38**:979–91.
24. Blaak R, Lehmann S, Likos CN. Charge-induced conformational changes of dendrimers. *Macromolecules* 2008;**41**:4452–8.
25. Tanis I, Karatasos K. Molecular dynamics simulations of polyamidoamine dendrimers and their complexes with linear poly(ethylene oxide) at different pH conditions: static properties and hydrogen bonding. *Phys Chem Chem Phys* 2009;**11**:10017–28.
26. Quintana A, Raczka E, Piehler L, Lee I, Myc A, Majoros IJ, et al. Design and function of a dendrimer-based therapeutic nanodevice targeted to tumor cells through the folate receptor. *Pharm Res* 2002;**19**:1310–6.
27. Shi X, Wang S, Meshinchi S, Antwerp MEV, Bi X, Lee I, et al. Dendrimer-entrapped gold nanoparticles as a platform for cancer-cell targeting and imaging. *Small* 2007;**3**:1245–52.
28. Shi X, Lee I, Baker Jr. JR. Acetylation of dendrimer-entropped gold and silver nanoparticles. *J Mater Chem* 2008;**18**:586–93.
29. Shi X, Wang SH, Lee I, Shen M, Baker Jr. JR. Comparison of the interalization of targeted dendrimers and dendrimer-entrapped gold nanoparticles into cancer cells. *Biopolymers* 2009;**91**:936–42.
30. Kandasamy SK, Lee H, Larson RG. Computational simulations of dendrimers. In: Majoros IJ, Baker Jr. JR, editors. *Dendrimer-based nanomedicine*. Chicago: Pan Stanford; 2009.
31. Lee H, Larson RG. Coarse-grained molecular dynamics studies of the concentration and size dependence of fifth- and seventh-generation PAMAMA dendrimers on pore fomration in DMPC bilayer. *J. Phys Chem B* 2008;**112**:7778–84.
32. Kelly CV, Leroueil PR, Orr BG, Banaszak Holl MM, Andricioaei I. Poly(amidoamine) dendrimers on lipid bilayers. II: Effects of bilayer phase and dendrimer termination. *J Phys Chem B* 2008;**112**:9346–53.
33. Lee H, Larson RG. Coarse-grained molecular dynamics studies of the concentration and size dependence of fifth- and seventh-generation PAMAM dendrimers on pore formation in DMPC bilayer. *J Phys Chem B* 2008;**112**:7778–84.
34. Lee H, Larson RG. Molecular dynamics simulations of PAMAM dendrimer-induced pore for mation in DPPC bilayers with a coarse-graned model. *J Phys Chem B* 2006;**110**:18304–11.
35. Xiang T-X, Anderson BD. Liposomal drug transport: a molecular perspective form molecular dynamics simulations in lipid bilayers. *Adv Drug Deliv Rev* 2006;**58**:1357–78.
36. Marrink S-J, Berendsen HJC. Simulation of water transport through a lipid membrane. *J Phys Chem B* 1994;**98**:4155–68.
37. Tu K, Tobias DJ, Blasie JK, Klein ML. Molecular dynamics investigation of the structure of a fully hydrated gel-phase dipalmitoylphosphatidylcholine bilayer. *Biophys J* 1996;**70**:595–608.
38. Venable RM, Brooks BR, Pator RW. Molecular dynamics simulations of gel (LB1) phase lipid bilayers in conatant and constant surface area ensembles. *J. Phys Chem B* 2000;**112**:4822–32.
39. Snyder RG, Tu K, Klein ML, Mendelssohn R, Strauss HL, Sun W. Acyl chain conformation and packing in dipalmotoylphosphatidylcholine bilayers from MD simulation and IR spectroscopy. *J Phys Chem B* 2002;**106**:6273–88.
40. Prato M, Kosstarelos K, Blanco A. Functionalized carbon nanotubes in drug design and discovery. *Acc Chem Res* 2008;**41**:60–8.

41. Liu Z, Chen K, Davis C, Sherlock S, Cao Q, Chen X, et al. Drug delivery with carbon nanotubes for *in vivo* cancer treatment. *Cancer Res* 2008;**68**:6652–60.
42. Hilder TA, Hill JM. Modeling the encapsulation of the anticancer drug cisplatin into carbon nanotubes. *Nanotechnology* 2007;**18**:1–8.
43. Shi X, Kong Y, Gao H. Coarse grained molecular dynamics and theoretical studies of carbon nanotubes entering cell membranes. *Acta Mech Sin* 2008;**24**:161–9.
44. Kukowska-Latallo J, Candido KA, Cao Z, Nigavekar SS, Majoros IJ, Thomas TP, et al. Nanoparticle targeting of anticancer drug improves therapeutic response in animal model of human epithelial cancer. *Cancer Res* 2005;**65**:5317–24.
45. Majoros IJ, Baker Jr. JR, editors. *Dendrimer-based nanomedicine*. Singapore: Pan Stanford Publishing Pte Ltd. 2008.
46. Langer R. Biomaterials in drug delivery and tissue engineering: one laboratory's experience. *Acc Chem Res* 2000;**33**:94–101.
47. Majoros IJ, Mehta CB, Baker Jr. JR. Mathematical description of dendrimer structure. *J Comput Theor Nanosci* 2004;**1**:193–8.
48. Uhrich KE, Cannizzaro SM, Langer R, Shakesheff KM. Polymeric systems for controlled drug release. *Chem Rev* 1999;**99**:3181–98.
49. Kumar N, Langer RS, Domb AJ. Polyanhydrides: an overview. *Adv Drug Deliv Rev* 2002;**54**:889–910.
50. Allen C, Yu Y, Maysinger D, Eisenburg A. Polycaprolactone-b-poly(ethylene oxide) block copolymermicelles as a novel drug delivery vehicle for neurotrophic agent FK506 and L-685, 818. *Bioconj Chem* 1998;**9**:564–72.
51. Jhaveri MS, Rait A, Chung K-M, Trepel JB, Chang EH. Antisense oligonucleotides targeted to human alpha folate receptor inhibit breast cancer cell growth and sensitize the cells to doxorubicin treatment. *Mol Cancer Ther* 2004;**3**:1505–12.
52. Yu W, Pirollo KF, Rait A, Yu B, Xiang LM, Huang WQ, et al. A sterically stabilized immunolipoplex for systemic administration of a therapeutic gene. *Gene Ther* 2004;**11**:1434–40.
53. Xu L, Pirollo KF, Chang EH. Tumor-targeted p53-gene therapy enhances the efficacy of conventional chemo/radiotherapy. *J Control Release* 2001;**74**:115–28.
54. Foley S, Crowley C, Smaihi M, Bonfils C, Erlanger B, Seta P, et al. *Biochem Biophys Res Commun* 2002;**294**:116–9.
55. Ashcroft J, Tsyboulski D, Hartman K, Zakharian T. Fullerene (C60) immunoconjugates: interaction of water-soluble C60 derivatives with the murine anti-gp240 melanoma antibody. *Chem Commun* 2006;**28**:3004–6.
56. Yang J, Wang K, Driver J, Yang J, Barron A. The use of fullerene substituted phenylalanine amino acid as a passport for peptides through cell membranes. *Org Biomol Chem* 2007;**5**:260–6.
57. Zhang Z, Yang X, Zhang Y, Zeng B, Wang S, Zhu T, et al. Delivery of telomerase reverse transcriptase small interfering RNA in complex with positively charged single-walled carbon nanotubes suppresses tumor growth. *Clin Cancer Res* 2006;**2**:4933–9.
58. Zhang CY, Yeh H, Kuroki M, Wang T. Single-quantum-dot based DNA nanosensor. *Nat Mater* 2005;**4**:826–31.
59. Kloepfer JA, Mielke RE, Wong MS, Nealson KH, Stucky G, Nadeau JL. Quantum dots as strain- and metabolism-specific microbiological labels. *Appl Environ Microbiol* 2003;**69**:4205–13.
60. Ness JM, Akhtar RS, Latham CB, Roth KA. Combined tyramide signal amplification and quantum dots for sensitive and photostable immuno-fluorescence detection. *J Histochem Cytochem* 2003;**51**:981–7.
61. Wu X, Liu H, Haley KN, Treadway JA, Larson JP, Ge N, et al. Immunofluorescent labeling of cancer marker Her2 and other cellular targets with semiconductor quantum dots. *Nat Biotechnol* 2003;**21**:41–6.

62. Tada H, Higuchi H, Wanatabe T, Ohuchi N. In vivo real-time tracking of single quantum dots conjugated with monoclonal anti-HER2 antibody in tumors of mice. *Cancer Res* 2007;**67**:1138–44.
63. Huang X, El-Sayed IH, Qian W, El-Sayed MA. Cancer cell imaging and photothermal therapy in the near-infared region by using gold nanorods. *J Am Chem Soc* 2006;**128**:2115–20.
64. James WD, Hirsch LR, West JL, Neal PD, Payne JD. Application of INAA to the build-up and clearance of gold nanoshells in clinical studies in mice. *J Radioanal Nucl Chem* 2007;**271**:455–9.
65. Ojima I. Guided molecular missiles for tumor-targeting chemotherapy; case studies using the second-generation taxoids as warheads. *Acc Chem Res* 2008;**41**:108–11.
66. Majoros IJ, Williams CR, Baker Jr. JR. Current dendrimer applications in cancer diagnosis and therapy. *Curr Top Med Chem* 2008;**8**:1165–79.
67. Feazell RP, Nakayama-Ratchford N, Dai H, Lippard SJ. Soluble single-walled carbon nanotubes as longboat delivery systems for platinum(IV) anticancer drug design. *J Am Chem Soc* 2007;**129**:8438–9.
68. Dubowchik GM, Walker MA. Receptor-mediated and enzyme-dependent targeting of cytotoxic anticancer drugs. *Pharmcol Ther* 1999;**83**:67–123.
69. Dancey JE, Chen HX. Strategies for optimizing combinations of moleuclarly targeted anticancer agents. *Nat Rev Drug Discovery* 2006;**5**:649–59.
70. Majoros IJ, Myc A, Thomas T, Mehta CB, Baker JR. PAMAM dendrimer-based multifunctional conjugate for cancer therapy: synthesis, characterization, and functionality. *Biomacromolecules* 2006;**7**:572–9.
71. Thomas TP, Majoros IJ, Kotlyar A, Kukowska-Latallo JF, Bielinska A, Myc A, et al. Targeting and inhibition of cell growth by an engineered dendritic nanodevice. *J Med Chem* 2005;**48**:3729–35.
72. Nowell PC. The clonal evolution of tumor cell populations. *Science* 1976;**194**:23–8.
73. Majoros IJ, Thomas TP, Mehta CB, Baker JR. Poly(amidoamine) dendrimer-based multifunctional engineered nanodevice for cancer therapy. *J Med Chem* 2005;**48**:5892–9.
74. Etrych T, Mrkvan T, Rihova B, Ulbrich K. Star-shaped immunoglobulin-containing HPMA-based conjugates with doxorubicin for cancer therapy. *J Control Release* 2007;**122**:31–8.
75. Bareford LM, Swaan PW. Endocytic mechanisms for targeted drug delivery. *Adv Drug Deliv Rev* 2007;**59**:748–58.
76. Dubowchik GM, Walker MA. Receptor-mediated and enzyme-dependent targeting of cytotoxic anticancer drugs. *Pharmacol Ther* 1999;**83**:67–123.
77. Joshi A, Vance D, Rai P, Thiyagarajan A, Kane RS. The design of polyvalent therapeutics. *Chem Eur J* 2008;**14**:7738–47.
78. Hilgenbrink AR, Low PS. Folate receptor-mediated drug targeting: from therapeutics to diagnostics. *J Pharm Sci* 2005;**94**:2135–46.
79. Lee RJ, Low PS. Folate-mediated tumor cell targeting of liposome-entrapped doxorubicin in vitro. *Biochim Biophys Acta—Biomembr* 1995;**1233**:134–44.
80. Montet X, Funovics M, Montet-Abou K, Weissleder R, Josephson L. Multivalent effects of RGD peptides obtained by nanoparticle display. *J Med Chem* 2006;**49**:6087–93.
81. Shukla R, Thomas TP, Peters J, Kotlyar A, Myc A, Baker Jr. JR. Tumor angiogenic vasculature targeting with PAMAM dendrimer-RGD conjugates. *Chem Commun* 2005;5739–41.
82. Temming K, Lacombe M, Schaapveld RQJ, Orfi L, Kéri G, Poelstra K, et al. Rational design of RGD-albumin conjugates for targeted delivery of the VEGF-R kinase inhibitor PTK787 to angiogenic endothelium. *ChemMedChem* 2006;**1**:1200–3.
83. Chen Y, Foss CA, Byun Y, Nimmagadda S, Pullambhatla M, Fox JJ, et al. Radiohalogenated prostate-specific membrane antigen (PSMA)-based ureas as imaging agents for prostate cancer. *J Med Chem* 2008;**51**:7933–43.

84. Qian ZM, Li H, Sun H, Ho K. Targeted drug delivery via the transferrin receptor-mediated endocytosis pathway. *Pharmacol Rev* 2002;**54**:561–87.
85. Shukla R, Thomas TP, Desai AM, Kotlyar A, Park SJ, Baker Jr. JR. HER2 specific delivery of methotrexate by dendrimer conjugated anti-HER2mAb. *Nanotechnology* 2008;**19**:295102.
86. Shukla R, Thomas TP, Desai AM, Kotlyar A, Park SJ, Baker Jr. JR. HER2 specific delivery of methotrexate by dendrimer conjugated anti-HER2 mAb. *Nanotechnology* 2008;**19**:295102.
87. Thomas TP, Shukla R, Kotlyar A, Liang B, Ye JY, Norris TB, et al. Dendrimer-epidermal growth factor conjugate displays superagonist activity. *Biomacromolecules* 2008;**9**:603–9.
88. Lee YC, Lee RT. Carbohydrate-protein interactions: basis of glycobiology. *Acc Chem Res* 1995;**28**:321–7.
89. Kiessling LL, Gestwicki JE, Strong LE. Synthetic multivalent ligands in the exploration of cell-surface interactions. *Curr Opin Chem Biol* 2000;**4**:696–703.
90. Mammen M, Choi SK, Whitesides GM. Polyvalent interactions in biological systems: implications for design and use of multivalent ligands and inhibitors. *Angew Chem Int Ed* 1998;**37**:2755.
91. Roy R. Syntheses and some applications of chemically defined multivalent glycoconjugates. *Curr Opin Struct Biol* 1996;**6**:692–702.
92. Choi Y, Thomas T, Kotlyar A, Islam MT, Baker JJR. Synthesis and functional evaluation of DNA-assembled polyamidoamine dendrimer clusters for cancer cell-specific targeting. *Chem Biol* 2005;**12**:35–43.
93. Lu Y, Low PS. Folate-mediated delivery of macromolecular anticancer therapeutic agents. *Adv Drug Deliv Rev* 2002;**54**:675–93.
94. Schnell JR, Dyson HJ, Wright PE. Strcuture, dynamics, and catalytic function of dihydrofolate. *Annu Rev Biophys Biomol Struct* 2004;**33**:119–40.
95. Chunduru S, Cody V, Luft J, Pangborn W, Appleman J, Blakley R. Methotrexate-resistant variants of human dihydrofolate reductase. Effects of Phe31 substitutions. *J Biol Chem* 1994;**269**:9547–55.
96. Tattersall MH, Brown B, Frei 3rd E. The reversal of methotrexate toxicity by thymidine with maintenance of antitumour effects. *Nature* 1975;**253**:198–200.
97. Myc A, Majoros IJ, Thomas TP, Baker JR. Dendrimer-based targeted delivery of an apoptotic sensor in cancer cells. *Biomacromolecules* 2007;**8**:13–8.
98. Kaminskas LM, Kelly BD, McLeod VM, Boyd BJ, Krippner GY, Williams ED, et al. Pharmacokinetics and tumor disposition of PEGylated, methotrexate conjugated Poly-l-lysine dendrimers. *Mol Pharm* 2009;**6**:1190–204.
99. Wu G, Barth RF, Yang W, Kawabata S, Zhang L, Green-Church K. Targeted delivery of methotrexate to epidermal growth factor receptor-positive brain tumors by means of cetuximab (IMC-C225) dendrimer bioconjugates. *Mol Cancer Ther* 2006;**5**:52–9.
100. Chau Y, Dang NM, Tan FE, Langer R. Investigation of targeting mechanism of new dextran-peptide-methotrexate conjugates using biodistribution study in matrix-metalloproteinase-overexpressing tumor xenograft model. *J Pharm Sci* 2006;**95**:542–51.
101. Bai KB, Lang O, Orban E, Szabo R, Kohidai L, Hudecz F, et al. Design, synthesis, and in vivo activity of novel drug delivery systems containing tufsin derivatives and methotrexate. *Bioconjug Chem* 2008;**19**:2260–9.
102. Han J, Lim S-J, Lee M-K, Kim C-K. Altered pharmacokinetics and liver targetability of methotrexate by conjugation with lactosylated albumins. *Drug Deliv* 2001;**8**:125–34.
103. Kohler N, Sun C, Wang J, Zhang M. Methotrexate-modified superparamagnetic nanoparticles and their intracellular uptake into human cancer cells. *Langmuir* 2005;**21**:8858–64.
104. Bai KB, Lng O, Orbn E, Szab R, Köhidai L, Hudecz F, et al. Design, synthesis, and in vitro activity of novel drug delivery systems containing tuftsin derivatives and methotrexate. *Bioconjug Chem* 2008;**19**:2260–9.

105. Kamen BA, Capdevila A. Receptor-mediated folate accumulation is regulated by the cellular folate content. *Proc Natl Acad Sci USA* 1986;**83**:5983–7.
106. Tomalia DA, Naylor AM, Goddard III WA. Starburst dendrimers: molecular-level control of size, shape, surface chemistry, topology, and flexibility from atoms to macroscopic matter. *Angew Chem Int Ed* 1990;**29**:138–75.
107. Schiff PB, Fant J, Horwitz SB. Promotion of microtubule assembly in vitro by taxol. *Nature* 1979;**277**:665–7.
108. Gueritte-Voegelein F, Guenard D, Lavelle F, Le Goff MT, Mangatal L, Potier P. Relationships between the structure of taxol analogs and their antimitotic activity. *J Med Chem* 1991;**34**:992–8.
109. Wang X, Li J, Wang Y, Cho KJ, Kim G, Gjyrezi A, et al. HFT-T, a targeting nanoparticle, enhances specific delivery of pacliltaxel to folate receptor-positive tumors. *ACS Nano* 2009;**3**:3165–74.
110. Lin Y-S, Tungpradit R, Sinchaikul S, An F-M, Liu D-Z, Phutrakul S, et al. Targeting the deliver of glycan-based paclitaxel prodrugs to cancer cells via glucose transporters. *J Med Chem* 2008;**51**:7428–41.
111. Pan J, Feng S-S. Targeted delivery of paclitaxel using folate-decorated poly(lactide)-vitamin E TPGS nanoparticles. *Biomaterials* 2008;**29**:2663–72.
112. Liang H-F, Chen S-C, Chen M-C, Lee P-W, Chen C-T, Sung H-W. Paclitaxel-loaded poly(I3-glutamic acid)-poly(lactide) nanoparticles as a targeted drug delivery system against cultured HepG2 cells. *Bioconjug Chem* 2006;**17**:291–9.
113. Cirilli M, Bachechi F, Ughetto G, Colonna FP, Capobianco ML. Interactions between morpholinyl anthracyclines and DNA: the crystal structure of a morpholino doxorubicin bound to d(CGTACG). *J Mol Biol* 1993;**230**:878–89.
114. Choi SK, Thomas T, Li M, Kotlyar A, Desai AM, Baker Jr. JR. Light-controlled release of caged doxorubicin from folate receptor-targeting PAMAM dendrimer nanoconjugates. *Chem Commun* 2010;**46**:2632–4.
115. Fan L, Li F, Zhang H, Wang Y, Cheng C, Li X, et al. Co-delivery of PDTC and doxorubicin by multifunctional micellar nanoparticles to achieve active targeted drug deliverty and overcome multidrug resistance. *Biomaterials* 2010;**31**:5634–42.
116. Grange C, Geninatti-Crich S, Esposito G, Alberti D, Tei L, Bussolati B, et al. Combined delivery and magnetic resonance imaging of neural cell adhesion molecule-targeted doxorubicin-containing liposomes in experimentally induced Kaposi's sarcoma. *Cancer Res* 2010;**70**:2180–90.
117. Zhang Y-F, Wang J-C, Bian D-Y, Zhang X, Zhang Q. Targeted delivery of RGD-modified liposomes encapsulating both combretastatin A-4 and doxorubicin for tumor therapy: In vitro and in vivo studies. *Eur J Pharm Biopharm* 2010;**74**:467–73.
118. Zhang L, Xia J, Zhao Q, Liu L, Zhang Z. Functional graphene oxide as a nanocarrier for controlled loading and targeted delivery of mixed anticancer drugs. *Small* 2010;**6**:537–44.
119. Xiong X-B, Ma Z, Lai R, Lavasanifar A. The therapeutic response to multifunctional polymeric nano-conjugates in the targeted cellular and subcellular delivery of doxurubicin. *Biomaterials* 2010;**31**:757–68.
120. Zhong S, Bhattacharya S, Chan W, Jasti B, Li X. Leucine-aspartic acid-valine sequence as targeting ligand and drug carrier for doxurubicin delivery to melanoma cells: In vitro cellular uptake and cytotoxicity studies. *Pharm Res* 2009;**26**:2578–87.
121. Dhar S, Liu Z, Thomale J, Dai H, Lippard SJ. Targeted single-wall carbon nanotube-mediated Pt(IV) prodrug delivery using folate as a homing device. *J Am Chem Soc* 2008;**130**:11467–76.
122. Dhar S, Gu FX, Langer R, Farokhzad OC, Lippard SJ. Targeted delivery of cisplatin to prostate cancer cells by aptamer functionalized Pt(IV) prodrug-PLGA-PEG nanoparticles. *Proc Natl Acad Sci USA* 2008;**105**:17356–61.
123. Lee RJ, Wang S, Low PS. Measurement of endosome pH following folate receptor-mediated endocytosis. *Biochim Biophys Acta—Mol Cell Res* 1996;**1312**:237–42.

124. Yang J, Chen H, Vlahov IR, Cheng J-X, Low PS. Characterization of the pH of folate receptor-containing endosomes and the rate of hydrolysis of internalized acid-labile folate-drug conjugates. *J Pharmacol Exp Ther* 2007;**321**:462–8.
125. Sutherland MSK, Sanderson RJ, Gordon KA, Andreyka J, Cerveny CG, Yu C, et al. Lysosomal trafficking and cysteine protease metabolism confer target-specific cytotoxicity by peptide-linked anti-CD30-auristatin conjugates. *J Biol Chem* 2006;**281**:10540–7.
126. Thomas TP, Kukowska-Latallo JR. In: Majoros IJ, Baker Jr. JR, editors. *Dendrimer-based nanomedicine*. Hackensack: Pan Stanford; 2008. p. 175–207.
127. Etrych T, Mrkvan T, Říhová B, Ulbrich K. Star-shaped immunoglobulin-containing HPMA-based conjugates with doxorubicin for cancer therapy. *J Control Release* 2007;**122**:31–8.
128. Myc A. Preclinical antitumor efficacy evaluation of dendrimer-based methotrexate conjugates. *Anticancer Drugs* 2008;**19**:143–9.
129. Choi SK, Thomas T, Li M, Kotlyar A, Desai A, Baker Jr. JR. Light-controlled release of caged doxorubicin from folate receptor-targeting PAMAM dendrimer nanoconjugate. *Chem Commun* 2010;**46**:2632–4.
130. Lai P-S, Lou P-J, Peng C-L, Pai C-L, Yen W-N, Huang M-Y, et al. Doxorubicin delivery by polyamidoamine dendrimer conjugation and photochemical internalization for cancer therapy. *J Control Release* 2007;**122**:39–46.
131. Lu Y, Wu J, Wu J, Gonit M, Yang X, Lee A, et al. Role of formulation composition in folate receptor-targeted liposomal doxorubicin delivery to acute myelogenous leukemia cells. *Mol Pharm* 2007;**4**:707–12.
132. Pan XQ, Zheng X, Shi G, Wang H, Ratnam M, Lee RJ. Strategy for the treatment of acute myelogenous leukemia based on folate receptor beta-targeted liposomal doxorubicin combined with receptor induction using all-trans retinoic acid. *Blood* 2002;**100**:594–602.
133. Weissleder R. Scaling down imaging: molecular mapping of cancer in mice. *Nat Rev Cancer* 2002;**2**:1–8.
134. Maeda H, Wu J, Sawa T, Matsumura Y, Hori K. Tumor vacular permeability and the EPR effect in macromolecular therapeutics: a review. *J Control Release* 2000;**65**:271–84.
135. Maeda H. The enhanced permeability and retention (EPR) effect in tumor vaculature: the key role of tumor-selective macromolecular drug targeting. *Adv Enzyme Regul* 2001;**41**:189–207.
136. Lee CC, Gillies ER, Fox ME, Guillaudeu SJ, Frachet JMJ, Dy EE, et al. A single dose of doxorubicin-functionalized bowtie dendrimer cures mice bearing C-26 colon carcinomas. *Proc Natl Acad Sci USA* 2006;**103**:16649–54.
137. Kresse M, Wagner S, Pfefferer D, Lawaczeck R, Elate V, Semmler W. Targeting of ultrasmall superparamagnetic iron oxide (USPIO) particles to tumor cells in vivo by using tranferrin receptor pathways. *Magn Reson Med* 1998;**40**:236–42.
138. Wunderbaldinger P, Josephson LJ, Weissleder R. Tat peptide directs enhanced clearance and hepatic permeability of magnetic nanoparticles. *Bioconjug Chem* 2002;**13**:264–8.
139. Farokhzad OC, Jon S, Khademhosseini A, Tran TN, Lavan DA, Langer R. Nanoparticle-aptamer bioconjugates: a new approach for targeting prostate cancer cells. *Cancer Res* 2004;**64**:7668–72.
140. Herr JK, Smith JE, Medley CD, Shangguan D, Tan W. Aptamer-conjuaged nanoparticles for selective collection and detection of cancer cells. *Anal Chem* 2006;**78**:2918–24.
141. Yigit MV, Mazumdar D, Kim HK, Lee JH, Odintsov B, Lu Y. Smart "turn-on" magnetic resonance contrast agents based on aptamer-functionalized superparamagnetic iron oxide nanoparticles. *ChemBioChem* 2007;**8**:1675–8.
142. Weissleder R, Kelly K, Sun EY, Shtatland T, Josephson L. Cell-specific targeting of nanoparticles by multivalent attachment of small molecules. *Nat Biotechnol* 2005;**23**:1418–23.
143. Zhang Y, Sun Y, Xu X, Zhu H, Huang L, Qi Y, et al. Synthesis, biodistribution, and microsingle photon emission computed tomography (SPECT) imaging study of technetium-99 m labeled

PEGylated dendrimer poly(amidoamine) (PAMAM)-folic acid conjugates. *J Med Chem* 2010;**53**:3262.

144. Swanson SD, Kukowska-Latallo JF, Patri AK, Chen C, Ge S, Cao Z, et al. *Int J Nanomedicine* 2008;**3**:201.
145. Cheng Z, Thorek DLJ, Tsurkas A. Gadolinium-conjugated dendrimer nanoclusters as a tumor-targeted TI magnetic resonance imaging contrast agent. *Angew Chem Int Ed* 2010;**49**:346.
146. Lauro R, Platania A, Liberatore C, Reda G, Spinelli C. Biochemical profile of essential aterial hypertension. Indications for a targeted therapy: experience with propranolol. *Clin Ter* 1978;**85**:19–25.
147. Foulds L. Tumor progression. *Cancer Res* 1957;**17**:355–66.
148. Olson ES, Jiang T, Aguilera TA, Nguyen QT, Ellies LG, Scadeng M, et al. Activatable cell penetrating peptides linked to nanoparticles as dual probes for in vivo fluorescence and MR imaging of proteases. *PNAS* 2010;**107**:4311.
149. Boswell CA, Eck PK, Regino CAS, Bernardo M, Wang KJ, Milenic DE, et al. Synthesis, characterization, biological evaluation of integrin a nb 3-targeted PAMAM dendrimer. *Mol Pharm* 2008;**5**:527–39.
150. Patri AK, Myc A, Beals J, Thomas TP, Bander NH, Baker Jr. JR. Synthesis and in vitro testing of J591 antibody-dendrimer conjugates for targeted prostate cancer therapy. *Bioconjug Chem* 2004;**15**:1174.
151. Kerr JF, Wyllie AH, Currie AR. Apoptosis: a basic biological phenomenon with wide-ranging implications in tissue kinetics. *Br J Cancer* 1972;**26**:239–57.
152. Zurgil N, Shafran Y, Fixler D, Deutsch M. Analysis of early apoptotic events in individual cells by fluorescence intensity and polarization measurements. *Biochem Biophys Res Commun* 2002;**290**:1573–82.
153. Finkel E. The mitochondrion: is it central to apoptosis? *Science* 2001;**292**:624–6.
154. Green DR, Reed JC. Mitochondria and apoptosis. *Science* 1998;**281**:1309–12.
155. Demoy M, Minko T, Kopeckova P, Kopecek J. Time- and concentration-dependent apoptosis and necrosis induced by free and HPMA copolymer-bound doxorubicin in human ovarian carcinoma cells. *J Control Release* 2000;**69**:185–96.
156. Heiskanen KM, Bhat MB, Wang HW, Ma J, Nieminen AL. Mitochondrial depolarization accompanies cytochrome c release during apoptosis in PC6 cells. *J Biol Chem* 1999;**274**:5654–8.
157. Strebel A, Harr T, Bachmann F, Wernli M, Erb P. Green fluorescent protein as a novel tool to measure apoptosis and necrosis. *Cytometry* 2001;**43**:126–33.
158. Liu J, Bhalgat M, Zhang C, Diwu Z, Hoyland B, Klaubert DH. Fluorescent molecular probes V: a sensitive caspase-3 substrate for fluorometric assays. *Bioorg Med Chem Lett* 1999;**9**:3231–6.
159. Hug H, Los M, Hirt W, Debatin KM. Rhodamine 110-linked amino acids and peptides as substrates to measure caspase activity upon apoptosis induction in intact cells. *Biochemistry* 1999;**38**:13906–11.
160. Zhang HZ, Kasibhatla S, Guastella J, Tseng B, Drewe J, Cai SX. N-Ac-DEVD-N′-(Polyfluorobenzoyl)-R110: novel cell-permeable fluorogenic caspase substrates for the detection of caspase activity and apoptosis. *Bioconjug Chem* 2003;**14**:458–63.
161. Sgonc R, Gruber J. Apoptosis detection: an overview. *Exp Gerontol* 1998;**33**:525–33.
162. Span LF, Pennings AH, Vierwinden G, Boezeman JB, Raymakers RA, de Witte T. The dynamic process of apoptosis analyzed by flow cytometry using Annexin-V/propidium iodide and a modified in situ end labeling technique. *Cytometry* 2002;**47**:24–31.
163. Steff AM, Fortin M, Arguin C, Hugo P. Detection of a decrease in green fluorescent protein fluorescence for the monitoring of cell death: an assay amenable to high-throughput screening technologies. *Cytometry* 2001;**45**:237–43.

164. Finucane DM, Bossy-Wetzel E, Waterhouse NJ, Cotter TG, Green DR. Bax-induced caspase activation and apoptosis via cytochrome c release from mitochondria is inhibitable by Bcl-xL. *J Biol Chem* 1999;**274**:2225–33.
165. Kelly KJ, Sandoval RM, Dunn KW, Molitoris BA, Dagher PC. A novel method to determine specificity and sensitivity of the TUNEL reaction in the quantitation of apoptosis. *Am J Physiol Cell Physiol* 2003;**284**:C1309–18.
166. Liu J, Lu Y. FRET study of a trifluorophore-labeled DNAzyme. *J Am Chem Soc* 2002;**124**:15208–16.
167. Berney C, Danuser G. FRET or no FRET: a quantitative comparison. *Biophys J* 2003;**84**:3992–4010.
168. Stauffer SR, Hartwig JF. Fluorescence resonance energy transfer (FRET) as a high-throughput assay for coupling reactions. Arylation of amines as a case study. *J Am Chem Soc* 2003;**125**:6977–85.
169. Luo KQ, Yu VC, Pu Y, Chang DC. Application of the fluorescence resonance energy transfer method for studying the dynamics of caspase-3 activation during UV-induced apoptosis in living HeLa cells. *Biochem Biophys Res Commun* 2001;**283**:1054–60.
170. Gopich I, Szabo A. Fluorophore-quencher distance correlation functions from single-molecule photon arrival trajectories. *J Phys Chem B* 2005;**109**:6845–8.
171. Yusuke Niino KH, Oka Kotar. Simultaneous live cell imaging using dual FRET sensors with a single excitation light. *PLoS ONE* 2009;**4**:e6036.

Applications of Molecular Imaging

Craig J. Galbán,[*] Stefanie Galbán,[†] Marcian E. Van Dort,[*] Gary D. Luker,[*] Mahaveer S. Bhojani,[*] Alnawaz Rehemtulla,[†] and Brian D. Ross[*]

[*]*Department of Radiology, University of Michigan, Center for Molecular Imaging, Ann Arbor, Michigan, USA*

[†]*Department of Radiation Oncology, University of Michigan, Center for Molecular Imaging, Ann Arbor, Michigan, USA*

Progress in Molecular Biology
and Translational Science, Vol. 95
DOI: 10.1016/S1877-1173(10)95009-6

1877-1173/10 $35.00

Today molecular imaging technologies play a central role in clinical oncology. The use of imaging techniques in early cancer detection, treatment response, and new therapy development is steadily growing and has already significantly impacted on clinical management of cancer. In this chapter, we overview three different molecular imaging technologies used for the understanding of disease biomarkers, drug development, or monitoring therapeutic outcome. They are (1) optical imaging (bioluminescence and fluorescence imaging), (2) magnetic resonance imaging (MRI), and (3) nuclear imaging (e.g., single-photon emission computed tomography (SPECT) and positron emission tomography (PET)). We review the use of molecular reporters of biological processes (e.g., apoptosis and protein kinase activity) for high-throughput drug screening and new cancer therapies, diffusion MRI as a biomarker for early treatment response and PET and SPECT radioligands in oncology.

I. Optical Imaging

A. Introduction

The integration of genetically encoded imaging reporters into cells and animals provides a unique opportunity to monitor molecular, biochemical, and cellular pathways *in vivo* as modifications of post- or cotranslational events such as phosphorylation or glycosylation can be monitored in real time. In this section, we discuss the use of such reporters for target validation and dose/schedule optimization and for identification of lead compounds from a library using cell-based, high-throughput screening (HTS). We further review the use of transgenic animals expressing tissue-specific fluorescent or bioluminescent reporters as a unique tool for studying cellular processes such as transcription or apoptosis in living cells and animals.

B. Clinical Significance for Imaging Kinase Activity

Dysregulation and mutation of kinases have been reported to play a causal role in many human diseases such as cancer, rheumatoid arthritis, cardiovascular and neurological disorders, asthma, and psoriasis.[1–6] Molecular profiling of pathological samples and corroborative development in bioinformatics in recent years have led to the identification of potential biomarkers.[7–11] Among the biomarkers identified, protein kinases have gained much attention as potential therapeutic targets. Thus, it is not surprising that the first molecularly targeted drugs approved by the Food Drug Administration are tyrosine kinase inhibitors specific to abl (the Abelson protooncogene), PDGFR (platelet-derived growth factor receptor), Erlotinib, Avastin, Lapatinib, and Herceptin.[12] However, other than Imatinib, most kinase inhibitors such as Erlotinib,

Lapatinib, Cetuximab, ABX-EGF, and Bevacizumab have relatively modest activity as a single agent.[13,14] This suggests the use of multitargeted kinase inhibitors or the combination with radiation therapy to be more beneficial for cancer therapy than a single agent. There is an increasing need for the preclinical evaluation of drug efficacies and also the optimization of dosing and scheduling of such agents. For instance the combinatorial treatment of gemcitabine and gefitinb was determined efficacious in a preclinical study of head and neck cancer but only when gefitinib was administered prior to gemcitabine.[15] The ability to test drug interaction and agent combination in preclinical models will facilitate drug development and serve to provide for rational does/schedule development for clinical trials. We believe that the use of targeted agents in oncology will significantly benefit from molecular imaging techniques in preclinical studies with the goal of identifying most efficacious agents modulating signaling pathways in a targeted manner. The increased development of such preclinical model systems has already begun to provide us with the ability to screen large numbers of molecules and will ensure the translation of the most efficacious drug or drug combination into the clinic in the future.

C. Imaging of Post- and Cotranslational Events

Screening and identification of new kinase inhibitors from a library of chemical compounds require robust methods. Western blotting with phospho-specific antibodies and *in vitro* kinase assays with radioisotopes are commonly used to assess kinase activity. However, these methods are invasive, require large numbers of animals, and only provide a snapshot of kinase activity at specific time points. Additionally, some of these assays are suitable only for *in vitro* studies and may not reflect the *in vivo* kinase activity or drug specificity. Recent discoveries in the field of optical imaging have overcome such limitations. The field of molecular imaging encompasses the noninvasive visual representation of biological processes at the cellular and molecular level in the whole organism and the modalities and instrumentation to support the visualization and measurement of these processes.[16] These imaging technologies are an attempt to bridge the gap between discovery of causal disease markers and identification of their inhibitor for potential therapeutic use.

We have recently developed a bioluminescent reporter to monitor activity of the serine/threonine kinase Akt/PKB,[17] whose expression profile has been linked to tumor initiation, progression, and also resistance to cancer therapy. This reporter was based on conformation-dependent complementation of firefly luciferase, wherein a monomeric reporter gene is split into two separate inactive components. The reporter activity is restored and can be measured when the two components are brought into close proximity. Fields and colleagues utilize this strategy first and developed the yeast two hybrid system,[18] which is based on protein complementation of GAL4, a transcriptional activator. Since then the system has been used successfully to identify novel protein–protein interactions. Yet it has limited utility in the context of a living cell or animal. Subsequently a number of molecular reporters routinely used in mammalian biology were engineered for complementation studies. These include a plethora of fluorescent proteins and bioluminescent enzymes (firefly luciferase and renilla luciferase), β-galactosidase, dihydrofolate reductase (DHFR), and TME-1 β-lactamase.[19–24]

The bioluminescence reporter approach has emerged as a useful technique for small animal imaging using complementation assays. In the presence of ATP and oxygen, luciferase modifies its substrate luciferin by releasing photons. The ATP-dependent firefly (*Photinus pyralis*) is most commonly used for *in vivo* imaging due to the fact that 30% of its light generated emits at a spectra above 600 nm, a region where the signal attenuation by the absorption and scattering properties of mammalian tissue is minimal.[25,26] Hence compared with fluorescence imaging wherein the excitation light excites other fluorescent molecules in the body, thus creating background/autofluoresence, BLI is superior especially for deep tissue imaging.

Previously Luker and colleagues developed a luciferase complementation system, by which N-terminal and C-terminal luciferase fragments were fused with FRB of the mammalian target of rapamycin and FK506-binding protein12 (FKBP), respectively.[27] Interaction between FRB–N-Luc/C-Luc–FKBP upon single-site binding of rapamycin to FKBP reconstituted luciferase activity in an FK506-competitive manner. Later the investigators used this strategy to demonstrate the phosphorylation-dependent interaction between human Cdc25C and 14-3-3ε *in vivo.*[27]

To create a prototype kinase reporter for reporting changes in kinase activity, we adapted the previously mentioned complementation system further. In particular, we flanked an Akt consensus substrate peptide and the phosphorylated amino acid binding domain from FHA2 with an amino-(N-Luc) and carboxyl-(C-Luc) terminal domain of the firefly luciferase reporter molecule (Fig. 1)[17] to develop a bioluminescent Akt reporter (BAR).

In the presence of Akt kinase activity, phosphorylation of the Akt consensus substrate sequence results in its binding to the FHA2 domain, thus sterically preventing reconstitution of a functional luciferase protein. In the absence of Akt kinase activity, release of this steric constraint allows reconstitution of the luciferase enzyme resulting in detectable bioluminescence. The advantage of this novel bioluminescent reporter is that (a) it can be adapted for monitoring other kinase activities and (b) it allows imaging in living cells and animals in a quantitative, dynamic, and noninvasive manner.

D. Preclinical Application of Kinase Reporters

Applications of a kinase reporter include studying signaling pathways in biological systems as well as testing drug efficacies for preclinical studies. We used the BAR reporter in cell culture to evaluate several inhibitors; an Akt inhibitor

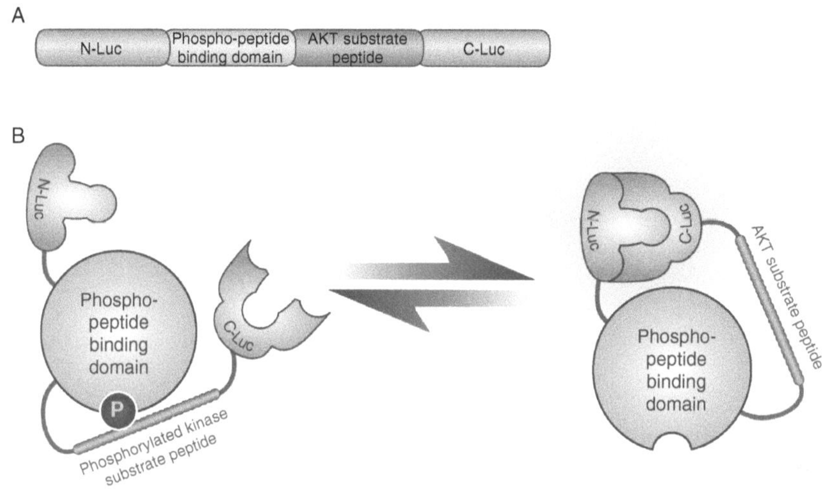

FIG. 1. Schematic representation of the Akt kinase reporter (BAR). (A) The chimeric Akt kinase reporter is a fusion of the N-Luc and C-Luc luciferase protein linked to an Akt peptide domain containing the Akt substrate sequence and a yeast FHA2 phospho-Ser/Thr binding domain. (B) Phosphorylation of the Akt substrate peptide in the reporter results in interaction with the FHA2 phospho-peptide binding domain leading to steric constraints on the N-Luc and C-Luc. Inhibition of Akt kinase activity results in decreased binding of phospho-peptide and the binding domain which enables reconstitution of luciferase and restored bioluminescence.

(API-2) along with a PI-3K inhibitor (Perifosine). Increased bioluminescence activity was observed in a time- and dose-dependent manner with both inhibitors indicating that BAR provides a surrogate for Akt activity in terms of quantity and dynamics. Further we confirmed Akt kinase inhibition with conventional western blotting using phosphor-specific antibodies against Akt.[17] Bioluminescent kinase reporters may additionally aid in investigating signaling events that are upstream of the particular kinase and thus impinge on its activity. When BAR reporter-expressing cells were treated with EGF, changes in BAR bioluminescent activity were detected thus confirming its application for pathway analysis.[17] Our results indicated that activation of EGFR, which has previously been reported to feed into the Akt cascade, can be monitored by Akt bioluminescent imaging. To test the use of BAR as a surrogate marker for EGFR signaling, different cell lines were treated with a known EGFR inhibitor, Erlotinib. Differential activation of the BAR reporter was observed in Erlotinib-sensitive and Erlotinib-resistant cell lines, thus confirming its use in sensing specific upstream signals.[17]

As mentioned previously, a significant advantage of using bioluminescent kinase reporters lies in their potential to monitor signaling pathways in live animals, and therefore providing a unique understanding of pharmacokinetics and bioavailability of specific drugs. For example, when animals with glioma xenografts were treated with 20 mg/kg of the Akt inhibitor AP1-2, peak inhibition was observed 12 hours after treatment, yet when treated with 40 mg/kg, BLI signals were detected for up to 24 h after treatment initiation (Fig. 2). Interestingly when animals were treated with Perifosine, Akt kinase inhibition was detected by BLI at 2 h after treatment and remained elevated for 7 days. This observation was confirmed by previously published data demonstrating high plasma concentration of the drug for up to 7 days after treatment.

In summary such studies establish a preclinical application for BLI reporters in identifying drug–target interaction in cells and living animals. Furthermore can these reporter assays be adapted for their *in vivo* use of dose and schedule optimization and identification of efficacious combinatorial treatments.

E. Modification of Bioluminescent Kinase Reporter for Enhanced Sensitivity

To increase reporter sensitivity we tested the hypothesis that subcellular localiation may influence kinases specificity. Distinct subcellular localization harbor variable kinase concentration and thus may result in improved reporter activity. Akt, for example, is recruited to the plasma membrane by PI-3 kinase-generated D3-phosphorylated phosphoinositides which bind to the Akt PH domain and induce its translocation.[28,29] Kinase-1, which depends on phosphoinositides colocalizes at the same time and phosphorylates Akt within the activation loop.[28,29] We hypothesized that constructing a membrane-targeted BAR would increase its sensitivity. By fusing 10 amino-terminal residues of Lyn kinase responsible for myristoylation and palmitoylation to BAR[30] (Fig. 3) sensitivity of the MyrPalm-BAR reporter was doubled compared with BAR (unpublished data). Thus, subcellular localization of the investigated kinase needs to be considered when designing the next generation reporters.

F. Molecular Imaging of N-Linked Glycosylation

N-linked glycosylation (NLG) is a complex biosynthetic process that regulates maturation of proteins through the secretory pathway. This cotranslational modification is regulated by a series of enzymatic reactions, which results in the transfer of a core glycan from the lipid carrier to a protein substrate. Previous work has shown that inhibition of (NLG) *in vitro*[31] reduces protein levels of receptor tyrosine kinases (RTKs) (i.e., EGFR, ErbB2, Erbb3, and IGF-1R) commonly overexpressed in many cancers, resulting in decreased signaling through both dominant and redundant RTK pathways. Hence inhibiting NLG has become a feasible strategy for cancer therapy[31] and has also been shown to have radiosensitizing effects as demonstrated for other RTK targeting agents in glioblastoma.[32–35] Despite validation of therapeutic NLG

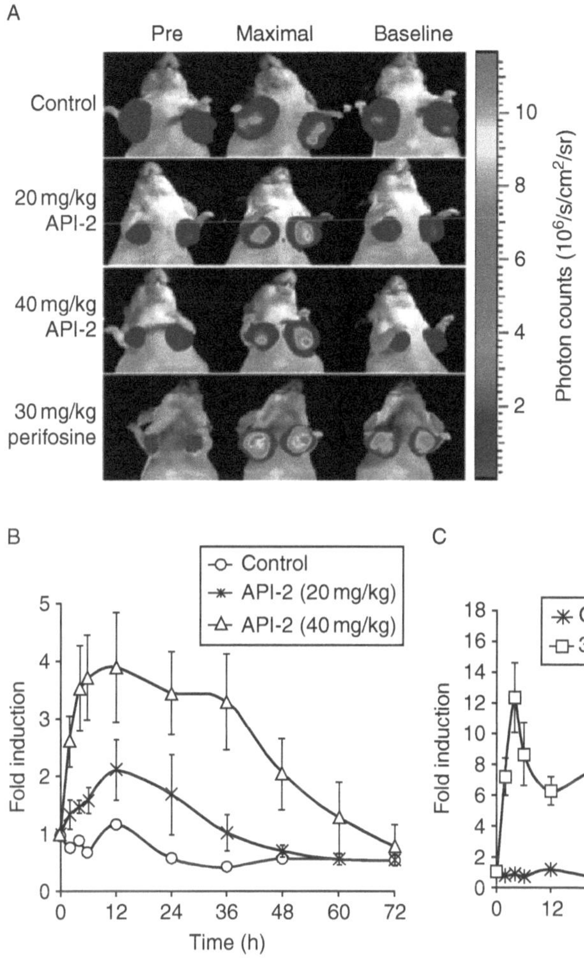

FIG. 2. Imaging of Akt kinase inhibiton in live animals. (A) Mice transplanted with D54 cells expressing the Akt reporter were treated with vehicle control (20% DMSO in PBS), API-2 (20 or 40 mg/kg), or Perifosine (30 mg/kg). Representative images are shown of mice prior to treatment or during maximal luciferase signal upon treatment (Max), and posttreatment. (B) Tumor-specific bioluminescence activity of D54 cells stably expressing BAR in mice, treated with either the vehicle control (20% DMSO in PBS) or API-2 (20 or 40 mg/kg), was monitored at various times. Fold induction of signal intensity over pretreatment values was plotted as mean ± SEM for each of the groups. (C), Bioluminescence activity in tumor-bearing mice before and after treatment with 30 mg/kg perifosine was plotted as fold induction over pretreatment values (± SEM) for each of the groups (from Ref. 17, with permission).

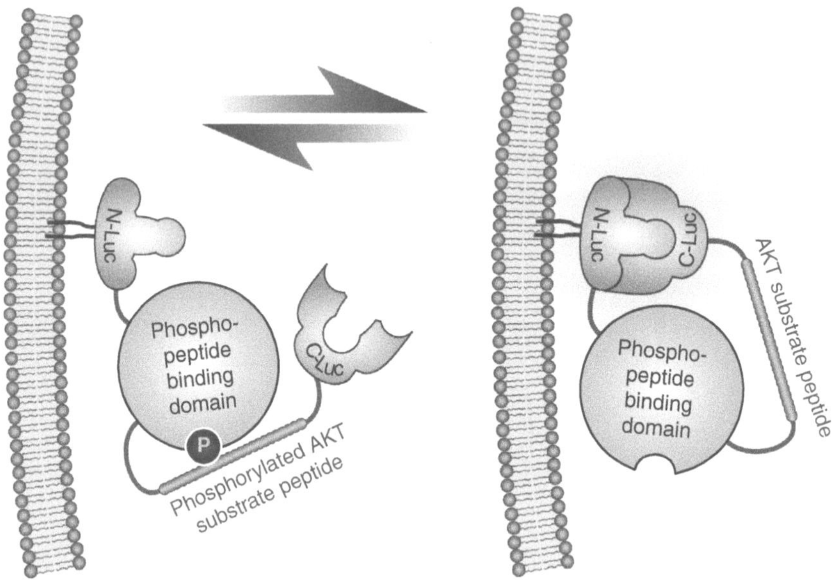

FIG. 3. Membrane-targeted bioluminescent Akt reporter. MyrPalm-BAR is generated by adding 10 amino acids from the N-terminus of Lyn kinase to BAR plasmid. The proposed basis of reporter activity for the MyrPalm-BAR reporter remains the same as that for BAR alone, which is described in Fig. 1. This involves Akt-dependent phosphorylation of the Akt peptide domain that results in its interaction with the FHA2 domain. In this form (Akt-ON), the reporter has minimal bioluminescence activity (Light-OFF). In the absence of Akt activity (Akt-OFF), association of the N-Luc and C-Luc domains restores bioluminescence activity (Light-ON).

inhibition *in vitro*, its evaluation as a potential target for cancer therapy is required *in vivo*. We have recently developed a model system that allows us to monitor NLG *in vivo* by using bioluminescent imaging techniques.[36] We utilized a modified luciferase reporter (ER-LucT) to monitor the cotranslational transfer of glycan precursors from its lipid precursor to consensus NLG sites (NXS/T) within the Luciferase reporter (Fig. 4A).[36] Glycosylation of Luc in the ER disrupted the ability of this enzyme to se luciferin and ATP as substrates and therefore had low bioluminescent activity. On the contrary, inhibition of NLG (and loss of the added glycan moiety) enhanced bioluminescent activity.[36] After *in vitro* validation, we tested this reporter in D54 glioma xenografts. With the use of this novel ER-LucT reporter, we demonstrated by noninvasively imaging these tumors that inhibition of NLG correlated with a decrease in RTK protein levels and tumor regression.[36] Using this molecular imaging approach, we further determined efficacious *in vivo* doses of the GlcNAc-1-phosphotransferase inhibitor, and tunicamycin, which blocks

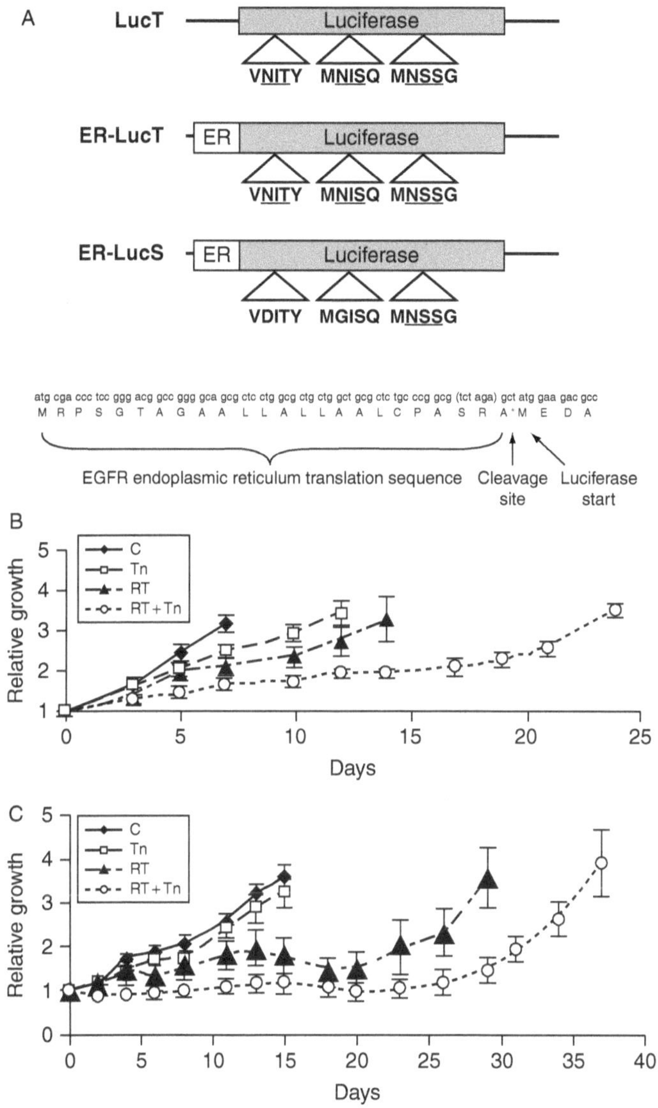

FIG. 4. Schematic diagram illustrating the NLG reporter design. (A) The ER translation sequence of the epidermal growth factor receptor is fused to the N-terminus of the Luc enzyme. Constructs contain either three (ER-LucT) sites for N-linked glycosylation or a single (ER-LucS) site. (B) Mice bearing wild-type D54 or (C) U87-MG xenograft tumors were randomized into four treatment groups: control, tunicamycin (Tn, 0.75 mg/kg), radiation (IR, 5 daily fractions of 2 Gy), and the combination of Tn with IR. Data points represent relative tumor growth compared to the tumor volume on Day 0, and error bars report the standard error (from Ref. 36, with permission).

N-glycan precursor biosynthesis. In summary, this reporter allowed for the determination that NLG inhibition in D54 and U87MG glioma xenograft tumors is therapeutically beneficial since the combination of NLG inhibiton and radiation therapy led to a significant tumor regression compared to NLG inhibition and radiation therapy given as single therapies (Fig. 4B, C).[36]

G. Clinical Significance for Imaging Proteases

Biological processes such as cleavage of nascent polypeptide chains, post-translational cleavage of inactive enzymes to yield functional enzymes, and proteolytic degradation of enzyme are regulated through the actions of proteases. Proteolytic processing is modulated both temporally and positionally, and thus contributes to protein activation and subcellular localization. Enzymatic cleavage plays a critical role in several physiological processes ranging from embryogenesis, hormone maturation, immunity, blood clotting, pathogenesis of viral and bacterial diseases to programmed cell death. Proteases are very tightly controlled and in the case of caspases even determine the cell's fate due to their critical role in coordinating proliferation and programmed cell death, which is essential for normal physiology. Dysregulation of signaling pathways culminating in caspase activation often results in diseases such as AIDS, neurodegenerative disorders, myelodysplastic syndromes, ischemia/reperfusion injury, autoimmune disease and cancer. Thus, quantitative noninvasive imaging of proteases is of importance for monitoring disease progression, and for screening and validation of experimental therapeutic agents.

H. Imaging of Enzymatic Activity

Programmed cell death culminates in the selective activation of caspases (cysteine-aspartic proteases) followed by the cleavage of specific target proteins and is initiated either by death receptors (extrinsic) or intrinsically by inhibitors of cellular pathways, such as staurosporine.[37–40] The converging point of this complex protease cascade of both intrinsic and extrinsic apoptotic pathways is the activation of caspase-3, which cleaves key proteins leading to the concomitant appearance of apoptotic morphology.[40–42]

To develop a molecular imaging tool, which would report on caspase-3 activity yet exhibit low levels of background bioluminescent activity when expressed in mammalian cells, we utilized our previously described chimeric luciferase reporter.[43] The fusion of an estrogen receptor (ER) regulatory domain to both the N- and C-terminus of luciferase had successfully silenced any bioluminescent activity of this enzyme (Fig. 5). The design of a novel apoptosis reporter included the protease cleavage site for caspase-3 (DEVD) at the junction of the luciferase and ER domains allowing protease-mediated activation of the reporter upon separation from the silencing domain (i.e., ER). When tested in cells undergoing apoptosis, caspase-3-dependent cleavage of

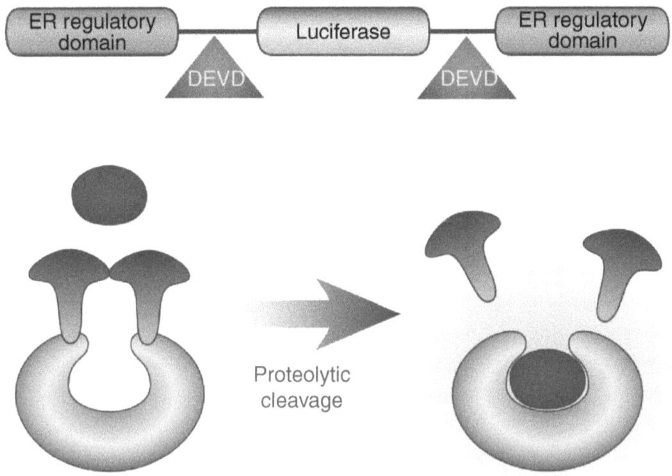

FIG. 5. Bioluminescent imaging reporter for apoptosis. Depicted is a recombinant reporter created by fusing residues 281–599 of the modified mouse estrogen receptor (ER) sequence to the N- and C-terminal ends of luciferase, respectively, yet separated by DEVD sequence on both ends. Caspase-3-dependent cleavage of the ER domains from the luciferase of the ER–DEVD–Luc–DEVD–ER reporter restores luciferase activity.

the reporter occurred and resulted in the restoration of luciferase activity.[43] Furthermore, the use of this bioluminescent reporter in a xenograft model enabled us to image caspase-3 activation noninvasively upon TRAIL treatment. It further provided insight into its potential application as a unique tool for evaluating therapeutic efficacy of experimental agents alone or in combination and for dose and schedule optimization. Using a combinatorial treatment regimen of 5-fluorouracil and TRAIL/Apo2L in a glioma xenograft model (D54 cells), the use of our caspase-3 reporter indeed allowed determination of increased efficacy of this drug combination versus each drug alone.[44]

I. Development of Caspase-3 Reporter Variants

Three additional variations of the caspase-3 reporter were developed to improve signal-to-noise ratios:

(A) Construction of a recombinant protein wherein Peptide A and B, a pair of peptides that had been reported to possess strong affinity for each other, were fused to N-Luc (ANLuc) and C-Luc (BCLuc), respectively with an intervening caspase-3 cleavage site.[27,45,46] This chimeric luciferase reporter had significantly reduced background luciferase activity as the

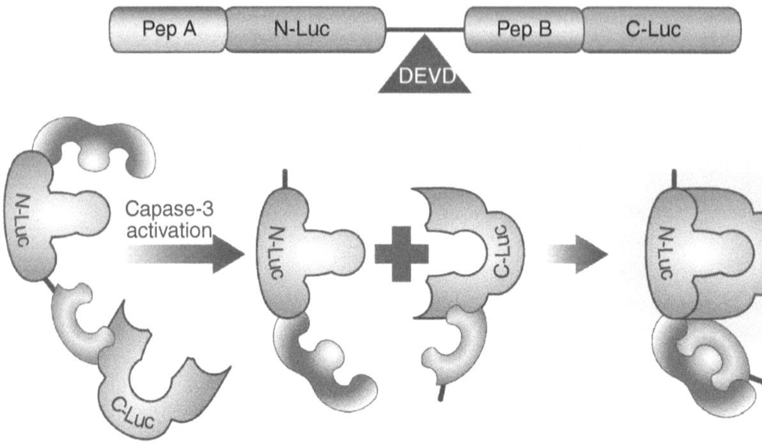

FIG. 6. Noninvasive imaging of caspase-3 utilizing a split-luciferase reporter strategy. (A) Schematic representation of the bioluminescent caspase-3 reporter (AN-Luc-BC-Luc). Apoptosis imaging reporter constitutes the split luciferase (N-Luc and C-Luc) domains fused to interacting peptides, pepA and pepB, with an intervening caspase-3 cleavage motif. (B) Upon induction of apoptosis, the reporter molecule is proteolytically cleaved by caspase-3 at the DEVD motif. This cleavage enables interaction between pepANLuc and pepBCLuc, thus reconstituting luciferase activity.

N-Luc and C-Luc were unable to complement when expressed as a fusion protein (Fig. 6). Recently published data with this optimized apoptosis reporter demonstrates that this is a highly sensitive, dynamic, and quantitative system for the detection of caspase-3 activity both *in vitro* and *in vivo*.[47] It further allowed *in vivo* optimization of dose, combination, and schedule of novel therapies in a dynamic, noninvasive manner.[47]

(B) To minimize potential oligomerization of the reporter, two ER regulatory domains were added which inhibited the interaction of the substrate with the enzyme and replaced the monomeric luciferase with the tetrameric β-galactosidase which hindered the oligomerization[48,49] (Fig. 7). Further advantages of using a β-galactosidase-based protease sensing reporter are the substrate variety available that are fluorogenic, paramagnetic, radioactive, or chemiluminescent and the increased stability of β-Galactosidase offering a robust and multimodality molecular imaging technology.[50–53] The use of this ER–*LacZ*–ER reporter for *in vivo* imaging in transgenic animals is described elsewhere in this chapter.

(C) A nonbioluminescent reporter which contained a single-chain antibody (harboring signal peptide, HA and myc tags, and a transmembrane domain), a Golgi retention signal and caspase-3 recognition and

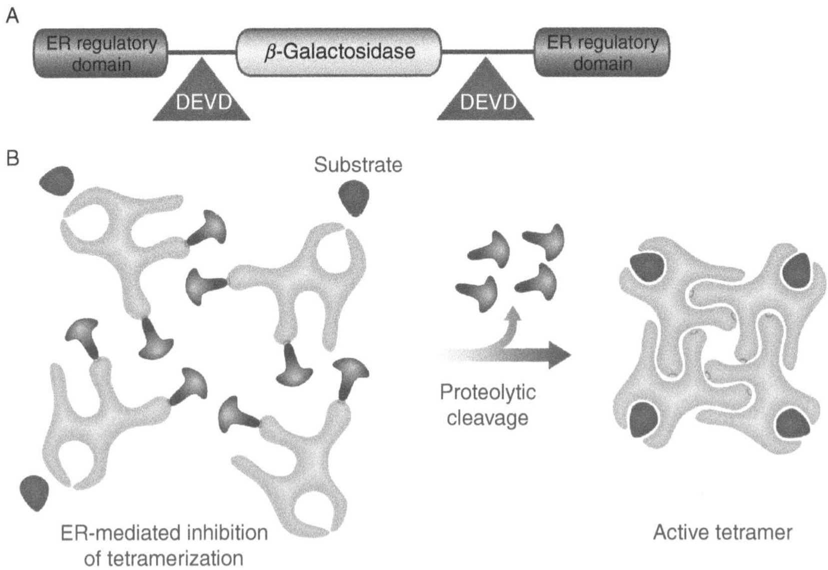

FIG. 7. Tetrameric LacZ-based reporter for monitoring cell death. (A) Recombinant beta-galactosidase-based apoptosis reporter was created by fusing ER at both N- and C-terminus of LacZ separated by DEVD sequence on both ends. (B) Addition of the ER to both end of LacZ had lead to inhibitory effect on the activity of ER–DEVD–LacZ–DEVD–ER: It inhibited the tetramerization of the enzyme, which is critical for its activity, leading to increased signal-to-noise ratio.

cleavage sequences (for details see Ref. 16) who also constructed cells which undergo cell death actively translocate this chimeric single-chain antibody, normally residing in the Golgi bodies, to the cell surface. Since this process is caspase-3 dependent, dying cells which contain the antibody on its cell surface were visualized (Fig. 8). Not only may this strategy enable a true three-dimensional imaging of apoptosis but it may also allow the visualization of tumor cells undergoing apoptosis by other imaging modalities when coupled with nanoparticles embedded with contrast agents.

J. Imaging of Organelle-Specific Proteases

Versatile and sensitive assay systems to monitor the activity of proteases involved in the maturation of secretory proteins have been lacking. These proteases reside in the trans-Golgi network (TGN), where they proteolytically process newly formed proteins from the endoplasmic reticulum before packaging into secretory vesicles. Studying TGN protease biology noninvasively

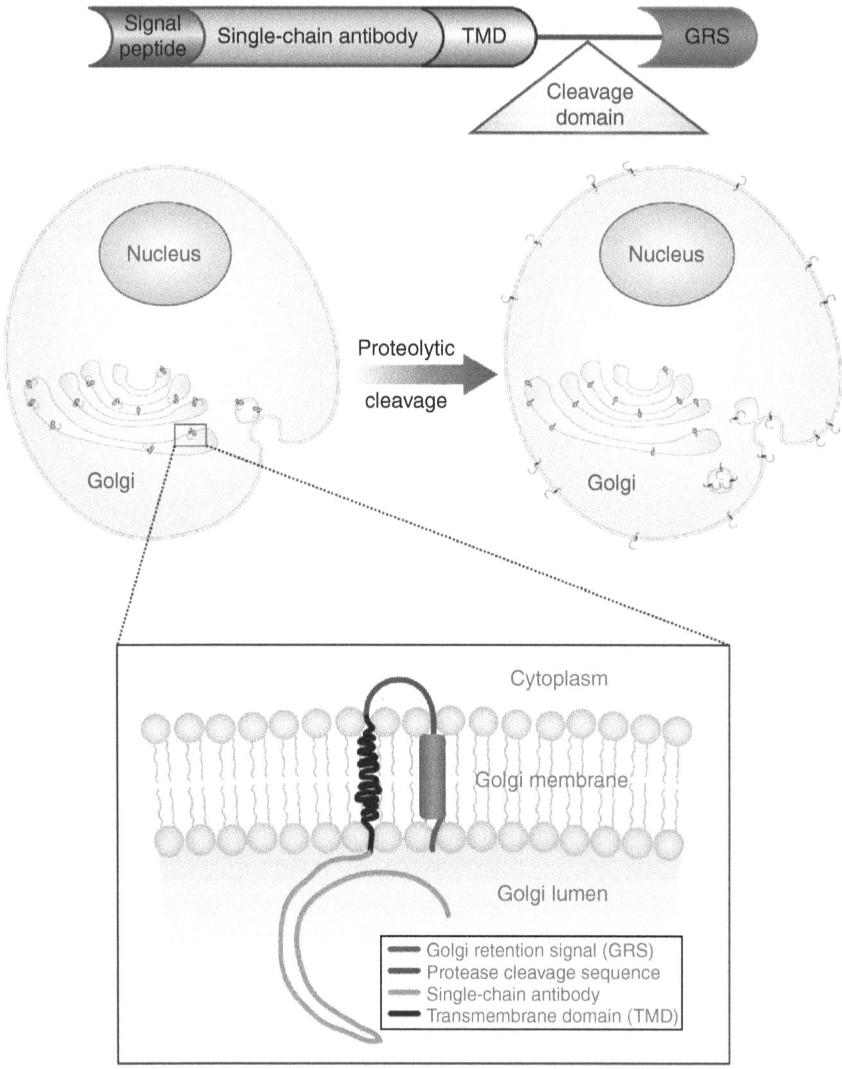

FIG. 8. Strategy for imaging of apoptosis based on conditional expression of single-chain antibody. (A) Constitutive imaging construct (CIC) was constructed from a single-chain antibody, signal peptide, HA and myc tags, and a transmembrane domain. When expressed in cells, this fusion protein localizes to the cell surface. (B) The inducible imaging construct (IIC) was CIC engineered to contain a Golgi retention signal, separated by caspase-3 recognition and cleavage sequence. The DEVD sequence was placed between the transmembrane domain and a Golgi retention signal to retain caspase-3 cleavage sequence in the cytoplasm (see inset). When expressed in cells, this chimeric protein localizes to Golgi bodies. Induction of apoptosis leads to caspase-3 activation and cleavage of the chimeric protein resulting in its translocation to cell surface.

would allow the preservation of the unique intracellular TGN environment (low pH, high Ca^{2+}) that is otherwise perturbed by commonly used biochemical methods. Such strategies are further advantageous for discovering novel pharmaceutical agents that can traverse plasma and Golgi membranes and retain inhibitory activity within the microenvironment of TGN. TGN-residing proteases include carboxypeptidases, prohormone convertase (PC) family members, and β-site amyloid precursor protein (APP)-cleaving enzyme (BACE) family members.[54–57]

In an effort to develop a noninvasive reporter for TGN-residing proteases, we used three domains: a secreted alkaline phosphatase (SEAP), a Golgi protease-specific recognition and cleavage site, and the cytoplasmic and transmembrane domains from BACE function to retain the reporter within the TGN (Fig. 9). This GRAP reporter localizes to TGN until it is cleaved by a specific TGN protease after which SEAP is secreted into the extracellular medium. Thus, SEAP levels present in the media are indicative of intracellular TGN protease activity. Decreases in SEAP levels signify a loss of protease activity and allows positive identification of protease inhibitors.

The TGN-enriched protease Furin plays a critical role in processing a myriad of proteins such as serum proteins (proalbumin), coagulation factors (pro-von Willebrand factor), growth factors, hormones, cell surface receptors (insulin proreceptor), and matrix metalloproteases (stromolysin-3 and MT1-MMP) and yet it is also utilized by a number of pathogens such as HIV-1, ebola, and avian influenza and virulent bacterial pathogens such as anthrax, pseudomonas, and diphtheria.

Thus, enhanced processing by Furin and other proteases has been linked to the development of several diseases, such as Alzheimer's disease (AD), arthritis, and cancer. Remarkably, recent data suggest that the acquisition a glycoprotein effectively cleaved by Furin may extend the ability of the newly discovered H5N1 avian influenza virus thus contributing to its pathogenicity. The activation of the toxin anthrax through Furin cleavage is another example where one may envision benefits for inhibiting such protease activity.

To obtain the ability for noninvasive monitoring Furin activity, we included a 10-amino acid recognition and cleavage site (GLSARNRQKR↓) of the furin substrate Stromolysin-3 (ST3) in our previously developed reporter and named it GRAPfurin. Overexpression of Furin in cells resulted in an observed increase in processing of GRAPfurin reporter as detected by western blotting and as an increase in SEAP activity in the media.

The specificity of this system was shown by construction of GRAPfurinmut in which the furin target recognition and cleavage sequence was mutated to GLSAANAQAA↓, rendering this reporter nonresponsive to Furin proteolytic activity. This led to a reduction in processed protein in both the lysate and extracellular medium and a concomitant decrease in SEAP activity in

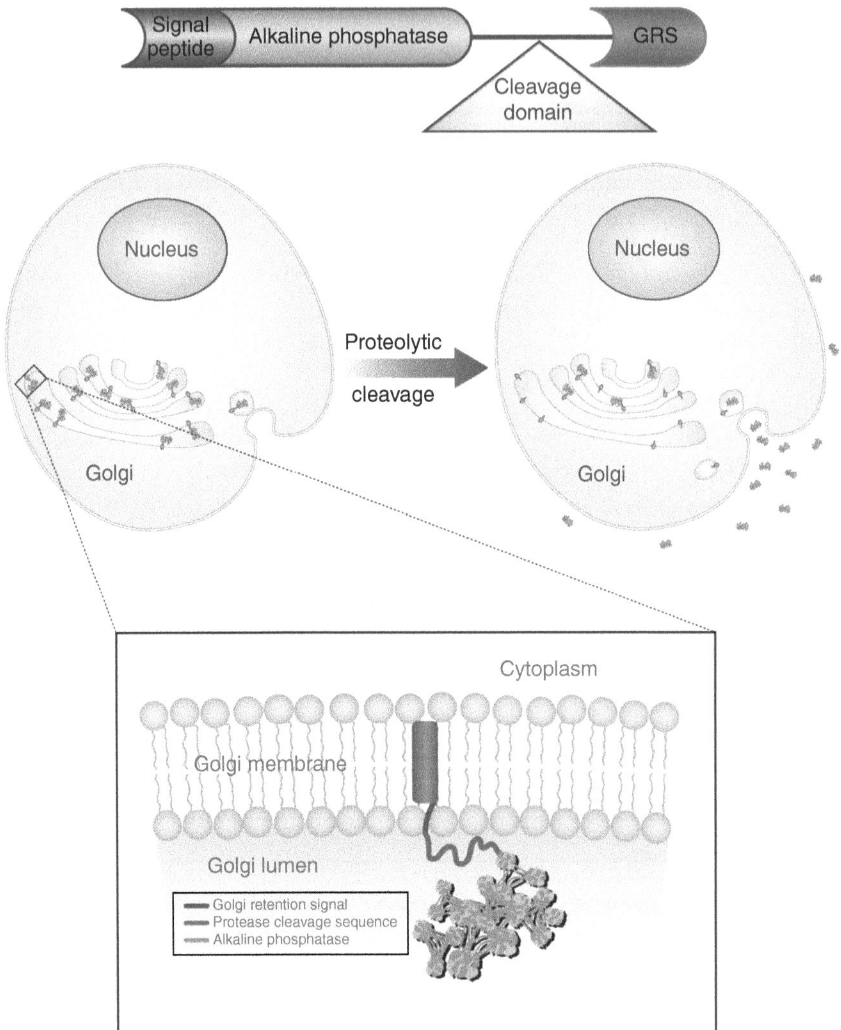

FIG. 9. TGN protease imaging reporter. (A) Fusion protein consisting of three functional domains, a secreted alkaline phosphatase (SEAP), a Golgi protease-specific recognition and cleavage site, and a Golgi retention signal from BACE that retains the reporter within the TGN, was created. This reporter localizes to TGN until it is cleaved by a specific TGN protease, after which SEAP is secreted into the extracellular media. The protease-specific recognition and cleavage sequence is located within the lumen of the Golgi to respond to specific TGN proteases residing in this compartment of the Golgi bodies (see inset).

media.[58,59] Additionally, a notable decrease in SEAP activity was observed when CHO-GRAPfurin cells were treated with 25 μM of the Furin inhibitor, dec-RVKR-CMK, whereas similar treatment with dec-RVKR-CMK caused no decrease in SEAP activity in the control CHO cell line expressing SEAP constitutively.[59] In summary, using this technology, monitoring of Furin protease activity is now possible using molecular imaging.[59]

K. Use of Molecular Reporters for High-Throughput Drug/Target Screening

The use of Furin inhibitors such as α1-PDX, D6R, D9R, and dec-RVKR-CMK as pharmaceutical agents is unfortunately hampered by their large size, lack of stability, and/or toxicity. To date, the only nonprotein/peptide inhibitor of Furin is a naturally occurring neoandrographolide and its succinoyl ester derivatives, with IC50 values ranging from high micromolar to low millimolar values. Thus, there is an urgent need for identifying novel Furin inhibitors with desired characteristics in terms of toxicity, solubility, and ability to interact with the protein target in its appropriate subcellular compartment. Such drug discovery can be achieved by modifying above described cell-based assays for HTS of compound and siRNA libraries. Utilizing this approach, identification of Furin inhibitors utilizing our GRAPfurin reporter from a screen of 39,000 molecules was accomplished. CCG 8294, one of the major hits in the high-throughput assay, has shown promise as a Furin inhibitor with high efficacy in cells and has also shown inhibition of Furin-mediated processing of polypeptides within the secretory pathway.[59]

This platform may also be used for imaging of kinase activity. These cell-based screens have unique advantages as only compounds interacting with the target in the correct cellular compartment and under normal cellular physiological conditions of that subcellular compartment (pH, concentrations of specific ions, etc.) would be identified. In addition, because the assay involves live cells, the reporter enables monitoring of the kinase in question in the context of other signaling pathways. Lastly, in contrast to other cell-based reporter screens, which are fraught with false positives, the kinase reporter such as BAR is a "gain-of-function assay" wherein the inhibition of kinase activity results in increase in bioluminescence. For example, compounds that are cytotoxic (and thus result in loss of signal) or those that inhibit the reporter directly (e.g., luciferase inhibitors) may show up as false positives in traditional kinase assays but not with the BAR reporter platform. Such carefully designed screening methodologies would enable one to narrow down the number of hits to a smaller group of "true positives."

In summary, we have developed imaging reporters for monitoring activities of kinases, proteases, and programmed cell death and thus provide imaging tools for HTS of targeted inhibitors, and pro- and antiapoptotic compounds and validation of such.

L. Design of Transgenic Reporter Mice for Optical Imaging

Transgenic reporter mice contain transgenes, which typically consist of a promoter driving the expression of a fluorescent or bioluminescent reporter gene. The advantage of such reporter mice lies in the ability to follow the activation of specific promoters and transcription factors or even enzymatic activity in real time by utilizing optical imaging techniques. Biological processes such as transcription or apoptosis can be monitored both over time and in all organs of the animal. The combination of the Cre-loxP system with the use of transgenic reporter genes allows tissue-specific Cre recombination and thereby activation of the fluorescent or bioluminescent reporter. The recent development of a reporter mouse, wherein global deletion of the loxP-flanked EGFP by germ line Cre recombination leads to the expression of Luciferase and β-GAL, exemplifies this concept.[60] This reporter mouse allows for noninvasive imaging of targeted Cre activation *in vitro* or in living animals and will be useful for future studies examining these events.

A wide range of human disorders involves the inappropriate regulation of NF-κB, including cancer, neurodegenerative diseases, asthma, and inflammatory diseases. With the development of transgenic mice reporting on NF-κB activity, key processes regulating NF-κB-dependent transcription can now be identified *in vivo*.[61] Another signaling molecule involved in the pathogenesis of many human diseases, including cancer, fibrotic disorders, and neurodegeneration is TGF-β. Transgenic mice containing a Smad-responsive luciferase reporter construct (SBE-luc) are useful instruments to assess Smad2/3 signaling activity. Smad 2/3 are anchor proteins important for downstream TGF-β signaling. The use of SBE-luc mice allows tissue-specific detection of TGF-β pathway activation in response to systemic endotoxin challenges or brain injury.[62]

Optical imaging plays also an important role in the preclinical assessment of drug target interaction and hence will continue to enhance our ability to fight cancer. The development of a transgenic mouse reporting on hydroxylase activity is an excellent example of such use.[63] The transcription factor hypoxia-inducible factor (HIF) consisting of a labile α subunit and stable β subunit is important for the cell's adaptation to hypoxia. Under conditions of normal oxygen tension (normoxia), members of the EGLN family hydroxylate HIFα subunits on conserved prolyl residues. This signals the pVHL E3 containing complex to polyubiquitinate HIFα subunits resulting in its degradation. Under hypoxic conditions, or in the absence of pVHL, the alpha subunits accumulate in the cell and engage in transcriptional activation of genes involved in acute or chronic adaptation to hypoxia. With the fusion of a luciferase to the HIF1α region that binds pVHL in and oxygen-dependent manner, Safran and colleagues[63] generated a transgenic mouse, which can report on EGLN hydroxylase activity. The ROSA26 ODD-Luc mouse expresses the oxygen-dependent domain of Hif1α subunit fused to luciferase protein (ODD-Luc) and responds to changes in oxygen tension. Moreover, this mouse has been used to monitor the action of small molecule inhibitors of HIF prolyl hydroxylase activity.[63]

Apoptosis is an essential process for the maintenance of normal physiology. Dysregulation of cell death has been defined as a hallmark of carcinogenesis. Thus, not only will monitoring apoptosis noninvasively in living animals provide a unique insight into its function in normal and disease processes, but it will also allow the assessment of preclinical drug efficacy. We have recently developed a novel reporter mouse, carrying a β-Galactosidase gene flanked by two regulatory domains of the estrogen receptor and intervening Asp-Glu-Val-Glu (DEVD) sequences which is controlled by the skin-specific keratin 5 promoter (kRT5).[48] The structure of the reporter is shown in Fig. 10A. The activation of the excecutioner caspase (C-3) results in the cleavage of the reporter at the C-3 recognition site (DEVD), which was visualized by using a near-infrared fluorescent substrate of β-galactosidase (DDAOG, (9H-{1,3-dichloro-9,9-dimethylacridin-2-one-7-yl} β-D-galactopyranoside).[48] Immunohistochemical analysis of skin tissue from transgenic animals revealed the presence of β-galactosidase immunoreactivity in epidermal cells, which corresponded to cells that had KRT5-positive staining (Fig. 10C). Control animals failed to show β-galactosidase-specific immunoreactivity (Fig. 10B). To investigate whether the reporter in the transgenic animals was conditionally activated in response to an apoptotic stimulus, mice were UV-irradiated and fluorescence imaging was performed upon administration of the DDAOG substrate.[48] A significant increase in DDAO fluorescence was observed at 24 h in UV-irradiated animals compared (Fig. 11) with control animals.[48] The analysis of skin samples from nontransgenic as well as unirradiated (transgenic CON) and UV-irradiated (transgenic UV) transgenic animals revealed the presence of the 190-kDa ER-LACZ-ER polypeptide in the transgenic animal. To validate the imaging studies, immunohistological studies using an antibody specific to active caspase-3 were accomplished to demonstrate the presence of apoptotic activity within UV-irradiated mouse skin samples. No significant staining was observed in unirradiated animals, whereas UV-treated animals had significant levels of active caspase 3 positivity within the epidermal cells. Finally, skin sections from an untreated control mouse and a UV-irradiated mouse were stained using 5-bromo-4-chloro-3-indolyl-b-D-galadctopyranoside to identify cells that possessed both active β-galactosidase and an antibody specific to active caspase-3.[48] Untreated control cells had

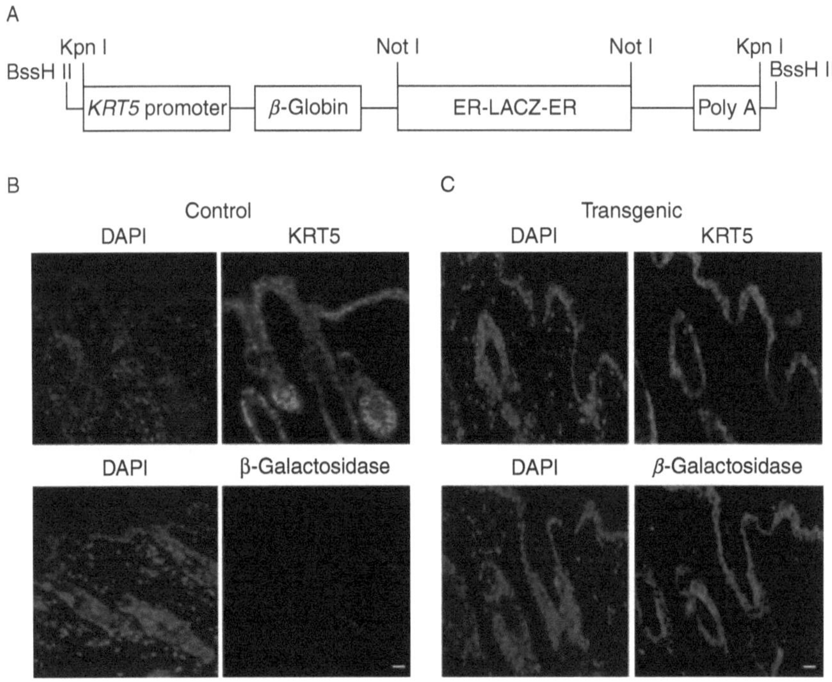

FIG. 10. Schematic of the transgenic apoptosis reporter construct. (A) The transgene depicted contains the ER-LACZ-ER coding sequence (see Fig. 7) under the transcriptional control of the keratin 5 (KRT5) promoter. B & C, Immunohistochemical analysis of formalin-fixed, paraffin-embedded dorsal skin sections from nontransgenic (control) (B) and transgenic (C) mice (ER-LACZ-ER transgenic). Presence of β-galactosidase protein and KRT5 was detected using the appropriate antisera. DAPI (4,6-diamidino-2-phenylindole) staining was used to identify nuclei. In control animals, KRT5 staining but not b-galactosidase staining was observed, whereas in transgenic animals, KRT5 staining and β-galactosidase staining was detected in similar populations within the section. Bar = 40 μm (from Ref. 48, with permission).

no significant staining for active caspase-3 and low β-galactosidase activity.[48] These results reveal colocalization of the β-galactosidase activity with activation of caspase-3, thus directly correlating detection of the fluorescent signal with apoptosis using this molecular imaging reporter system.[48]

In summary enzymatic activation of the reporter during apoptosis enabled us to monitor β-galactosidase activity noninvasively in living cells in a dose- and time-dependent manner. Transgenic animals provide the ability to image apoptosis in the skin, and thus will enable unique insights into the role of apoptosis in skin biology, for example, wound healing and UV-light-induced DNA damage and melanoma to be obtained *in vivo*.[48]

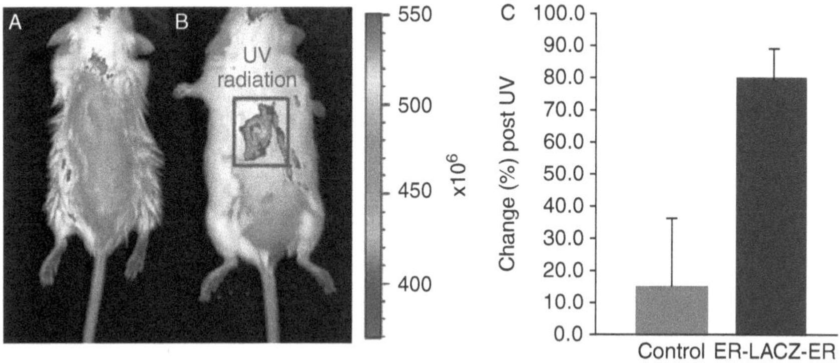

FIG. 11. Imaging of UV-induced apoptosis in transgenic animals. Transgenic mice were shaved and UVB-irradiated (B) or mock irradiated (A). The irradiated area is outlined as a red square. After irradiation, mice were injected with the fluorescent substrate DDAOG (9H-{1,3-dichloro-9,9-dimethylacridin-2-one-7-yl} β-D-galactopyranoside) and imaged using a Xenogen IVIS system. As compared with the control animals, the UVB-treated animals showed a significant fluorescent signal 24 h after radiation. (C) Control animals (open bar) showed an approximately 10% change in fluorescence, whereas UV-irradiated animals (solid bar) had an 80% mean increase in fluorescence activity. Error bars represent ± SD, $n = 5$ animals per group (from Ref. 48, with permission). (For interpretation of the references to color in this figure legend, the reader is referred to the Web version of this chapter.)

In conclusion, the development of transgenic reporter mice and their increasing use for preclinical drug/target identification, validation and dosing, and schedule optimization will contribute in many ways to improving drug-development.

II. Magnetic Resonance Imaging

A. Introduction

Magnetic resonance imaging (MRI) has proved to be one of the most important advances in the radiologic diagnosis of oncology patients. With high spatial resolution and soft tissue contrast, MRI allows precise noninvasive radiographic measurements of tumor location and size. Rapid advancements in functional MR technologies have facilitated rapid growth and widespread availability of clinical MR scanners. Routine MR acquisition sequences provide the capability of generating images of fundamental biophysical, physiologic, metabolic, or functional properties of tissues. This allows for characterization of tissue perfusion,[64] vascular permeability,[65,66] tissue oxygenation,[67] cellular status,[68] cellular density,[69] and microstructural organization,[70,71] all of which are used in research and clinical studies. Development and validation of MRI biomarkers, capable of monitoring the biology and behavior of tumors, are being investigated for their efficacy at predicting outcomes in the clinical management of individual cancer patients. This is of great interest since standard risk factors currently used cannot account for the variable and unpredictable treatment responses of patients with a similar risk profiles. This chapter highlights several key emerging functional and molecular imaging approaches as they are applied to translational imaging.

B. Diffusion Magnetic Resonance Imaging

Molecular imaging commonly refers to imaging techniques capable of measuring biologic processes at the cellular and molecular level. An example of an imaging readout serving as a biomarker for a cellular event, such as enzymatic expression from a targeted gene, qualifies as a molecular imaging modality. DW-MRI is sensitive to molecular water interactions, resulting from thermal motion, that occur at the cellular level. Water molecules are typically the signal source therefore, water mobility is probed in DW-MRI. In pure water, temperature is the only significant modulator of molecular mobility, and in fact, diffusion MRI has been used to measure temperature noninvasively.[72] However, water mobility in cancer tissue is strongly impacted by biologic factors on the cellular level, making DW-MRI a unique diagnostic tool.[73,74] Indeed, DWI is readily available and increasing in its use in clinical practice owing to its exquisite sensitivity to cellular status, cytotoxic edema, cellular density, and cellular organization of tissues.[75–78] The objective of this section is to provide a broad overview of basic methods and applications of diffusion MRI as applied to cancer imaging.

1. Principles Involved in Diffusion Imaging of Cancer

By acquiring water signal using a spin echo sequence in the presence of bipolar time-dependent field gradients, Stejskal and Tanner[79] measured water diffusivity, which was first demonstrated in 1965. They showed that water signal attenuates at a rate proportional to the diffusivity of the molecule. Diffusion-weighted MRI was not applied to *in vivo* systems until the 1980s,[74,80,81] Since then, DW-MRI has found wide applications as a research tool for studying biological systems resulting in reviews on the technical aspects and consensus biomarker recommendations using diffusion imaging.[77,82–83] Molecular diffusion is the thermally driven random translational motion of molecules in media commonly referred to as *Brownian motion*. Key factors that influence the mobility of a diffusing molecule include media viscosity, temperature, and its molecular mass. Unlike, magnetization-related process such as T1 and T2 relaxation times which drive conventional MRI contrast diffusion measurements obtained by DW-MRI are unaffected by field strength. This allows for the noninvasive and consistent quantification (image) of water diffusion values spatially *in vivo*. Diffusion values are obtained in part through the use of magnetic field gradients which "encode" the initial locations of constituent water molecules in the tissue. Following a brief interval, the same gradients "decode" the molecular locations. The decoding of water molecules that undergo displacement during the time interval is incomplete resulting in the loss of signal through spin dephasing. The extent of dephasing increases proportionally to the distance translated between encode/decode diffusion gradient pulses. A larger loss of signal will be observed in highly mobile water molecules compared to immobile water in more hindered/cellular tissue environments, which result in relatively strong signal on diffusion-weighted images. The extent of signal loss at various diffusion gradient settings provides the means for calculating molecular mobility in complex systems, such as tumor tissue. However, water within tumor tissue maybe bound to macromolecules through hydrogen bonding[84] or compartmentalized and separated by semipermeable membranes. Thus, the concept of a single diffusion coefficient is not valid but rather a spectrum of diffusion values. As such diffusion measurements are typically reported as an "apparent diffusion coefficient" (ADC) when performing diffusion-sensitive sequences on tissues.[74,78] ADC measurements can be used to assess a myriad of effects that impede molecular motions including cell membrane integrity, cell density, interactions with macromolecules, as well as processes that enhance mobility via active transport, convective motion, and perfusion.

Pure water at body temperature has a diffusion coefficient of approximately 3×10^{-3} mm^2/s, which results in a displacement distance of 0.03 mm, or 30 μm, in 50 ms, which is on the order of the typical MR time interval used clinically. Tumor cells as well as other structures such as membranes, organelles, myelin layers, and macromolecules span smaller dimensions than those displaced by pure water. Thus, a water molecule will likely encounter many interactions with cellular or subcellular entities over this measurement interval. Transient association of water with obstacles residing within tumor tissue effectively reduces water mobility to an ADC lower than free water diffusion. The greater the bulk density of structures within a tumor tissue that hinders water mobility, the lower the ADC value for that tumor. For this reason, ADC is considered a noninvasive imaging biomarker of cellularity or cell density. However, if two tissues have different ADC values, the lower ADC tissue may not necessarily have the greater number of cells per unit volume. Other factors such as cell size, relative sizes compartment volumes, and membrane permeability also affect water mobility and ADC. For a given tissue, ADC is useful as an indicator of the relative cellularity, such as in the evolution of tumor over time following therapy. Alterations in the cellular makeup of the tumor due to disease or intervention, as well as changes in cellular organization or integrity of cellular elements, are available for study by DW-MRI.

Accurate diffusion measurement is attainable in spite of the presence of physiologic motions. A single-shot echo-planar imaging (EPI) approach[85] has become the standard MR sequence in clinical studies as its rapid acquisition speed allows the entire set of echoes for an image to be collected within one single scan period, essentially eliminating bulk tissue motion. However, images generated by EPI are sensitive to other artifacts such as image distortion and signal loss as a result of magnetic susceptibility. These limitations aside, EPI combined with diffusion-sensitization gradient pulses is the most commonly used clinical sequence for DW-MRI.

2. Diffusion Imaging in Tissue Characterization

Tumor ADC maps generated from DW-MRI data aid in defining solid enhancing tumor, noncontrast enhancing lesions, peritumoral edema, and necrotic or cystic regions from normal surrounding tissue. Progressively increasing ADC values have been widely observed in dense cellular tumors to necrotic cysts which is consistent with known histological properties of tumors. The ADCs for highly cellular dense tumors are $0.6–0.8 \times 10^{-3}$ mm^2/s,[86] whereas the ADCs for solid enhancing high-grade glioma span from 0.8 to 1.3×10^{-3} mm^2/s.[86] ADC values of edematous brain are in the range of $1.3–1.4 \times 10^{-3}$ mm^2/s, and a necrotic tumor core typically has an ADC of $1.8–2.4 \times 10^{-3}$ mm^2/s. ADC measurements obtained by diffusion MRI has also been reliably shown in abdominal organs and tumors within organ sites such as kidney, liver, and pancreas.[87] ADC values have also been evaluated in colorectal hepatic metastases.[88–89] This ability to reliably obtain ADC measurements in internal organs has allowed the investigation of pretreatment ADC as a predictive biomarker of chemotherapeutic response of hepatic metastatic lesions from colorectal cancer.[89] It was observed that significantly high mean pretreatment ADC values were found in metastatic lesions that responded to chemotherapy, which may have implications for future development of individualized therapy.

Diffusion measurements have also been investigated for differentiating benign and malignant lesions in liver, breast, and prostate, where increased cellularity of malignant lesions hinders water motion in a reduced extracellular space.[90] Whole-body diffusion MRI has recently been reported for screening malignancies throughout the body.[91] This approach of acquiring whole-body diffusion images has been demonstrated on freely breathing patients. Figure 12 shows whole-body MRI, including DW-MRI of a 60-year-old male with stage III diffuse large B-cell lymphoma. Coronal T1-weighted and slab maximum intensity projection (MIP) DW-MRI images reveal lymph node involvement on both sides of the diaphragm (see arrows). The highest lymph node-to-background contrast was observed in the whole-body DW-MRI. Whole-body diffusion MRI for applications in tumor detection and monitoring of treatment response will continue to be an active area of investigation.

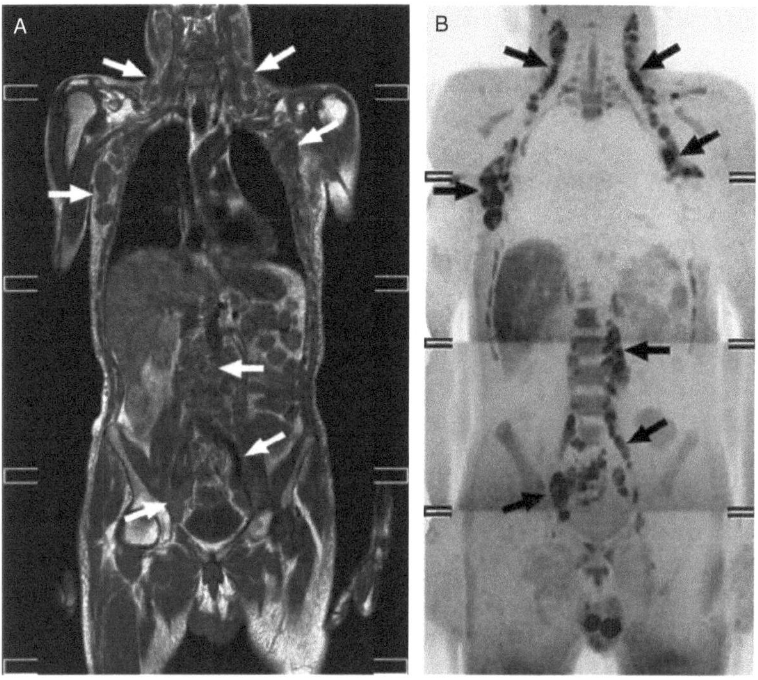

FIG. 12. Whole-body anatomical and DW-MR images of a 60-year-old male with stage III diffuse large B-cell lymphoma. (A) Coronal T1-weighted and (B) maximum intensity projection DW-MRI obtained with a *b*-value of 1000 s/mm^2 show lymph node involvement on both sides of the diaphragm (arrows). Note that whole-body DW-MRI provides superior lymph node-to-background contrast. (Figure kindly provided by Dr. Thomas Kwee, Department of Radiology, University Medical Centre Utrecht, Utrecht.)

3. Diffusion Imaging in Tumor Grading

Investigation into differentiating between tumor type and grade has also been done using DW-MRI and diffusion tensor MRI (DT-MRI) in adult as well as pediatric populations. Preliminary results using diffusion MRI for detecting pancreatic adenocarcinoma have been reported with high sensitivity and specificity.[92] Diffusion MRI has also provided useful diagnostic information for discriminating poorly differentiated from undifferentiated carcinomas[93] and benign from malignant salivary gland tumors.[94] Several studies have also observed high ADC values in low-grade astrocytoma, whereas low ADC values were reported in high-grade malignant glioma. These findings reflect a more hindered diffusion with increasing tumor cellularity.[69,95] Although tumor anisotropy has been investigated, it remains uncertain whether tumor type and grade can be differentiated by anisotropy indices derived from DT-MRI.

Tumor cytoarchitecture is predominantly random; therefore, negligible anisotropy is observed in tumor. In addition, normal tissue anisotropy depends heavily on its location in the brain,[96] which implies that the contrast of tumor to normal background, as depicted by anisotropy, will depend on lesion location. There is justifiable optimism that anisotropy will be valuable in assessing the effect of tumor on normally unidirectional white matter structures. Displacement and compression of white matter tracks by tumor mass effects as well as destruction of track organization by tumor infiltration have been documented by anisotropy-based DT-MRI, which suggests that this technology may have a role in presurgical planning.[97–99]

4. Diffusion Imaging to Assess Tumor Cellularity and Treatment Response

It is traditionally viewed that as cellular density increases, the added tortuosity reduces water mobility. The inverse relationship between ADC and cellular density has been noted by several groups.[69,98,100,101] A recently proposed biphasic model relating ADC values to cellularity assumes that water resides in two pools within tissue, a fast diffusion and a slow diffusion pool.[84] The slow diffusion pool is proposed to consist of a water layer trapped by electrostatic forces, that is, hydrogen bonding, of the protein membranes and associated cytoskeleton. The fast diffusion pool is thought to belong to water compartmentalized in intra- and extracellular spaces, which are slower than free water. Based on either the traditional, water resides in intra- and extracellular compartments, or biphasic diffusion models water diffusion is expected to decrease during cell swelling or cell proliferation and increase during treatment-induced loss of cellular viability or density. Whatever the specific underlying mechanism governing water diffusion in tissue, the fact remains that tumor diffusion values increase as tumor tissue progresses from a solid, cellular lesion to an acellular, necrotic tumor during successful cytotoxic therapy. This characteristic of tumor water diffusion values provides a key opportunity to use this quantifiable ADC parameter as a sensitive biomarker for detecting the underlying changes of tumor cytoarchitecture associated with treatment.[102]

Because treatment-induced cellular changes precede gross volumetric changes in tumor size, diffusion MRI can provide early detection of changes in tumor structure. Preclinical and clinical cancer studies could thus utilize this imaging biomarker as an early response indicator. Sixteen years of research in preclinical studies have supported this notion that diffusion MR can be used to noninvasively detect cellular changes associated with treatment-induced cell killing in animal models.[44,69,98,100,103–108] The key findings in these studies were that changes in ADC values precede changes in tumor volume regression and were treatment-independent and dose-dependent, all supporting the claim that this imaging biomarker may indeed be used as an early surrogate for treatment outcome.

Tumor burden is typically assessed between pretreatment scans and those obtained weeks to months after the conclusion of a therapeutic protocol.[109,110] Clinical studies have found that early changes in tumor ADC correlate with a delayed clinical response to therapy.[101,111–146] In general, a significant difference in the mean ADC between responders and nonresponders to therapy has been reported, as well as a linear correlation between the relative change in ADC and the normalized change in tumor volume.[124] The correlation of increasing ADC with a positive clinical outcome was also observed in head and neck squamous cell carcinomas.[140] Presented in Fig. 13 are the corresponding pretreatment and

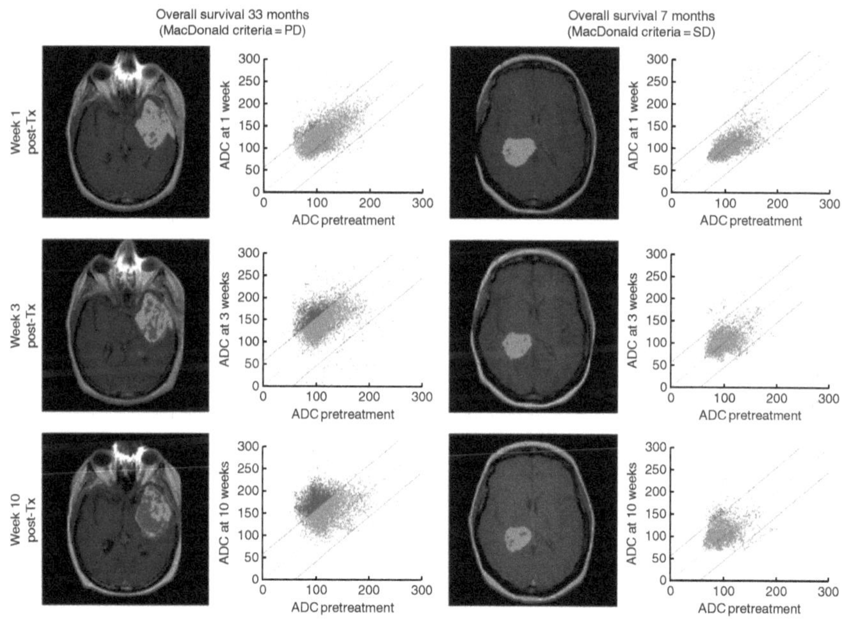

FIG. 13. Representative functional diffusion map (fDM) analysis at 1, 3, and 10 weeks of clinically (left panels) responding and (right panels) nonresponding patients to fractionated radiation therapy. Presented images are single slices of the T1-weighted contrast-enhanced scans at each time point with a pseudocolor overlay of the fDM. Red voxels indicate regions with a significant rise in apparent diffusion coefficient (ADC) at each time point compared with pretreatment, green regions had unchanged ADC, and blue voxels indicate areas of significant decline in ADC. The scatter plots display data for the entire tumor volume and not just for the depicted slice at each time point, with the pretreatment ADC on the x-axis and posttreatment ADC on the y-axis. The central red line represents unity, and the flanking blue lines represent the 95% CIs (from Ref. 138, with permission). (For interpretation of the references to color in this figure legend, the reader is referred to the Web version of this chapter.)

3 week post treatment initiation T1-weighted and ADC maps with corresponding histograms of a patient treated by nonsurgical organ preservation therapy. The diffusion histogram reflects a broad ADC distribution with some areas exhibiting a very low ADC consistent with high cellularity (mean ADC of 120×10^{-5} mm^2/s). Following 3 weeks of treatment, the nodal tumor showed a negligible increase in tumor volume, whereas the ADC had increased by 28% (153×10^{-5} mm^2/s). Given the large nodal disease and minimal volumetric response, this patient underwent a clinically necessitated cervical lymph node dissection which revealed no evidence of residual disease (complete responder) and was alive and free of disease 35 months from the completion of treatment. DW-MRI of a breast cancer patient who has undergone two cycles of neoadjuvant therapy revealed an increase in the tumor diffusion values.[119] This indicates a reduction in the tumor cell density as a result of treatment but, no significant reduction in tumor size was observed. A significant decrease in tumor volume was noted prior to the second treatment cycle. Although initial increases in tumor ADC values during treatment are typically associated with cell death, a drop in ADC values has been observed later within the tumor, even to pretreatment levels. This trend in ADC may indicate tumor regrowth or fibrosis. This is supported by findings of lower ADC values in contrast-enhancing portions of recurrent high-grade gliomas compared with those obtained in patients with radiation injury and necrosis where higher diffusion values are observed in necrotic regions of osteosarcomas.[113,147]

A major confounding factor in assigning a single indicator for patient tumor response is tumor heterogeneity. A given lesion often contains wide gradations of viable cellularity and necrosis resulting in a nonuniform response of tumor subregions to treatment. Whole-tumor histogram-based analysis of ADC values is one approach to addressing spatial heterogeneity to response. Nevertheless, the extent of regional changes may be underestimated by whole-tumor averages. To deal with intrinsic heterogeneity of diffusion values within a tumor, an alternative approach for image analysis has been proposed, referred to as *functional diffusion mapping* (fDM).[121] A key component of fDM is spatial alignment, that is, registration, of all three-dimensional image sets into a common geometrical framework. Once aligned, changes in diffusion values are determined on a voxel-by-voxel basis from pretreatment, mid-treatment, and posttreatment ADC maps. To visualize changes in ADC within the tumor, tumor volumes are segmented into three categories representing voxels for which ADC (1) increased by a specified threshold (red voxels), (2) decreased (blue voxels), and (3) did not change outside this threshold range (green voxels). Illustrated in Fig. 14 are a series of fDM color overlays and corresponding voxel-wise scatter plots of ADC pretreatment versus at 1, 3, and 10 weeks posttreatment initiation for a therapeutically responding and nonresponding patient. The relative volume of tumor that exhibited a significant increase in ADC, shown as red voxels in the color overlay and scatter plots, is used as a biomarker to predict treatment outcome.

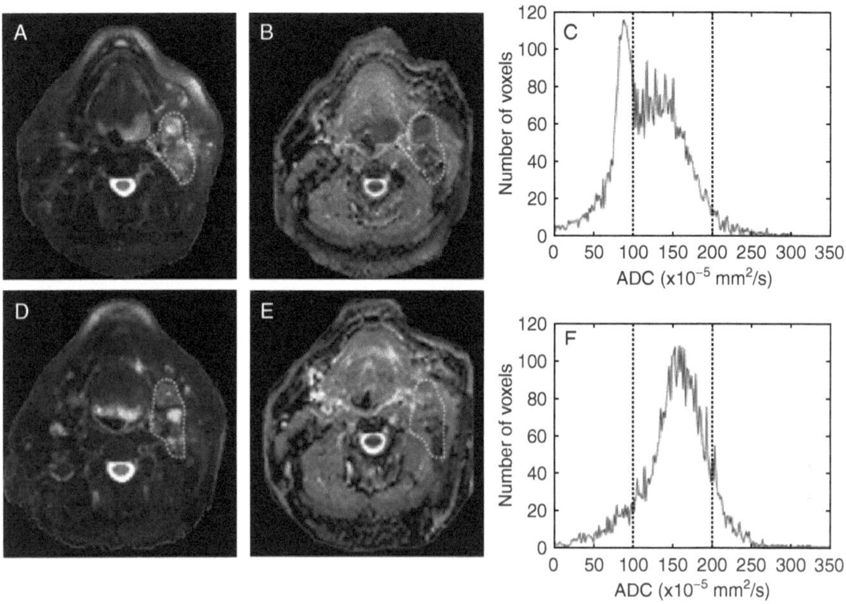

FIG. 14. Pre- (top row) and 3 week posttreatment initiation (bottom row). (A) T2-weighted, (B) apparent diffusion coefficient (ADC) maps with corresponding (C) ADC histograms from the whole-tumor of a patient treated for HNSCC of the left base of the tongue. The tumor is outlined by the yellow contour line. Mean ADC values increased from 120 to 156 $\times 10^{-5}$ mm^2/s following treatment initiation of a nonsurgical organ preservation therapy. This patient was subsequently determined to be a complete responder to therapy (from Ref. 140, with permission).

It was found in previous work that the relative volume of tumor that exhibited a significant increase in ADC values as determined by fDM measured at 3 weeks into treatment was predictive of radiographic response measured at 10 weeks.[121,139] Moreover, tumor assessment by fDM at 3 weeks into treatment provided an early indicator of the eventual clinical responses of disease time to progression and overall survival in patients with malignant glioma.[148] Preliminary work was performed to determine the feasibility of clinically translating the fDM imaging biomarker for use in quantifying bone tumor response in a patient treated for metastatic prostate cancer to the bone.[149] DW-MRI was performed prior to treatment and again at 2 and 8 weeks posttreatment initiation to quantify changes in tumor diffusion values. Three metastatic lesions were identified for fDM analysis, of which two are presented in Fig. 15. All tumors demonstrated early changes in diffusion values at 2 weeks and continued to increase further by week 8 posttreatment initiation. Compared to the percent change in tumor mean ADC, the fDM analysis offered improved sensitivity over the histogram-based approach when

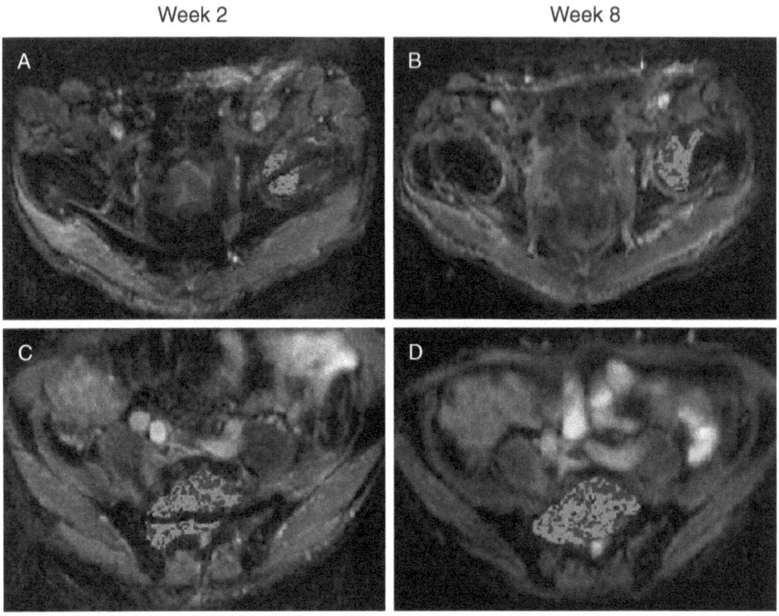

FIG. 15. Presented is the fDM analysis in a patient with metastatic prostate cancer to the bone treated with hormone therapy. Areas with increased ADC (red voxels), decreased ADC (blue voxels), and areas where ADC did not change significantly (green voxels) are visually apparent. fDM analysis of the femoral head and sacral lesions at (A, C) 2 and (B, D) 8 weeks after treatment initiation revealed distinct regions of red voxels signifying areas with significant increases in ADC ($> 26 \times 10^5$ mm^2/s). (Reprint from Ref. 141, with permission from Neoplasia.) (For interpretation of the references to color in this figure legend, the reader is referred to the Web version of this chapter.)

tumors have heterogeneous changes over time. A strong correlation in fDM was observed with patient's prostate-specific antigen (PSA) levels, which is suggestive of patient response. In contrast, CT, bone scans, and anatomic MRI images obtained posttreatment were not useful for the assessment of treatment efficacy. In general, quantification of tumor response by fDM may provide a standardized approach to treatment response assessment using diffusion MRI. Extension of this image postprocessing approach for other tumor types is possible. While further validation of DW-MRI measurements as a biomarker for early treatment response is needed, recent studies have shown promising results.

5. Summary

MRI methods such as DW-MRI and DT-MRI, which are based on tissue biophysical properties, are rapidly being incorporated into routine imaging protocols for improving diagnosis, characterization, and management of cancer patients. In the future, these imaging techniques when combined with other physiology-based methods, such as MR permeability and magnetic resonance spectroscopy (MRS) metabolite mapping, as well as excellent anatomic images, are anticipated to improve tumor diagnosis, biopsy guidance, pretreatment and presurgical planning, and the assessment of early therapeutic efficacy in individuals. Research continues in determining how best to use diffusion information to positively impact on patient management.

C. Permeability Magnetic Resonance Imaging

An active area for treating tumors is the development of drugs that target the vascular support network of the tumor. In response to this growing field, imaging biomarkers for tumor angiogenesis are under investigation for detection and quantification of pharmacodynamic drug activity. Dynamic contrast-enhanced magnetic resonance imaging (DCE-MRI) is presently being applied for assessment of tumor pathophysiology and treatment response against radiotherapy as well as antiangiogenic and vascular disruption agents. This MRI technique is rapidly progressing as a noninvasive imaging-based biomarker for drug efficacy studies in clinical trials.[150–166]

1. Principles of DCE-MRI

DCE-MRI is an imaging technique used to investigate microvascular structure and function by recording the pharmacokinetics of injected low-molecular-weight contrast agents as they pass through the tumor vasculature. Following an intravenous bolus injection of a paramagnetic contrast agent, the contrast agent enters the tumor arterioles, passes through capillary beds, and finally drains via the veins within the tumor. Commonly used is a gadolinium-based contrast agent which shortens the T1 relaxation time of blood resulting in a concentration-dependent spatially varying enhancement of signal (contrast) on T1-weighted images. Collection of rapidly acquired serial T1-weighted images prior and early postinjection of the contrast agent provides the initial area under the contrast-agent concentration time curve, which can be analyzed for kinetic information. Signal enhancement within the tumor depends on a variety of physiological factors such as tissue perfusion, arterial input function, capillary surface area, capillary permeability, and the volume of the extracellular extravascular space (EES). Imaging data are typically analyzed from a defined region of interest that may encompass all or part of the tumor. The data are fit to a compartmental model of tumor microvasculature that generates parameters that describe the behavior of the contrast agent time-dependent concentration curve, which represents a combination of flow, blood volume, vessel permeability, and EES. Standardized terms for the kinetic variables within the model are commonly used in these studies.[167] The two-compartment model regards the EES and plasma as the two compartments that are well mixed with contrast agent and which have a constant permeability. Transport between these compartments is determined by the parameter K^{trans} (volume transfer constant between the blood plasma and the EES) and k_{ep} (rate constant between the EES and the blood plasma). The EES fractional volume (v_e) is related to K^{trans} and k_{ep} via the equation, $v_e = (K^{trans}/k_{ep})$. Although a primary end point used in clinical trials, changes in K^{trans} may represent different physiological processes in different individuals within a patient population (e.g., a reduction of K^{trans} could represent reduced permeability, reduced blood flow, or a combination of the two). There are a variety of analytical approaches used for postprocessing and analysis of DCE-MRI data. To date, there remains no consensus on a recommended model for deriving the volume transfer coefficient of contrast between the blood plasma and the EES (K^{trans}) and the size of the EES (v_e), as well as should descriptive parameters such as the initial area under the contrast-agent concentration time curve be used for in assessing antiangiogenic and vascular disrupting agents in clinical trials.[168]

2. Dynamic Contrast-Enhanced Magnetic Resonance Imaging for Detection of Residual Disease

DCE-MRI following therapy has been proposed to aid in detecting residual disease or early recurrence, which may be difficult to detect in tissue regions exposed to radiotherapy. DCE-MRI measurements in tumor treated with radiotherapy have been investigated in the cervix, lung, head and neck, and bladder tumors wherein high enhancement was associated with an increase in local recurrence and poor survival.[169–172] Following treatment completion, a contrast agent enhancement pattern that persists or has returned early has been attributed to viable tumor cells. It has been shown in cervical cancer that DCE-MRI enhancement obtained early in therapy was associated with early recurrence and poor survival.[172]

DCE-MRI also shows promise as a noninvasive imaging technique for determining the malignancy of a tumor. As presented in Fig. 16, a large tumor is observed in a series of T1-weighted contrast-enhanced MR images, which show strong signal enhancement over time postinjection of contrast. Using the serial images, tumor hemodynamics can be quantified. These values, such as permeability constant, K^{trans}, and extracellular/extravascular fraction, v_e, have been shown to suggest the extent of tumor malignancy. In a study involving head and neck tumor patients, the presence of lesion enhancement was found to have a positive correlation with viable tumor cells in postradiation surgical specimens.[169] Finally, in meningioma patients examined using DCE-MRI following radiotherapy, pharmacokinetic analysis revealed a decrease in the exchange rate constant in patients who responded relative to nonresponders. Overall, although

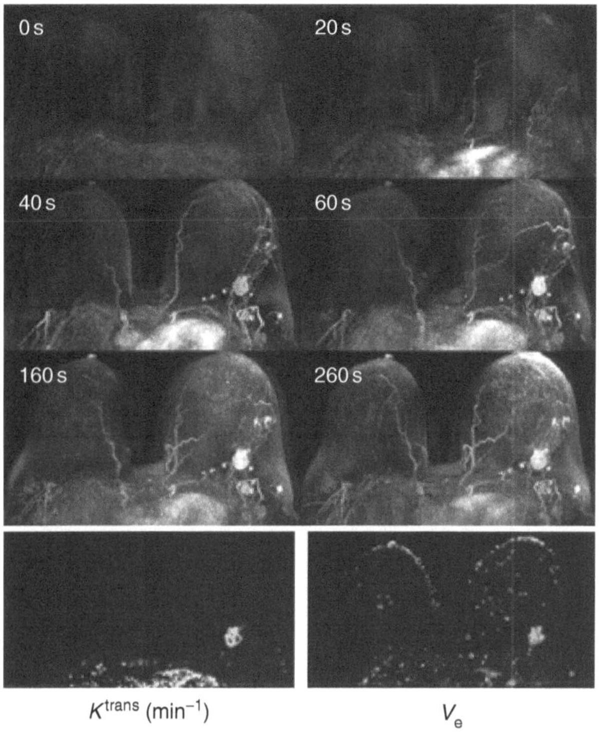

FIG. 16. Demonstration of Dynamic Contrast Enhance (DCE)-MRI shown as maximum intensity projections (MIPs) and mid-tumor axial images of the modeled permeability constant K^{trans} and extracellular/extravascular fraction v_e. Note marked intratumoral heterogeneity of vasculature denoting high malignancy.

there is emerging evidence to the potential use of DCE-MRI for detecting posttreatment residual disease, further studies are required in order to adequately validate this approach for routine clinical use.

3. Dynamic Contrast-Enhanced Magnetic Resonance Imaging as Predictor of Treatment Response

DCE-MRI is capable of providing anatomic and physiological information using conventional clinical MRI sequences. These sequences are incorporated into standard imaging protocols that can be used for treatment assessment. Many studies have evaluated the prognostic value of DCE-MRI in assessing treatment response to radiotherapy, antivascular, and antiangiogenic therapies.[150,151] Most studies consisted of small cohort single-center phase I trials, although a few phase II trials have also incorporated DCE-MRI.[173] In addition, the efficacy of antiangiogenic drugs has been demonstrated using DCE-MRI in clinical trials. For example, a colorectal cancer treated with a vascular disrupting agent, combretastatin, resulted in a rapid tumor vascular shutdown within 4 h following the first dose.[174] Quantitative maps of area under gadolinium contrast medium curve at 60 s revealed significant decreases in permeability. The permeability constant (K^{trans}) and extracellular/extravascular space (v_e) had also revealed that the vascular shutdown was induced by this drug with negligible morphologic change in tumor.

DCE-MRI biomarkers have been used as early indicators of efficacy, dose, and outcome as well as assist in defining the biologically active and maximum tolerated doses. However, few trials have demonstrated that DCE-MRI measurements correlate with clinical outcome measures. Thus, observed changes in DCE-MRI biomarkers should not be considered a surrogate of a successful outcome measure in randomized phase III trials. Advancement of DCE-MRI as a diagnostic/predictive indicator of treatment response will require added attention to the design of a clinical protocol where the acquisition of the imaging data must be at well-conceived scanning interval after administration of the agent. For example, an agent that causes vascular disruption in the tumor may require image time points pretreatment, 4- to 6-h, and 24-h posttreatment initiation. In comparison with a trial involving an antiangiogenic agent (or radiotherapy), DCE-MRI may be more sensitive to treatment-induced effects at longer interscan intervals of days to weeks or months in order to reach maximal change in the DCE-MRI biomarkers.

Another area for potential improvement of the employment of DCE-MRI as a predictive biomarker of tumor response to treatment may be in the choice of the specific imaging biomarker(s) or parameters selected for a given therapeutic intervention. Moreover, the criteria for selecting regions of interest within the tumor along with the specific pharmacokinetic model used to analyze the imaging data, which both provide for the quantification of the DCE-MRI biomarkers, may impact the final results. Alternative methods for image data analysis may have a significant role in DCE-MRI, especially in tumors with significant regions of heterogeneity.

4. Summary

As with any biomarker, the routine application of DCE-MRI as a noninvasive biomarker of tumor angiogenesis and response to therapy requires validation through statistical correlation with traditional clinical outcome measures (i.e., radiologic response, overall survival). Currently, there is a lack of clinical data correlating changes in quantified DCE-MRI biomarkers with outcome measures. Thus, this measurements adoption as a surrogate end point in drug efficacy studies is limited. New contrast agents, higher magnetic fields (7 and 9.4 T), and image acquisition and analysis tools are currently in development, which may help to improve the prognostic value of DCE-MRI in cancer trials.

III. Nuclear Imaging

A. Introduction

Positron emission tomography (PET) and single-photon emission computed tomography (SPECT) are nuclear imaging techniques used to map physiological and biological processes in humans and animals following the administration of radiolabeled tracers. A unique advantage of PET and SPECT imaging techniques is their potential for detecting disease-related biochemical and physiologic abnormalities prior to the appearance of anatomical changes which are visualized by conventional imaging modalities such as CT and MRI. PET uses radioisotopes that decay via emission of positrons, whereas, SPECT radioisotopes decay by electron capture and/or gamma emission.

Table I lists some of the most commonly used PET and SPECT radioisotopes and their physical data. The short half-lives of the positron-emitters carbon-11, nitrogen-13, and oxygen-15 dictates that radioligand synthesis with these isotopes can only be accomplished in close proximity to a cyclotron. On the other hand, radioisotopes such as fluorine-18, copper-64, indium-111, iodine-123, and iodine-124 are sufficiently long-lived to allow transportation from regional commercial sites. Additionally, the radioisotopes gallium-68, copper-62, and technetium-99m can be conveniently obtained from an

TABLE I
COMMONLY USED PET AND SPECT RADIOISOTOPES

Isotope	Imaging mode	Production method	Half-life	Decay mode(s)
^{11}C	PET	Cyclotron	20.4 min	β+ (99+ %)
^{13}N	PET	Cyclotron	10 min	β+ (100%)
^{15}O	PET	Cyclotron	2.03 min	β+ (99.9%)
^{18}F	PET	Cyclotron	110 min	β+ (97%) EC (3%)
^{124}I	PET	Accelerator	4.2 days	EC (74.4%) β+ (25.6%)
^{68}Ga	PET	Generator	68.3 min	β+ (90%) EC (10%)
^{62}Cu	PET	Generator	9.73 min	β+ (98%) EC (2%)
^{64}Cu	PET	Reactor	12.7 h	β+EC
^{99m}Tc	SPECT	Generator	6.02 h	IT
^{111}In	Gamma scintigraphy	Accelerator	2.8 days	EC X-ray
^{123}I	SPECT	Accelerator	13.3 h	EC

β+ = positron emission, EC = electron capture, IT = isomeric transition.

in-house generator. At the present time, clinical SPECT imaging is more frequently conducted than PET imaging due to its cost-effectiveness and the greater availability of SPECT scanners at most clinical centers.

In PET, positrons (positively charged electrons, $\beta+$) ejected from the nucleus during radioactive decay travel a few millimeters in tissue, after which, they undergo annihilation by collision with electrons. Each annihilation event releases two γ-ray photons of equal energy (511 keV) in opposite trajectories (180° apart). PET scanners utilize the simultaneous detection of these two photons (coincidence detection) to precisely locate the source of the annihilation event. Subsequently, the event data is processed by computers to reconstruct the spatial distribution of the annihilation events. SPECT scanners on the other hand, use collimators (lead shields containing narrow parallel holes) to acquire only those photons that have a parallel trajectory. Thus, the original path of the detected photon can be linearly extrapolated from knowledge of the collimators orientation. Coincidence detection is significantly more efficient than collimation at recording annihilation events as the latter approach results in discarding a high percentage of useful emitted photons. Thus, PET provides a much better sensitivity (2–3 orders of magnitude), quantitation capability and spatial resolution than SPECT.

The process of developing a useful nuclear imaging radiotracer for biological imaging has several requirements that can pose special challenges. Incorporation of the radionuclide (including its chelating functionality in some cases) in a target ligand should have a negligible effect on its binding affinity. Radioligand binding sites (receptor, enzyme, etc.,) usually exist in low concentration (micromolar to nanomolar). Thus, the specific activity of the radioligand should be sufficiently high to represent a high radiative emission from a very small quantity (mass) of radiodiagnostic to avoid producing a pharmacologic effect. Due to the constraints of working with a short radioisotope half-life, the overall synthetic strategy for radioligand preparation should be short, the individual reaction steps rapid and high yielding, and the entire process should be adaptable to microscale manipulation. From an *in vivo* standpoint, the radioligand should display low or negligible nonspecific binding so as to provide a high target-to-background signal and the *in vivo* kinetics of radioligand target uptake and washout should be compatible with the half-life of the radioisotope. Additionally, the radioligand should not be extensively metabolized and metabolites, if present, should not compete with the radioligand at its intended binding site. Despite these rigorous requirements, many radioligands have been developed that display demonstrated clinical utility for biological imaging (e.g., [^{18}F]fluoro-deoxyglucose ([^{18}F]FDG), [^{18}F]FLT, radiolabeled somatostatin analogs, etc.).

Currently, [^{18}F]FDG (Fig. 17); a radioligand marker for tumor glucose metabolism), is the workhorse of PET, reportedly used in at least 90% of human PET studies. The majority of these studies have been in oncology

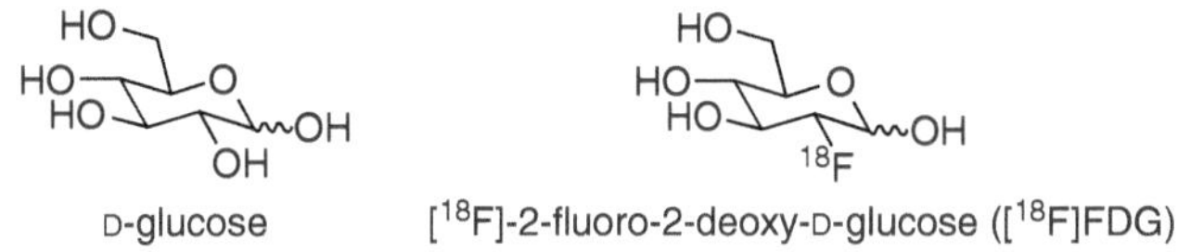

FIG. 17. Structure of D-glucose and [^{18}F]FDG.

where [^{18}F]FDG PET is the primary method used for detection and staging of many cancers.[175] However, [^{18}F]FDG is not tumor specific and is known to accumulate in many benign inflammatory processes leading to false-positive interpretation.[176] The past decade has also seen the investigation and validation of several alternative radioligands to [^{18}F]FDG that target specific aspects of tumor biology. These targets include molecular biomarkers such as growth factor receptors, protein kinases, specific receptor overexpression or biological events such as angiogenesis, apoptosis, hypoxia, and tumor proliferation. This review highlights recent developments in PET and SPECT radioligand development in the area of oncology and CNS imaging with a major focus on its application in oncology.

B. Radioligands for Imaging Angiogenesis

Angiogenesis, the formation of new blood vessels through capillary sprouting from preexisting vasculature, plays a key role in the growth and metastatic potential of solid tumors.[177,178] Tumor growth beyond a 1–2 mm^3 volume requires an independent vasculature for the cellular supply of oxygen and nutrients and removal of waste products.[179] Consequently, tumors that outgrow their existing blood supply frequently display oxygen deficiency (hypoxia) which can trigger the secretion of various pro-angiogenic growth factors, such as, vascular endothelial growth factors (VEGFs) for initiating new blood vessel growth.[177,180] Binding of VEGFs to the VEGF family of receptors (VEGFR) initiates a signaling cascade that promotes the proliferation, migration, and survival of endothelial cells, ultimately leading to angiogenesis.[181,182] The angiogenic effects of the VEGF family are believed to be primarily mediated through VEGF-A. To date, VEGF-A (also referred to as VEGF) and its receptors are the most characterized signaling pathways in developmental and tumor angiogenesis.

Alternative splicing of RNA has revealed the existence of at least seven different molecular isoforms for VEGF-A, comprising, 121, 145, 148, 165, 189, or 206 amino acids.[183] The angiogenic actions of VEGF-A are mediated primarily via two closely related endothelium-specific receptor tyrosine kinases (VEGFR-1 and VEGFR-2).[184,185] All of the VEGF-A isoforms bind to both

VEGFR-1 and VEGFR-2, of which, VEGFR-2 is the major mediator of the mitogenic, angiogenic, and permeability enhancing effects of VEGF-A.[185] VEGFRs are overexpressed in a variety of solid tumors with overexpression of VEGFR-2 or VEGF-A in particular, serving as poor prognostic markers.[181,186]

PET imaging of VEGFR expression *in vivo* was first demonstrated using $VEGF_{121}$ (a molecular isoform of VEGF) radiolabeled with ^{64}Cu.[187] Radiolabeling was achieved via ^{64}Cu chelation to a DOTA-$VEGF_{121}$ conjugate (DOTA is an abbreviation for 1,4,7,10-tetraazacyclododecane-*N*,*N*′,*N*″,*N*‴-tetraacetic acid). *In vivo* evaluation of ^{64}Cu-DOTA-$VEGF_{121}$ using microPET imaging of athymic nude mice bearing U87MG human glioblastoma xenografts showed rapid and high specific accumulation of the radioligand in small U87MG tumors (16% injected dose per gram [ID/g]) at 4 h postinjection. Larger tumors showed significantly lower uptake (1–3% ID/g). Differences in tumor localization between large and small tumors showed a good correlation with tumor VEGF receptor expression (VEGR2). *In vivo* VEGFR2 specificity of the radioligand was also confirmed by pharmacological blocking experiments and *ex vivo* studies (immunofluorescence staining, western blot analysis). This study also demonstrated the dynamic nature of VEGFR expression during tumor expression within the same animal model. Subsequently, these authors also reported on the development of a ^{64}Cu-labeled vasculature-targeting fusion toxin (VEGF121/rGel) composed of a $VEGF_{121}$ linked recombinant plant toxin gelonin construct (rGel) for multimodality imaging and therapy of glioblastoma.[188] High sustained tumor accumulation and tumor-to-target ratios were demonstrated by this radioligand in mice bearing glioblastoma xenografts up to 48 h postinjection. Based on the pharmacokinetic information obtained from the imaging studies, therapeutic administration of VEGF121/rGel to mice bearing orthotopic U87MG glioblastomas revealed specific tumor neovasculature damage by histological analysis after a 4 dose treatment regimen.[188]

Myocardial infarction (MI) is known to activate many biological pathways including VEGF/VEGFR signaling. PET imaging studies conducted in a rat model of MI with ^{64}Cu-DOTA-$VEGF_{121}$ revealed a 3–4 higher myocardial uptake of radioactivity for up to 2 weeks following infarction compared to controls.[189] In addition, PET imaging of VEGFR expression with ^{64}Cu-DOTA-$VEGF_{121}$ in a rat stroke model showed peak tracer uptake in the stroke border zone at approximately 10 days postsurgery indicating neovascularization as confirmed by histopathology studies.[190]

Chan and coworkers have reported on the synthesis and evaluation in tumor-bearing mice of an ^{111}In-labeled recombinant VEGF isoform $VEGF_{165}$ (^{111}In-hn-Tf-VEGF) as a tumor angiogenesis marker.[191] $VEGF_{165}$ was fused through a flexible polypeptide linker to the *n*-lobe of human transferrin. The latter construct permitted labeling of the radioligand with ^{111}In at a site remote from the VEGF receptor-binding domain. In radioligand stability studies,

^{111}In-hn-Tf-VEGF demonstrated a moderate loss of ^{111}In to transferrin in human plasma *in vitro* over a 72 h period (21.3 ± 3.4% per day). Radioligand biodistribution studies and whole-body gamma camera imaging were conducted in athymic mice bearing subcutaneous U87MG human glioblastoma xenografts. ^{111}In-hn-Tf-VEGF displayed tumor and blood radioactivity accumulations of 6.7 ± 1.1%ID/g and 1.6 ± 0.4%ID/g, respectively, at 72 h postinjection. Coadministration of a 100-fold excess of VEGF led to a 15-fold decrease in tumor uptake of radioactivity. High uptake of radioactivity was also observed in liver (45.5 ± 7.5%ID/g), kidneys (39.4 ± 7.0%ID/g) and spleen (35.6 ± 4.4%ID/g) at this time interval. The authors present evidence to indicate that uptake of radioactivity in these organs is due to ^{111}In-hn-Tf-VEGF and not due to ^{111}In-transferrin via transchelation of ^{111}In from the radioligand to transferrin.

An indirect approach to angiogenesis imaging has focused on radioligands targeting the $\alpha_v\beta_3$ class of cell adhesion molecule integrins.[192] Integrin $\alpha_v\beta_3$ receptors are significantly up-regulated in endothelial cells during angiogenesis but not in mature vessels or nonneoplastic epithelium.[193–195] Integrin $\alpha_v\beta_3$ is also expressed in a variety of tumor cells, including melanoma, late-stage glioblastoma, ovarian, breast, and prostate cancer.[196] The ability to visualize and quantify integrin $\alpha_v\beta_3$ expression *in vivo* would allow for appropriate selection of patients for anti-integrin treatment and also monitor treatment efficacy in such patients.

Radioligand development for $\alpha_v\beta_3$ imaging has focused primarily on small RGD peptide antagonists.[197] The tripeptide sequence motif, arginine-glycine-aspartate (RGD), is found in proteins of the extracellular matrix. Many integrins, including $\alpha_v\beta_3$, link the intracellular cytoskeleton of cells with the extracellular matrix via recognition and binding to this RGD motif. Wu and coworkers have reported on the enhanced $\alpha_v\beta_3$ receptor binding characteristics of dimeric and multimeric RGD peptides over monomeric peptides which has been attributed to an increased local concentration of RGD domains at the receptor vicinity (polyvalency effect).[198] Accordingly, several [^{18}F]- and [^{64}Cu]-labeled, dimeric, and tetrameric RGD peptide analogs have been recently synthesized and evaluated by this group for integrin-targeted imaging in lung, brain, and breast cancer.[198–200] As an example, microPET imaging studies with a dimeric RGD peptide coupled to 4-[^{18}F]Fluorobenzoate {[^{18}F]-FB-E[c(RGDyK)]$_2$} showed predominantly renal excretion and twice as much tumor uptake in the same animal model as the monomeric analog [^{18}F]-FB-c(RGDyK).[201] Binding potentials derived from tracer kinetic modeling studies showed good correlation with tumor integrin expression levels as measured by SDS-PAGE/autoradiography in the six tumor models tested.[201]

Matrix metalloproteinases (MMPs), a family of zinc-and calcium-dependent endopeptidases, facilitate endothelial cell migration during angiogenesis via the enzymatic degradation of connective tissue.[202] Consequently, several MMP-specific peptides as well as small-molecule inhibitors (MMPI) have been radiolabeled with ^{125}I, ^{99m}Tc, ^{111}In, ^{64}Cu, ^{11}C, or ^{18}F for PET or SPECT imaging of angiogenesis.[203,204] However, their utility for angiogenesis imaging has yet to be demonstrated in most cases due to their low tumor targeting and *in vivo* instability.

C. Radioligands for Imaging Apoptosis

Apoptosis (programmed cell death) plays a critical role in the homeostasis of multicellular organisms. Initiation of apoptotic cell death leads to activation of a family of cysteine proteases (caspases) which act as central executioners of the apoptotic process.[39] Radiation as well as anticancer drug treatment can induce apoptosis in tumor cells. Consequently, imaging methods that provide information on the rate and extent of apoptosis are of interest in monitoring the efficacy of anticancer treatment. A vast majority of the work on apoptosis-targeted radioligands has focused on Annexin V and its derivatives.[205] Annexin V is a member of the calcium and phospholipid binding superfamily of Annexin proteins that displays selective, nanomolar affinity ($K_d \sim 0.5$–7 nM) toward phosphatidylserine (PS) residues. Induction of apoptosis results in a rapid externalization of PS from the inner leaflet of the plasma membrane to its outer surface.[206] Accordingly, Annexin-mediated imaging of PS has been extensively investigated for identifying cells at the early stages of apoptosis. Annexin V and its derivatives have been radiolabeled with a wide variety of radioisotopes including radioiodine (^{123}I, ^{124}I, ^{125}I), ^{18}F, ^{99m}Tc, ^{111}In, ^{11}C, and ^{64}Cu.[207] Tait and coworkers have compared the apoptosis-specific liver uptake of several [^{99m}Tc]-labeled Annexin V derivatives prepared by amine-directed modification with that labeled site-specifically at the N-terminus.[208] A clear improvement was seen for site-specific labeling as compared to amine-directed modification. Use of [^{99m}Tc]-labeled hydrazinenicotinamide-annexin imaging for assessment of response to chemotherapy has also been reported.[209] The reported imaging protocol was able to distinguish responders from nonresponders with 94% accuracy (16/17 patients) and a sensitivity and specificity of 86% and 100%, respectively.

Radiolabeled caspase substrates, inhibitors, and monoclonal antibodies targeted to human Annexin V have also been reported as alternative approaches to apoptosis imaging.[210,211] An excellent review on these developments has been recently published by Lahorte and coworkers.[207]

More recent approaches toward apoptosis imaging radioligand development have focused on small-molecule probes that target the early stages of the apoptotic cascade as exemplified by the 4-[^{18}F]Fluorobenzyltriphenylphosphonium cation ([^{18}F]-FBnTP) and the [^{18}F]-labeled malonic acid analog ([^{18}F]-ML-10) (Fig. 18).[212,213] The latter radioligand which is capable of distinguishing between apoptotic and necrotic cells has shown promising initial results in several small-scale clinical trials.[214]

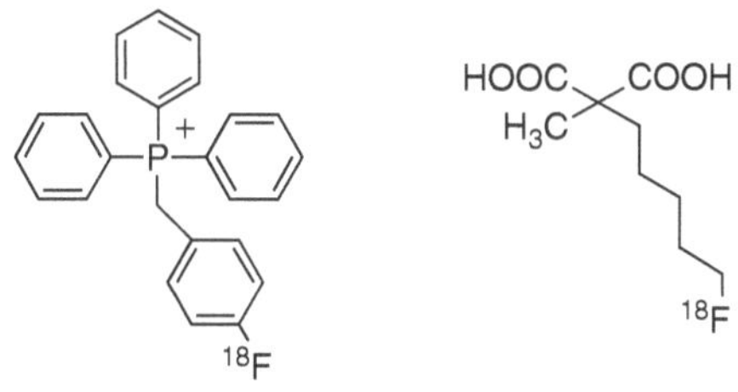

[^{18}F]BnTP [^{18}F]ML-10

FIG. 18. Structure of radioligands for imaging apoptosis.

D. Radioligands for Imaging Hypoxia

The growth of solid malignant tumors is frequently accompanied by tissue hypoxia due to outgrowth of its blood supply. Hypoxia in tumor cells leads to resistance to both radiation and anticancer treatment.[215] Noninvasive imaging methods for identification and quantitation of tumor hypoxia status could play a central role in predicting and monitoring treatment response. Initial approaches to hypoxia imaging focused on radiolabeled 2-nitroimidazole derivatives, a class of hypoxia-activated prodrugs.[216] These bioreductive prodrugs undergo reductive metabolism in the hypoxic environment of tumors thereby releasing toxic metabolites that can lead to cell damage.[217] Among the 2-nitroimidazole class of radioligands, [^{18}F]fluoromisonidazole ([^{18}F]FMISO) is currently the most widely used clinical PET hypoxia tracer (Fig. 19).[218]

More recently, Rischin *et al.* have shown that [^{18}F]FMISO PET imaging is a useful method for identifying head and neck cancer patients most likely to benefit from treatment with the hypoxic cell cytotoxin, tirapazamine in a chemoradiotherapy regimen.[219] In an extensive study involving 73 patients with head and neck cancer, Rajendran *et al.* have shown that [^{18}F]FMISO PET imaging is a predictor of survival prior to radiation therapy.[220] A ^{18}F-labeled 2-nitroimidazole derivative, fluoroazomycin arabinoside ([^{18}F]FAZA; Fig. 19) that displays enhanced *in vivo* stability to enzymatic cleavage has been described by Piert and colleagues.[221] Studies indicate that [^{18}F]FAZA may be superior to [^{18}F]FMISO for hypoxia imaging due to its superior biokinetic profile.[221] These authors have also shown that [^{18}F]FAZA imaging has predictive value for the determination of radiotherapy success when used in combination with tirapazamine.[222] Copper-64 labeled Cu-Diacetyl-bis(N^4-methylthiosemicarbazone) (^{64}Cu-ATSM) has also been proposed as a PET hypoxia imaging agent. Differences in tumor type selectivity for this radiotracer, however, raises questions regarding its use as a universal PET hypoxia marker.[223]

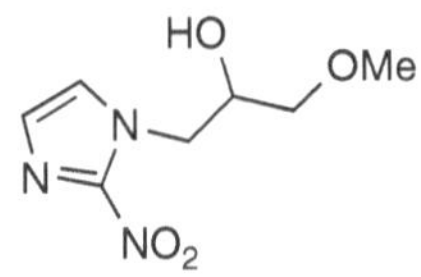

Misonidazole (MISO)

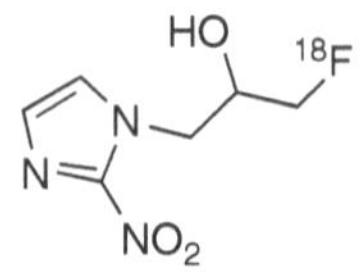

[^{18}F]fluoromisonidazole ([^{18}F]FMISO)

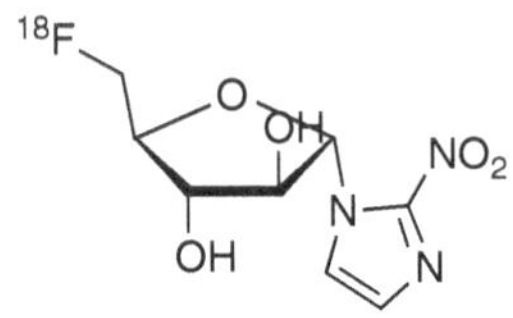

[^{18}F]fluoroazomycin arabinoside ([^{18}F]FAZA)

FIG. 19. Structure of misonidazole (MISO), [^{18}F]FMISO, and [^{18}F]FAZA.

E. Radioligands for Imaging EGF Receptors

The search for radioligands that target the ErbB family of RTKs is an active area of research. This receptor family includes four members: epidermal growth factor receptor (EGFR/ErbB1/HER1), HER2 (ErbB2/neu), HER3 (ErbB3), and HER4 (ErbB4).[224,225] Overexpression of these receptors, particularly HER1 and HER2, has been documented in many epithelial cancers including breast, nonsmall cell lung (NSCLC), ovarian, and bladder cancer.[226,227] It has also been shown that such overexpression is frequently associated with a poor prognosis.[228–230]

Preclinical evaluation of a ^{68}Ga-labeled, recombinant human epidermal growth factor DOTA conjugate (^{68}Ga-DOTA-hEGF) has been reported for HER1 imaging.[231] *In vitro* studies with ^{68}Ga-DOTA-hEGF conducted on EGFR-expressing cell lines, U343 glioma and A431 cervical carcinoma, demonstrated high-affinity binding (2 nM), rapid internalization of radioligand and good retention of radioactivity. Radioligand biodistribution in mice bearing A431 tumor xenografts showed a tumor-to-blood ratio of 4.5 at 30 min postinjection (2.7% ID/g in tumor). Interestingly, tumor uptake was dependent on the specific activity of the radioligand: a twofold increase in tumor uptake was observed with a 10-fold lower specific activity material. Tumors were clearly visualized by microPET imaging in a tumor-bearing mouse although the kinetics of tumor uptake of radioactivity was slow compared to that of liver and kidney.

An ^{111}In-labeled human EGF-diethylenetriaminepentaacetic acid (DTPA) conjugate (^{111}In-DTPA-hEGF) is under investigation for future Phase 1 clinical trials as a radiotherapeutic for breast cancer.[232] Preclinical pharmacokinetic, toxicologic, and dosimetry studies conducted in mice and rabbits with ^{111}In-DTPA-hEGF showed no acute toxicity in female BALB/c mice at 42 times the maximum planned human dose. Highest uptake of radioactivity was seen in liver (41.3% ID/g) and kidney (18.6% ID/g) at 1 h postinjection, although these values had decreased to 4.5–4.9% at 72 h. The radiotracer showed fast blood clearance following intravenous injection and no morphologic changes were seen by light microscopy in 19 sampled tissues.

Cetuximab (Erbitux; ImClone Systems, Inc.), a chimeric anti-EGFR monoclonal antibody, is an FDA-approved drug for the treatment of advanced metastatic colorectal cancer. Cai and coworkers describe microPET imaging with a ^{64}Cu-labeled DOTA-cetuximab conjugate (^{64}Cu-DOTA-cetuximab) in seven xenograft tumor mouse models.[233] Uptake of radioactivity for ^{64}Cu-labeled DOTA-cetuximab was similar in major organs and tissues in all seven of the tested tumor models (U87MG human glioblastoma, PC-3 human prostate carcinoma, CT-26 murine colorectal carcinoma, HCT-8, HCT-116, SW620 human colorectal carcinoma and MDA-MB-435 human breast cancer). High uptake of radioactivity was seen for the high EGFR-expression U87MG and PC-3 tumors (13.2%ID/g and 12.8% ID/g, respectively) at 48 h postinjection. Tumor radioactivity uptake determined by microPET imaging showed a good correlation with EGFR-expression levels as measured by western blot analysis.

An Affibody dimer, His$_6$-(Z$_{HER2:4}$)$_2$, has been recently described as a high-affinity HER2/neu ligand.[234] Orlova and coworkers have prepared ^{99m}Tc- and ^{125}I-labeled His$_6$-(Z$_{HER2:4}$)$_2$ and compared their biodistribution in tumor-bearing BALB/c *nu/nu* mice.[235] Significantly, higher levels of radioactivity were observed in tumor compared to liver for the ^{125}I-labeled ligand. These studies indicate that the radioiodinated ligand may be more suitable than the corresponding [^{99m}Tc]-labeled ligand for imaging tumor HER2 expression levels, particularly in liver. A ^{99m}Tc-labeled Affibody compound MAG3-(Z$_{HER2:342}$) labeled via a MAG3 chelator has also been recently reported by this group.[236] Smith-Jones and colleagues have used mouse microPET imaging with a ^{68}Ga-labeled F(ab′)$_2$ fragment of herceptin (^{68}Ga-DOTA-F(ab′)$_2$-herceptin) to image HER2 downregulation after heat shock protein (Hsp90) inhibition.[237] The Ansamycin antibiotic, geldanamycin is known to cause HER2 degradation via inhibition of the Hsp90 chaperone protein. PET imaging was conducted on mice bearing BT474 breast tumor xenografts with ^{68}Ga-DOTA-F(ab′)$_2$-herceptin and [^{18}F]-fluorodeoxyglucose demethoxygeldanamycin (17-AAG). A significant decrease of HER2 expression was seen within 24 h of 17-AAG treatment with ^{68}Ga-DOTA-F(ab′)$_2$-herceptin imaging. In contrast, tumor uptake of [^{18}F]FDG (a marker of glycolysis) was unchanged. The authors conclude that PET imaging with the HER2 radioligand, ^{68}Ga-DOTA-F(ab′)$_2$-herceptin is superior to [^{18}F]FDG imaging for evaluating tumor response to 17-AAG therapy.

Blakenberg and colleagues have recently described a new generation of radiotracers based on a recombinant human EGF expressing a cysteine-containing fusion tag. Incorporation of the cysteine-fusion tag allowed for site-specific conjugation with either ^{99m}Tc or ^{64}Cu for imaging EGFR expression with SPECT or PET, respectively.[238] Tumor accumulation of the radiotracers were demonstrated in orthotopic MDA231luc breast carcinoma by PET and SPECT imaging and also in the spontaneous mouse lung carcinoma using SPECT in a bitransgenic mouse model.[238]

F. Radioligands for Imaging Somatostatin Receptors

The utilization of radiolabeled peptide analogs of the hormone somatostatin for the diagnostic imaging and therapy of neuroendocrine tumors (NET) has had notable success.[239,240] Somatostatin exists in two isoforms: a short peptide having 14 amino acids, and a second peptide with 28 amino acids, both of which bind with high affinity to five receptor subtypes (sst_1–sst_5).[241] A majority of malignant tumors (e.g., NET, small cell lung cancer, malignant lymphoma and breast tumors) overexpress multiple sst receptor subtypes relative to nontumor tissues, of which the sst_2 subtype is frequently more predominantly expressed.[242] Since these G-protein coupled receptors undergo internalization on ligand binding they are uniquely suited for radionuclide imaging. Internalization of the receptor–radioligand complex allows for extended tumor retention times which could enhance diagnostic sensitivity due to improved tumor-to-background ratios. Receptor–radioligand complex internalization could also be an important advantage in targeted radiotherapy applications.[243] A recent elegant study by Cescato and colleagues describes the use of new immunocytochemical methods to quantitatively measure sst_2, sst_3, and sst_5 receptor subtype internalization induced by a variety of somatostatin analogs in human embryonic kidney 293 (HEK293) cells expressing these subtypes.[244]

Somatostatin has a short plasma half-life (approximately 2 min) and therefore is unsuitable for radioligand development.[245] The discovery of the short octapeptide somatostatin analog, octreotide, which displays a superior pharmacological profile, played a major role in the development of radioligands for somatostatin receptor imaging.[246] Octreotide which displays high affinity for sst_2 ($IC_{50} = 2$ nM) has a plasma half-life of 1.7 h and higher metabolic stability than somatostatin.[247] Subsequently, [^{111}In]-labeled DTPA conjugated octreotide (Octreoscan; Mallinkrodt Medical) was developed and introduced for scintigraphic imaging of sst-expressing NET. Octreoscan is currently the gold standard for the localization, staging, and management of NET.[248] However, for imaging purposes, use of the ^{111}In radionuclide has certain disadvantages, including high cost, limited availability, less than optimum image quality and elevated patient radiation dose.[248] Consequently, several newer somatostatin analogs labeled with single-photon emitters (^{99m}Tc, ^{123}I) or positron-emitters (^{68}Ga, ^{18}F, ^{64}Cu) for SPECT or PET application, respectively, have been described.[249–254] A recent review describes the progress of functional imaging of NET using PET.[255]

Current work in sst radioligand development has been directed primarily toward modulation of radioligand subtype specificity or evaluation of different radionuclides for improved tumor-targeted radiotherapy.[247] Ginj *et al.* have reported on novel ^{111}In-labeled DOTA-conjugated octreotide analogs that display high-affinity binding (1.4–13 nM) to several sst receptor subtypes (sst_2, sst_3, and sst_5).[243] Animal biodistribution studies showed high, specific uptake of radioactivity in sst receptor-expressing tumors (AR4-2J) in a rat model. Both radiopeptides were more efficiently internalized than [^{111}In-DOTA,Tyr^3]-octreotide. The authors propose that the high-affinity, broad sst specificity of these radioligands will be advantageous for the diagnosis and targeted radiotherapy of a broader range of sst-expressing tumors.

In an extensive study involving 84 patients, Gabriel and coworkers have compared the utility of a new PET somatostatin analog, ^{68}Ga-D-Phe^1-Tyr^3-octreotide DOTA conjugate (^{68}Ga-DOTA-TOC), for NET imaging with that of SPECT and CT.[250] ^{68}Ga-DOTA-TOC was designed by replacement of the Phe^3 residue in the corresponding octapeptide by Tyr which leads to increased hydrophilicity and improved kidney clearance.[256,257] This modification also provides enhanced affinity for human sst_2.[256,258] Gabriel *et al.* compared PET imaging with ^{68}Ga-DOTA-TOC to SPECT imaging with the following radioligands: (^{111}In-DOTA-TOC) and ^{99m}Tc-labeled hydrazinonicotinyl-Tyr3-octreotide (^{99m}Tc-HYNIC-TOC).[250] CT imaging was also performed on each patient. Comparison of the three imaging modalities revealed an accuracy of 96% for PET imaging, which was significantly higher than that of CT (63%) and SPECT (58%). In addition, ^{68}Ga-DOTA-TOC imaging results were true-positive in 32 patients whose SPECT results were false-negative and it was able to detect more lesions than either SPECT or CT. Moreover, PET detected more metastatic tumors (lymph node, bone, and liver) than SPECT thus permitting more accurate disease staging. The authors conclude that PET imaging with ^{68}Ga-DOTA-TOC in conjunction with CT is superior to SPECT in the clinical diagnosis of NET.

Radiolabeled somatostatin analogs that incorporate beta-emitting (^{90}Y, ^{177}Lu) or alpha-emitting (^{213}Bi) radioisotopes for targeted tumor radiotherapy are under active investigation.[259,260] Preliminary, preclinical data suggest that the radionuclide ^{90}Y (high energy, pure beta-emitter; Emax = 2.25 MeV) may be more effective for treating larger tumors while the use of ^{177}Lu (low energy, beta-emitter; Emax = 0.497 MeV) leads to fewer relapses when treating smaller lesions.[261,262] These developments have been recently reviewed.[247]

G. Radioligands for Imaging Tumor Cell Proliferation

Tritium-labeled thymidine ([^{3}H]thymidine) has been shown to be rapidly incorporated into newly synthesized DNA. Hence [^{3}H]thymidine has been the gold standard for many years for assessing cell proliferation in cell culture and animal studies.[263] [^{11}C]Thymidine was subsequently developed as a PET radioligand for monitoring cell proliferation *in vivo*. However, general use of this radioligand is hampered by its rapid catabolism and short half-life.[264] Two PET radioligands that were subsequently developed to address these limitations include the [^{18}F]-labeled analogs, 3′-deoxy-3′-[^{18}F]fluorothymidine ([^{18}F]FLT) and 1-(2′-deoxy-2′-[^{18}F]fluoro-beta-D-arabinofuranosyl)thymine ([^{18}F]FMAU) (Fig. 20).[1] Of these, [^{18}F]FLT has been the most studied to date.[175] [^{18}F]FLT is taken up by cells and phosphorylated by thymidine kinase 1 (TK-1) with subsequent trapping within the cell.[263] Intracellular [^{18}F]FLT retention can thus be used as a measure of cellular TK-1 activity. Since TK-1 enzyme activity closely parallels the DNA synthesis pathway, cellular retention of [^{18}F]FLT is also a measure of cellular proliferation.

Leyton and coworkers have published several reports describing the use of PET imaging of [^{18}F]FLT uptake as a surrogate marker for early *in vivo* quantitative imaging of drug-induced changes in cell proliferation. In one example, PET imaging with [^{18}F]FLT was used for noninvasive measurement

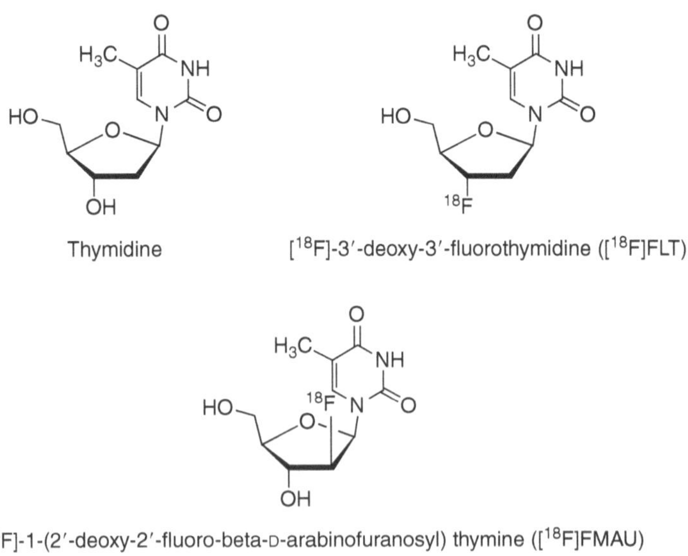

FIG. 20. Structure of [^{18}F]FLT and [^{18}F]FMAU.

of the biological activity of a novel histone deacetylase inhibitor (LAQ824).[265] Histone deacetylase inhibitors have been shown to cause growth inhibition of cancer cells *in vitro* and in animal models.[266] Treatment of mice bearing HCT116 colon-carcinoma xenografts with LAQ824 resulted in a dose-dependent decrease in [^{18}F]FLT tumor uptake. Dose-dependent decreases of tumor TK1 and protein levels were also observed with LAQ824 pretreatment.

In a second study, [^{18}F]FLT PET imaging was found to be superior to [^{18}F]FDG PET imaging for quantitative measurement of tumor cell proliferation following cisplatin treatment.[267] Cisplatin is a widely used chemotherapeutic drug for cervical, lung, bladder, and prostate cancer.[268] Decreased [^{18}F]FLT tumor uptake was seen in mice bearing radiation-induced fibrosarcoma 1 (RIF-1) tumors subject to cisplatin treatment, despite a lack of change in tumor size. The decrease in [^{18}F]FLT uptake was associated with a decrease in cell proliferation determined by immunohistochemical analysis. [^{18}F]FLT PET imaging was also successful in quantification of the activity of an administered replicating oncolytic viral vector (*dl*922-947) in mice bearing IGROVI ovarian carcinoma xenografts.[269]

Waldherr and colleagues have used [^{18}F]FLT PET imaging to measure the effect of the ErbB-selective kinase inhibitor PK1-166 on tumor cell proliferation in SCID mice having ErbB1-overexpressing A431 xenograft tumors.[270] Treatment with PK1-166 markedly lowered tumor [^{18}F]FLT uptake within 48 h of drug exposure and led to a 79% decrease of [^{18}F]FLT uptake within a week of treatment. [^{18}F]FLT PET imaging has also been used to monitor decreased tumor proliferation in a murine squamous cell carcinoma following radiation therapy.[271]

Taken together, these studies provide good initial evidence that [^{18}F]FLT PET imaging may be useful for monitoring therapeutic response early in the course of treatment. However, present studies suggest that [^{18}F]FDG may be more useful for tumor detection and staging.[272]

H. Radioligands for Imaging β-Amyloid

AD is a progressive neurodegenerative disorder characterized by irreversible impairment of memory and cognitive function and behavior changes. The pathological features of the disease are the presence of intracellular neurofibrillary tangles (NFT) and extracellular amyloid plagues consisting of dense aggregates of β-amyloid (Aβ).[273] Aβ is a 39–43 amino acid metalloprotein resulting from the proteolytic cleavage of APP by β- and γ-secretases. Although the precise mechanism of neuronal death in AD remains unknown, it is widely accepted that progressive accumulation of Aβ in brain due an imbalance between its production and removal is central to AD pathogenesis. Currently, postmortem analysis of brain tissue remains the only definitive method of AD diagnosis. For these reasons, intense efforts have been directed toward the development of PET and SPECT radioligands for the diagnostic imaging of amyloid deposits in AD.

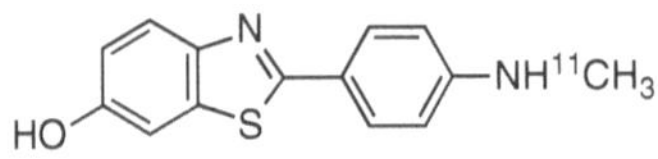

[^{11}C]PIB

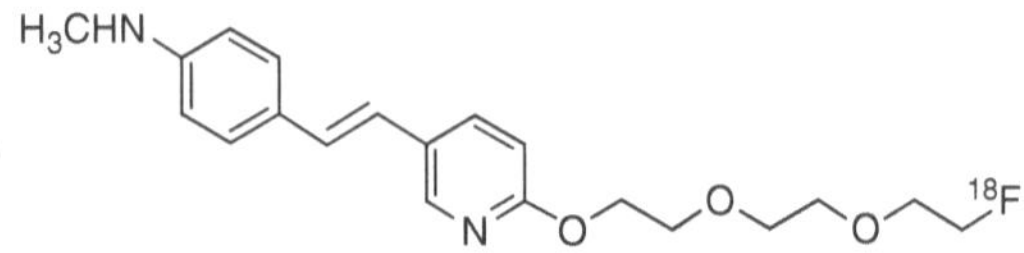

[^{18}F]AV-45

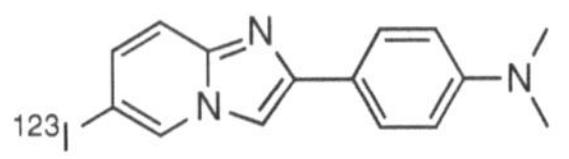

[^{123}I]IMPY

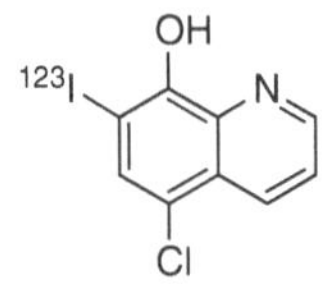

[^{123}I]CQ

FIG. 21. Structure of radioligands for imaging beta-amyloid.

Prominent among these is [^{11}C]PIB (Fig. 21), a carbon-11 labeled derivative of the fluorescent dye Thioflavin T routinely used to quantify beta-sheet fibrilization *in vitro*. [^{11}C]PIB displays high *in vitro* binding affinity for Aβ plagues (Ki = 4.3 nM) and a high initial brain uptake (7.0% ID/g at 2 min) followed by a rapid clearance from normal mouse brain (0.6% ID/g at 30 min) following i.v. administration. Initial clinical studies conducted with [^{11}C]PIB in patients with mild AD showed marked retention in areas of association cortex, which are known to contain large amounts of amyloid deposits in AD, with respect to control subjects (Fig. 22).[274] An F-18 labeled stilbene derivative ([^{18}F]AV45) developed at the University of Pennsylvania is also under clinical evaluation for imaging of AD with PET.[275] [^{18}F]AV45 displays rapid reversible binding characteristics which permit subject scanning at 45 min postinjection and several studies have confirmed its ability to discriminate between AD and age-matched controls.[276] Several [^{123}I]-labeled SPECT Aβ radioligands as exemplified by [^{123}I]IMPY and [^{123}I]CQ are also under evaluation due to the advantage of more widespread application.[277] [^{123}I] IMPY displayed selective labeling of Aβ plaques in *ex vivo* autoradiography studies using a double transgenic (PSAPP) mouse model of AD.[278]

I. Summary and Conclusion

The ability of PET and SPECT imaging to noninvasively image physiological and biochemical processes has opened up numerous applications in clinical as well as basic scientific research. Additionally, PET and SPECT imaging could play an important role in many aspects of the drug development process. At the preclinical stage, noninvasive imaging studies with radiolabeled investigational

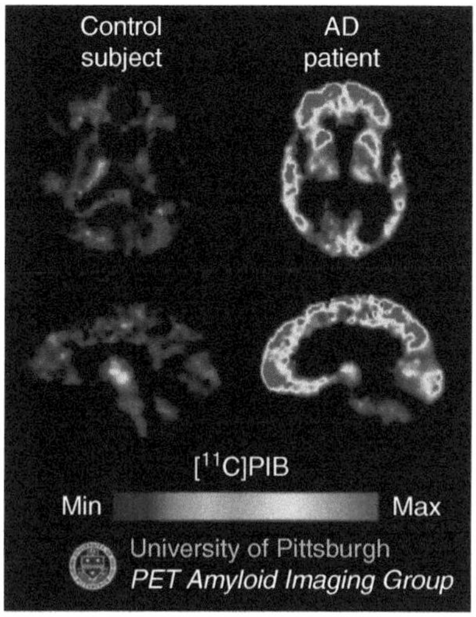

FIG. 22. Comparative PET images of [^{11}C]PIB binding in an AD and control subject. (Image courtesy of http://www.thinkgene.com.)

drugs can provide vital proof-of-concept information to aid proper candidate selection. These studies would allow noninvasive determination of a drug's pharmacokinetic behavior and target versus nontarget accumulation. Importantly, such studies could provide information early in the development process if a drug is, in fact, reaching its target and also identify potential toxicity issues. Drug analogs under investigation could also be screened *in vivo* in animal models against a validated radioligand selective for the same biological target (e.g., receptor, enzyme binding site, etc.) to provide direct measures of receptor occupancy. Such data could be used to design optimal dosing and timing schedules in clinical trials thereby improving their efficiency and cost-effectiveness.

At the clinical level, microdosing studies with suitable radioligands could provide useful metabolic and toxicology data prior to the conducting of classical Phase I trials. The FDA Critical Path Initiative has emphasized the development of validated imaging-based biomarkers as an important tool for streamlining the drug development process. For example, in oncological drug development, clinically validated radioligand probes that image downstream biological effects of cancer treatment such as apoptosis, angiogenesis, and decreased tumor cell proliferation could be useful surrogate markers for monitoring the therapeutic response of new oncology drugs in development.

[^{18}F]FDG PET imaging currently plays an increasingly important role particularly in cancer treatment by virtue of its ability to diagnose, stage, and assess tumor response to chemotherapy and chemoradiotherapy. It is highly likely that the future availability of validated molecularly targeted radioligands will likewise play a major role in therapeutic drug development. Such achievements will greatly advance the ultimate goal of personalized medical treatment in providing the right drug to the right patient.

Acknowledgments

We thank Steven Kronenberg for designing the figures and the members of the Center for Molecular Imaging (CMI) for critical reading of the manuscript. The authors acknowledge support from the following National Institutes of Health grants (PO1 CA85878, P50 CA93990, and R24 CA83099) during the preparation of this review.

References

1. Blume-Jensen P, Hunter T. Oncogenic kinase signalling. *Nature* 2001;**411**(6835):355–65.
2. Westra J, Limburg PC. p38 mitogen-activated protein kinase (MAPK) in rheumatoid arthritis. *Mini Rev Med Chem* 2006;**6**(8):867–74.
3. Ben-Bassat H. Biological activity of tyrosine kinase inhibitors: novel agents for psoriasis therapy. *Curr Opin Investig Drugs* 2001;**2**(11):1539–45.
4. Blease K. Targeting kinases in asthma. *Expert Opin Investig Drugs* 2005;**14**(10):1213–20.
5. Kumar R, Singh VP, Baker KM. Kinase inhibitors for cardiovascular disease. *J Mol Cell Cardiol* 2007;**42**(1):1–11.
6. Mueller BK, Mack H, Teusch N. Rho kinase, a promising drug target for neurological disorders. *Nat Rev Drug Discov* 2005;**4**(5):387–98.
7. Alizadeh A, et al. The lymphochip: a specialized cDNA microarray for the genomic-scale analysis of gene expression in normal and malignant lymphocytes. *Cold Spring Harb Symp Quant Biol* 1999;**64**:71–8.
8. Margalit O, et al. Microarray-based gene expression profiling of hematologic malignancies: basic concepts and clinical applications. *Blood Rev* 2005;**19**(4):223–34.
9. Morris DS, et al. Integrating biomedical knowledge to model pathways of prostate cancer progression. *Cell Cycle* 2007;**6**(10):1177–87.
10. Nishiu M, et al. Microarray analysis of gene-expression profiles in diffuse large B-cell lymphoma: identification of genes related to disease progression. *Jpn J Cancer Res* 2002;**93**(8):894–901.
11. Tschoep K, et al. Gene expression profiling in sarcomas. *Crit Rev Oncol Hematol* 2007;**63**(2):111–24.
12. Radford IR. Imatinib. Novartis. *Curr Opin Investig Drugs* 2002;**3**(3):492–9.
13. Melnikova I, Golden J. Targeting protein kinases. *Nat Rev Drug Discov* 2004;**3**(12):993–4.
14. Melnikova I, Golden J. Apoptosis-targeting therapies. *Nat Rev Drug Discov* 2004;**3**(11):905–6.
15. Chun PY, et al. Synergistic effects of gemcitabine and gefitinib in the treatment of head and neck carcinoma. *Cancer Res* 2006;**66**(2):981–8.
16. Bhojani MS, et al. Imaging of proteolytic activity using a conditional cell surface receptor. *Mol Imaging* 2006;**5**(2):129–37.
17. Zhang L, et al. Molecular imaging of Akt kinase activity. *Nat Med* 2007;**13**(9):1114–9.
18. Fields S, Song O. A novel genetic system to detect protein–protein interactions. *Nature* 1989;**340**(6230):245–6.

19. Luker KE, Piwnica-Worms D. Optimizing luciferase protein fragment complementation for bioluminescent imaging of protein–protein interactions in live cells and animals. *Methods Enzymol* 2004;**385**:349–60.
20. Michnick SW, et al. Universal strategies in research and drug discovery based on protein-fragment complementation assays. *Nat Rev Drug Discov* 2007;**6**(7):569–82.
21. Remy I, Ghaddar G, Michnick SW. Using the beta-lactamase protein-fragment complementation assay to probe dynamic protein–protein interactions. *Nat Protoc* 2007;**2**(9):2302–6.
22. Remy I, Michnick SW. A highly sensitive protein–protein interaction assay based on Gaussia luciferase. *Nat Methods* 2006;**3**(12):977–9.
23. Remy I, Michnick SW. Application of protein-fragment complementation assays in cell biology. *Biotechniques* 2007;**42**(2) p. 137, 139, 141 passim.
24. Stefan E, et al. Quantification of dynamic protein complexes using Renilla luciferase fragment complementation applied to protein kinase A activities in vivo. *Proc Natl Acad Sci USA* 2007;**104**(43):16916–21.
25. Contag CH, Ross BD. It's not just about anatomy: in vivo bioluminescence imaging as an eyepiece into biology. *J Magn Reson Imaging* 2002;**16**(4):378–87.
26. Greer III LF, Szalay AA. Imaging of light emission from the expression of luciferases in living cells and organisms: a review. *Luminescence* 2002;**17**(1):43–74.
27. Luker KE, et al. Kinetics of regulated protein–protein interactions revealed with firefly luciferase complementation imaging in cells and living animals. *Proc Natl Acad Sci USA* 2004;**101**(33):12288–93.
28. Cantley LC. The phosphoinositide 3-kinase pathway. *Science* 2002;**296**(5573):1655–7.
29. Chan TO, Rittenhouse SE, Tsichlis PN. AKT/PKB and other D3 phosphoinositide-regulated kinases: kinase activation by phosphoinositide-dependent phosphorylation. *Annu Rev Biochem* 1999;**68**:965–1014.
30. Kovarova M, et al. Structure–function analysis of Lyn kinase association with lipid rafts and initiation of early signaling events after Fcepsilon receptor I aggregation. *Mol Cell Biol* 2001;**21**(24):8318–28.
31. Contessa JN, et al. Inhibition of N-linked glycosylation disrupts receptor tyrosine kinase signaling in tumor cells. *Cancer Res* 2008;**68**(10):3803–9.
32. Chakravarti A, Loeffler JS, Dyson NJ. Insulin-like growth factor receptor I mediates resistance to anti-epidermal growth factor receptor therapy in primary human glioblastoma cells through continued activation of phosphoinositide 3-kinase signaling. *Cancer Res* 2002;**62**(1):200–7.
33. Damiano V, et al. Cooperative antitumor effect of multitargeted kinase inhibitor ZD6474 and ionizing radiation in glioblastoma. *Clin Cancer Res* 2005;**11**(15):5639–44.
34. Dote H, et al. ErbB3 expression predicts tumor cell radiosensitization induced by Hsp90 inhibition. *Cancer Res* 2005;**65**(15):6967–75.
35. Lammering G, et al. Radiosensitization of malignant glioma cells through overexpression of dominant-negative epidermal growth factor receptor. *Clin Cancer Res* 2001;**7**(3):682–90.
36. Contessa JN, et al. Molecular imaging of N-linked glycosylation suggests glycan biosynthesis is a novel target for cancer therapy. *Clin Cancer Res* 2010;**16**(12):3205–14.
37. Ameisen JC. On the origin, evolution, and nature of programmed cell death: a timeline of four billion years. *Cell Death Differ* 2002;**9**(4):367–93.
38. Bhojani MS, Rossu BD, Rehemtulla A. TRAIL and anti-tumor responses. *Cancer Biol Ther* 2003;**2**(4 Suppl. 1):S71–8.
39. Hengartner MO. The biochemistry of apoptosis. *Nature* 2000;**407**(6805):770–6.
40. Nicholson DW, Thornberry NA. Apoptosis. Life and death decisions. *Science* 2003;**299**(5604):214–5.

41. Nicholson DW, Thornberry NA. Caspases: killer proteases. *Trends Biochem Sci* 1997;**22** (8):299–306.
42. Shi Y. Apoptosome: the cellular engine for the activation of caspase-9. *Structure* 2002;**10** (3):285–8.
43. Laxman B, et al. Noninvasive real-time imaging of apoptosis. *Proc Natl Acad Sci USA* 2002;**99** (26):16551–5.
44. Lee KC, et al. Noninvasive molecular imaging sheds light on the synergy between 5-fluorouracil and TRAIL/Apo2L for cancer therapy. *Clin Cancer Res* 2007;**13**(6):1839–46.
45. Thormeyer D, et al. Characterization of lacZ complementation deletions using membrane receptor dimerization. *Biotechniques* 2003;**34**(2):346–50, 352–5.
46. Zhang Z, Zhu W, Kodadek T. Selection and application of peptide-binding peptides. *Nat Biotechnol* 2000;**18**(1):71–4.
47. Coppola JM, Ross BD, Rehemtulla A. Noninvasive imaging of apoptosis and its application in cancer therapeutics. *Clin Cancer Res* 2008;**14**(8):2492–501.
48. Khanna D, et al. A transgenic mouse for imaging caspase-dependent apoptosis within the skin. *J Invest Dermatol* 2010;**130**(7):1797–806.
49. Matthews BW. The structure of E. coli beta-galactosidase. *C R Biol* 2005;**328**(6):549–56.
50. Celen S, et al. Synthesis and evaluation of 18F- and 11C-labeled phenyl-galactopyranosides as potential probes for in vivo visualization of LacZ gene expression using positron emission tomography. *Bioconjug Chem* 2008;**19**(2):441–9.
51. Olesen CE, et al. Novel methods for chemiluminescent detection of reporter enzymes. *Methods Enzymol* 2000;**326**:175–202.
52. Smith AD, Trempe JP. Luminometric quantitation of photinus pyralis firefly luciferase and Escherichia coli beta-galactosidase in blood-contaminated organ lysates. *Anal Biochem* 2000;**286**(1):164–72.
53. Van Poucke SO, Nelis HJ. Rapid detection of fluorescent and chemiluminescent total coliforms and Escherichia coli on membrane filters. *J Microbiol Methods* 2000;**42**(3):233–44.
54. Molloy SS, et al. Bi-cycling the furin pathway: from TGN localization to pathogen activation and embryogenesis. *Trends Cell Biol* 1999;**9**(1):28–35.
55. Reznik SE, Fricker LD. Carboxypeptidases from A to z: implications in embryonic development and Wnt binding. *Cell Mol Life Sci* 2001;**58**(12–13):1790–804.
56. Vassar R. Beta-secretase (BACE) as a drug target for Alzheimer's disease. *Adv Drug Deliv Rev* 2002;**54**(12):1589–602.
57. Vassar R, et al. Beta-secretase cleavage of Alzheimer's amyloid precursor protein by the transmembrane aspartic protease BACE. *Science* 1999;**286**(5440):735–41.
58. Coppola JM, et al. A small-molecule furin inhibitor inhibits cancer cell motility and invasiveness. *Neoplasia* 2008;**10**(4):363–70.
59. Coppola JM, et al. Identification of inhibitors using a cell-based assay for monitoring Golgi-resident protease activity. *Anal Biochem* 2007;**364**(1):19–29.
60. Ishikawa TO, Herschman HR. Conditional bicistronic Cre reporter line expressing both firefly luciferase and beta-galactosidase. *Mol Imaging Biol* 2010.
61. Carlsen H, et al. In vivo imaging of NF-kappa B activity. *J Immunol* 2002;**168**(3):1441–6.
62. Lin AH, et al. Global analysis of Smad2/3-dependent TGF-beta signaling in living mice reveals prominent tissue-specific responses to injury. *J Immunol* 2005;**175**(1):547–54.
63. Safran M, et al. Mouse model for noninvasive imaging of HIF prolyl hydroxylase activity: assessment of an oral agent that stimulates erythropoietin production. *Proc Natl Acad Sci USA* 2006;**103**(1):105–10.
64. Rosen BR, et al. Perfusion imaging with NMR contrast agents. *Magn Reson Med* 1990;**14** (2):249–65.

65. Tofts PS, Kermode AG. Measurement of the blood-brain barrier permeability and leakage space using dynamic MR imaging. 1. Fundamental concepts. *Magn Reson Med* 1991;**17**(2):357–67.
66. Schwickert HC, et al. Contrast-enhanced MR imaging assessment of tumor capillary permeability: effect of irradiation on delivery of chemotherapy. *Radiology* 1996;**198**(3):893–8.
67. Su FC, et al. Temporal resolving power of perfusion- and BOLD-based event-related functional MRI. *Med Phys* 2004;**31**(1):154–60.
68. Warach S, et al. Acute human stroke studied by whole brain echo planar diffusion-weighted magnetic resonance imaging. *Ann Neurol* 1995;**37**(2):231–41.
69. Guo AC, et al. Lymphomas and high-grade astrocytomas: comparison of water diffusibility and histologic characteristics. *Radiology* 2002;**224**(1):177–83.
70. Pierpaoli C, Basser PJ. Toward a quantitative assessment of diffusion anisotropy. *Magn Reson Med* 1996;**36**(6):893–906.
71. Basser PJ, Pierpaoli C. Microstructural and physiological features of tissues elucidated by quantitative-diffusion-tensor MRI. *J Magn Reson B* 1996;**111**(3):209–19.
72. Gellermann J, et al. Methods and potentials of magnetic resonance imaging for monitoring radiofrequency hyperthermia in a hybrid system. *Int J Hyperthermia* 2005;**21**(6):497–513.
73. Sorensen AG, et al. Hyperacute stroke: evaluation with combined multisection diffusion-weighted and hemodynamically weighted echo-planar MR imaging. *Radiology* 1996;**199**(2):391–401.
74. Le Bihan D, et al. Separation of diffusion and perfusion in intravoxel incoherent motion MR imaging. *Radiology* 1988;**168**(2):497–505.
75. Moseley ME, et al. Diffusion-weighted MR imaging of anisotropic water diffusion in cat central nervous system. *Radiology* 1990;**176**(2):439–45.
76. Le Bihan D, et al. Diffusion and perfusion magnetic resonance imaging in brain tumors. *Top Magn Reson Imaging* 1993;**5**(1):25–31.
77. Le Bihan D. Molecular diffusion nuclear magnetic resonance imaging. *Magn Reson Q* 1991;**7**(1):1–30.
78. Moseley ME, et al. Diffusion-weighted MR imaging of acute stroke: correlation with T2-weighted and magnetic susceptibility-enhanced MR imaging in cats. *AJNR Am J Neuroradiol* 1990;**11**(3):423–9.
79. Stejskal E, Tanner J. Spin diffusion measurements: spin echoes in the presence of a time-dependent field gradient. *J Chem Phys* 1965;**42**(1):288–92.
80. Thomsen C, Henriksen O, Ring P. In vivo measurement of water self diffusion in the human brain by magnetic resonance imaging. *Acta Radiol* 1987;**28**(3):353–61.
81. Merboldt KD, et al. MRI of "diffusion" in the human brain: new results using a modified CE-FAST sequence. *Magn Reson Med* 1989;**9**(3):423–9.
82. Bammer R. Basic principles of diffusion-weighted imaging. *Eur J Radiol* 2003;**45**(3):169–84.
83. Padhani AR, et al. Diffusion-weighted magnetic resonance imaging as a cancer biomarker: consensus and recommendations. *Neoplasia* 2009;**11**(2):102–25.
84. Le Bihan D. The 'wet mind': water and functional neuroimaging. *Phys Med Biol* 2007;**52**(7):R57–90.
85. Edelman RR, Wielopolski P, Schmitt F. Echo-planar MR imaging. *Radiology* 1994;**192**(3):600–12.
86. Provenzale JM, et al. Peritumoral brain regions in gliomas and meningiomas: investigation with isotropic diffusion-weighted MR imaging and diffusion-tensor MR imaging. *Radiology* 2004;**232**(2):451–60.
87. Yoshikawa T, et al. ADC measurement of abdominal organs and lesions using parallel imaging technique. *AJR Am J Roentgenol* 2006;**187**(6):1521–30.

88. Koh DM, et al. Colorectal hepatic metastases: quantitative measurements using single-shot echo-planar diffusion-weighted MR imaging. *Eur Radiol* 2006;**16**(9):1898–905.
89. Koh DM, et al. Predicting response of colorectal hepatic metastasis: value of pretreatment apparent diffusion coefficients. *AJR Am J Roentgenol* 2007;**188**(4):1001–8.
90. Charles-Edwards EM, deSouza NM. Diffusion-weighted magnetic resonance imaging and its application to cancer. *Cancer Imaging* 2006;**6**:135–43.
91. Takahara T, et al. Diffusion weighted whole body imaging with background body signal suppression (DWIBS): technical improvement using free breathing, STIR and high resolution 3D display. *Radiat Med* 2004;**22**(4):275–82.
92. Ichikawa T, et al. High-b value diffusion-weighted MRI for detecting pancreatic adenocarcinoma: preliminary results. *AJR Am J Roentgenol* 2007;**188**(2):409–14.
93. Sumi M, Ichikawa Y, Nakamura T. Diagnostic ability of apparent diffusion coefficients for lymphomas and carcinomas in the pharynx. *Eur Radiol* 2007;**17**(10):2631–7.
94. Eida S, et al. Apparent diffusion coefficient mapping of salivary gland tumors: prediction of the benignancy and malignancy. *AJNR Am J Neuroradiol* 2007;**28**(1):116–21.
95. Bulakbasi N, et al. The added value of the apparent diffusion coefficient calculation to magnetic resonance imaging in the differentiation and grading of malignant brain tumors. *J Comput Assist Tomogr* 2004;**28**(6):735–46.
96. Shimony JS, et al. Quantitative diffusion-tensor anisotropy brain MR imaging: normative human data and anatomic analysis. *Radiology* 1999;**212**(3):770–84.
97. Mori S, et al. Brain white matter anatomy of tumor patients evaluated with diffusion tensor imaging. *Ann Neurol* 2002;**51**(3):377–80.
98. Chenevert TL, McKeever PE, Ross BD. Monitoring early response of experimental brain tumors to therapy using diffusion magnetic resonance imaging. *Clin Cancer Res* 1997;**3**(9):1457–66.
99. Brunberg JA, et al. In vivo MR determination of water diffusion coefficients and diffusion anisotropy: correlation with structural alteration in gliomas of the cerebral hemispheres. *AJNR Am J Neuroradiol* 1995;**16**(2):361–71.
100. Lyng H, Haraldseth O, Rofstad EK. Measurement of cell density and necrotic fraction in human melanoma xenografts by diffusion weighted magnetic resonance imaging. *Magn Reson Med* 2000;**43**(6):828–36.
101. Chenevert TL, et al. Diffusion magnetic resonance imaging: an early surrogate marker of therapeutic efficacy in brain tumors. *J Natl Cancer Inst* 2000;**92**(24):2029–36.
102. Chenevert TL, Sundgren PC, Ross BD. Diffusion imaging: insight to cell status and cytoarchitecture. *Neuroimaging Clin N Am* 2006;**16**(4):619–32, viii–ix.
103. Ross BD, et al. Evaluation of cancer therapy using diffusion magnetic resonance imaging. *Mol Cancer Ther* 2003;**2**(6):581–7.
104. Lee KC, et al. Prospective early response imaging biomarker for neoadjuvant breast cancer chemotherapy. *Clin Cancer Res* 2007;**13**(2 Pt 1):443–50.
105. Lee KC, et al. Dynamic imaging of emerging resistance during cancer therapy. *Cancer Res* 2006;**66**(9):4687–92.
106. Lee KC, et al. An imaging biomarker of early treatment response in prostate cancer that has metastasized to the bone. *Cancer Res* 2007;**67**(8):3524–8.
107. Hamstra DA, et al. The use of 19F spectroscopy and diffusion-weighted MRI to evaluate differences in gene-dependent enzyme prodrug therapies. *Mol Ther* 2004;**10**(5):916–28.
108. Lee KC, et al. Fusion of the HSV-1 tegument protein vp22 to cytosine deaminase confers enhanced bystander effect and increased therapeutic benefit. *Gene Ther* 2006;**13**(2):127–37.
109. Therasse P, et al. New guidelines to evaluate the response to treatment in solid tumors. European Organization for Research and Treatment of Cancer, National Cancer Institute of the United States, National Cancer Institute of Canada. *J Natl Cancer Inst* 2000;**92**(3):205–16.

110. Sorensen AG, et al. Comparison of diameter and perimeter methods for tumor volume calculation. *J Clin Oncol* 2001;**19**(2):551–7.
111. Yoshida S, et al. Initial experience of diffusion-weighted magnetic resonance imaging to assess therapeutic response to induction chemoradiotherapy against muscle-invasive bladder cancer. *Urology* 2010;**75**(2):387–91.
112. Yankeelov TE, et al. Integration of quantitative DCE-MRI and ADC mapping to monitor treatment response in human breast cancer: initial results. *Magn Reson Imaging* 2007;**25**(1):1–13.
113. Uhl M, et al. Osteosarcoma: preliminary results of in vivo assessment of tumor necrosis after chemotherapy with diffusion- and perfusion-weighted magnetic resonance imaging. *Invest Radiol* 2006;**41**(8):618–23.
114. Tomura N, et al. Diffusion changes in a tumor and peritumoral tissue after stereotactic irradiation for brain tumors: possible prediction of treatment response. *J Comput Assist Tomogr* 2006;**30**(3):496–500.
115. Theilmann RJ, et al. Changes in water mobility measured by diffusion MRI predict response of metastatic breast cancer to chemotherapy. *Neoplasia* 2004;**6**(6):831–7.
116. Sun YS, et al. Locally advanced rectal carcinoma treated with preoperative chemotherapy and radiation therapy: preliminary analysis of diffusion-weighted MR imaging for early detection of tumor histopathologic downstaging. *Radiology* 2010;**254**(1):170–8.
117. Schubert MI, et al. Diffusion-weighted magnetic resonance imaging of treatment-associated changes in recurrent and residual medulloblastoma: preliminary observations in three children. *Acta Radiol* 2006;**47**(10):1100–4.
118. Rhee TK, et al. Tumor response after yttrium-90 radioembolization for hepatocellular carcinoma: comparison of diffusion-weighted functional MR imaging with anatomic MR imaging. *J Vasc Interv Radiol* 2008;**19**(8):1180–6.
119. Pickles MD, et al. Diffusion changes precede size reduction in neoadjuvant treatment of breast cancer. *Magn Reson Imaging* 2006;**24**(7):843–7.
120. Okuma T, et al. Assessment of early treatment response after CT-guided radiofrequency ablation of unresectable lung tumours by diffusion-weighted MRI: a pilot study. *Br J Radiol* 2009;**82**(984):989–94.
121. Moffat BA, et al. Functional diffusion map: a noninvasive MRI biomarker for early stratification of clinical brain tumor response. *Proc Natl Acad Sci USA* 2005;**102**(15):5524–9.
122. Marugami N, et al. Early detection of therapeutic response to hepatic arterial infusion chemotherapy of liver metastases from colorectal cancer using diffusion-weighted MR imaging. *Cardiovasc Intervent Radiol* 2009;**32**(4):638–46.
123. Mardor Y, et al. Monitoring response to convection-enhanced taxol delivery in brain tumor patients using diffusion-weighted magnetic resonance imaging. *Cancer Res* 2001;**61**(13):4971–3.
124. Mardor Y, et al. Early detection of response to radiation therapy in patients with brain malignancies using conventional and high b-value diffusion-weighted magnetic resonance imaging. *J Clin Oncol* 2003;**21**(6):1094–100.
125. Liu Y, et al. Diffusion-weighted imaging in predicting and monitoring the response of uterine cervical cancer to combined chemoradiation. *Clin Radiol* 2009;**64**(11):1067–74.
126. Lin YC, et al. Significant temporal evolution of diffusion anisotropy for evaluating early response to radiosurgery in patients with vestibular schwannoma: findings from functional diffusion maps. *AJNR Am J Neuroradiol* 2010;**31**(2):269–74.
127. Liapi E, et al. Assessment of response of uterine fibroids and myometrium to embolization using diffusion-weighted echoplanar MR imaging. *J Comput Assist Tomogr* 2005;**29**(1):83–6.
128. Kremser C, et al. Preliminary results on the influence of chemoradiation on apparent diffusion coefficients of primary rectal carcinoma measured by magnetic resonance imaging. *Strahlenther Onkol* 2003;**179**(9):641–9.

129. Koh DM, et al. Reproducibility and changes in the apparent diffusion coefficients of solid tumours treated with combretastatin A4 phosphate and bevacizumab in a two-centre phase I clinical trial. *Eur Radiol* 2009;**19**(11):2728–38.
130. Kim S, et al. Diffusion-weighted magnetic resonance imaging for predicting and detecting early response to chemoradiation therapy of squamous cell carcinomas of the head and neck. *Clin Cancer Res* 2009;**15**(3):986–94.
131. Kim SH, et al. Locally advanced rectal cancer: added value of diffusion-weighted MR imaging in the evaluation of tumor response to neoadjuvant chemo- and radiation therapy. *Radiology* 2009;**253**(1):116–25.
132. Kato H, et al. Head and neck squamous cell carcinoma: usefulness of diffusion-weighted MR imaging in the prediction of a neoadjuvant therapeutic effect. *Eur Radiol* 2009;**19**(1):103–9.
133. Kamel IR, et al. Functional MR imaging assessment of tumor response after 90Y microsphere treatment in patients with unresectable hepatocellular carcinoma. *J Vasc Interv Radiol* 2007;**18**(1 Pt 1):49–56.
134. Kamel IR, et al. The role of functional MR imaging in the assessment of tumor response after chemoembolization in patients with hepatocellular carcinoma. *J Vasc Interv Radiol* 2006;**17** (3):505–12.
135. Jain R, et al. Imaging response criteria for recurrent gliomas treated with bevacizumab: role of diffusion weighted imaging as an imaging biomarker. *J Neurooncol* 2010;**96**(3):423–31.
136. Jacobs MA, Herskovits EH, Kim HS. Uterine fibroids: diffusion-weighted MR imaging for monitoring therapy with focused ultrasound surgery—preliminary study. *Radiology* 2005;**236** (1):196–203.
137. Hayashida Y, et al. Monitoring therapeutic responses of primary bone tumors by diffusion-weighted image: Initial results. *Eur Radiol* 2006;**16**(12):2637–43.
138. Harry VN, et al. Diffusion-weighted magnetic resonance imaging in the early detection of response to chemoradiation in cervical cancer. *Gynecol Oncol* 2008;**111**(2):213–20.
139. Hamstra DA, et al. Evaluation of the functional diffusion map as an early biomarker of time-to-progression and overall survival in high-grade glioma. *Proc Natl Acad Sci USA* 2005;**102** (46):16759–64.
140. Galban CJ, et al. A feasibility study of parametric response map analysis of diffusion-weighted magnetic resonance imaging scans of head and neck cancer patients for providing early detection of therapeutic efficacy. *Transl Oncol* 2009;**2**(3):184–90.
141. Dzik-Jurasz A, et al. Diffusion MRI for prediction of response of rectal cancer to chemoradiation. *Lancet* 2002;**360**(9329):307–8.
142. Dirix P, et al. Dose painting in radiotherapy for head and neck squamous cell carcinoma: value of repeated functional imaging with (18)F-FDG PET, (18)F-fluoromisonidazole PET, diffusion-weighted MRI, and dynamic contrast-enhanced MRI. *J Nucl Med* 2009;**50**(7):1020–7.
143. Deng J, et al. Diffusion-weighted MR imaging for determination of hepatocellular carcinoma response to yttrium-90 radioembolization. *J Vasc Interv Radiol* 2006;**17**(7):1195–200.
144. Cui Y, et al. Apparent diffusion coefficient: potential imaging biomarker for prediction and early detection of response to chemotherapy in hepatic metastases. *Radiology* 2008;**248** (3):894–900.
145. Byun WM, et al. Diffusion-weighted MR imaging of metastatic disease of the spine: assessment of response to therapy. *AJNR Am J Neuroradiol* 2002;**23**(6):906–12.
146. Eccles CL, et al. Change in diffusion weighted MRI during liver cancer radiotherapy: preliminary observations. *Acta Oncol* 2009;**48**(7):1034–43.
147. Hein PA, et al. Diffusion-weighted imaging in the follow-up of treated high-grade gliomas: tumor recurrence versus radiation injury. *AJNR Am J Neuroradiol* 2004;**25**(2):201–9.

148. Hamstra DA, et al. Functional diffusion map as an early imaging biomarker for high-grade glioma: correlation with conventional radiologic response and overall survival. *J Clin Oncol* 2008;**26**(20):3387–94.
149. Lee KC, et al. A feasibility study evaluating the functional diffusion map as a predictive imaging biomarker for detection of treatment response in a patient with metastatic prostate cancer to the bone. *Neoplasia* 2007;**9**(12):1003–11.
150. Zahra MA, et al. Dynamic contrast-enhanced MRI as a predictor of tumour response to radiotherapy. *Lancet Oncol* 2007;**8**(1):63–74.
151. O'Connor JP, et al. DCE-MRI biomarkers in the clinical evaluation of antiangiogenic and vascular disrupting agents. *Br J Cancer* 2007;**96**(2):189–95.
152. Ah-See ML, et al. Early changes in functional dynamic magnetic resonance imaging predict for pathologic response to neoadjuvant chemotherapy in primary breast cancer. *Clin Cancer Res* 2008;**14**(20):6580–9.
153. Baar J, et al. A vasculature-targeting regimen of preoperative docetaxel with or without bevacizumab for locally advanced breast cancer: impact on angiogenic biomarkers. *Clin Cancer Res* 2009;**15**(10):3583–90.
154. Batchelor TT, et al. AZD2171, a pan-VEGF receptor tyrosine kinase inhibitor, normalizes tumor vasculature and alleviates edema in glioblastoma patients. *Cancer Cell* 2007;**11**(1):83–95.
155. Bauerle T, et al. Imaging anti-angiogenic treatment response with DCE-VCT, DCE-MRI and DWI in an animal model of breast cancer bone metastasis. *Eur J Radiol* 2010;**73**(2):280–7.
156. Jain RK, et al. Biomarkers of response and resistance to antiangiogenic therapy. *Nat Rev Clin Oncol* 2009;**6**(6):327–38.
157. Jarnagin WR, et al. Regional chemotherapy for unresectable primary liver cancer: results of a phase II clinical trial and assessment of DCE-MRI as a biomarker of survival. *Ann Oncol* 2009;**20**(9):1589–95.
158. Kamoun WS, et al. Edema control by cediranib, a vascular endothelial growth factor receptor-targeted kinase inhibitor, prolongs survival despite persistent brain tumor growth in mice. *J Clin Oncol* 2009;**27**(15):2542–52.
159. Kim S, et al. Prediction of response to chemoradiation therapy in squamous cell carcinomas of the head and neck using dynamic contrast-enhanced MR imaging. *AJNR Am J Neuroradiol* 2010;**31**(2):262–8.
160. Lockhart AC, et al. Phase I study of intravenous vascular endothelial growth factor trap, aflibercept, in patients with advanced solid tumors. *J Clin Oncol* 2010;**28**(2):207–14.
161. Meyer T, et al. A phase I trial of radioimmunotherapy with 131I-A5B7 anti-CEA antibody in combination with combretastatin-A4-phosphate in advanced gastrointestinal carcinomas. *Clin Cancer Res* 2009;**15**(13):4484–92.
162. Sorensen AG, et al. A "vascular normalization index" as potential mechanistic biomarker to predict survival after a single dose of cediranib in recurrent glioblastoma patients. *Cancer Res* 2009;**69**(13):5296–300.
163. van Laarhoven HW, et al. Phase I clinical and magnetic resonance imaging study of the vascular agent NGR-hTNF in patients with advanced cancers (European Organization for Research and Treatment of Cancer Study 16041). *Clin Cancer Res* 2010;**16**(4):1315–23.
164. Vriens D, et al. Chemotherapy response monitoring of colorectal liver metastases by dynamic Gd-DTPA-enhanced MRI perfusion parameters and 18F-FDG PET metabolic rate. *J Nucl Med* 2009;**50**(11):1777–84.
165. Wong CI, et al. Phase I and biomarker study of ABT-869, a multiple receptor tyrosine kinase inhibitor, in patients with refractory solid malignancies. *J Clin Oncol* 2009;**27**(28):4718–26.
166. Zahra MA, et al. Semiquantitative and quantitative dynamic contrast-enhanced magnetic resonance imaging measurements predict radiation response in cervix cancer. *Int J Radiat Oncol Biol Phys* 2009;**74**(3):766–73.

167. Tofts PS, et al. Estimating kinetic parameters from dynamic contrast-enhanced T(1)-weighted MRI of a diffusable tracer: standardized quantities and symbols. *J Magn Reson Imaging* 1999;**10**(3):223–32.
168. Leach MO, et al. The assessment of antiangiogenic and antivascular therapies in early-stage clinical trials using magnetic resonance imaging: issues and recommendations. *Br J Cancer* 2005;**92**(9):1599–610.
169. Semiz Oysu A, et al. Dynamic contrast-enhanced MRI in the differentiation of posttreatment fibrosis from recurrent carcinoma of the head and neck. *Clin Imaging* 2005;**29**(5):307–12.
170. Ohno Y, et al. Prognostic value of dynamic MR imaging for non-small-cell lung cancer patients after chemoradiotherapy. *J Magn Reson Imaging* 2005;**21**(6):775–83.
171. Dobson MJ, et al. The assessment of irradiated bladder carcinoma using dynamic contrast-enhanced MR imaging. *Clin Radiol* 2001;**56**(2):94–8.
172. Boss EA, et al. Post-radiotherapy contrast enhancement changes in fast dynamic MRI of cervical carcinoma. *J Magn Reson Imaging* 2001;**13**(4):600–6.
173. Wedam SB, et al. Antiangiogenic and antitumor effects of bevacizumab in patients with inflammatory and locally advanced breast cancer. *J Clin Oncol* 2006;**24**(5):769–77.
174. Goh V, Padhani AR, Rasheed S. Functional imaging of colorectal cancer angiogenesis. *Lancet Oncol* 2007;**8**(3):245–55.
175. Shields AF. Positron emission tomography measurement of tumor metabolism and growth: its expanding role in oncology. *Mol Imaging Biol* 2006;**8**(3):141–50.
176. Metser U, Even-Sapir E. Increased (18)F-fluorodeoxyglucose uptake in benign, nonphysiologic lesions found on whole-body positron emission tomography/computed tomography (PET/CT): accumulated data from four years of experience with PET/CT. *Semin Nucl Med* 2007;**37**(3):206–22.
177. Carmeliet P. Angiogenesis in health and disease. *Nat Med* 2003;**9**(6):653–60.
178. Sturk C, Duont D. Angiogenesis. In: Tannock IF, et al. editors. *The basic science of oncology*. New York: McGraw-Hill; 2005. p. 231–48.
179. Bergers G, Benjamin LE. Tumorigenesis and the angiogenic switch. *Nat Rev Cancer* 2003;**3**(6):401–10.
180. Safran M, Kaelin Jr. WG. HIF hydroxylation and the mammalian oxygen-sensing pathway. *J Clin Invest* 2003;**111**(6):779–83.
181. Ferrara N. Vascular endothelial growth factor: basic science and clinical progress. *Endocr Rev* 2004;**25**(4):581–611.
182. Hicklin DJ, Ellis LM. Role of the vascular endothelial growth factor pathway in tumor growth and angiogenesis. *J Clin Oncol* 2005;**23**(5):1011–27.
183. Renner W, Pilger E. Simultaneous in vivo quantitation of vascular endothelial growth factor mRNA splice variants. *J Vasc Res* 1999;**36**(2):133–8.
184. Sato Y, et al. Properties of two VEGF receptors, Flt-1 and KDR, in signal transduction. *Ann N Y Acad Sci* 2000;**902**:201–5 discussion 205–207.
185. Ferrara N. The role of VEGF in the regulation of physiological and pathological angiogenesis. *Exs* 2005;(94):209–31.
186. Rudlowski C, et al. Prognostic significance of vascular endothelial growth factor expression in ovarian cancer patients: a long-term follow-up. *Int J Gynecol Cancer* 2006;**16**(Suppl. 1):183–9.
187. Cai W, et al. PET of vascular endothelial growth factor receptor expression. *J Nucl Med* 2006;**47**(12):2048–56.
188. Hsu AR, et al. Multimodality molecular imaging of glioblastoma growth inhibition with vasculature-targeting fusion toxin VEGF121/rGel. *J Nucl Med* 2007;**48**(3):445–54.
189. Rodriguez-Porcel M, et al. Imaging of VEGF receptor in a rat myocardial infarction model using PET. *J Nucl Med* 2008;**49**(4):667–73.

190. Cai W, et al. Positron emission tomography imaging of poststroke angiogenesis. *Stroke* 2009;**40**(1):270–7.
191. Chan C, et al. A human transferrin-vascular endothelial growth factor (hnTf-VEGF) fusion protein containing an integrated binding site for (111)In for imaging tumor angiogenesis. *J Nucl Med* 2005;**46**(10):1745–52.
192. Danen EH. Integrins: regulators of tissue function and cancer progression. *Curr Pharm Des* 2005;**11**(7):881–91.
193. Hood JD, Cheresh DA. Role of integrins in cell invasion and migration. *Nat Rev Cancer* 2002;**2**(2):91–100.
194. Kumar CC. Integrin alpha v beta 3 as a therapeutic target for blocking tumor-induced angiogenesis. *Curr Drug Targets* 2003;**4**(2):123–31.
195. Ruoslahti E. Specialization of tumour vasculature. *Nat Rev Cancer* 2002;**2**(2):83–90.
196. Cai W, et al. In vitro and In vivo Characterization of 64Cu-Labeled AbegrinTM, a Humanized Monoclonal Antibody against Integrin {alpha}v{beta}3. *Cancer Res* 2006;**66**(19):9673–81.
197. Cai W, Chen X. Anti-angiogenic cancer therapy based on integrin alphavbeta3 antagonism. *Anticancer Agents Med Chem* 2006;**6**(5):407–28.
198. Wu Y, et al. microPET imaging of glioma integrin {alpha}v{beta}3 expression using (64)Cu-labeled tetrameric RGD peptide. *J Nucl Med* 2005;**46**(10):1707–18.
199. Chen X, et al. Integrin alpha v beta 3-targeted imaging of lung cancer. *Neoplasia* 2005;**7**(3):271–9.
200. Cai W, et al. A thiol-reactive 18F-labeling agent, N-[2-(4-18F-fluorobenzamido)ethyl]maleimide, and synthesis of RGD peptide-based tracer for PET imaging of alpha v beta 3 integrin expression. *J Nucl Med* 2006;**47**(7):1172–80.
201. Zhang X, et al. Quantitative PET imaging of tumor integrin alphavbeta3 expression with 18F-FRGD2. *J Nucl Med* 2006;**47**(1):113–21.
202. Pendas AM, et al. Diet-induced obesity and reduced skin cancer susceptibility in matrix metalloproteinase 19-deficient mice. *Mol Cell Biol* 2004;**24**(12):5304–13.
203. Wagner S, et al. Molecular imaging of matrix metalloproteinases in vivo using small molecule inhibitors for SPECT and PET. *Curr Med Chem* 2006;**13**(23):2819–38.
204. Medina OP, et al. Radionuclide imaging of tumor xenografts in mice using a gelatinase-targeting peptide. *Anticancer Res* 2005;**25**(1A):33–42.
205. Blankenberg FG. Recent advances in the imaging of programmed cell death. *Curr Pharm Des* 2004;**10**(13):1457–67.
206. Corsten MF, et al. Counting heads in the war against cancer: defining the role of annexin A5 imaging in cancer treatment and surveillance. *Cancer Res* 2006;**66**(3):1255–60.
207. Lahorte CM, et al. Apoptosis-detecting radioligands: current state of the art and future perspectives. *Eur J Nucl Med Mol Imaging* 2004;**31**(6):887–919.
208. Tait JF, et al. Improved detection of cell death in vivo with annexin V radiolabeled by site-specific methods. *J Nucl Med* 2006;**47**(9):1546–53.
209. Rottey S, et al. Sequential 99mTc-hydrazinonicotinamide-annexin V imaging for predicting response to chemotherapy. *J Nucl Med* 2006;**47**(11):1813–8.
210. Haberkorn U, et al. Investigation of a potential scintigraphic marker of apoptosis: radioiodinated Z-Val-Ala-DL-Asp(O-methyl)-fluoromethyl ketone. *Nucl Med Biol* 2001;**28**(7):793–8.
211. Bauer C, et al. 131I-labeled peptides as caspase substrates for apoptosis imaging. *J Nucl Med* 2005;**46**(6):1066–74.
212. Reshef A, et al. Small-molecule biomarkers for clinical PET imaging of apoptosis. *J Nucl Med* 2010;**51**(6):837–40.
213. Madar I, et al. Detection and quantification of the evolution dynamics of apoptosis using the PET voltage sensor 18F-fluorobenzyl triphenyl phosphonium. *J Nucl Med* 2009;**50**(5):774–80.

214. Cohen A, et al. From the Gla domain to a novel small-molecule detector of apoptosis. *Cell Res* 2009;**19**(5):625–37.
215. Harrison LB, et al. Impact of tumor hypoxia and anemia on radiation therapy outcomes. *Oncologist* 2002;**7**(6):492–508.
216. Chapman JD. Hypoxic sensitizers–implications for radiation therapy. *N Engl J Med* 1979;**301** (26):1429–32.
217. Denny WA. Prodrug strategies in cancer therapy. *Eur J Med Chem* 2001;**36**(7–8):577–95.
218. Rajendran JG, Mankoff DA. Beyond detection: novel applications for PET imaging to guide cancer therapy. *J Nucl Med* 2007;**48**(6):855–6.
219. Rischin D, et al. Prognostic significance of [18F]-misonidazole positron emission tomography-detected tumor hypoxia in patients with advanced head and neck cancer randomly assigned to chemoradiation with or without tirapazamine: a substudy of Trans-Tasman Radiation Oncology Group Study 98.02. *J Clin Oncol* 2006;**24**(13):2098–104.
220. Rajendran JG, et al. Tumor hypoxia imaging with [F-18] fluoromisonidazole positron emission tomography in head and neck cancer. *Clin Cancer Res* 2006;**12**(18):5435–41.
221. Piert M, et al. Hypoxia-specific tumor imaging with 18F-fluoroazomycin arabinoside. *J Nucl Med* 2005;**46**(1):106–13.
222. Beck R, et al. Pretreatment 18F-FAZA PET Predicts Success of Hypoxia-Directed Radiochemotherapy Using Tirapazamine. *J Nucl Med* 2007;**48**(6):973–80.
223. Yuan H, et al. Intertumoral differences in hypoxia selectivity of the PET imaging agent 64Cu (II)-diacetyl-bis(N4-methylthiosemicarbazone). *J Nucl Med* 2006;**47**(6):989–98.
224. Hynes NE, Lane HA. ERBB receptors and cancer: the complexity of targeted inhibitors. *Nat Rev Cancer* 2005;**5**(5):341–54.
225. Yarden Y, Sliwkowski MX. Untangling the ErbB signalling network. *Nat Rev Mol Cell Biol* 2001;**2**(2):127–37.
226. Baselga J, Hammond LA. HER-targeted tyrosine-kinase inhibitors. *Oncology* 2002;**63**(Suppl. 1):6–16.
227. Nahta R, Hortobagyi GN, Esteva FJ. Growth factor receptors in breast cancer: potential for therapeutic intervention. *Oncologist* 2003;**8**(1):5–17.
228. Selvaggi G, et al. Epidermal growth factor receptor overexpression correlates with a poor prognosis in completely resected non-small-cell lung cancer. *Ann Oncol* 2004;**15**(1):28–32.
229. Witton CJ, et al. Expression of the HER1-4 family of receptor tyrosine kinases in breast cancer. *J Pathol* 2003;**200**(3):290–7.
230. Memon AA, et al. The relation between survival and expression of HER1 and HER2 depends on the expression of HER3 and HER4: a study in bladder cancer patients. *Br J Cancer* 2006;**94**(11):1703–9.
231. Velikyan I, et al. Preparation and evaluation of (68)Ga-DOTA-hEGF for visualization of EGFR expression in malignant tumors. *J Nucl Med* 2005;**46**(11):1881–8.
232. Reilly RM, et al. Preclinical pharmacokinetic, biodistribution, toxicology, and dosimetry studies of 111In-DTPA-human epidermal growth factor: an auger electron-emitting radiotherapeutic agent for epidermal growth factor receptor-positive breast cancer. *J Nucl Med* 2006;**47**(6):1023–31.
233. Cai W, et al. Quantitative PET of EGFR expression in xenograft-bearing mice using (64)Cu-labeled cetuximab, a chimeric anti-EGFR monoclonal antibody. *Eur J Nucl Med Mol Imaging* 2007;**34**(6):850–8.
234. Steffen AC, et al. In vitro characterization of a bivalent anti-HER-2 affibody with potential for radionuclide-based diagnostics. *Cancer Biother Radiopharm* 2005;**20**(3):239–48.
235. Orlova A, et al. Comparative in vivo evaluation of technetium and iodine labels on an anti-HER2 affibody for single-photon imaging of HER2 expression in tumors. *J Nucl Med* 2006;**47** (3):512–9.

236. Engfeldt T, et al. Imaging of HER2-expressing tumours using a synthetic Affibody molecule containing the (99m)Tc-chelating mercaptoacetyl-glycyl-glycyl-glycyl (MAG3) sequence. *Eur J Nucl Med Mol Imaging* 2007;**34**(5):722–33.
237. Smith-Jones PM, et al. Early tumor response to Hsp90 therapy using HER2 PET: comparison with 18F-FDG PET. *J Nucl Med* 2006;**47**(5):793–6.
238. Levashova Z, et al. SPECT and PET imaging of EGF receptors with site-specifically labeled EGF and dimeric EGF. *Bioconjug Chem* 2009;**20**(4):742–9.
239. Mariani G, Erba PA, Signore A. Receptor-mediated tumor targeting with radiolabeled peptides: there is more to it than somatostatin analogs. *J Nucl Med* 2006;**47**(12):1904–7.
240. Schillaci O. Somatostatin receptor imaging in patients with neuroendocrine tumors: not only SPECT? *J Nucl Med* 2007;**48**(4):498–500.
241. Bodei L, Paganelli G, Mariani G. Receptor radionuclide therapy of tumors: a road from basic research to clinical applications. *J Nucl Med* 2006;**47**(3):375–7.
242. Reubi JC, Waser B. Concomitant expression of several peptide receptors in neuroendocrine tumours: molecular basis for in vivo multireceptor tumour targeting. *Eur J Nucl Med Mol Imaging* 2003;**30**(5):781–93.
243. Ginj M, et al. Preclinical evaluation of new and highly potent analogues of octreotide for predictive imaging and targeted radiotherapy. *Clin Cancer Res* 2005;**11**(3):1136–45.
244. Cescato R, et al. Internalization of sst2, sst3, and sst5 receptors: effects of somatostatin agonists and antagonists. *J Nucl Med* 2006;**47**(3):502–11.
245. Scarpignato C, Pelosini I. Somatostatin analogs for cancer treatment and diagnosis: an overview. *Chemotherapy* 2001;**47**(Suppl. 2):1–29.
246. Lamberts SW, et al. Somatostatin-receptor imaging in the localization of endocrine tumors. *N Engl J Med* 1990;**323**(18):1246–9.
247. Forrer F, et al. Neuroendocrine tumors. Peptide receptor radionuclide therapy. *Best Pract Res Clin Endocrinol Metab* 2007;**21**(1):111–29.
248. Storch D, et al. Evaluation of [99mTc/EDDA/HYNIC0]octreotide derivatives compared with [111In-DOTA0, Tyr3, Thr8]octreotide and [111In-DTPA0]octreotide: does tumor or pancreas uptake correlate with the rate of internalization? *J Nucl Med* 2005;**46**(9):1561–9.
249. Guggenberg EV, et al. Radiopharmaceutical development of a freeze-dried kit formulation for the preparation of [99mTc-EDDA-HYNIC-D-Phe1, Tyr3]-octreotide, a somatostatin analog for tumor diagnosis. *J Pharm Sci* 2004;**93**(10):2497–506.
250. Gabriel M, et al. 68Ga-DOTA-Tyr3-octreotide PET in neuroendocrine tumors: comparison with somatostatin receptor scintigraphy and CT. *J Nucl Med* 2007;**48**(4):508–18.
251. Krenning EP, et al. Somatostatin receptor scintigraphy with [111In-DTPA-D-Phe1]- and [123I-Tyr3]-octreotide: the Rotterdam experience with more than 1000 patients. *Eur J Nucl Med* 1993;**20**(8):716–31.
252. Henze M, et al. Characterization of 68Ga-DOTA-D-Phe1-Tyr3-octreotide kinetics in patients with meningiomas. *J Nucl Med* 2005;**46**(5):763–9.
253. Meisetschlager G, et al. Gluc-Lys([18F]FP)-TOCA PET in patients with SSTR-positive tumors: biodistribution and diagnostic evaluation compared with [111In]DTPA-octreotide. *J Nucl Med* 2006;**47**(4):566–73.
254. Anderson CJ, et al. 64Cu-TETA-octreotide as a PET imaging agent for patients with neuroendocrine tumors. *J Nucl Med* 2001;**42**(2):213–21.
255. Mottaghy FM, Reske SN. Functional imaging of neuroendocrine tumours with PET. *Pituitary* 2006;**9**(3):237–42.
256. Reubi JC, et al. Affinity profiles for human somatostatin receptor subtypes SST1-SST5 of somatostatin radiotracers selected for scintigraphic and radiotherapeutic use. *Eur J Nucl Med* 2000;**27**(3):273–82.

257. Heppeler A, et al. Receptor targeting for tumor localisation and therapy with radiopeptides. *Curr Med Chem* 2000;**7**(9):971–94.
258. Rufini V, Calcagni ML, Baum RP. Imaging of neuroendocrine tumors. *Semin Nucl Med* 2006;**36**(3):228–47.
259. Forrer F, et al. Treatment with 177Lu-DOTATOC of patients with relapse of neuroendocrine tumors after treatment with 90Y-DOTATOC. *J Nucl Med* 2005;**46**(8):1310–6.
260. van Essen M, et al. Effects of therapy with [177Lu-DOTA0, Tyr3]octreotate in patients with paraganglioma, meningioma, small cell lung carcinoma, and melanoma. *J Nucl Med* 2006;**47**(10):1599–606.
261. de Jong M, et al. Combination radionuclide therapy using 177Lu- and 90Y-labeled somatostatin analogs. *J Nucl Med* 2005;**46**(Suppl 1):13S–7S.
262. De Jong M, et al. Somatostatin receptor-targeted radionuclide therapy of tumors: preclinical and clinical findings. *Semin Nucl Med* 2002;**32**(2):133–40.
263. Shields AF, et al. Imaging proliferation in vivo with [F-18]FLT and positron emission tomography. *Nat Med* 1998;**4**(11):1334–6.
264. Mankoff DA, Shields AF, Krohn KA. PET imaging of cellular proliferation. *Radiol Clin North Am* 2005;**43**(1):153–67.
265. Leyton J, et al. In vivo biological activity of the histone deacetylase inhibitor LAQ824 is detectable with 3'-deoxy-3'-[18F]fluorothymidine positron emission tomography. *Cancer Res* 2006;**66**(15):7621–9.
266. McLaughlin F, La Thangue NB. Histone deacetylase inhibitors open new doors in cancer therapy. *Biochem Pharmacol* 2004;**68**(6):1139–44.
267. Leyton J, et al. Early detection of tumor response to chemotherapy by 3'-deoxy-3'-[18F] fluorothymidine positron emission tomography: the effect of cisplatin on a fibrosarcoma tumor model in vivo. *Cancer Res* 2005;**65**(10):4202–10.
268. Boulikas T, Vougiouka M. Recent clinical trials using cisplatin, carboplatin and their combination chemotherapy drugs (review). *Oncol Rep* 2004;**11**(3):559–95.
269. Leyton J, et al. Quantifying the activity of adenoviral E1A CR2 deletion mutants using renilla luciferase bioluminescence and 3'-deoxy-3'-[18F]fluorothymidine positron emission tomography imaging. *Cancer Res* 2006;**66**(18):9178–85.
270. Waldherr C, et al. Monitoring antiproliferative responses to kinase inhibitor therapy in mice with 3'-deoxy-3'-18F-fluorothymidine PET. *J Nucl Med* 2005;**46**(1):114–20.
271. Yang YJ, et al. Use of 3'-deoxy-3'-[18F]fluorothymidine PET to monitor early responses to radiation therapy in murine SCCVII tumors. *Eur J Nucl Med Mol Imaging* 2006;**33**(4):412–9.
272. Yap CS, et al. Evaluation of thoracic tumors with 18F-fluorothymidine and 18F-fluorodeoxyglucose-positron emission tomography. *Chest* 2006;**129**(2):393–401.
273. Masters C. Neuropathology of Alzheimer's disease. In: Burns A, O'Brien J, Ames D, editors. *Dememtia*. London: Hodder Arnold; 2005. p. 393–407.
274. Klunk WE, et al. Imaging brain amyloid in Alzheimer's disease with Pittsburgh Compound-B. *Ann Neurol* 2004;**55**(3):306–19.
275. Lin K-J, et al. Whole-body biodistribution and brain PET imaging with [18F]AV-45, a novel amyloid imaging agent—a pilot study. *Nucl Med Biol* 2010;**37**:497–508.
276. Choi SR, et al. Preclinical Properties of 18F-AV-45: a PET imaging agent for mapping beta-amyloid plagues in the brain. *J Nucl Med* 2009;**50**:1887–94.
277. Ono M. Development of positron-emission tomography/single-photon emission computed tomography imaging probes for in vivo detection of beta-amyloid plaques in Alzheimer's brains. *Chem Pharm Bull (Tokyo)* 2009;**57**(10):1029–39.
278. Kung MP, et al. Characterization of IMPY as a potential imaging agent for beta-amyloid plaques in double transgenic PSAPP mice. *Eur J Nucl Med Mol Imaging* 2004;**31**(8):1136–45.

Cancer Epigenetics

Wendell Weber

Department of Pharmacology, University of Michigan Medical School, Ann Arbor, Michigan, USA

Epigenetic studies reveal important insights into cancer biology. This chapter presents a broad picture of how epigenetic changes in health and disease and in response to environment contribute to carcinogenesis, and how findings from newer approaches and emergent technologies may extend these observations. Consideration is given to biological insights drawn from studies of epigenomic patterns in cancer cells, and the influence of epigenomic profiling on diagnosis, therapy, and prognosis. The chapter begins with a brief overview of the concepts and foundations on which epigenetics is built and concludes with comments on prospects for the future of cancer epigenetics.

I. Introduction

Cancers are often said to be diseases of development,[1] or to be genetic disorders arising from mutations in DNA sequences that cause disruption and disorganization of genomes,[2] and more recently, to be disorders arising from a combination of genetic and epigenetic aberrations.[3] Epigenetics initially

Progress in Molecular Biology
and Translational Science, Vol. 95
DOI: 10.1016/S1877-1173(10)95010-2

referred mainly to developmental phenomena[4] but in modern biology, the term is applied more broadly to signify a relation to gene action including all heritable changes in gene expression and chromatin structure that are not encoded in the DNA sequence itself.

Aberrant gene function and altered patterns of gene expression are key features of cancer, and as we have learned more about genomes, we see the potential importance of epigenetic processes, particularly those that result in silencing key regulatory genes, in alterations that occur in the earliest stages of cancer involving pathways of growth and differentiation of stem cells, and in the therapeutic targeting of these defects.[5–7] That epigenetic control of gene expression may respond to environmental factors in a manner distinct from effects of genetics on gene expression is of considerable interest.[8] Currently, it is reasonable to expect that cancer represents a group of heterogeneous disorders that is driven mainly by combinations of genetic and epigenetic abnormalities.

The primary objective of this chapter is to describe the relevance of epigenetic changes to the initiation and progression of cancer, and how recent findings from newer approaches and emergent technologies extend these observations. Consideration is given to biological insights drawn from studies of epigenomic patterns in cancer cells and to the influence of epigenomic profiling on diagnosis, therapy, and prognosis. The chapter begins with a brief historical introduction followed by an account of the experimental foundations of the field and concludes with remarks concerning prospects for cancer epigenetics of the future.

II. First, a Little History

As far back as the eighteenth century, scientists debated whether or not acquired traits were heritable. "Naturalists" favored the argument that such traits were heritable, while "geneticists" believed that heritability arose only through natural selection. In the nineteenth century, cytologists were aware of the curious assortment of densely staining agglomerations that are present in cell nuclei of various plants and animals. Modern insight began with the suggestion of Emil Heitz, a German cytologist, that these structures had certain genetic attributes and were related to chromosomes. Heitz recognized two classes of chromosomal material, euchromatin which underwent a typical cycle of condensation and unraveling, and heterochromatin which maintained its compactness in the nucleus. In his review of heterochromatin published nearly 40 years later, Spencer Brown saw the investigation of chromatin as "one of the most challenging and diffuse in modern biology."[9] By the time of Brown's review in 1966, the repressive action of chromatin on gene action had been recognized, and the two states of chromatin were viewed as a visible guide to gene action during evolution and development.

It will be recalled that in 1910, Thomas Hunt Morgan, who had chosen *Drosophila* (fruit fly) for his studies of heredity, had formulated a revolutionary chromosome theory of heredity, and had provided the first elegant proof that chromosomes must contain genes.[10] Following the work of Morgan and his students, Avery and colleagues had demonstrated in the 1940s that DNA, not protein, was the genetic material, and in 1953, Watson and Crick had demonstrated that the double helix of DNA was the molecular basis of heredity. During the 1960s, the informational foundations of modern genetics were established when the genetic code was deciphered by Marshall Nirenberg, Severo Ochoa, and others.[11] Thus, in a span of little more than half a century, the views of geneticists had rapidly gained ground while those of naturalists were in the decline.

Then, a sequence of events began with Edwin Southern's landmark article "Detection of specific sequences among DNA fragments separated by gel electrophoresis" published in 1975.[12] Southern's article was the first of several newly invented methods for manipulating DNA and other nucleotides for determining the composition of genomic DNA that were widely adopted for identifying genetic lesions of medical importance (reviewed in Chapter 1; Weber[13]). From 1975 to the present, molecular biological approaches in all forms dominated biological research, and the extensive record of research that defines the genomic revolution is inseparably associated with the range of innovative technologies that completely transformed life science research.

For more than 50 years, genomic research has been driven largely by the central dogma of molecular biology that "genes beget RNA which in turn begets protein."[14] Though this model has been suitable for the development of molecular biology, it has suffered from inadequacies in explaining transmission of hereditary information as newer data have come to light. The idea of unidirectional gene expression that was implicit in the central dogma as originally formulated was negated by the discovery of reverse transcriptase. The predictive value of the genotype was further confounded when it became apparent that some genes encode just one protein, while others encode more than one protein, and still others do not encode any protein. Posttranslational protein modifications added another unexplained twist. Identification of previously unknown pathway components illustrated the complexity of cellular events, and the recognition that gene expression could be altered at the translational, transcriptional, and posttranscriptional levels necessitated a wider view of phenotypic expression. Thus, the concepts embodied in the original version of the central dogma formulated in the 1950s required expansion to accommodate new knowledge affecting transmission of heritable information (Fig. 1).

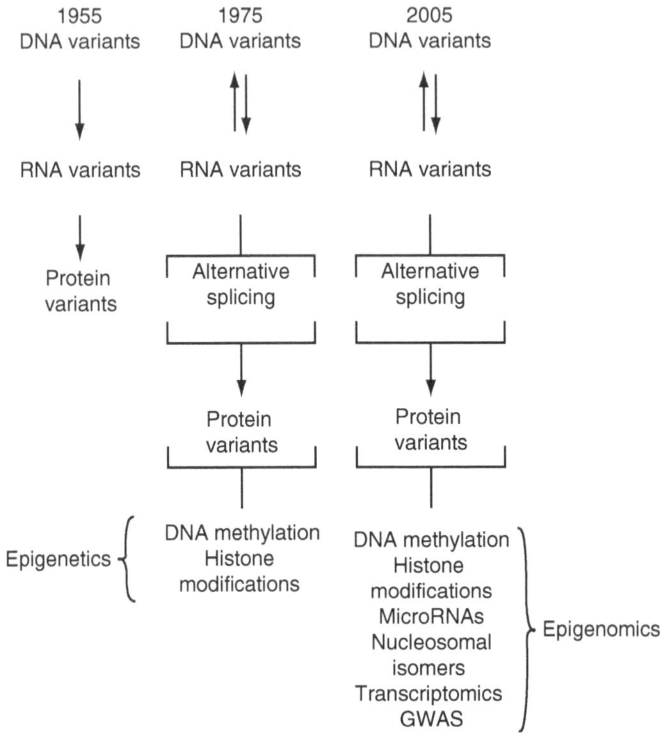

FIG. 1. The expanding dogma of molecular biology.

III. Epigenetic Patterns in Normal Cells

Cells transmit information via two distinct routes, genetic and epigenetic. Transmission of genetic information is based on the DNA code, and individual variation in the information transmitted is attributed to changes in the DNA sequence. Transmission of epigenetic information, on the other hand, is based on gene expression, and variation in gene expression is determined in turn by the effects of a set of epigenetic marks (covalent modifications) on DNA and on chromatin without modifying the DNA sequence. Methyl groups as well as other small molecules (acetyl, phosphoryl, etc.), and small noncoding RNAs can all serve as epigenetic marks of DNA and chromatin. Chromatin is a polymeric complex packaged with DNA to form the nucleosome which comprises a histone octamer that includes the core histones H2A, H2B, H3, and H4 interconnected by H1 histones. Histones are small (11,000–21,000 molecular weight) basic proteins that bind noncovalently with acidic DNA to form nucleosomes. The nucleosome, which constitutes the basic building block of chromatin, is a dynamic participant in the regulation of virtually all chromosomal processes, including transcription, replication, and DNA repair. Combinations of epigenetic marks on DNA and chromatin work together with additional enzymes and nonhistone coadaptor molecules to modify and remodel chromatin during gene expression. It is the addition and removal of these covalent modifications on histone residues and remodeling of the nucleosome that give chromatin its dynamic character.

The covalent modifications on DNA and chromatin are reversible and they can change during development, with age, and under stressful environmental conditions. They can also change the state of chromatin to increase the chance of genetic change, possibly in a heritable way, by mutation, recombination, and the movement of jumping genes. Furthermore, a genetic change in two regions with identical DNA sequences is likely to differ if they have different epigenetic marks. As a rule, inactive genes usually have more compact chromatin, whereas active genes tend to exhibit a more open configuration of chromatin favoring the interplay between genetic and epigenetic components. Wherever DNA is less condensed, it is more vulnerable to disruption and aberrant regulation because it is more accessible to mutagens and to enzymes implicated in repair and recombination. The idea that certain diseases such as cancer may be caused by such nascent changes in the genome adds to their significance.

Perhaps the main function of the epigenome, and its capacity to respond with reversible modifications, is to serve under normal circumstances as a functional intermediary between the heritable genome (the genomic DNA sequence) and the stresses imposed by the environment. Such an interface could promote epigenetic modifications that result in a more stable, homeostatic profile of gene expression. This interpretation is similar to that suggested by Feinberg[5] as a metastable state that results in a new cellular set point with a new range of gene expression patterns.

A. The Epigenetic Experimental Framework: Tools for Dissecting Epigenetic Pathways and Networks

Epigenomic research over the last three decades has highlighted the roles of DNA methylation, histone modification, genomic imprinting, and noncoding RNAs in modulating gene expression patterns from early human development to adulthood. Rapid progress has been made in quantifying, mapping and characterizing these events, and this section considers first several established methodologies that have been used for analyzing these events (reviewed by van Steensel[15,16] and by Havlis and Trbusek[17]) and concludes with brief descriptions of selected technologies of epigenetic interest that have been developed recently.

1. Established Technologies Widely Used in Epigenetic Research

Lack of a suitable technique to decode genome-wide DNA methylation patterns with high resolution hampered progress in epigenetics until Susan Clark and coworkers developed the bisulfite conversion and sequencing protocol.[18] Clark's method uses sodium bisulfite to convert cytosine residues to uracil in single-stranded DNA under conditions whereby 5-methylcytosine (5mC) is nonreactive. All the cytosine residues remaining in the sequence represent previously methylated cytosines in the genome. The converted DNA is amplified with specific primers and sequenced. Bisulfite sequencing is straightforward and efficient. In the event that a site is only partially methylated, the method has the added advantage of enabling determination of the proportion of cells that are methylated. The important features of several widely used methods that employ 5mC as a marker are summarized below and in Table I. For a summary of these methodologies, see the review by Havlis and Trbusek[17] and additional citations in Table I.

Transcription factors, nucleosomes, chromatin-modifying proteins, and epigenetic marks together form extremely complex regulatory networks. During the past 10 years, two microarray approaches have been developed for genome-wide mapping of the binding sites of regulatory proteins and the distributions of methylation patterns and histone modifications.[16] In the first of these, chromatin immunoprecipitation (ChIP) is combined with microarray detection (chip). ChIP-chip was first used to map DNA-binding proteins on a genome-wide scale, but it was also applied to map other phenomena such as histone modifications and nucleosome distribution. The second method, called the DamID, utilizes an entirely different principle. In this case, a transcription factor or chromatin-binding protein of interest is fused to DNA adenine methyltransferase (Dam). When this fusion protein is expressed, Dam will be targeted to binding sites of its fusion partner, resulting in methylation of adenines in DNA nearby the binding sites. To identify these sites, the methylated regions are either purified or selectively amplified from genomic DNA, fluorescently labeled, and hybridized to a microarray. Because adenine methylation does not occur endogenously in mammals, the binding sites of targeted methylation can be derived from the microarray signals. Additional details and the merits and drawbacks of the two methods are considered elsewhere.[16]

Subsequently, a variety of sequencing protocols have been developed to analyze ChIP samples (reviewed by Schones and Zhao[30]). Most of these protocols combine ChIP with serial analysis of gene expression (SAGE), serial analysis of chromatin occupancy (SACO), the genome-wide mapping technique (GMAT), and ChIP combined with paired-end ditag sequencing (ChIP-PET). The combination of ChIP with massively parallel sequencing

TABLE I

Tools for Monitoring DNA Methylation

Method	Detection	DNA amount needed	Sensitivity	Comments	Selected references
High performance thin layer chromatography	Scintillation	5 μg	20 fmol	Simple, low cost, rapid, good for large-scale screening	19
High performance liquid chromatography	Optical—UV, scintillation, fluorescence, MS	< 1 μg	400 fmol	Quantitative, reproducible, sensitive	20
Capillary electrophoresis	MS	< 1 μg	100 fmol	Automation possible, high sample throughput	21
Immunoassay	Fluorescence	NA	1.5 fmol	Spatial resolution on metaphase chromosomes previously stained by the Giemsa method	22
Modification-sensitive restriction enzymes (MSRE)	Gel electrophoresis, Southern blot	> 5 μg	NA	Methylation site-specific	23,24
Bisulfite sequencing	Gel electrophoresis	10 ng	2.5 fmol	Sensitive, easy, best for analysis of different sequences in a small number of samples	18,25,26
Bisulfite sequenceing + chloroaldehyde	Fluorescence	10 ng	175 fmol	Slow and chloroaldehyde is toxic, does not require extensive purification of DNA	27
Combined bisulfite restriction analysis (COBRA)	Gel electrophoresis	1 μg	125 fmol	Rapid, sensitive, quantitative and can be used with paraffin sections	28
Methylation-sensitive single nucleotide primer extension (MS-SNuPE)	Gel electrophoresis	5 ng	500 fmol	Avoids MSRE and is automatable. Target sequence should contain only A, C, and T, while primer should contain only A, G, and T	29

(ChIP-Seq) permits surveillance of the genome in shorter time and will probably disclose new aspects of biology. Schones and Zhao review applications of these techniques for profiling of DNA methylation, histone modifications, nucleosome positioning, and chromatin accessibility and summarize some of the more interesting results that have been obtained. For example, they consider how meaningful biological phenomena can be extracted from an analysis of large data sets when conducting genome-wide experiments. They point out that metazoan genomes take account of the three-dimensional architecture in regulating gene expression so that regulatory factors for transcription do not function solely by binding target sites in proximity to a gene, but may rely instead on long-range interactions among DNA regions spanning as much as 100 kb and even across chromosomes depending on the developmental stage of particular cell types. Current data support the idea that DNA methylation, histone modifications, nucleosome location, noncoding RNAs, DNA-binding proteins, and the three-dimensional architecture are not independent elements of functional epigenomes but influence each other during the dynamic regulation of cellular differentiation or pathological states.

2. Recent Advances in Epigenetic Technologies

Technologies of special interest to epigenetics research that have been developed recently for dissecting human proteome-wide chromatin marks, for genome-wide studies of the human methylomes, for the construction of a microchip for genome-wide profiling of microRNAs, and for assessing the transcriptomics of cells are described in this section.

3. An Epigenome Microarray Platform for Dissecting Proteome-Wide Chromatin Marks[31]

At the molecular level, histone marks can act as ligands for domains found on regulatory proteins of chromatin. Insight into how these domains influence chromatin activities has come from identification and characterization of methyl-lysine effectors. These marks are believed to create a distinct molecular architecture on histones that is recognizable by specialized binding domains such as chromodomains (CDs). For example, components of transcriptional repressive protein complexes such as heterochromatin protein 1 (HP1) contain CDs that allow them to recognize the appropriate transcriptional repressive mark, histone 3 trimethylated at lysine 9 (H3K9me3). Similarly, components of a transcriptional activation complex can recognize histone 3 trimethylated at lysine 4 (H3K4me3) found on several modules associated with transcriptional activation. However, H3K4me3 is also a ligand for complexes with very different activities, such as transcriptional repression and recombination. Thus, the biological outcomes of histone marks depend on both their location and the repertoire of effectors that have access to those regions. The proteins (or protein domains) that recognize histone modifications in this context are termed "effectors" or "readers."

Previously, Gozani and colleagues had demonstrated the utility of a histone peptide microarray to characterize methyl-lysine functions for the plant homeodomain (PHD) fingers present within the yeast proteome[32] More recently, they[31] described the development, validation, and application of a human epigenome peptide microarray platform (HEMP) for high-throughput identification of ligands for effector molecules. They have tested the integrity of individual peptide features by probing this platform with modification-specific antibodies and known chromatin effector domains. They have screened a large library of Royal Domain family members and identified three modules, the CD of MPP8 ($MPP8_{CD}$), and the tudor domains (TD) of TDRD7 ($TDRD7_{TD}$), and JMJ2C ($JMJ2C_{TD}$). This technology will facilitate dissection of chromatin signaling networks and could contribute to the unraveling of epigenetic mechanisms.

4. Genome-Wide Study of Human DNA Methylomes Showing Widespread Epigenomic Differences[33]

The study by Lister and coworkers[33] described below are the first genome-wide, single-base-resolution maps of methylated cytosines in a mammalian genome from two genomes: human embryonic stem cells and fetal fibroblasts. Genome-wide studies of mammalian DNA methylation have previously been performed, but they were limited by low resolution, sequence-specific bias, or by analyzing only a small fraction of the genome. In their more comprehensive study, Lister and coworkers found widespread differences in the composition and patterning of cytosine methylation between the two genomes. Nearly one-fourth of all methylation identified in embryonic stem cells was in a non-CG context, suggesting that embryonic stem cells may use different methylation mechanisms to affect gene regulation. Methylation in non-CG contexts also showed enrichment in gene bodies and depletion in protein-binding sites and enhancers. Non-CG methylation disappeared upon induced differentiation of the embryonic stem cells, and was restored in induced pluripotent stem cells. Hundreds of differentially methylated regions (DMRs) proximal to genes involved in pluripotency and differentiation and widespread reduced methylation levels in fibroblasts associated with lower transcriptional activity were identified. They suggest that future studies should explore the prevalence of non-CG methylation in diverse cell types, including variation throughout differentiation and its potential reestablishment in induced pluripotent states.

5. A Microchip for Genome-Wide MicroRNA Profiling[34]

To assist in their expression profiling of microRNA signatures in B cell chronic lymphocytic leukemias (CLLs),[35,36] Carlo Croce's group developed an miRNA microchip and established detection methodology to investigate alterations in expression of all known miRNAs in human cancer.[34] They used this microarray to determine tissue-specific miRNA expression signatures for human and mouse miRNAs. MiRNA oligo probe design, miRNA microfabrication, target preparation, array hybridization, data analysis, and other technical details are described.[34] The microchip contains gene-specific 40-mer oligonucleotide probes generated from human and mouse precursor miRNA collected from the Sanger database or from published papers. They tested this platform by comparing it with a panel of human normal tissues and identified specific miRNA expression signatures for each tissue type. Based on these signatures, hemopoietic tissues cluster together in a group distinct from that of nonhemopoietic tissues showing that miRNA expression profiles differ with cell- and tissue-type, suggesting that abnormal cell/tissues also have distinctive miRNA expression profiles.

6. RNA-Seq: A Tool for Assessing Transcriptomics[37]

The transcriptome is the complete set of transcripts in a cell. Understanding the transcriptome is essential for identifying and interpreting the functional elements of the genome and for understanding development and disease. A variety of technologies, including hybridization and sequence-based approaches, have been developed. Hybridization-based approaches are high throughput and relatively inexpensive (except for tiling microarrays), have limitations because they rely on existing knowledge about genome sequence, suffer high background levels due to cross hybridization, and a small dynamic range of detection because of background as well as saturation of signals. Recently, a new high-throughput DNA sequencing method, termed RNA-Seq (RNA sequencing), has been applied to several species, including mouse and human cells. The advantages of this technique have been demonstrated by Wang and coworkers for analysis of eukaryotic transcriptomes.[37]

A typical RNA-Seq experiment is conducted as follows: long RNAs are first converted into a library of cDNA fragments through either RNA fragmentation or DNA fragmentation. Sequencing adaptors are subsequently added to each cDNA fragment and a short sequence is obtained from each cDNA using high-throughput sequencing technology. The resulting sequence reads are aligned with the reference genome or transcriptome, and reads are classified as three types: exonic, junctions, and poly(A) end. These three types of reads are used to generate a base resolution expression profile for each gene (see Fig. 1 in the article by Wang *et al.*[37])

The benefits, challenges, and new transcriptomic insights are reviewed by Wang and colleagues[37] and by Schones and colleagues.[30] As the cost of sequencing methods falls, RNA-Seq is likely to replace microarrays for many applications that involve determination of the composition and dynamics of cellular transcriptomes.[30, 37–39]

While the major short-term goal of sequencing is to determine the roles of DNA methylation, histone modification, genomic imprinting, and noncoding RNAs in modulating gene expression, the long-term goal of much of the sequencing work is to establish a reference epigenome in health and disease by sequencing different tissues[40] as part of the Human Epigenome Project (http://www.epigenome.org/index.php). This information is intended to support the creation of future epigenome projects of disease genomes such as the Cancer Epigenome Project. Human malignancies, for example, are characterized by cell- and tissue-specific alterations in aberrant patterns of DNA methylation, histone modification, and genomic imprinting, as well as in disruptions of microRNA regulation, as will be described. There are also indications of individual epigenomic differences highlighted by studies of DNA methylation patterns performed on monozygotic twins[41,42] that will necessitate resequencing of epigenomes from both healthy and diseased individuals.

B. Epigenetic Inheritance in Normal Cellular Processes

Epigenetic inheritance was hardly recognized until the mid-1970s, because developmental biologists had been more interested in how cells differentiated and acquired their specialized roles rather than how they remembered their differentiated state and transmitted it to their progeny. In tissues and organs of humans, there are at least 200 different types of cells. They all contain essentially the same genetic information, yet their size, shape, and behavior are markedly different. These differences are brought about by various mechanisms of epigenetic control rather than genetic differences.

We are familiar with how genes are made up of DNA, RNA, proteins, and other molecules to form a tightly coiled package called chromatin. Chromatin is chemically marked as we have said, and these marks will determine whether genes are to be translated into protein, or directed to perform developmental or regulatory functions. Attachment of a methyl group to cytosine, a nucleotide molecule of DNA, in or near the promoter sequence of a gene is a common way of doing this. Extensive cytosine methylation of the promoter will turn off the gene during development or during postdevelopmental life. Another important epigenetic mechanism of gene expression involves covalent modification of histone tails. Histone tails are regions of the histone protein that protrude from the chromosomes that are available for modification by acetyl, methyl, phosphoryl, and other small chemical groups. These modifications are important for chromatin organization as well as for the conduct of cellular processes that require access to the DNA template including gene transcription, DNA repair, and replication. A third important epigenetic mechanism for regulating gene expression is genomic imprinting. Imprinting somehow enables the gene to remember whether it was inherited from its father or mother, and only the imprinted version from a specified parent will be expressed. However, should the imprinted allele be defective, as may occur through pathologic loss of imprinting (LOI), human disease, including cancer, can result even though the other parental allele is normal.

While the genetic sequence is the same for every cell in the body during a person's lifetime, each cell has its own characteristic set of marks that define its epigenome. These marks change with developmental stage,[43,44] sex,[45] age,[1,45] and may also change in response to environmental toxins,[46–50] and stress.[51] The epigenotype of individuals is thus more complex than the genotype because it amounts to the sum of all cellular epigenotypes.[45,52,53] This complexity also makes screening for epigenetic marks and aberrant epigenetic regulation more difficult than genetic screening. Another crucial feature of epigenetic mechanisms is that they are remembered by the cell when it divides so that the daughter cell follows the same patterns of development and maturation as those of the parent cell.[40] Harmful changes to the epigenome, on the other hand, can result in various human diseases, including cancer.

The involvement of DNA methylation in cell differentiation and gene function was initially proposed in the 1960s by Scarano,[54] amplified by Riggs[55] and by Holliday and Pugh[56] in 1975, further substantiated by Razin and Riggs[57] in 1980, and reviewed by Razin and Kantor in 2005.[58] This section summarizes the sequence of the main events that occurred during this period and how fundamental epigenetic mechanisms operate to control gene expression during normal cellular processes.

1. DNA Methylation

During the 1960s, in studies of enzymatic synthesis of DNA ongoing in Arthur Kornberg's laboratory, Josse noted that the frequency of cytosine in vertebrate genomes was unusually low compared to that expected from the overall base composition of DNA.[59] But the significance of this curiosity was not apparent until Grippo in Scarano's laboratory[60] found that 5mC of sea urchin embryos was the only methylated base in DNA. Rollin Hotchkiss had actually noted the presence of methylcytosine in calf thymus DNA almost 20 years earlier, which he had called "epicytosine".[61] Additionally, Grippo *et al.* observed that 90% of 5mC occurred in the form of CpG dinucleotide doublets (CCGG). Subsequently, Scarano[54] called attention to the instability of 5mC suggesting that it would deaminate spontaneously to yield TpG + CpA. In follow-up studies, Salser *et al.* in 1977,[62] and Bird in 1980[63] provided evidence affirming Scarano's suggestion. Making use of studies based on restriction enzymes that were capable of distinguishing methylated DNA from unmethylated DNA, Bird also observed that DNA methylation within the animal kingdom ranged from undetectable levels in arthropods to low levels in echinoderms to high levels in vertebrates.

In 1975, articles by Riggs[55] and Holliday and Pugh[56] independently suggested that DNA methylation was somehow related to control of gene expression in mammalian cells. Riggs pointed out that DNA regulation in eukaryotes had not been considered in the light of accumulating evidence of changes in gene regulation in *Escherichia coli* involving DNA methylases, while Holliday and Pugh believed that methylation of CpG doublets was exceptional because it occurred much less frequently than expected from the overall base composition. Holliday and Pugh also speculated that methylation of cytosine in DNA might serve to regulate gene expression.

These ideas greatly advanced the understanding of cell memory and epigenetic inheritance by generating a lot of interest in studying the biochemistry and genetics of cytosine methylation in higher (eukaryotic) organisms (reviewed in Razin and Riggs[57]). In 1977, Christian and colleagues demonstrated a correlation between hypomethylation of DNA and expression of globin genes in Friend erythroleukemia cells from which they postulated a link between hypomethylation of DNA and gene expression. A more convenient and generally applicable approach involving restriction enzymes became possible when it was found that Msp I recognized the same sequence as Hpa II (CCGG) but cuts the DNA regardless of the methylation state of the internal cytosine of the CpG dinucleotide. In 1978, Waalwijk and Flavell[64] provided the first convincing evidence for tissue-specific methylation patterns in the rabbit β-globin gene region (reviewed in Razin and Riggs[57]).

Going into the 1980s, there was a general consensus that methylation of cytosines within CpG dinucleotide doublets was an established characteristic of genomic DNA, and that deficiency of CpG doublets in genomic doublets in vertebrate genomes was due to instability of 5mC through its mutation to thymine. It was also agreed that the distribution of CpGs in vertebrate genomic DNA was not random. Further comparisons across nonvertebrate and vertebrate genomes revealed that 98% of DNA was methylated at very low levels but the remaining 2% of methylated DNA was concentrated within regions of high-density methylation and their existence was particularly evident in DNA of vertebrates. Gardiner-Garden and Frommer[65] proposed that these regions be defined as "CpG islands" after a large-scale analysis of their length, nucleotide composition, frequency, and location relative to the transcription unit of associated genes.

Throughout much of the 1980s, attempts to define the significance of DNA methylation continued. It had been demonstrated in several model systems that transposed elements such as infectious proviruses were rendered harmless by methylation; from which it had been inferred that DNA methylation prevented damage by "selfish" foreign elements by suppressing their capacity to disrupt gene structure and function. Bird[66–88] believed this to be the ancestral function of invertebrate DNA methylation. He also thought that vertebrates had retained this ancestral function, and had adapted methylation at CpG residues as a suppressor of endogenous promoters of genes. More recent information has confirmed several of these ideas as DNA methylation tends to occur predominantly in large repetitive genomic regions, including rDNA, satellite sequences, centromeric repeats, and parasitic elements, and endogenous retroviruses.[69,70]

In vertebrate genomes, DNA methylation involves two dynamically regulated pathways. Approximately 70% of CpG residues are methylated, most of which occurs during the S-phase of the cell cycle, whereas a similar proportion of genes that possess CpG islands are mostly unmethylated. The pattern of genomic methylation found in adult eukaryotic tissues is explained by a combination of two distinct processes—*de novo* methylation and maintenance methylation. *De novo* methylation refers to enzymatic transfer of the methyl group to unmethylated cytosines of CpG dinucleotides, a process that occurs mainly in the early embryo. In contrast, maintenance methylation converts a "hemimethylated CpG" (a CpG in which only one strand of DNA is methylated) into a symmetrically methylated form. At the next round of replication, hemimethylated CpG is formed; hence the pattern of methylation in the parent nucleus is transmitted to the daughter cell by only one strand of the DNA double helix. The hemimethylated CpGs are rapidly converted to symmetrically methylated forms by maintenance methylation. Initially, DNA methylation patterns, once established, were thought to be faithfully maintained at each cell division, but it is now evident that the methylation state at any one CpG site is not always maintained. Apparently, an interplay between *de novo* methylation and demethylation at each cell division gives rise to a heterogeneous pattern of methylation for any one molecule and Riggs and colleagues[71] have estimated that failure of maintenance occurs at approximately 5% per CpG site per cell division (reviewed by Clark[72]).

Enzymatic methylation of the C5-carbon of cytosine in a DNA strand yields 5-methyl-2′-deoxycytidine monophosphate, a reaction that is catalyzed by families of methyltransferases (EC 2.1.1 MTs). Eukaryotic DNA methyltransferase (DNMT) was first cloned and sequenced from mouse erythroleukemia cells in 1988 by Bestor *et al.*[73] Currently, there are four members of the DNMT family: Dnmt1,[74] Dnmt2,[75] Dnmt3,[76] and Dnmt3L. Three of these are implicated in the establishment and maintenance of genomic methylation patterns in mice and/or humans. DNMT1 is capable of both *de novo* and maintenance methylation at CpG sites, and can also maintain methylation of some non-CpG sites. Yoder and Bestor,[75] Van den Wyngaert and colleagues,[77] and Herman *et al.*[78] identified members of the DNMT2 family; Okano and colleagues identified two members of the Dnmt3 family, Dnmt3a, and Dnmt3b in mice. Dnmt3a and Dnmt3b both show equal activity toward hemimethylated and unmethylated DNA; but the expression pattern plus substrate selectivity suggested that Dnmt3a and Dnmt3b might encode *de novo* methyltransferases.[76,79] The human homologs, DNMT3A and DNMT3B, are highly homologous to the mouse genes. The fourth family member, Dnmt3L, belongs to the Dnmt3 family by virtue of its sequence homology to the PHD-like motif. Dnmt3L is essential for the establishment of maternal genomic imprints and, because it lacks a methyltransferase domain, is more likely to regulate methylation rather than acting as an enzyme that methylates DNA. It interacts with Dnmt3a and Dnmt3b to facilitate methylation of retroposons.[80]

As evidence about methylation of DNA accumulated, Razin and Riggs[57] recognized that DNA methylation offered an attractive explanation for control of gene expression, but experimental support for that idea was elusive. Adrian Bird and coworkers postulated that access of transcription factors to methylated sites on DNA were physically blocked, perhaps by unknown proteins. The first methyl-CpG-binding protein was discovered accidentally by Adrian Bird and coworkers who were attempting to identify factors that bind to unmethylated DNA and would function to protect CpG islands from DNA methylation. Subsequently, they identified and characterized several methyl-CpG-binding proteins.[81–85] MeCP2, an important member of the methyl-CpG-binding proteins, contains both a methyl-CpG-binding domain (MBD) and a transcriptional repressor domain (TRD). Bird and coworkers[84] proposed that MeCP2 could bind methylated DNA in the context of chromatin and they suggested this protein contributed to the long-term silencing of gene activity. Currently, two major families of methyl-CpG-binding proteins have been identified in mammals: MBDs and Kaiso-like proteins (Table II).[86]

By the early 1990s, almost a decade had elapsed since the existence of CpG islands was first appreciated. By that time, the number and genomic distribution of these short regions of genomic DNA in the human and mouse genomes had been determined. Their structure, particularly the presence of CpG dinucleotides, allowed them to be distinguished from the rest of the genome, facilitating the isolation of their associated genes. Some of their distinctive properties were recognized. Methylation of CpG islands, for example, appeared to be important in gene silencing in such processes as X-inactivation, imprinting, and possibly in cancer.[70] With these findings in hand, questions regarding the function of DNA methylation and the relation of histone acetylation and DNA methylation to chromatin remodeling and to repression of gene activity were beginning to yield to experimental scrutiny. Acetylation of conserved lysines on the amino terminals of the core histones was shown to be an important mechanism by which chromatin structure is altered. Histone

TABLE II
A PARTIAL LIST OF PROTEINS THAT ARE DEREGULATED IN CANCEROUS PROCESSES

Protein category	Epigenetic protein	Status in cancer	Cancer type
DNMTs	DNMT1	Mutation/overexpression	Colon/multiple
	DNMT3A	Overexpression	Multiple
	DNMT3B	Overexpression	Multiple
Methyl-binding proteins	MeCP2	Overexpression, mutation	Multiple
	MBD1	Overexpression, mutation	Multiple
	MBD2	Overexpression, mutation	Multiple
	MBD3	Overexpression, mutation	Multiple
	MBD4	Mutation	Colon, stomach, endometrium
Histone acetylases	P300	Mutations, translocations, deletions	Multiple
	CBP	Mutations, translocations, deletions	Multiple
	pCAF	Mutations	Colon
	MOZ	Translocations	Hematological
	MORF	Translocations	Hematological, uterine, leiomyoma
	Tip60	Underexpression	Multiple
Histone deacetylases	HDAC1	Overexpression/underexpression/mutation	Multiple/colon
	HDAC2	Overexpression/mutation	Multiple, colon, gastric, endometrial
	HDAC3	Overexpression	Colon
	HDAC4	Overexpression/underexpression/mutation	Prostate, breast; colon/breast
	HDAC5	Underexpression	Colon, AML
	HDAC6	Overexpression	Breast, AML
	HDAC7	Overexpression	Colon
	HDAC8	Overexpression	Colon
	SIRT1	Overexpression/underexpression	Multiple/colon
	SIRT2	Underexpression/deletion	Glioma
	SIRT3	Overexpression	Breast
	SIRT4	Underexpression	AML
	SIRT7	Overexpression	Breast, thyroid carcinoma

Modified from Miremadi *et al.*[109] and Ellis *et al.*[110]

acetylation was associated with an open chromatin conformation allowing for gene transcription, while histone deacetylation maintained the chromatin in the closed, nontranscribed state. Aided by the tools of molecular biology, investigators had learned how CpG dinucleotides were targeted for methylation, and how the patterns of methylation were read, maintained, and in most cases, faithfully transmitted from one generation to the next.

2. Synergistic Effects of Histone Deacetylation and Methylation in Transcriptional Gene Silencing

Nan *et al.*[87] and Jones *et al.*[88] were first to demonstrate that transcriptional silencing involves the cooperation of DNA histone acetylation and chromatin modification. They reported independently that a methylcytosine guanine dinucleotide-binding protein, previously identified as MeCP2,[83,89] resides in a complex with several histone deacetylases (HDACs). MeCP2 is an abundant protein that contains both an MBD and a transcriptional repression domain (TRD), and MeCP2 binds to chromosomes at sites known to contain methylated DNA.[83,89] The complex also includes Sin3A, a corepressor in other deacetylation-dependent silencing processes, plus several unidentified proteins. Nan *et al.* and Jones *et al.* also demonstrated that the methylation-dependent transcriptional silencing could be reversed by the specific histone deacetylase inhibitor, trichostatin, and that histone deacetylation was guided to specific chromatin domains by genomic methylation patterns. Transcriptional silencing in both instances relied on histone deacetylation. Bestor[90] suggested that deacetylation favored greater ionic interactions between the positively charged N-terminal histone tails and the negatively charged phosphate backbone of DNA that could interfere with binding of transcriptional factors to their specific DNA sequences. He also suggested that deacetylation might lead to compaction of the chromatin through favorable interactions between adjacent nucleosomes. These findings pointed to a direct causal relationship between DNA acetylation and chromatin modification in methylation-dependent transcriptional silencing.

Shortly after Nan *et al.*[87] and Jones *et al.*[88] had established the cooperation of histone acetylation and chromatin modification in transcriptional gene silencing, Ng and colleagues[91] found that HeLa cells deficient in MeCP2 were still capable of repressing transcription. They concluded that MeCP2 was probably not the sole connection between DNA methylation and transcriptional silencing. In a follow-up study reported in 2001, Tamaru and Selker[92] identified a previously unknown gene, *dim-5*, was required for methylation of lysines of chromatin histone 3 tails in the fungus *Neurosparra crassa*. In characterizing the gene, Tamaru and Selker found they had accidentally generated a stop codon in a distinctive 130 amino acid sequence motif known as the evolutionarily conserved SET domain. On mapping the mutant gene, they found it was located in a region homologous to histone methyltransferases. The region demonstrated that the gene in *Drosophila* and various other organisms, including mammals, was required for heterochromatin formation; they further demonstrated that recombinant DIM-5 protein specifically methylated histone 3, and that replacements of lysine 9, with either a leucine or an arginine, caused loss of DNA methylation. Tamaru and Selker concluded that in addition to DNA acetylation, methylation-dependent transcriptional silencing depends on methylation of histone.

Nakayama and colleagues[93] then provided evidence that a conserved lysine residue, lysine 9 of histone 3 (H3 Lys^9 or H3K9), was preferentially methylated at heterochromatic regions of fission yeast (*Schizosaccharomyces pombe*), and that modifications of histone tails were linked to heterochromatin assembly. Nakayama *et al.* proposed that histone deacetylases and histone methyltransferases cooperate to establish a "histone code" that would result in self-propagating heterochromatin assembly. Assuming that certain transacting proteins that affect silencing (Clr4/SUV39H1 and Swi6/HP1) are conserved and that H3Lys9 methylation occurs in higher eukaryotes, Nakayama *et al.* predicted that a similar mechanism might be responsible for higher order chromatin assembly in humans as well as in yeast.

3. Chromatin

Kornberg's proposal in 1974[94] that chromatin structure was based on a repeating unit of eight histone molecules and about 200 DNA base pairs have provided the basis for chromatin research since then. The X-ray structure of the repeating unit, the nucleosome, was solved and subsequent research revealed its biological significance. A total of five types of histones were observed as components of the nucleosome, and powerful protease inhibitors led to discovery of an H3–H4 tetramer and H2A–H2B dimer. These histone oligomers could be recombined to generate X-ray patterns of chromatin and the organizing principle of the nucleosome, a histone octamer and its mode of interaction with DNA could be induced.

4. Normal Chromatin Organization

Structural studies of the mammalian cell demonstrated that the genomic DNA molecule of humans, which is approximately 1.7 m long in its extended conformation, is folded and compacted into a 5-μm nucleus in cells.[94,95] In all eukaryotes, nuclear DNA is associated with chromatin in a package that permits it to be replicated and transcribed.[94–98] Histone proteins packaged with DNA to form the nucleosome are the basic building block of chromatin. Each nucleosome is formed of approximately 146 base pairs of DNA wrapped around a histone octamer core particle containing one H3–H4 tetramer and two H2A and H2B dimers,[94,99] or alternatively, of histone variants. Repeating nucleosome cores are assembled into higher order structures which are stabilized by the linker H1 histone.[95,100] In addition to packaging and compaction of DNA, nucleosomes participate in virtually all chromosomal processes, including transcription, replication, and DNA repair as well as construction of the kinetochore and centromere, and in telomere maintenance.[98]

The histones that comprise the nucleosome contain evolutionarily conserved N-terminal tails. Across species, histones are among the most invariant proteins known. They can act on chromatin structure by altering the net charge on the histone tail to reduce histone–DNA binding, or alternatively, specific modified residues or combinations of residues can form sites for nonhistone proteins that in turn can influence chromatin structure and function[101]). Because each modification represents a dynamic balance between the effects of the modifying enzymes and because many, if not all, enzymes depend upon, or are influenced by, metabolites or components present in the intracellular and extracellular environment, the nucleosome can serve as a finely tuned sensor of the metabolic state of the cell and the composition of the environment.

Functionally, histone modifications have been divided into those involved in the establishment of global chromatin environments, and those involved in the coordination of DNA-based biological tasks.[102] Histone modifications partition the genome into distinct domains such as euchromatin in which DNA is accessible to transcription, and heterochromatin, in which chromatin is inaccessible for transcription. To coordinate DNA-based functions, histone modifications guide unraveling of chromatin to execute a specific function. Operationally, the presence of modifications on histones function either by disrupting chromatin contacts in which DNA is packaged or by coordinating the recruitment of enzyme complexes and nonhistone proteins to manipulate DNA. In this way, histone modifications and the histone code have the potential to act cooperatively to alter local functions as in gene transcription, or genome-wide functions, as in DNA replication, repair, and chromosome condensation. The two categories of function resulting from histone modifications are described by Kouzarides.[102] Pathways and mechanisms that reproduce chromatin organization in the wake of DNA replication and repair are discussed elsewhere by Groth[103] and Vaissiere and Herceg.[104]

5. Chromatin Modifications and Gene Expression

A striking feature of the core histones, particularly of their tails, is the large number of posttranslational modified residues they possess,[102] all of which are believed to play an important role in diverse cellular processes that require access to the DNA template. The 23–35 residues[99,105] of the amino termini tails of all histones protrude from the nucleosome core and are accessible to histone acetyltransferases, histone deacetylases, histone methyltransferases, kinases, and other enzymes that attach or reverse these modifications, a feature crucial to the role of the nucleosome in transcriptional regulation. At least eight different reversible modifications are found including acetylation, methylation, ubiquination, and sumoylation of lysine residues, methylation and deimination of arginine residues, phosphorylation of serine and threonine residues, and ADP ribosylation of glutamate residues at specific sites on histone tails. We have most information about acetylation, methylation, and phosphorylation.[102,105,106] These modifications have led to the "histone code," a hypothesis based on the idea that distinct histone modifications on one or more tails are read by other proteins and thereby dictate specific downstream events.[105]

For purposes of transcription, modifications are divisible into those that lead either to activation or to repression. Acetylation, methylation, phosphorylation, and ubiquination are involved in activation, whereas methylation, ubiquination, sumoylation, deimination, and proline isomerization are involved in repression. It is likely, however, that any given modification has the potential to activate or repress transcription under different conditions. For example, methylation at H3K36 and H3K9 are activating modifications when in the coding region but are repressing when in the promoter.[102]

Histone modification by acetylation is almost invariably associated with activation of transcription because this modification partially neutralizes the positive charge of histones to reduce the affinity of histone proteins for DNA and chromatin packaging is relaxed. The connection between acetylation and transcription had been suspected since the pioneering studies of Allfrey in the 1960s[107], although this relation remained uncertain until the yeast Gcn5 protein, a positive transcription regulator of many genes, was demonstrated to have acetylase (HAT) activity[108] (reviewed by Kornberg[94]). Although recombinant Gcn5 protein acetylates histones in the free state, it fails to do so in nucleosomes. This lack of activity of the natural substrate resulted in the discovery of the SAGA complex, so-called for its content of additional proteins affecting transcriptional activation and promoter function. A human counterpart of the SAGA complex containing the acetylase PCAF has been described[109,110] (Table II). Generally, these enzymes modify more than one lysine, but some do show limited specificity. Most acetylation sites are those more accessible to modification.

Deacetylation, the reversal of acetylation, is usually associated with repression and silencing.[111] The connection between deacetylation and repression was most clearly demonstrated by the isolation of a human histone deacetylase, HDAC1, whose sequence is highly similar to that of a yeast negative regulatory protein, reduced potassium dependency (Rpd3). All additional acetylases, which have been identified in yeast and human cells, occur in multiprotein complexes that have important functional consequences: the complexes can deacetylate histones in nucleosomes while the isolated deacetylase cannot, and the complexes contain other proteins previously implicated in transcriptional repression and chromosome interactions[94] (Table II).

Histone methylation can be either an activating or repressing mark. For example, methylation on H3K4, H3K36, and H3K79 activates transcription, whereas methylation on H3K9, H3K27, and H4K20 represses transcription. Furthermore, the methylation degree on a specific residue as well as the location of the methylated histone within the nucleosomes affects the transcriptional process.[112] Three classes of histone methylating enzymes are recognized: SET domain lysine methyltransferases, non-SET-domain lysine methyltransferases, and arginine methyltransferases. Improved understanding of histone methylation has shown that this epigenetic mark is dynamically regulated. For instance, hypermethylation of CpG islands in gene promoter regions is associated with dimethylation of histone H3 at lysine 9, deacetylation of the same residue, trimethylation of histone 3 at lysine 27, and loss of the transcriptional activating mark H3K4me2 (reviewed in Jacinto and Esteller[113]).

Originally, histone demethylation was a contentious issue and was initially believed to be irreversible until the subsequent detection of histone demethylating enzymes[114] (reviewed in Spanhoff *et al.*[106]). Unlike the deacetylases,[111] histone methyltransferases are usually specific for the arginine or lysine they target. An important part of the specificity of lysine demethylases concerns the state of methylation they act on. Their selectivity for mono-, di- and trimethylated lysines provides a measure of control of lysine methylation. However, present information is too fragmentary to describe the function of these enzymes precisely. Information concerning the effect of other histone modifications on gene transcription including phosphorylation, deimination, deubiquination, ADP ribosylation, and proline isomerization is also limited and fragmentary (reviewed by Kouzarides[102]).

Histone modifications are also implicated in DNA repair, DNA replication, and chromosome condensation, although the information is of limited scope and fragmentary (reviewed by Kouzarides[102]). With respect to DNA repair, one of the earliest recognized responses to DNA damage is the phosphorylation of histone variant γ-H2AX in mammalian cells. Two phosphorylation sites on this histone participate in repair of double-strand breaks via nonhomologous end joining. In human cells, mono- and dimethyl forms of H4K20 appear to be implicated in repair of radiation-induced DNA damage. In yeast, and in the presence of DNA damage, acetylation of H3K56 has been implicated in genome stability and DNA replication. Another histone acetylase of yeast that acetylates H4K12 is recruited to sites of DNA repair. With respect to DNA replication, findings in *Drosophila* suggest that the acetyltransferase, HBO1, is required for S-phase initiation and fixing of replication origins. With respect to chromosome condensation, there is evidence for involvement of phosphorylation and acetylation. Condensation and decondensation of chromatin are important during the cell cycle. Two phosphorylations may be important in mitosis, the first at H4S10 (serine 10) and the second at H3T3 (threonine 3).

Since extensive studies had established that acetylation of histone tails was primarily associated with gene activation, while methylation, depending on its position and state, was associated with either repression or activation, investigators proposed that multiple modifications of histones might act in a combinatorial or sequential fashion to specify distinct chromatin states in accord with the histone code hypothesis.[105] However, there have been only isolated reports in support of this idea until Wang and coworkers[115] systematically analyzed genome-wide histone modifications of acetylations and methylations in human CD4+ cells. Wang *et al.* identified a common modification module detected at gene promoters consisting of 17 modifications (H2A.Z, H2BK5ac, H2BK12ac, H2BK20ac, H2BK120ac, H3K4ac, H3K4me1, H3K4me2, H3K4me3, H3K9ac, H3K9me1, H3K18ac, H3K27ac, H3K36ac, H4K5ac, H4K8ac, and H4K91ac).

Genes associated with this module had higher expression, and addition of more modifications to this module is associated with further increases in gene expression. The data of Wang *et al.* suggested that these acetylations and methylations act cooperatively to prepare chromatin for transcriptional activation. Wang *et al.* could also classify modifications into three classes according to expression patterns. Class I expression patterns contained H3K27me3 and correlated with low expression. This class also contained H3K4me1/2/3, H3K9me1, and H2A.Z but no acetylations. The patterns containing only H3K4me3 or no modification also belonged to this class. Class II patterns contained H3K36me3 or the modification backbone consisting of the 17 modifications described above, or the backbone plus H4K16ac which correlated with intermediate gene expression. Class III showed the highest expression, and it included H2BK5me1, H4K16ac, H4K20me1, and H3K79me1/2/3 in addition to the modification backbone. Further analysis suggested that genes involved in cellular physiology and metabolism were enriched in the class III patterns, consistent with their housekeeping roles. In contrast, many genes involved in development, cell–cell signaling, and synaptic transmission were enriched in the inactive class I patterns, consistent with their not being required for mature T-cell function. Results concerning enhancers of transcription suggested that there are distinctive patterns at different enhancers, but no significant correlations were found between modification patterns at enhancers and gene expression.

6. Synergistic Effects of Chromatin Modification and Remodeling

Early research on chromatin focused on the packaging and compaction of DNA by nucleosomes,[94] which led to the idea that chromatin might be a relatively static unit. But, as noted above, more recent studies indicate that nucleosomes themselves are dynamic participants in virtually all chromosomal processes, including transcription, replication, and repair (reviewed by Saha *et al.*[98]and by Jones and Baylin[7]). The dynamic properties of nucleosomes are primarily due to the action of nucleosome-modifying and remodeling complexes. Nucleosome-modifying complexes add or remove covalent modifications at particular residues of histone proteins that are recognized by transcriptional regulators and other factors, whereas chromatin remodeling complexes restructure, mobilize, and eject nucleosomes to regulate access to the DNA. Each remodeller affects the structure of nucleosomes and arrays of nucleosomes in a distinct manner, perhaps because different remodelers use unrelated mechanisms to restructure the nucleosome. All remodellers, however, require ATP hydrolysis for their remodeling functions, and all contain an ATPase domain that is highly similar to those present in known DNA translocases. Modifying and remodeling complexes cooperate to regulate access to the DNA and together, they guide the recruitment of transcriptional regulators to particular loci and give chromatin its dynamic character by modifying the covalent attachments on lysine residues of histone tails of the octamer histones, and by mobilizing nucleosomes to alternative positions along the DNA, or, at times, by replacing a canonical core histone (e.g., H2A) with a variant histone (e.g., H2AZ or H2AX).

Remodellers of eukaryotes have evolved into several families: SWI/SNF, ISWI, NURD/Mi-2/CHD, INO80, and SWR1. Currently, the two best studied families of chromatin remodelers, SWI/SNF and ISWI, have provided insight into the remodeling process that performs specific chromatin tasks. Together, these specialized remodellers mediate many biological processes by establishing or altering regional properties of chromatin (see Table I in the review by Saha *et al.*[98]for a list of biological functions). Genome-wide analysis techniques have improved markedly and have increased understanding of chromatin regulation. Newer methods such as combining ChIP-SAGE or with massively parallel sequencing (ChIP-Seq) have provided insight and new appreciation of the ATP-dependent remodellers in development and their underlying mechanisms (for discussion, see Ho and Crabtree[116]).

Recent studies shed light on how epigenetic information controls DNA repair.[117] Double-strand breaks are among the most damaging lesions of DNA. They are constantly produced by various genotoxic endogenous and exogenous (environmental) agents. If left unrepaired, these lesions can lead to cell death or mutations in oncogenes, tumor-suppressor genes, or DNA repair genes resulting in genomic instability, oncogenic transformation, and development of disease, including cancer.[113] Cells have evolved mechanisms to repair such lesions that may vary according to the type of damage incurred. Among eukaryotes, two major, highly conserved pathways, homologous recombination and nonhomologous end joining, have evolved for repairing double-strand breaks.

Repair of double-strand breaks through either of these pathways is a complicated, dynamically regulated process. First, compacted chromatin must be relaxed to allow repair machinery to reach the damaged DNA. Then, cellular mechanisms involving modifying and remodeling complexes alter the structure of chromatin, although the way repair factors achieve this and how repair is coordinated with transcription and other processes are not known. Only recently have various molecular players including chromatin-modifying and remodeling complexes been associated with DNA repair. These activities include ATP-dependent nucleosome (chromatin remodeling) and posttranslational histone modifications. Exchange of histone variants into nucleosomes around break sites is an additional mechanism that may facilitate DNA repair, and once DNA repair is completed, additional enzymatic activities are needed to restore chromatin structure.

Loizou and colleagues present a stepwise model that aids in the understanding of the interplay between chromatin-modifying/remodeling complexes during the repair of double-strand breaks.[117] Their model assumes that cells can utilize the activities of histone-modifying processes and remodeling complexes needed to achieve the repair. The description in Box 1 is an adaptation of the pattern of events that Loizou *et al.* describe pictorially and in the accompanying legend to their Fig. 1.[117]

BOX 1

THE INTERPLAY OF CHROMATIN-MODIFYING/REMODELING COMPLEXES IN REPAIR OF DOUBLE-STRAND BREAKS (ADAPTED FROM LOIZOU *ET AL.*[117])

Step 1. In response to a double-strand break, the MRN (MRE11-RAD50-NBS1) complex and ATM kinase are recruited to the DNA break site. Activated ATM kinase phosphorylates the histone H2A variant (H2AX) over a large region facilitating the recruitment of early response proteins such as MDC1.

Step 2. Step 1 is followed by recruitment of the TRRAP/TIP60 HAT complex that acetylates core histones around the break site. Histone acetylation unwinds chromatin and/or serves as a binding platform to facilitate recruitment of remodeling complexes, such as INO8O and SWR1, and late repair proteins such as RAD51 and BRCA1. The presence of INO8O may facilitate the eviction or sliding of nucleosomes in the vicinity of the break to allow resection of the 5′-3′ strand and generation of a 3′ single-strand DNA (ssDNA) overhang. This maneuver allows RAD51 and BRCA1 to stimulate double-strand break to be repaired through homologous recombination.

Step 3: After double-strand break repair, dephosphorylation of the incorporated or evicted γ-H2AX may be mediated by Pph3 and PP2A, respectively.

Step 4. Deacetylation of histone occurs to restore chromatin after the DNA break is repaired.

Note: Some of the chromatin/remodeling mechanisms may act in a species and DNA repair type-specific manner.

7. MicroRNAs

MicroRNAs (miRNAs) are small, noncoding RNAs about 22 nucleotides long that are processed by Dicer from precursors with a characteristic hairpin secondary structure. Ambros and coworkers[118] present specific criteria for expression and biogenesis that are required for the identification and annotation of miRNAs so that they can be reliably distinguished from other RNAs such as small interfering RNAs. As none of the criteria on its own is sufficient for a candidate gene to be annotated as a miRNA, evidence for both their expression and biogenesis is required for reliable annotation. Profiles of different cell types and tissues indicate that expression patterns of miRNAs are cell type-dependent and tightly associated with cell differentiation and development. They are highly conserved, and they play crucial roles in important regulatory processes, including gene expression during development, proliferation, differentiation, apoptosis, and stress response.

After the initial discovery of the small RNA *lin-4* gene locus in 1993 in developing worm larvae by Lee, Ambros, and colleagues[119] (reviewed by Ambros[120]), several years elapsed before Fire, Mello, and associates recognized that these noncoding, nonmessenger RNAs possess potent and specific interference with gene expression.[121] While Fire, Mello and colleagues were attempting to use antisense RNAs to inhibit gene expression in *Caenorhabditis elegans* worms, they tested the double-strand RNA mixture as a silencer of gene expression. They found to their surprise that the double-stranded RNA mixture was at least an order of magnitude more potent than were sense or antisense RNAs alone (reviewed by Hannon[122]). The effects of this interference were evident in both the injected animals and their progeny. Only a few molecules of injected double-stranded RNA were required per affected cell, suggesting that there could be a catalytic or amplification component in the interference process. They recognized that genetic interference by double-stranded RNA could conceivably be used more generally by the organism as a tool for physiological gene silencing.

miRNAs genes are expressed in all metazoan organisms studied so far, including mammals.[120] These genes represent about 1% of the genome and are among the more abundant gene-regulatory molecules in animal cells. Each microRNA is derived from a gene that is dedicated to the production of a particular RNA about 22 nucleotides long that has hundreds of targets. It is estimated that 30% of genes are regulated by at least one microRNA, some of which pair with mRNAs of protein-coding genes to produce posttranslational repression through a mechanism involving the RNA interference machinery (reviewed in Bartel and coworkers [123,124]). Most animal miRNAs are imprecisely complementary to their mRNA targets, and they inhibit protein synthesis through a mechanism that preserves the stability of the mRNA target.[120]

miRNAs are transcribed by RNA polymerase II as lengthy hairpin primary structures called primicroRNAs. PrimicroRNAs are processed into the nucleus by RNAse III Drosha premicro RNAs, 70–100 nucleotides long.[125,126] These molecules are transported to the cytoplasm by Exportin 5 where an additional step mediated by RNAse III Dicer generates a double-stranded RNA of about 22 nucleotides named miR/miR. One of these strands is incorporated into the microRNA-containing RNA-induced silencing complex (miRISC); the other strand is believed to be degraded. The miRISC regulates gene expression posttranscriptionally, binding to the 3′ untranslated region (3′-UTR) through partial complementarity. At the same time, the complex leads to mRNA degradation and translation inhibition (reviewed in Iorio and Croce[125] and by Guil and Esteller[126]).

The first set of miRNA/target searches regarding miRNA biology pointed toward control of cell fate as a common theme for the activity of miRNAs. For example, the lists of predicted insect miRNA targets seemed to be enriched in genes encoding transcription factors, but also included genes with diverse functions not directly related to gene expression. Predicted targets in *Drosophila* of miR-277, for instance, included genes for the biochemical pathway for catabolism of leucine, isoleucine, and valine, and strongly suggested that miR-277 could regulate this pathway at several points.[120] While many outstanding questions remained, the genetic analysis of miRNAs was beginning to reveal the range of functions that these RNAs might have in control of animal development and physiology. As the program of expression of miRNAs is dependent on cell type and tightly associated with cell differentiation and development, expression of aberrant miRNAs appeared to be highly likely (reviewed in Iorio *et al.*[125]) as discussed below (see IV.C).

MiRNA is a comparatively new model in regulatory biology[122] and the mechanistic complexity of the process and its biological ramifications are only beginning to be appreciated. The technique has been harnessed for the analysis of gene function in several diverse systems, including plants, fungi, and metazoans, but its use in mammals has lagged somewhat. The first indication that miRNA could induce gene silencing in mammals came from observations in early mouse embryos and numerous mammalian cell lines, but silencing in these systems was transient. By utilizing long, hairpin dsRNAs, Paddison and coworkers[127] succeeded in creating stable gene silencing substantially increasing the power of miRNA as a genetic tool. The ability to create permanent cell lines with stable "knockdown" phenotypes extended the utility of miRNA in several ways, one of which is its application to epigenetics research.

8. Genomic Imprinting

Genomic imprinting, also referred to as gametic or parental imprinting, is another distinctive development bearing on the initiation of certain cancers that surfaced from studies of chromosome biology. Several wide-ranging phenomena such as X-inactivation, position effect variegation in *Drosophila*, and genomic imprinting are attributed to epigenetic regulation. Most of these changes in gene expression are mediated by DNA methylation of cytosines at CpG dinucleotide islands, coupled with modifications in chromatin of core histone tails. It results in parent-of-origin, monoallelic expression of genes, and is involved in the pathogenesis of several human conditions, including cancer and neurological disorders (reviewed by Das *et al.*[128]).

In 1949, Barr and Bertram[129] demonstrated an anatomical distinction between somatic cells of males and females, easily visible at metaphase under an ordinary microscope that could be used to sort tissues and individuals into two groups according to gender without prior knowledge of sex. In 1959, Ohno *et al.*[130] explained that the pair of X chromosomes in female cells was unlike each other because one of the pair remained extended during mitosis while the other assumed a condensed state forming the "sex chromatin body" that Barr identified. As a follow-up to Ohno's study in the 1960s, the sex chromatin body was characterized by Lyon[131] at the cellular level using X-linked markers of coat color of mice and at the genetic level by Beutler and colleagues[132] using X-linked markers of G6PD of human red blood cells to show that X expression of these markers was a mosaic in normal females. Lyon and Beutler *et al.* concluded independently that each female cell became a mosaic consisting of one inactive and one active X chromosome by a process of random inactivation. On average, half the cells of females have the maternal chromosome active and half have the paternal X chromosome active.

While the studies of Lyon and Beutler *et al.* were in progress, Crouse[133] identified a strange form of chromosomal behavior in the mealy bug *Sciara*. Embryos that were initially triploid, having two copies of the paternal gamete and a single copy of the maternal gamete, inactivate one or both paternal copies but always retain the maternal copy. She proposed that "the chromosome which passes through the male germ line acquires an imprint that results in behavior exactly opposite to the imprint conferred on the same chromosome by the female germ line." This was the earliest definition of a gametic imprint resulting in a functional difference between parental chromosomes.

Investigators had speculated on various models to explain genomic imprinting,[55,56] but the molecular basis of this epigenetic modification was uncertain until Barton *et al.*[134] and McGrath and Solter[135] presented experimental evidence that functional differences existed between maternal and paternal alleles. They showed that mouse embryos derived from purely maternal, or purely paternal, genomes failed to develop beyond implantation demonstrating that maternal and paternal genomes were both required for normal embryonic development. By the 1990s, numerous studies of mouse genes known to be imprinted suggested cytosine methylation to be a part of the imprinting mechanism.[136–141] These models focused primarily on the idea that parental imprinting resulted from gamete-specific marks that were established to achieve sex-specific gene expression patterns in mature gametes.[141,142] Increasing evidence supported the idea that stable chromatin modifications were controlled by small segments of methylated DNA a few kilobases long. These segments appeared to occur at a small number of chromosomal locations that showed different levels of DNA methylation of maternal and paternal alleles (reviewed by Whitelaw and Garrick[141]).

Reinhard Stöger and colleagues were first to demonstrate in 1993 an example of a primary gametic imprint that was differentially methylated. Building on a previous study in Stöger's laboratory that the *Igfr2* locus was imprinted, Stöger *et al.* searched *Igfr2* for the presence of parental-specific methylation modifications. (Igfr2 is an insulin-related protein that is expressed in rodents at high levels during embryonic and fetal development but at low levels in adults. This somatic growth factor enhances placental nutrient exchange of glucose for fetal growth and its impairment restricts fetal growth.) They identified two DMRs: region 1 contained the transcription start site and was methylated only on the silent paternal chromosome; region 2 contained in an intron and was methylated only on the expressed maternal chromosome. However, methylation of region 1 was acquired after fertilization, while methylation of region 2 was inherited from the female gamete. These data indicated that the expressed locus in region 2 carried a potential imprinting signal and implied that methylation was necessary for expression of the *Igfr2* gene.

Imprinted genes represent a small subset of the approximately 20,000 autosomal genes in the human genome. They are involved in embryonic, fetal, and placental development, cell proliferation, and adult behavior, and faulty imprinting is linked to cancer as well as obesity, diabetes, neurodevelopmental, and various behavioral disorders. Recently, Luedi and coworkers[143] developed a statistical model that identifies potentially imprinted genes, and also predicts the parental allele from which they are expressed. Of 23,788 annotated autosomal mouse genes, their model identified 600 (2.5%) to be potentially imprinted, 64% of which are predicted to exhibit maternal expression. Luedi and colleagues also applied and extended their model to identify imprinted genes in the human genome.[144] They predicted 156 imprinted genes of 20,770 (0.75%) annotated autosomal genes in the human genome. However, the overlap in the repertoires of imprinted genes in humans and mice was only 32%, emphasizing a marked species difference in imprinting,[144] calling into question the significance of human cancer risk assessments based solely on nonprimate animal studies. A list of known imprinted genes is available at the website http://www.geneimprint.com; http://www.geneimprint.com/site/genes-by-status.

IV. Epigenetic Patterns in Cancer

The development of cancer is closely tied to genetic instability combined with clonal expansion of cells that have accumulated an advantageous set of genetic and epigenetic aberrations. Instability of genetic origin may arise from point mutations of DNA sequences, chromosomal rearrangements, DNA dosage abnormalities, and perturbations of microsatellite sequences. Instability of epigenetic origin was initially thought to result from aberrant patterns of DNA methylation, faulty imprinting, or histone modifications in chromatin, but more recent advances suggest that the causes of epigenetic instability, particularly dysregulation of processes resulting in silencing of regulatory genes, should be expanded to include virtually every component of chromatin including changes in nucleosomal architecture and noncoding RNAs. Most commonly, the types of epigenetic change observed in cancer cells are increases in methylation of CpG islands within gene promoter regions and deacetylation with or without methylation of histone proteins. These abnormalities may act alone or work together to alter the functions or expression of cellular components, and they may occur at any stage in the development of cancer (for recent reviews, see Jones and Baylin,[7] Esteller,[145] Guil and Esteller,[126] Esteller,[6] Clark,[72] Cheung *et al.*,[146] McCabe *et al.*,[147] and Clouaire and Stancheva[86]).

A. Abnormal DNA Methylation in Cancer

Aberrant patterns in DNA methylation provided early hints of epigenetic dysregulation in human cancers. Hypomethylation at both the individual gene and globally was the first of these patterns to be reported.[148,149] Cancer cells typically exhibit hypomethylation of intergenic regions that usually contain the majority of a cell's methylcytosine content. As a consequence, transposable elements may be activated that contribute to genomic instability and chromosomal rearrangements, both of which may lead to further cancer-related events.

At the same time, cancer cells may exhibit hypermethylation, particularly of CpG islands at gene promoters.[150] Promoter transcriptional silencing of hypermethylation of genes involving important cellular pathways is a prominent feature of many major human tumor types.[151] In cancer cells, hypermethylation is a key event in the carcinogenic process, contributing to all of the typical hallmarks that result from transcriptional silencing of tumor-suppressor genes.[113,151] It turns genes off that should be on (tumor suppressors, DNA repair), and vice versa (oncogenes, invasion, and metastasis). In addition, hypermethylation is often accompanied by global hypomethylation and this combination could affect cancer cells to a greater extent than coding region deletions or mutations which are relatively rare.

Since the discovery of altered methylation in human cancer, many studies and reviews have focused on the hypermethylation of genes of specific interest[151–158] and pathways, processes or regions assumed to be of functional importance.[113,151,159] For example, in an extensive study by Esteller and coworkers[151], a total of 12 genes were analyzed, including tumor-suppressor genes (*p16*^{INK4a}, *p15*^{INK4b}, *p73*, *APC*, and *BRCA1*), DNA repair genes (*hMLH1*, *GSTP1*, and *MGMT*), and genes related to metastasis and invasion (*CDH1*, *TIMP3*, and

DAPK), all of which had been rigorously characterized. Each gene had been characterized for abnormal silencing in cancer in DNA from over 600 primary tumor samples representing 15 major tumor types. The data showed that promoter hypermethylation is a feature of each of the 15 tumor types. Additionally, unique tumor profiles exhibited simultaneous inactivation of several pathways by aberrant methylation for the tumor types: that is, a colorectal tumor might have disruption of cell cycle, of DNA repair, and of a metastasis-related process by hypermethylation of $p16^{INK4a}$, *hMLH1*, and *TIMP-3*, respectively. In addition, a mammary tumor could accomplish similar objectives by silencing $p16^{INK4a}$, *BRCA1*, and *CDH1*. In the cases cited, these epigenetic lesions occur in the absence of a genetic lesion, and they are often early events in the natural history of the cancer. That the spectrum of epigenetic alterations described provides a powerful system of biomarkers for developing molecular detection strategies for many forms of human cancer is also of diagnostic interest. A more recent compilation shows that DNA hypermethylation can occur in many genes involved in different biochemical pathways that are related to tumor development and progression (reviewed by Cheung *et al.*[146], see Table I). In addition to the tumor suppressor, DNA repair, and metastasis genes cited above,[151] these include cell-cycle genes plus genes that regulate apoptosis, detoxification, hormone response, Ras signaling, and Wnt signaling.

Cheung *et al.*[146] also discuss genes and genomic regions which are often associated with oncogenes and are reactivated by hypomethylation. C-Myc, a transcription factor that acts as an oncogene, is often reported hypomethylated in hepatocellular carcinoma, leukemia, and gastric carcinoma, and is often associated with cancers of the bladder, colorectum, and breast. Many other genes are found to be hypomethylated and reactivated including PSG in testicular germ cell cancer, WNT5A, CRIP1, and S100P in prostate cancer, L1 adhesion molecule in colorectal cancer, and the cancer/testis antigen gene, XAGE-1, in gastric cancers, but their role in oncogenesis is not fully understood. Cheung and colleagues also describe the role of global hypomethylation of repetitive sequences such as Line 1, Alu sequences, and transposable elements in promoting genomic instability in various cancers.

Recently, Irizarry and coworkers[160] conducted a study that raises a question regarding the location of alterations that are responsible for the colon cancer-related differential methylation. Irizarry *et al.* performed a genome-wide study to determine the relationship between DNA methylation changes (gain and loss) in cancer versus that in normal differentiation. They asked (1) where are the DNA methylation changes located that distinguish tissue types; (2) where are DNA methylation alterations in cancer compared to those in matched normal mucosa; and (3) what is the functional role of each of these methylation changes. Irizarry *et al.* found that most methylation alterations in colon cancer did not occur in promoters, and also not in CpG islands, but in sequences up to 2 kb distant (which they term "CpG island shores"). They found that CpG island shore methylation was strongly related to gene expression, and that it was highly conserved in mouse, discriminating tissue types regardless of species of origin. Irizarry *et al.* also found there was overlap (45–65%) of the location of colon cancer-related methylation changes with those of normal tissues, with hypermethylation enriched closer to the associated CpG islands, and hypomethylation enriched further from the associated CpG islands and resembling that of noncolon normal tissues. They concluded that methylation changes in cancer are at sites that vary normally in tissue differentiation. They state that their findings are consistent with the epigenetic progenitor model of cancer,[161] which proposes that epigenetic alterations affecting tissue-specific differentiation are the predominant mechanism by which epigenetic changes cause cancer.

In an effort to improve understanding of the causes and global patterns of methylation patterns, Toyota and colleagues examined the methylation status of CpG islands in a panel of 50 primary colorectal cancers and 15 adenomas.[162] They found that a majority of CpG islands methylated in colon cancer were also methylated in a subset of normal colonic cells as an age-related consequence of incremental hypermethylation. In contrast, methylation of the cancer-specific clones was found exclusively in a subset of colorectal cancers which appeared to exhibit a CpG island methylator phenotype (CIMP). The CIMP+ tumors included the majority of sporadic colorectal cancers with microsatellite instability related to methylation of the mismatch repair gene hMLH1. The data suggested to Toyota *et al.* the existence of a pathway in colorectal cancer that was responsible for the risk of mismatch repair-positive sporadic tumors.[162]

Feinberg and coworkers believed that special significance attaches to loss of imprinting (LOI) in cancer and they sought to determine the mechanism by which this epigenetic change might enhance the risk of initiation and progression of carcinogenesis. In the first of two papers, Cui *et al.*[163] found that LOI of the insulin-like growth factor II (*IGF2*) gene, a feature of many human cancers, occurred in about 10% of the normal human population.[163] LOI in this segment of the population increased the risk of colorectal cancer about a 3.5–5-fold, suggesting that faulty imprinting was related to the risk of cancer. In the second paper, Sakatani *et al.*[164] created a mouse model to investigate the mechanism by which LOI of *Igf2* contributed to intestinal cancer.[164] They knew from the work of others that imprinting of *Igf2* was regulated by a DMR upstream of the nearby untranslated *H19* gene, and that deletion of the DMR would lead to biallelic expression (LOI) of *Igf2* in the offspring. To model intestinal neoplasia, they used *Min* mice with an *Apc* mutation with or without a maternally inherited deletion, that is, with or without LOI, and they designed the model to mimic closely the human situation where LOI caused only a modest increase in *IGF2* expression. They created their model of *Igf2*

LOI by crossing female heterozygous carriers of a deletion in $H19^{+/-}$ with male heterozygous carriers of $Apc^{+/Min}$. Their results in offspring of this cross showed that LOI mice developed twice as many intestinal tumors as control litter mates. Their results also showed a shift toward a less differentiated normal intestinal epithelium. In a comparative study of human tissues, a similar shift in differentiation was seen in the normal colonic mucosa of humans with LOI. These observations suggested to Feinberg and associates that loss/impairment of normal parental imprinting might interfere with cellular differentiation and thereby increase the risk of cancer. In more general terms, they concluded that mutation of a cancer gene (*APC*) and an epigenetically imposed delay in cell maturation might act synergistically to initiate tumor development.[164]

The development of genome-wide methylation technologies has expanded understanding of DNA methylation patterns in normal and cancerous cells. Studies using these techniques have confirmed that the repetitive portion of the genome of normal cells is heavily methylated and most CpG islands are unmethylated, while cancer cells exhibit widespread loss of intergenic DNA methylation with gain of methylation at many gene-associated CpG islands. They have also generated new information about the DNA methylation patterns. For example, within the DNA methylome of individual tumors, about 1–10% of CpG islands are aberrantly hypomethylated. One study found that almost 5% of gene-associated CpG islands are methylated and that a fraction of these normally methylated CpG islands becomes hypomethylated and transcriptionally active in cancer cells. Promoter-associated CpG islands are not the only islands affected by aberrant DNA methylation, as some CpG islands located within 3′ ends of genes and in intergenic regions exhibit hypermethylation in cancer cells. Whether, and to what extent, methylation affects expression of these nonpromoter regions is unclear. Analysis of several genes with hypermethylated 3′ CpG islands showed, however, increased gene expression, suggesting a new function for DNA methylation in this location. These findings indicate that methylation patterns may have unanticipated effects on gene expression and cellular function than previously believed in association with cancer (reviewed by McCabe *et al.*[147]). The review by McCabe and colleagues also draws attention to various hypotheses regarding relationships between the DNA methylomes of different tumor types and describes potential mechanisms to explain the occurrence of aberrant hypermethylation revealed in these genome-wide studies.[147]

Exogenous insults may initiate hypomethylation of genomic DNA via DNA damage pathways which may predispose cells to development of cancer. These insults include dietary methyl donor deficiency, UV irradiation and chemicals, and bacterial infection. An area of concern where environmental conditions may influence epigenetic programming is the use of assisted reproductive technologies. Children born through these technologies have an increased frequency, ninefold greater than the general public, of developing Beckwith–Wiedemann syndrome that has an associated increase risk of embryonal tumors, particularly Wilm's tumor. Children born of this technology also have a higher incidence of retinoblastomas (reviewed in Wilson[165] and in Dean[166]). The underlying mechanisms of tumor induction through these processes remain speculative.

B. Aberrant Chromatin Modification and Remodeling in Cancer

Once transcriptional gene silencing in normal cellular processes was associated with methylation of CpG islands and conformational changes in chromatin involving histone deacetylation as described above[87,88,91–93] (see III.B.2), the focus of research shifted to include mechanisms by which chromatin modifications control gene activity in both normal cells and cancer cells. Covalent modifications of histones that can control gene activity are foremost among these.

A major function of histone modifications of chromatin, heretofore referred to as the histone code, is to rearrange the chromatin environment in a fashion that is either permissive or repressive of gene transcription. Fraga and coworkers[167] were among the first to profile posttranslational histone modifications by histone H4, in a comprehensive panel of normal tissues, cancer cell lines, and primary tumors. They demonstrated that cancer cells overall lost monoacetylation at H4-Lys16 (H4K16ac) and trimethylation of histone H4-Lys20 (H4K20me3). They showed in a mouse model of multistage carcinogenesis that these changes appeared early and accumulated during the tumorigenic process. These changes were also associated with hypomethylation of DNA repetitive sequences of cancer cells. Fraga *et al.* interpreted the global loss of monoacetylation and trimethylation of histone 4 as hallmarks of human cancer cells. Specific lysine residues such as lysine 9 in histone 3 (H3K9) or lysine 27 in histone 3 (H3K27) were found to participate in transcriptional gene silencing.[168]

More recently, other investigators have shown that overall, cancer cells exhibit a global decrease in H4K20me2/3, H3K9me2, and H4 acetylation, particularly at H4K16 (reviewed by McCabe *et al.*[147]). The loss of H4K16ac and H4K20me2/3 is primarily from the repetitive fraction of the genome, occurs in premalignant lesions, and increases during tumor progression. The loss of DNA methylation, H3K9me2 and H4K20me3 concerns global dysregulation of transcriptional repression in cancer cells, and may promote tumorigenesis through de-repression of exogenous repetitive elements such as transposons or miRNA, impaired DNA damage response, and chromosomal instability. It will be recalled that in normal cells, an open chromatin structure marked by hyperacetylation of histones H3 and H4 and di and trimethylation of histone H3 at lysine 4 (H3K4me2/3) constitutes a permissive region for transcription, whereas repressed regions exhibit a compact chromatin structure that lacks H3/H4 acetylation and H3K4 methylation, and is enriched instead in repressive modifications, di- and trimethylation of H3K9 (H3Kme2/3), trimethylation of H3K27 (H3K27me3), and trimethylation of H4K20 (H4K20me3).

Epigenetic mechanisms controlling transcription of genes involved in cell differentiation, proliferation, and survival are often targets for deregulation in the development of malignancy. The proteins responsible for the alterations characteristic of the cancer epigenome are the enzymes that catalyze DNA methylation, the proteins that bind methylated DNA at promoters and contribute to silencing, and the chromatin modifier enzymes that catalyze histone acetylation, deacetylation, methylation, and demethylation. The deregulation of these epigenetic modifiers has been characterized in many malignancies, and the disruption of a number of histone-modifying proteins, by mutations, deletions, or over- and underexpression is supportive of the critical role of these effectors in carcinogenesis.[109,110] A partial list of these proteins is presented in Table II. A more complete list is contained in reviews by Miremadi *et al.*[109]and Ellis *et al.*[110] Extensions of this work, now widely accepted, have shown that DNA cytosine methylation and histone modifications are intimately linked to nucleosomal remodeling in cancer cells and that the interplay between all three of these processes which results in permanent silencing of cancer-relevant genes, may be deregulated in cancer (reviewed by Jones and Baylin[7]).

C. MicroRNA Dysregulation in Cancer

Much of the more recent research on microRNAs (abbreviated *miR* and miRNA in the following discussion) has attempted to gain a better understanding of how these noncoding RNAs function in both normal and pathological states. It has been shown that miRNAs have a predilection toward targeting developmental genes, and it is well accepted that miRNAs are fundamental to the regulation of proliferation, differentiation, and apoptosis during normal development. Furthermore, alterations in the expression of miRNAs are seen in a variety of pathological processes, including cancer. Aberrant miRNA expression has been demonstrated in virtually every cancer type studied, including breast cancer, ovarian cancer, pancreatic cancer, non-small cell lung cancer, leukemia, and brain tumors, as reviewed by Turner *et al.*,[169] Guil and Esteller,[126] Iorio and Croce[125], and Lujambio and Esteller.[170] miRNA expression can be altered in cancer through a variety of mechanisms such as chromosomal changes, epigenetic defects, genetic mutations, and alterations in the machinery involved in miRNA biogenesis. MiRNAs can serve as biomarkers, and altered expression profiles demonstrate that they are key regulators of carcinogenesis. There is also accumulating evidence that they are involved in cell-cycle checkpoint regulation, and they have been associated with tumor progression and metastatic potential in addition to their role in cancer formation. The following examples demonstrate some of the major avenues of importance in the pathobiology of miRNAs.

The first link between microRNA genes and cancer was found in 2002 by Carlo Croce and coworkers. Croce's laboratory was attempting to identify tumor suppressors at chromosome 13q14 that might be involved in the pathogenesis of CLL. Deletions at chromosome 13q14 occur in approximately 50% of CLLs, while loss of heterozygosity (LOH) in this region occurs in approximately 70% of CLLs. They found, however, that this region did not contain a protein-coding tumor suppressor, but two *miR* genes, *miR-15a* and *miR-16-1*, which are expressed in the same region. Study of a large collection of CLLs showed knockdown or knockout of *miR-15a* and *miR-16-1* in 69% of CLLs. Croce and colleagues speculated that this event might be playing an important role, perhaps the initiation of a very early event, in the pathogenesis of CLL.

In pursuit of these initial observations, Croce's group found through mapping all known microRNA genes in the human genome that many are located in regions involving chromosomal alterations, such as deletion and amplification, in a variety of different tumors. They and others have since assessed global expression of microRNA genes in normal and diseased tissues and conducted profiling studies to determine the extent of microRNA dysregulation in human cancer and to determine whether microRNA profiling might be a tool suitable for assessing classification and prognosis of human cancers (reviewed in Iorio and Croce[125]). They found, for example, that profiling of different cell types and tissues indicated that the pattern of expression of microRNAs was cell type- and tissue-specific in various tumors, including CLL, acute myelocytic leukemia, lymphoma, multiple myeloma, breast cancer, lung cancer, and hepatocellular carcinoma. In a large study of indolent versus aggressive CLL, Croce *et al.* found 13 microRNAs capable of distinguishing between indolent and aggressive CLL.[171] Additionally, they found a germ line $C \rightarrow T$ homozygous substitution mutation in *pri-miR-16-1* in two patients. Both patients had a substantial reduction (15% and 40%) in the expression of *miR-16-1*. This was important because previous data indicated that *miR-16-1* and *miR-15a* behaved as tumor-suppressor genes in CLL and because LOH combined with a germ line mutation is characteristic, according to Knudson's model, of inactivation of a tumor-suppressor gene. The presence of pathogenic mutations in the *miR-15a-miR-16-1* cluster (as well as various mutations in other microRNAs) indicated that this class of genes is involved in CLL,[35] and that at least some microRNAs can function as tumor-suppressor genes.

As to microRNAs in solid cancers, breast cancer was in 2005 the first solid tumor to be profiled for microRNA expression. In 2005, Iorio *et al.* described the first microRNA signature characteristic of a solid tumor identifying 13 microRNAs that discriminated breast cancer tissue from normal tissue with

100% accuracy.[125] In this study, *miR-21*, overexpressed in breast carcinoma, mediates cell survival and proliferation directly by targeting several oncosuppressors, including *PTEN*, *PDCD4*, and *TPM1*. In addition, *miR-21* has been associated with advanced clinical stage, lymph node metastasis, and poor prognosis, and has also been found overexpressed in a variety of other cancers, for example, glioblastoma, ovary, and lung.

As happens with protein-coding genes, an aberrant pattern of methylation of CpG islands near or within microRNA genes can result in dysregulated miRNA expression and ultimately in pathogenic alterations including cancer (reviewed in Guil and Esteller[126]). There are numerous reports showing that miRNA genes are subject to hypermethylation and hypomethylation in both a tumor- and tissue-specific manner. The study by Iorio *et al.* in 2007,[172] for example, reveals a number of miRNA hypomethylated genes, including *miR-21*, *miR-203*, and *miR-205*, that are aberrantly upregulated in ovarian cancer.

The silencing of *miR-223* in leukemias illustrates in detail how individual miRNAs can suffer altered expression in cancer through dysregulation of chromatin modifiers.[126] *MiR-223*, a highly specific regulator of myelopoiesis, is inhibited in primary leukemias, and this repression may underlie the block in myeloid differentiation that occurs in cancer. Transcription of *miR-223* is under direct control of the oncogenic fusion protein AML1/ETO, the product of the most frequent chromosomal rearrangement in leukemias. Expression of this fusion protein in cancer drives histone deacetylation and DNA methylation of the *miR-223* gene, resulting in heterochromatic silencing of *miR-223*. As a consequence, H3 and H4 histones become deacetylated, and a small CpG island present on the core promoter region of *miR-223* close to the DNA region around the AML1-binding site is hypermethylated. Newly methylated CpGs act as binding sites for the DNA-methyl-CpG-binding protein MeCP2. All these changes in chromatin modifications depend on the presence of AML1/ETO, and they illustrate how an aberrantly formed chromatin remodeling complex may control the transcriptional silencing of a differentiation-associated miRNA gene upon the onset of cancer.

Lujambio and coworkers have shown through a pharmacological approach that miRNAs can play an important role in cancer metastasis by epigenetic mechanisms (reviewed in Lujambio and Esteller[170]). They measured the miRNA expression levels of three metastatic cell lines, treated or not treated with the DNA demethylating agent, 5-aza-2′-deoxycytidine, using a miRNA expression-profiling method. Treatment with this agent induced loss of DNA methylation associated with a release of miRNA gene silencing. They discovered five hypermethylated miRNAs exhibiting cancer-specific methylation, *miR148a*, three members of the *miR-9* family, and the *miR-34b/c* cluster. Restoration of expression of two of these methylated miRNAs, *miR-148a* and the *miR-34b/c* cluster, affected invasion capacity, both *in vitro* and *in vivo*. They also showed that the epigenetic silencing of these miRNAs mediated the activation of oncogenic and metastatic genes including *E3F3*, *C-MYC*, and *CDK6*, for *miR-34b/c* and the *TGIF2* for *miR-148a*. In human primary tumors, they showed that the *miR-34b/c* methylation was significantly correlated with oncogenic target upregulation, meaning that these oncogenes are targeted *in vivo*, and that the epigenetic silencing of the miRNA leads to their upregulation in cancer patients. Most importantly, these miRNAs were significantly more methylated in the primary tumors that gave rise to metastasis, highlighting the importance of the *in vivo* role of these miRNAs in suppressing tumor dissemination. These findings have implications for therapeutic possibilities for epigenetic drugs that can act on metastasis-related genes and miRNAs by restoring their expression.

D. Aberrant Genomic Imprinting in Cancer

Imprinted genes are implicated in many aspects of development, such as fetal and placental growth, cell proliferation, and adult behavior, and it is not surprising that abnormal expression of these genes is associated with numerous human disorders, including cancer. Certain aberrations of human pregnancy show that LOI plays an important role in embryogenesis (Table III). For example, ovarian dermoid cysts arise from LOI, resulting in benign cystic tumors that contain two maternal chromosomes and no paternal chromosomes[173,174], whereas hydatidiform moles contain a completely androgenic genome through LOI with two paternal chromosomes and no maternal chromosome.[175] Numerous reports of other tumors are associated with preferential loss of a particular parental chromosome. Examples include acute neuroblastoma (maternal chromosome 1p36 and paternal chromosome 2), Wilm's tumor (maternal chromosome 11p15.5), rhabdomyosarcoma (maternal chromosome 11p15.5), and sporadic osteosarcoma (maternal chromosome 13) (reviewed by Falls *et al.*[186]).

The role of defective imprinting in cancer is well illustrated by the occurrence of Wilm's tumors in association with the Beckwith–Wiedemann syndrome (BWS). This syndrome is a model for understanding the epigenetics of cancer as a family disorder caused by epigenetic changes in several genes. It maps to chromosome 11p15 and is characterized by generalized overgrowth of body parts including hemihypertrophy, macroglossia, and visceromegaly. A defect in imprinting was first suspected when preferential maternal transmission of mutations was observed in some BWS families.[187] Ten to twenty percent of BWS individuals are predisposed to embryonal tumors, most frequently to Wilms' tumors and adenocortical carcinomas. Among BWS patients that do not have cytogenetic abnormalities, the most common molecular event is the biallelic expression of *IGF2* due to LOI. In such instances, 70% of Wilms' tumors were found to exhibit biallelic *IGF2* expression that is thought to link

TABLE III

Human Tumors Due to Faulty Genomic Imprinting

Human disorder	Salient clinical features	Molecular pathology	Reference
Benign dermoid ovarian teratomas	Tumors contain many tissue types but no placental trophoblast	LOI results in tumors with two maternal chromosomes and no paternal contribution	173
Hydatidiform moles	Placental-derived extraembryonic tumors	LOI causes tumors with two paternal chromosomes with no maternal contribution	175,176
Wilms' tumors	Nephroblastoma of childhood	LOI causes preferential loss of maternal alleles on chromosome 11p15	177–180
Embryonal rhabdomyosarcoma	Tumors of striated muscle		179
Beckwith–Wiedemann syndrome	Pre and postnatal overgrowth, macroglossia, and other organomegaly, childhood tumors such as Wilms' tumor of the kidney, hypoglycemia, hemihypertrophy, and other minor complications	LOI results in biallelic expression of *IGF2* (80%), silencing or mutation of *H19* (35%), and silencing of *CDKN1C* (12%).[181]	182
Neuroblastoma	Childhood neural crest tumor	In a series of 13 neuroblastomas, loss of heterozygosity (LOH) results in loss of maternal 1p36 occurred in at least 10 cases, and two with loss of paternal alleles, 10 of which showed *N*-myc amplification.	183
Acute childhood leukemia	Bone marrow cells from patients with the infant seven syndrome showed various findings: three had loss of maternal alleles and 5/5 with MDS had loss of pad loss of paternal alleles	Findings suggest that imprinting of genes on chromosome 7, within bands q31-q36 may be important in myelodysplastic syndromes (MDS) and acute myeloid leukemia	184
Sporadic osteosarcoma	A total of 13 osteosarcoma cases were used to identify the parental origin of the lost chromosome or chromosome segment	Findings in this series indicate preferential loss of the maternal chromosome. This indicates that the initial event in the origin of this tumor occurred preferentially on the paternally derived chromosome 13	185

tumorigenesis directly to aberrant imprinting. Inactivation of *H19* genes was also present in a number of these cases, suggesting that the biallelic *IGF2* expression is coupled with *H19* inactivation (reviewed by Falls *et al.*[186]). Falls and colleagues point out that many other malignancies show LOI at the IGF2 locus.

They believe that deregulation of IGF2 imprinting is mechanistically involved in the development of a variety of childhood and adult tumors (see Table II in the review by Falls *et al.*[186]). It should also be remembered that because imprinting results in monoallelic expression, an imprinted tumor-suppressor gene would be expected to increase cancer susceptibility, since the inactivation of the remaining allele would eliminate tumor-suppressor function. *WT1*, *p57KIP2*, and *M6P/GF2R* represent examples of imprinted tumor-suppressor genes.

V. Epigenetic Therapies for Cancer

Silencing of key nonmutated genes such as tumor-suppressor genes and mismatch repair genes is a common event in cancer progression[152,162,188–194] including cancers of hematological origin.[195–197] Methylation and demethylation of CpG islands located in promoter regions of cancer cell genes and modifying enzymes of chromatin involving histone deacetylation are reversible, interacting processes associated with transcriptional silencing. Encouraged by the possibility that reversal of these processes could be important in preventing or reversing the disease phenotype, these processes have become therapeutic targets in the treatment of cancer.[106,198–202] Numerous preclinical and clinical trials have resorted to the treatment of various hemoglobinopathies, myelodysplastic, and leukemic syndromes with demethylating agents, histone demethylating agents, or the combined manipulation of cytosine methylation and histone acetylation. Agents used in these trials include older (5-azacytidine, 2-deoxy-5-azacytidine, or decitabine) and newer (MG98, an antisense DNMT1 inhibitor), demethylating agents, and older (sodium butyrate, sodium phenyl butyrate) and newer (trichostatin, suberoylanilide hydroxamic acid, and depsipeptide) HDAC inhibitors[200,201]

A. Methyltransferase Inhibitors and Demethylating Agents

One possible approach to promote expression of genes abnormally silenced by methylation is through inhibition of DNMTs, or alternatively, by agents capable of demethylating DNA.[106,199,202] These approaches have been studied in hematological and myeloid disorders although the data are limited. For example, in 1982, Ley *et al*. reported that the treatment of a patient with severe β-thallasemia with 5-azacytidine as a demethylating agent resulted in selective increases in γ-globin synthesis and hemoglobin F. Measurement of pretreatment methylation levels compared to posttreatment levels revealed hypomethylation of bone marrow DNA in regions near the γ-globin and the ϵ-globin genes.[203] Subsequently, several studies examined the use of demethylating agents such as 5-aza-2′-deoxycytidine (decitabine) in the treatment of another heritable hemoglobinopathy, sickle cell anemia. Treatment of this disorder with 2-deoxy-5-azacytidine led to significant increases in hemoglobin F and γ-globin that attained a maximum after 4 weeks of treatment and persisted for 2 weeks before falling below 90% of the maximum.[204] The mechanism of the therapeutic effect was not entirely clear but may have been caused by low pretreatment levels of methylation of the γ-globin gene and altered differentiation of stem cells induced by 2-deoxy-5-azacytidine.

Evidence also points to hypermethylation in the pathogenesis of the myelodysplastic syndromes. Patients with these disorders usually die from bone marrow failure or transformation to acute leukemia—standard care for this disorder is supportive. In one reported instance, the cyclin-dependent kinase inhibitor, $p15^{INK4b}$, was progressively hypermethylated and silenced in high-grade myelodysplasias, and treatment with 2-deoxy-5-azacytidine resulted in a decrease in *p15* promoter methylation and a positive clinical response in 9 of 12 myelodysplastic patients.[205] In another reported instance, 191 patients with high-risk myelodysplastic syndromes were treated with 5-azacytidine (dose 75 mg/m^2/day) for 7 days every 4 weeks. Statistically significant differences seen in the azacytidine group favored improved response rates, quality of life, reduced risk of leukemic transformation, and improved survival compared to supportive care.[206]

The potential reversal of epigenetic silencing by altering methylation levels with methyltransferase inhibitors or DNA demethylating agents has shown promise as a mode of therapy. In 2004, azacytidine was the first agent to receive FDA approval for treatment of several myelodysplastic syndrome subtypes. Cytidine analogs, such as 5-azacytidine and 5-aza-2′-deoxycytidine, achieve their therapeutic effects after a series of biochemical transformations. First, these agents are phosphorylated by a series of kinases to azacytidine triphosphate that is incorporated into RNA, disrupting RNA metabolism, and protein synthesis. Azacytidine diphosphate is reduced by ribonucleotide reductase to 5-aza-2′-deoxycytidine diphosphate, which is phosphorylated to triphosphate and incorporated into DNA. There it binds stoichiometrically to trap DNMTs and causes hypomethylation of replicating DNA.[206] Most methyltransferase inhibitors are, however, not specific for a particular methyltransferase, and several of them have unfavorable toxicity profiles, including severe nausea and vomiting.

There are newer agents under development that may improve the targeting of methylation. Among these, is MG98, a second-generation antisense oligonucleotide methyltransferase inhibitor that is specific for DNMT1.[199] MG98 produced dose-dependent reduction of DNMT1 and demethylation of the *p16* gene promoter and reexpression of p16 protein in tumor cell lines. A two-stage phase 2 trial was performed to assess antitumor activity of MG98 in patients with metastatic renal carcinoma, a solid tumor that has been shown to have hypermethylation of promoter regions of tumor-suppressor genes. The study was stopped after the first stage because neither the response nor the progression-free criteria for continuing to the second stage was met. Despite the negative results, the investigators believe that the rationale for further study of agents targeting DNA methylation in cancer should not diminish, and that future studies should attempt to assess target effects at the molecular level in cancers thought to be susceptible to this approach.

B. Histone Deacetylase Inhibitors

Acetylation of DNA-associated histones is linked to activation of gene transcription, whereas histone deacetylation is associated with transcriptional repression. Acute promyelocytic leukemia (APL) provides an excellent model to illustrate the modulation of gene transcription by acetylation and the therapeutic potential of histone deacetylase inhibitors. APL is a hematopoietic cancer that involves the retinoic acid receptor alpha (*RARα*) gene that maps to the long arm of chromosome 17q21. Ninety-five percent of APL cases arise from a translocation between chromosomes 15 and 17, (t15:17.q21) which leads to the formation of the fusion protein PML-RARα. PML-RARα results in a transcriptional block of the normal granulocytic differentiation pathway. RARα is a member of the nuclear hormone receptor family that acts as a ligand-inducible transcriptional activation factor by binding to retinoic acid response elements (RAREs) in a heterodimer with RXR, a related family of nuclear receptors. In the presence of a ligand (*all-trans retinoic acid*), the complex promotes transcription of retinoic acid responsive genes. In the absence of ligand, transcription is silenced by a multistep process involving recruitment of transcriptional regulators, corepressors, and nuclear receptor core repressors such as Sin3 to form a complex. Sin3, in turn, recruits a histone deacetylase that causes condensation of chromatin and prevents accessibility of transcriptional machinery to target genes. The presence of ligand (*all-trans retinoic acid*) induces a conformational change in RAR enabling the dissociation of the repressor complex and recruitment of coactivators (such as the *p160* family members). The coactivator molecules possess intrinsic histone acetylase activity that causes unwinding of DNA thereby facilitating transcription and promoting granulocyte differentiation. In an APL patient with a transcriptional block and refractoriness to *all-trans retinoic acid* resulting in a highly resistant form of APL, Warrell and colleagues showed that treatment with sodium butyrate, a histone deacetylase inhibitor restored sensitivity to the antileukemic effects of *all-trans retinoic acid*.[195]

Evaluation of sodium phenyl butyrate (buphenyl) has demonstrated its beneficial effect in treatment of other disorders, including the hemoglobinopathy β-thallesemia, as well as acute myelogenous leukemia and prostate cancer. Phenyl butyrate is one of the old generation of histone deacetylase inhibitors and presently additional inhibitors are being tested in clinical trials.[199] Among these, suberoylanilide hydroxamic acid has shown differentiating effects in a bladder cancer cell line. Another agent, depsipeptide, isolated from *Chrombacterium violaceum*, has been demonstrated to have potent cytotoxic activity through several different mechanisms, including histone deacetylase inhibition. This agent demonstrated activity against chronic myelogenous leukemia cells resulting in acetylation of histone H3 and H4 as well as expression of apoptotic proteins involving caspase pathways.[199–201]

C. Hypermethylation and Histone Deacetylation

The combined manipulation of histone acetylation and cytosine methylation in chromatin presents another strategy for gene-targeted therapy through epigenetic modification. These two epigenetic processes are linked, as was shown by Nan *et al.*[87] and Jones *et al.*[88] These authors showed that the repressive chromatin structure associated with dense methylation was also associated with histone deacetylation. Methylated DNA binds the transcriptional repressor, MeCP2, at the MBD which recruits the Sin 3A/histone deacetylase complex to form transcriptionally repressive chromatin. This process was reversed by trichostatin A, a specific inhibitor of histone deacetylases.

Since little was known about the importance of methylation relative to histone deacetylation in the inhibition of gene transcription, Cameron *et al.* examined this question.[207] They found that trichostatin alone did not reactivate several hypermethylated genes *MLH1*, *TIMP3*, *CDKN2B* (*INK4B*, *p15*), and *CDKN2A* (*INK4*, *p16*) under conditions that allowed reactivation of nonmethylated genes. These findings suggested that dense CpG island methylation in gene promoter regions was dominant over histone deacetylation in maintaining gene repression. They then induced partial CpG island demethylation by treatment with the demethylating agent, 5-aza-2′-deoxycytidine, in the presence or absence of histone deacetylase inhibition. They observed robust expression (fourfold increase) of the genes tested by combined drug treatment (trichostatin plus 5-aza-2′-deoxycytidine) in an experiment in which low-level reactivation was seen with 5-aza-2′-deoxycytidine treatment alone. These results indicated that histone deacetylation may not be needed to maintain a silenced transcriptional state, but histone deacetylase has a role in silencing when levels of DNA methylation are reduced. Bisulfite sequencing showed that the increase in gene expression brought about by the combination of the two drugs occurred with retention of extensive methylation in the genes tested. They also found that inhibition of deacetylase activity can induce gene expression without a large-scale change from repressive to accessible chromatin in agreement with the work of others. Taken together, the data suggested that decreased methylation is a prerequisite for transcription following histone deacetylase inhibition.

In experiments similar to those of Cameron *et al.*, Chiurazzi and colleagues examined the relative roles of methylation and histone deacetylation in silencing the *FMR1* gene in fragile-X syndrome.[208,209] Hypermethylation of CGG repeats in this disorder silences the *FMR1* gene to cause the absence of the FMR1 protein that subsequently leads to mental retardation. In their first paper, Chiurazzi *et al.* found that the demethylating agent 5-aza-2′-deoxycytidine partially restored FMR1 protein expression in B-lymphblastoid cell lines obtained from fragile-X patients confirming the role of *FMR1* promoter hypermethylation in the pathogenesis of fragile-X syndrome.[208] In their second paper, they found that combining 5-aza-2′-deoxycytidine with histone deacetylase inhibitors such as 4-phenylbutyrate, sodium butyrate, or trichostatin resulted in a 2–5-fold increase in FMR1 mRNA levels over that obtained with 5-aza-2′-deoxycytidine alone. The marked synergistic effect observed revealed that both histone hyperacetylation and DNA demethylation participate in regulating *FMR1* activity. These results may help pave the way for future attempts at pharmacologically restoring mutant *FMR1* activity *in vivo*.[209]

Methylation and histone deacetylation thus appear to act as layers for epigenetic silencing. Cameron *et al.* believe that one function of DNA methylation may be to firmly "lock" genes into a silenced chromatin state.[207] They suggested that this effect may be involved in transcriptional repression of methylated inactive X chromosomal genes and imprinted alleles. They proposed that to achieve maximal gene reactivation, it might be necessary to block simultaneously both DNA methylation and histone deacetylation, both of which are essential to the formation and maintenance of repressive chromatin.

The contributions of epigenetics to human disease and how to optimize its management are in their infancy. As we learn more about the proteins targeted by therapeutic agents, and the molecular interactions perturbed, rationally designed drugs and individualized therapy may be reasonable goals.[198,210] Various HDAC inhibitors seem to enhance the tumor response to ionizing radiation and thereby may protect normal tissues from radiation damage, and combinations of demethylating agents with HDAC inhibitors are also being studied with great interest.[211]

VI. Prospects for the Future of Cancer Epigenetics

After a slow start, epigenetics emerged from seemingly disparate observations in developmental and chromosomal biology to become a stand-alone discipline complementary to genetics. While discoveries of the hereditary nature and double helix of DNA define the molecular basis of modern heredity, epigenetic research demonstrated that all cellular differentiation from early human development to adulthood is guided and maintained by epigenetic mechanisms that allow stable propagation of gene expression from one generation to the next. The Encyclopedia of DNA Elements (ENCODE) project (2003–2011) and the US-NIH Epigenomics Roadmap Program (2008–2013) continue to define the constituents of the human genome.[212] These projects identify the genes (protein-coding and noncoding) and the patterns of DNA methylation, chromatin modification, genomic imprinting, and RNA modulation that determine whether genes are switched on or off in a given tissue. During the last two or three decades, epigenetic researchers have demonstrated that these mechanisms and interactions between them are grossly perturbed, resulting in inappropriate gene expression, inefficient DNA repair, and aberrant DNA replication and cell division, leading to the development of cancer and other human disorders. More recently, genome-wide association studies have revealed an astounding number of common DNA variations and are now revealing a multitude of epigenetic variations that cause cancer.

In normal cellular processes, about 95% of DNA methylation patterns are maintained through cell divisions to regulate the expression of genes that characterize differentiated cells. Exogenous inserted sequences such as transposons, parasitic, and viral elements are usually silenced by hypermethylation. Epigenetic information is stored in the amino acid residues in tails of core histones of chromatin and these residues are altered by various covalent modifications such as acetylation of lysine, methylation of lysine and arginine, phosphorylation of serine, and more. These modifications are reversible and they have special functions in gene translation, DNA replication, and DNA repair. Hypermethylation of CpG islands in gene promoter regions is associated with specific modifications such as dimethylation of histone H3 at lysine 9 (H3K9me2), deacetylation of this residue, trimethylation of H3 of lysine 27 (H3K27me3), and loss of the transcriptional activating mark, H3K4me2. Synergies that occur between DNA methylation and histone modifications in gene promoter regions suggest that DNA methylation is part of the normal epigenetic program that leads to transcriptional gene silencing. Profiling of different cell types and tissues indicates that the pattern of expression of miRNAs is cell type- and tissue-specific suggesting that every cellular process is likely to be regulated by miRNAs.

In contrast to epigenetic patterns in normal cellular processes, the patterns in cancerous cells are severely disorganized and disrupted. The resultant effects may be summarized as follows: (1) Tumor-suppressor genes often undergo silencing by aberrant CpG island hypermethylation that leads to cancer-related events. At the same time, many other regulatory genes of different cellular processes and pathways in multiple cancer types are also disrupted and undergo silencing through aberrant hypermethylation. (2) CpG island hypermethylation frequently occurs in conjunction with global hypomethylation, and in contrast to hypermethylation, hypomethylation is usually accompanied by reactivation of oncogenes and of exogenous inserted sequences. Failure to silence transposons and other exogenous inserted sequences can also lead to cancer-related events. (3) LOI also favors the development of cancer. (4) MiRNA expression can be disrupted by several mechanisms in human cancer: chromosomal abnormalities, mutations, defects in miRNA biogenesis machinery, and epigenetic changes such as altered DNA methylation miRNA expression profiling provide evidence for the association of miRNAs with the development and progression of cancer. An increasing number of studies indicate that miRNAs can function as oncogenes or as tumor-suppressor genes, depending on the cellular context and on the target genes they regulate.

Epigenomic research combined with genomic research has led to a fuller appreciation of the hereditary and environmental causes of cancer. Much of our knowledge of epigenetics stems from cancer-related studies of epigenetic phenomena, and today rapid progress is being made in quantifying, mapping, and characterizing these phenomena. Development of methods to identify variations in epigenetic phenomena associated with cell- and tissue-specific tumors must continue so that we may improve our understanding of their contribution to the development and progression of cancer. DNA microarray-based techniques including ChIP-chip have provided much valuable information, and newer, high-throughput protocols show potential to reveal additional features of the epigenome, particularly of the human epigenome. It is expected that ChIP-Seq will find broad applications in genome-wide mapping of DNA methylation, histone modifications, nucleosome positioning, the dynamics of long-range chromatin interactions, and other epigenetic processes and will reveal additional contributions to development and pathological conditions. Although many of the basic principles and complexities of epigenetic phenomena have been identified, the molecular mechanisms by which they are established and maintained are not clear so it is important to continue to develop tools and techniques to advance understanding of epigenome function and gene expression. Certain histone modifiers are proving to be attractive molecular targets for therapeutic intervention as highlighted by the number of drugs now in clinical trial that target histone methylation and acetylation (e.g., histone methylase and HDAC inhibitors) and by the fact that some of these inhibitors have received approval by the United States Food and Drug Administration (FDA).

References

1. Gilbert SF. *J Biosci* 2009;**34**:601–4.
2. Futreal PA, Kasprzyk A, Birney E, Mullikin JC, Wooster R, Stratton MR. *Nature* 2001;**409**:850–2.
3. Klein G. *Nature* 2005;**434**:150.
4. Holliday R. *Epigenetics* 2006;**1**:76–80.
5. Feinberg AP. *JAMA* 2008;**299**:1345–50.
6. Esteller M. *N Engl J Med* 2008;**358**:1148–59.
7. Jones PA, Baylin SB. *Cell* 2007;**128**:683–92.
8. Dolinoy DC. (McQueen CA, ed.), p. xxx, Elsevier Ltd, Amsterdam (2010), in press.
9. Brown SW. *Science* 1966;**151**:417–25.
10. Rubin GM, Lewis EB. *Science* 2001;**287**:2216–8.
11. Leder P. *Science* 2010;**327**:972.
12. Southern EM. *J Mol Biol* 1975;**98**:503–17.
13. Weber WW. *Pharmacogenetics.* New York: Oxford University Press; 2008.
14. Silverman PH. *Scientist* 2004;**18**:32–3.
15. van Steensel B, Henikoff S. *BioTechniques* 2003;**35**(346–4):356.
16. van Steensel B. *Nat Genet* 2005;**37**:S18–24.
17. Havlis J, Trbusek M. J Chromatogr B Analyt Technol Biomed. *Life Sci* 2002;**781**:373–92.
18. Clark SJ, Harrison J, Paul CL, Frommer M. *Nucleic Acids Res* 1994;**22**:2990–7.
19. Gowher H, Leismann O, Jeltsch A. *EMBO J* 2000;**19**:6918–23.
20. Havlis J, Madden JE, Revilla AL, Havel J. *J Chromatogr B Biomed Sci Appl* 2001;**755**:185–94.
21. Larsen LA, Christiansen M, Vuust J, Andersen PS. *Comb Chem High Throughput Screen* 2000;**3**:393–409.
22. Pfarr W, Webersinke G, Paar C, Wechselberger C. *BioTechniques* 2005;**38**:527–30.
23. Rousseau F, Heitz D, Biancalana V, Blumenfeld S, Kretz C, Boue J, et al. *N Engl J Med* 1991;**325**:1673–81.
24. Knox MR, Ellis TH. *Mol Genet Genomics* 2001;**265**:497–507.
25. Frommer M, McDonald LE, Millar DS, Collis CM, Watt F, Grigg GW, et al. *Proc Natl Acad Sci USA* 1992;**89**:1827–31.
26. Thomassin H, Oakeley EJ, Grange T. *Methods* 1999;**19**:465–75.
27. Oakeley EJ, Schmitt F, Jost JP. *BioTechniques* 1999;**27**(744–50):752.
28. Xiong Z, Laird PW. *Nucleic Acids Res* 1997;**25**:2532–4.
29. Gonzalgo ML, Jones PA. *Nucleic Acids Res* 1997;**25**:2529–31.
30. Schones DE, Zhao K. *Nat Rev Genet* 2008;**9**:179–91.
31. Bua DJ, Kuo AJ, Cheung P, Liu CL, Migliori V, Espejo A, et al. *PLoS One* 2009;**4**:e6789.
32. Shi X, Kachirskaia I, Walter KL, Kuo JH, Lake A, Davrazou F, et al. *J Biol Chem* 2007;**282**:2450–5.
33. Lister R, Pelizzola M, Dowen RH, Hawkins RD, Hon G, Tonti-Filippini J, et al. *Nature* 2009;**462**:315–22.
34. Liu CG, Calin GA, Meloon B, Gamliel N, Sevignani C, Ferracin M, et al. *Proc Natl Acad Sci USA* 2004;**101**:9740–4.

35. Calin GA, Liu CG, Sevignani C, Ferracin M, Felli N, Dumitru CD, et al. *Proc Natl Acad Sci USA* 2004;**101**:11755–60.
36. Calin GA, Dumitru CD, Shimizu M, Bichi R, Zupo S, Noch E, et al. *Proc Natl Acad Sci USA* 2002;**99**:15524–9.
37. Wang Z, Gerstein M, Snyder M. *Nat Rev Genet* 2009;**10**:57–63.
38. Pickrell JK, Marioni JC, Pai AA, Degner JF, Engelhardt BE, Nkadori E, et al. *Nature* 2010;**464**:768–72.
39. Montgomery SB, Sammeth M, Gutierrez-Arcelus M, Lach RP, Ingle C, Nisbett J, et al. *Nature* 2010;**464**:773–7.
40. Brena RM, Huang TH, Plass C. *Nat Genet* 2006;**38**:1359–60.
41. Fraga MF, Ballestar E, Paz MF, Ropero S, Setien F, Ballestar ML, et al. *Proc Natl Acad Sci USA* 2005;**102**:10604–9.
42. Javierre BM, Fernandez AF, Richter J, Al Shahrour F, Martin-Subero JI, Rodriguez-Ubreva J, et al. *Genome Res* 2010;**20**:170–9.
43. Hanson MA, Gluckman PD. *Basic Clin Pharmacol Toxicol* 2008;**102**:90–3.
44. Fan S, Zhang X. *Biochem Biophys Res Commun* 2009;**383**:421–5.
45. Murrell A, Rakyan VK, Beck S. *Hum Mol Genet* 2005;**14 Spec No 1**:R3–R10.
46. Dolinoy DC, Huang D, Jirtle RL. *Proc Natl Acad Sci USA* 2007;**104**:13056–61.
47. Liu Y, Lan Q, Siegfried JM, Luketich JD, Keohavong P. *Neoplasia* 2006;**8**:46–51.
48. Russo AL, Thiagalingam A, Pan H, Califano J, Cheng KH, Ponte JF, et al. *Clin Cancer Res* 2005;**11**:2466–70.
49. Kim DH, Nelson HH, Wiencke JK, Zheng S, Christiani DC, Wain JC, et al. *Cancer Res* 2001;**61**:3419–24.
50. Tessema M, Willink R, Do K, Yu YY, Yu W, Machida EO, et al. *Cancer Res* 2008;**68**:1707–14.
51. Heijmans BT, Tobi EW, Stein AD, Putter H, Blauw GJ, Susser ES, et al. *Proc Natl Acad Sci USA* 2008;**105**:17046–9.
52. Eckhardt F, Lewin J, Cortese R, Rakyan VK, Attwood J, Burger M, et al. *Nat Genet* 2006;**38**:1378–85.
53. Jaenisch R, Bird A. *Nat Genet* 2003;**33**(Suppl):245–54.
54. Scarano E. *Adv Cytopharmacol* 1971;**1**:13–24.
55. Riggs AD. *Cytogenet Cell Genet* 1975;**14**:9–25.
56. Holliday R, Pugh JE. *Science* 1975;**187**:226–32.
57. Razin A, Riggs AD. *Science* 1980;**210**:604–10.
58. Razin A, Kantor B. *Prog Mol Subcell Biol* 2005;**38**:151–67.
59. Josse J, Kaiser AD, Kornberg A. *J Biol Chem* 1961;**236**:864–75.
60. Grippo P, Iaccarino M, Parisi E, Scarano E. *J Mol Biol* 1968;**36**:195–208.
61. Hotchkiss RD. *J Biol Chem* 1948;**175**:315–32.
62. Salser, W. CSHL XLVII, 1977, 985–1003.
63. Bird AP. *Nucleic Acids Res* 1980;**8**:1499–504.
64. Waalwijk C, Flavell RA. *Nucleic Acids Res* 1978;**5**:4631–4.
65. Gardiner-Garden M, Frommer M. *J Mol Biol* 1987;**196**:261–82.
66. Bird AP. *Nature* 1986;**321**:209–13.
67. Bird AP, Taggart MH. *Nucleic Acids Res* 1980;**8**:1485–97.
68. Bird AP. *Cold Spring Harb Symp Quant Biol* 1993;**58**:281–5.
69. Yoder JA, Walsh CP, Bestor TH. *Trends Genet* 1997;**13**:335–40.
70. Cross SH, Bird AP. *Curr Opin Genet Dev* 1995;**5**:309–14.
71. Dutnall RN, Denu JM. *Nat Struct Biol* 2002;**9**:888–91.
72. Clark SJ. *Hum MolGenet* 2007;**16 Spec No 1**:R88–95.
73. Bestor T, Laudano A, Mattaliano R, Ingram V. *J Mol Biol* 1988;**203**:971–83.
74. Bestor TH. *Hum Mol Genet* 2000;**9**:2395–402.

75. Yoder JA, Bestor TH. *Hum Mol Genet* 1998;**7**:279–84.
76. Okano M, Xie S, Li E. *Nat Genet* 1998;**19**:219–20.
77. Van den Wyngaert I, Sprengel J, Kass SU, Luyten WH. *FEBS Lett* 1998;**426**:283–9.
78. Hermann A, Schmitt S, Jeltsch A. *J Biol Chem* 2003;**278**:31717–21.
79. Okano M, Bell DW, Haber DA, Li E. *Cell* 1999;**99**:247–57.
80. Deplus R, Brenner C, Burgers WA, Putmans P, Kouzarides T, de Launoit Y, et al. *Nucleic Acids Res* 2002;**30**:3831–8.
81. Boyes J, Bird A. *Cell* 1991;**64**:1123–34.
82. Boyes J, Bird A. *EMBO J* 1992;**11**:327–33.
83. Hendrich B, Bird A. *Mol Cell Biol* 1998;**18**:6538–47.
84. Meehan RR, Lewis JD, Bird AP. *Nucleic Acids Res* 1992;**20**:5085–92.
85. Cross SH, Meehan RR, Nan X, Bird A. *Nat Genet* 1997;**16**:256–9.
86. Clouaire T, Stancheva I. *Cell Mol Life Sci* 2008;**65**:1509–22.
87. Nan X, Ng HH, Johnson CA, Laherty CD, Turner BM, Eisenman RN, et al. *Nature* 1998;**393**:386–9.
88. Jones PL, Veenstra GJ, Wade PA, Vermaak D, Kass SU, Landsberger N, et al. *Nat Genet* 1998;**19**:187–91.
89. Nan X, Campoy FJ, Bird A. *Cell* 1997;**88**:471–81.
90. Bestor TH. *Nature* 1998;**393**:311–2.
91. Ng HH, Zhang Y, Hendrich B, Johnson CA, Turner BM, Erdjument-Bromage H, et al. *Nat Genet* 1999;**23**:58–61.
92. Tamaru H, Selker EU. *Nature* 2001;**414**:277–83.
93. Nakayama J, Rice JC, Strahl BD, Allis CD, Grewal SI. *Science* 2001;**292**:110–3.
94. Kornberg RD, Lorch Y. *Cell* 1999;**98**:285–94.
95. Luger K, Mader AW, Richmond RK, Sargent DF, Richmond TJ. *Nature* 1997;**389**:251–60.
96. Kimmins S, Sassone-Corsi P. *Nature* 2005;**434**:583–9.
97. Luo RX, Dean DC. *J Natl Cancer Inst* 1999;**91**:1288–94.
98. Saha A, Wittmeyer J, Cairns BR. *Nat Rev Mol Cell Biol* 2006;**7**:437–47.
99. Cheung P, Allis CD, Sassone-Corsi P. *Cell* 2000;**103**:263–71.
100. Horn PJ, Peterson CL. *Science* 2002;**297**:1824–7.
101. Turner BM. *Cell* 1993;**75**:5–8.
102. Kouzarides T. *Cell* 2007;**128**:693–705.
103. Groth A, Rocha W, Verreault A, Almouzni G. *Cell* 2007;**128**:721–33.
104. Vaissiere T, Herceg Z. *Cell Res* 2010;**20**:113–5.
105. Strahl BD, Allis CD. *Nature* 2000;**403**:41–5.
106. Spannhoff A, Hauser AT, Heinke R, Sippl W, Jung M. *ChemMedChem* 2009;**4**:1568–82.
107. Allfrey VG, Faulkner R, Mirsky AE. *Proc Natl Acad Sci USA* 1964;**51**:786–94.
108. Kuo MH, Zhou J, Jambeck P, Churchill ME, Allis CD. *Genes Dev* 1998;**12**:627–39.
109. Miremadi A, Oestergaard MZ, Pharoah PD, Caldas C. *Hum Mol Genet* 2007;**16 Spec No 1**: R28–49.
110. Ellis L, Atadja PW, Johnstone RW. *Mol Cancer Ther* 2009;**8**:1409–20.
111. Sengupta N, Seto E. *J Cell Biochem* 2004;**93**:57–67.
112. Santos-Rosa H, Schneider R, Bannister AJ, Sherriff J, Bernstein BE, Emre NC, et al. *Nature* 2002;**419**:407–11.
113. Jacinto FV, Esteller M. *Mutagenesis* 2007;**22**:247–53.
114. Paik WK, Kim S. *Biochem Biophys Res Commun* 1973;**51**:781–8.
115. Wang Z, Zang C, Rosenfeld JA, Schones DE, Barski A, Cuddapah S, et al. *Nat Genet* 2008;**40**:897–903.
116. Ho L, Crabtree GR. *Nature* 2010;**463**:474–84.

117. Loizou JI, Murr R, Finkbeiner MG, Sawan C, Wang ZQ, Herceg Z. *Cell Cycle* 2006;**5**:696–701.
118. Ambros V, Bartel B, Bartel DP, Burge CB, Carrington JC, Chen X, et al. *RNA* 2003;**9**:277–9.
119. Lee RC, Feinbaum RL, Ambros V. *Cell* 1993;**75**:843–54.
120. Ambros V. *Nature* 2004;**431**:350–5.
121. Fire A, Xu S, Montgomery MK, Kostas SA, Driver SE, Mello CC. *Nature* 1998;**391**:806–11.
122. Hannon GJ. *Nature* 2002;**418**:244–51.
123. Bartel DP. *Cell* 2004;**116**:281–97.
124. Bartel DP, Chen CZ. *Nat Rev Genet* 2004;**5**:396–400.
125. Iorio MV, Croce CM. *J Clin Oncol* 2009;**27**:5848–56.
126. Guil S, Esteller M. *Int J Biochem Cell Biol* 2009;**41**:87–95.
127. Paddison PJ, Caudy AA, Hannon GJ. *Proc Natl Acad Sci USA* 2002;**99**:1443–8.
128. Das R, Hampton DD, Jirtle RL. *Mamm Genome* 2009;**20**:563–72.
129. Barr ML, Bertram EG. *Nature* 1949;**163**:676–7.
130. Ohno S, Kaplan WD, Kinosita R. *Exp Cell Res* 1959;**18**:415–8.
131. Lyon MF. *Cytogenet Cell Genet* 1998;**80**:133–7.
132. Beutler E, YEH M, Fairbanks VF. *Proc Natl Acad Sci USA* 1962;**48**:9–16.
133. Crouse H. *Genetics* 1960;**45**:1429–43.
134. Barton SC, Surani MA, Norris ML. *Nature* 1984;**311**:374–6.
135. McGrath J, Solter D. *Cell* 1984;**37**:179–83.
136. Stoger R, Kubicka P, Liu CG, Kafri T, Razin A, Cedar H, et al. *Cell* 1993;**73**:61–71.
137. Reik W, Collick A, Norris ML, Barton SC, Surani MA. *Nature* 1987;**328**:248–51.
138. Sapienza C, Peterson AC, Rossant J, Balling R. *Nature* 1987;**328**:251–4.
139. Monk M. *Nature* 1987;**328**:203–4.
140. Barlow DP. *Trends Genet* 1994;**10**:194–9.
141. Whitelaw E, Garrick D. In: Ruvinsky A, Graves J, editors. *Mammalian Genomics*. Cambridge, MA: CABI; 2005. p. 179–200.
142. Reik W, Walter J. *Curr Opin Genet Dev* 1998;**8**:154–64.
143. Luedi PP, Hartemink AJ, Jirtle RL. *Genome Res* 2005;**15**:875–84.
144. Luedi PP, Dietrich FS, Weidman JR, Bosko JM, Jirtle RL, Hartemink AJ. *Genome Res* 2007;**17**:1723–30.
145. Esteller M. *Nat Rev Genet* 2007;**8**:286–98.
146. Cheung HH, Lee TL, Rennert OM, Chan WY. *Birth Defects Res C Embryo Today* 2009;**87**:335–50.
147. McCabe MT, Brandes JC, Vertino PM. *Clin Cancer Res* 2009;**15**:3927–37.
148. Feinberg AP, Vogelstein B. *Nature* 1983;**301**:89–92.
149. Feinberg AP, Gehrke CW, Kuo KC, Ehrlich M. *Cancer Res* 1988;**48**:1159–61.
150. Baylin SB, Hoppener JW, de Bustros A, Steenbergh PH, Lips CJ, Nelkin BD. *Cancer Res* 1986;**46**:2917–22.
151. Esteller M, Corn PG, Baylin SB, Herman JG. *Cancer Res* 2001;**61**:3225–9.
152. Baylin SB, Herman JG. *Trends Genet* 2000;**16**:168–74.
153. Jones PA, Laird PW. *Nat Genet* 1999;**21**:163–7.
154. Sakai T, Toguchida J, Ohtani N, Yandell DW, Rapaport JM, Dryja TP. *Am J Hum Genet* 1991;**48**:880–8.
155. Issa JP, Ottaviano YL, Celano P, Hamilton SR, Davidson NE, Baylin SB. *Nat Genet* 1994;**7**:536–40.
156. Lin X, Tascilar M, Lee WH, Vles WJ, Lee BH, Veeraswamy R, et al. *Am J Pathol* 2001;**159**:1815–26.
157. Braggio E, Maiolino A, Gouveia ME, Magalhaes R, Souto Filho JT, Garnica M, et al. *Int J Hematol* 2010;**91**:87–96.

158. Laird PW, Jaenisch R. *Hum Mol Genet* 1994;**3 Spec No**:1487–95.
159. Jones PA, Baylin SB. *Nat Rev Genet* 2002;**3**:415–28.
160. Irizarry RA, Ladd-Acosta C, Wen B, Wu Z, Montano C, Onyango P, et al. *Nat Genet* 2009;**41**:178–86.
161. Feinberg AP, Ohlsson R, Henikoff S. *Nat Rev Genet* 2006;**7**:21–33.
162. Toyota M, Ahuja N, Ohe-Toyota M, Herman JG, Baylin SB, Issa JP. *Proc Natl Acad Sci USA* 1999;**96**:8681–6.
163. Cui H, Cruz-Correa M, Giardiello FM, Hutcheon DF, Kafonek DR, Brandenburg S, et al. *Science* 2003;**299**:1753–5.
164. Sakatani T, Kaneda A, Iacobuzio-Donahue CA, Carter MG, Witzel SD, Okano H, et al. *Science* 2005;**307**:1976–8.
165. Wilson AS, Power BE, Molloy PL. *Biochim Biophys Acta* 2007;**1775**:138–62.
166. Dean W, Lucifero D, Santos F. *Birth Defects Res C Embryo Today* 2005;**75**:98–111.
167. Fraga MF, Ballestar E, Villar-Garea A, Boix-Chornet M, Espada J, Schotta G, et al. *Nat Genet* 2005;**37**:391–400.
168. Jenuwein T. *FEBS J* 2006;**273**:3121–35.
169. Turner JD, Williamson R, Almefty KK, Nakaji P, Porter R, Tse V, et al. *Neurosurg Focus* 2010;**28**:E3.
170. Lujambio A, Esteller M. *Cell Cycle* 2009;**8**:377–82.
171. Calin GA, Ferracin M, Cimmino A, Di Leva G, Shimizu M, Wojcik SE, et al. *N Engl J Med* 2005;**353**:1793–801.
172. Iorio MV, Visone R, Di Leva G, Donati V, Petrocca F, Casalini P, et al. *Cancer Res* 2007;**67**:8699–707.
173. Ohama K, Nomura K, Okamoto E, Fukuda Y, Ihara T, Fujiwara A. *Am J Obstet Gynecol* 1985;**152**:896–900.
174. Linder D, McCaw BK, Hecht F. *N Engl J Med* 1975;**292**:63–6.
175. Kajii T, Ohama K. *Nature* 1977;**268**:633–4.
176. Ohama K, Kajii T, Okamoto E, Fukuda Y, Imaizumi K, Tsukahara M, et al. *Nature* 1981;**292**:551–2.
177. Pal N, Wadey RB, Buckle B, Yeomans E, Pritchard J, Cowell JK. *Oncogene* 1990;**5**:1665–8.
178. Schroeder WT, Chao LY, Dao DD, Strong LC, Pathak S, Riccardi V, et al. *Am J Hum Genet* 1987;**40**:413–20.
179. Scrable H, Cavenee W, Ghavimi F, Lovell M, Morgan K, Sapienza C. *Proc Natl Acad Sci USA* 1989;**86**:7480–4.
180. Moulton T, Chung WY, Yuan L, Hensle T, Waber P, Nisen P, et al. *Med Pediatr Oncol* 1996;**27**:476–83.
181. Reik W, Maher ER. *Trends Genet* 1997;**13**:330–4.
182. Mannens M, Hoovers JM, Redeker E, Verjaal M, Feinberg AP, Little P, et al. *Eur J Hum Genet* 1994;**2**:3–23.
183. Caron H, van Sluis P, van Hoeve M, de Kraker J, Bras J, Slater R, et al. *Nat Genet* 1993;**4**:187–90.
184. Katz F, Webb D, Gibbons B, Reeves B, McMahon C, Chessells J, et al. *Br J Haematol* 1992;**80**:332–6.
185. Toguchida J, Ishizaki K, Sasaki MS, Nakamura Y, Ikenaga M, Kato M, et al. *Nature* 1989;**338**:156–8.
186. Falls JG, Pulford DJ, Wylie AA, Jirtle RL. *Am J Pathol* 1999;**154**:635–47.
187. Viljoen D, Ramesar R. *J Med Genet* 1992;**29**:221–5.
188. Plumb JA, Strathdee G, Sludden J, Kaye SB, Brown R. *Cancer Res* 2000;**60**:6039–44.
189. Ricciardiello L, Goel A, Mantovani V, Fiorini T, Fossi S, Chang DK, et al. *Cancer Res* 2003;**63**:787–92.

190. Lee WH, Morton RA, Epstein JI, Brooks JD, Campbell PA, Bova GS, et al. *Proc Natl Acad Sci USA* 1994;**91**:11733–7.
191. Esteller M, Hamilton SR, Burger PC, Baylin SB, Herman JG. *Cancer Res* 1999;**59**:793–7.
192. Esteller M, Toyota M, Sanchez-Cespedes M, Capella G, Peinado MA, Watkins DN, et al. *Cancer Res* 2000;**60**:2368–71.
193. Bastian PJ, Yegnasubramanian S, Palapattu GS, Rogers CG, Lin X, De Marzo AM, et al. *Eur Urol* 2004;**46**:698–708.
194. Lynch HT, de la Chapelle A. *N Engl J Med* 2003;**348**:919–32.
195. Warrell Jr. RP, He LZ, Richon V, Calleja E, Pandolfi PP. *J Natl Cancer Inst* 1998;**90**:1621–5.
196. Stirewalt DL, Radich JP. *Hematol* 2000;**5**:15–25.
197. Chim CS, Tam CY, Liang R, Kwong YL. *Cancer* 2001;**91**:2222–9.
198. Egger G, Liang G, Aparicio A, Jones PA. *Nature* 2004;**429**:457–63.
199. Gilbert J, Gore SD, Herman JG, Carducci MA. *Clin Cancer Res* 2004;**10**:4589–96.
200. Karberg S. *Cell* 2009;**139**:1029–31.
201. Friedrich MJ. *JAMA* 2010;**303**:213–4.
202. Spannhoff A, Sippl W, Jung M. *Int J Biochem Cell Biol* 2009;**41**:4–11.
203. Ley TJ, DeSimone J, Anagnou NP, Keller GH, Humphries RK, Turner PH, et al. *N Engl J Med* 1982;**307**:1469–75.
204. Koshy M, Dorn L, Bressler L, Molokie R, Lavelle D, Talischy N, et al. *Blood* 2000;**96**:2379–84.
205. Uchida T, Kinoshita T, Nagai H, Nakahara Y, Saito H, Hotta T, et al. *Blood* 1997;**90**:1403–9.
206. Silverman LR, Demakos EP, Peterson BL, Kornblith AB, Holland JC, Odchimar-Reissig R, et al. *J Clin Oncol* 2002;**20**:2429–40.
207. Cameron EE, Bachman KE, Myohanen S, Herman JG, Baylin SB. *Nat Genet* 1999;**21**:103–7.
208. Chiurazzi P, Pomponi MG, Willemsen R, Oostra BA, Neri G. *Hum Mol Genet* 1998;**7**:109–13.
209. Chiurazzi P, Pomponi MG, Pietrobono R, Bakker CE, Neri G, Oostra BA. *Hum Mol Genet* 1999;**8**:2317–23.
210. Issa JP, Garcia-Manero G, Giles FJ, Mannari R, Thomas D, Faderl S, et al. *Blood* 2004;**103**:1635–40.
211. Conley BA, Wright JJ, Kummar S. *Cancer* 2006;**107**:832–40.
212. Collins F. *Nature* 2010;**464**:674–5.

Molecular Targets and Clinical Cancer Risk Reductive Interventions

Madhuri Kakarala and
Dean E. Brenner

University of Michigan Medical Center and VA Medical Center, Ann Arbor, Michigan, USA

I. Defining Cancer Risk Reductive Intervention (Chemoprevention)

The term "chemoprevention" was first used by Sporn *et al.* to define the use of interventions ranging from diet modification, purified dietary extracts, to drugs for the purpose of blocking, reversing, or preventing the development of invasive cancer.[1–3] The term "chemoprevention," however, is imprecise as it does not define what disease we are preventing, and it alludes to the use of only chemicals rather than the wide variety of interventions under investigation such

Progress in Molecular Biology
and Translational Science, Vol. 95
DOI: 10.1016/S1877-1173(10)95011-4

1877-1173/10 $35.00

as diets, smoking cessation, exercise interventions. Moreover, the term carries negative connotations with the public, conjuring images of sick cancer patients suffering the toxicities of myelosuppression, fatigue, and alopecia so commonly associated with cytotoxic anticancer treatments called "chemotherapy." We prefer the term "cancer risk reductive intervention" (CRRI) over the term "chemoprevention" in order to precisely identify the broad range of strategies to reverse or delay carcinogenesis, including whole diet interventions, dietary extracts, drugs, behavior modification, smoking cessation, and others.

The carcinogenesis process occurs over decades and provides opportunities to intervene at many steps in transformation and cancer initiation. Since the carcinogenesis process requires a cascade of molecular changes over time, the interventions designed to reverse or delay these changes need to be diverse and target multiple mechanisms of carcinogenesis. In the development and clinical deployment of CRRIs, one must be mindful of therapeutic index, the toxicity, or side effects of a given intervention compared to the benefit derived from that intervention. CRRIs are intended for use in healthy individuals who have increased risk of developing cancer due to genetic predisposition or environmental exposures or both.[4,5] Successful deployment of CRRIs will reduce cancer-associated mortality or delay the age of mortality.

II. Cellular Transformational Molecular Events as Targets for CRRIs

The molecular events that define carcinogenesis, as outlined by Hanahan and Weinberg, can be categorized into six essential alterations in cell physiology: self-sufficiency in growth signals, insensitivity to growth-inhibitory signals, evasion of apoptosis, limitless replication potential, sustained angiogenesis, and tissue invasion and metastasis.[6] Accumulation of mutations or epigenetic alterations in key regulatory signaling pathways comprises the latest targets for CRRI development. Many such signaling intermediates have a common function in multiple organ sites (Fig. 1). The complexity and overlap of signal transduction pathways suggest that single targets may not be sufficient to optimally reduce the risk of cancer. Interventions to multiple pathways or targets may be required to arrest or reverse cellular carcinogenesis. Prominent interventions and their targets are listed in Table I.

III. Inherited Genetic Mutations (Cancer Susceptibility Syndromes)

High penetrance cancer susceptibility syndromes, while rare, provide important models for identifying key carcinogenesis-related molecular pathways and events. Perhaps more importantly, somatic mutations in known cancer

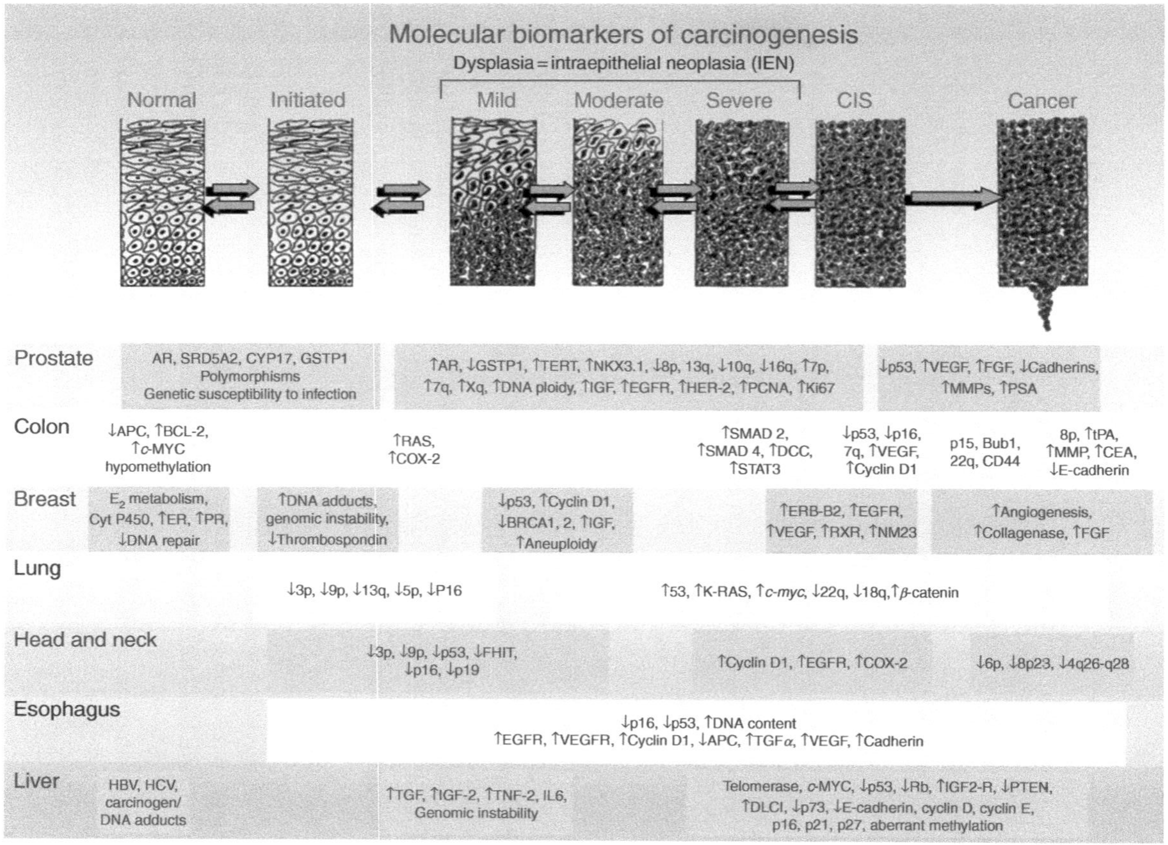

FIG. 1. *Genetic progression in major cancers*. Carcinogenesis is driven by genetic progression. This progression is marked by the appearance of molecular biomarkers in distinctive patterns representing accumulating changes in gene expression and correlating with changes in histologic phenotype as cells move from normal through the early stages of clonal expansion to dysplasia and finally to early invasive, locally advanced, and metastatic cancer. The figure shows candidate molecular biomarkers of genetic progression in seven target organs: prostate,[7–9] colon,[10,11] breast,[12,13] lung,[14–16] head and neck,[17–20] esophagus,[11,21] and liver.[22] (Figure and revised caption from Kelloff *et al*.[23]; published with permission from the American Association for Cancer Research.)

TABLE I

MOLECULAR MECHANISMS COMMON TO TRANSFORMING CELLS AND POTENTIAL PREVENTIVE INTERVENTIONS

Characteristics of neoplasia	Possible molecular targets
• Self-sufficiency in cell growth	Epidermal growth factor receptor, platelet-derived growth factor, MAP-kinase, PI3K.
• Insensitivity to antigrowth signals	SMADs, pRb, cyclin-dependent kinases, myc.
• Limitless replicative potential	hTERT, pRb, p53.
• Evading apoptosis	Bcl-2, BAX, caspases, Fas, tumor necrosis factor receptor, insulin growth factor/PI3K/AKT, mTOR p53, NF-κB, PTEN, *Ras*.
• Sustained angiogenesis	Vascular endothelial growth factor, basic fibroblast growth factor, integrins ($\alpha_v\beta_3$), thrombospondin-1, hypoxia-inducible factor-1α.
• Tissue invasion and metastases	Matrix metalloproteinases, MAP-kinase, E-cadherin.

PI3K, phosphoinositol-3-kinase; SMAD, drosophila protein, mothers against decapentaplegic gene and the *elegans* protein SMA; pRb, phosphorylate Rb protein; hTERT, human telomerase reverse transcriptase; mTOR, mammalian target of rapamycin; NFκB, nuclear factor kappa B; PTEN, phosphatase and tensin homolog; MAP, mitogen-activated protein. Adapted from Kelloff *et al.* and derived from Hanahan and Weinberg.[6,23]

susceptibility genes in sporadic tumors (e.g., mismatch repair gene mutations or methylation resulting in microsatellite instability in sporadic colon cancers) can have diagnostic and therapeutic implications.[24,25] Detailed knowledge of inherited genetic mutations and dysregulated pathways and their associated signaling intermediates have permitted the identification of targets and biomarker endpoints for efficacy testing for new CRRIs.[24–26] The normal functions and role of key tumor suppressor genes (e.g., APC, BRCA1 or 2, NF1, p53) in carcinogenesis have been elucidated through the more prevalent inherited genetic syndromes (hereditary nonpolyposis colon cancer, hereditary breast-ovarian cancer syndrome, neurofibromatosis type I, and familial adenomatous polyposis).[24–26]

As information regarding the functional consequences of these genetic syndromes has become available, new therapeutic and cancer risk reductive strategies have emerged. For example, BRCA1 and BRCA2 are each required for fundamental DNA repair by homologous recombination.[27] BRCA1- or BRCA2-mutated tumors, having lost one DNA repair pathway, are more reliant on other essential DNA repair pathways, especially base excision repair (BER).[28] In contrast, normal cells, even cells heterozygous for BRCA mutation, with intact homologous recombination, can tolerate the loss of BER.[28]

BER requires the activity of the enzyme poly ADPribose polymerase (PARP) 1.[29,30] Preclinical data showed that inhibition of PARP1 activity in a BRCA1 or BRCA2 homozygously deficient cell leads to cell death, but

BRCA1/2 wild-type or heterozygous cells are unaffected (synthetic lethality).[28] Among 60 patients with diverse advanced malignancies (nine breast cancers), 22 were BRCA mutation carriers and one individual with a family history consistent with BRCA mutation, who declined genetic testing. Marked clinical responses were seen, but exclusively in BRCA mutation carriers.[31,32]

IV. Special Features of CRRI Development

Cancer risk reduction poses unique challenges to clinical translation of interventions designed to modulate or inhibit molecular targets as compared to the interventions designed to treat a cancer diagnosis. First, therapeutic index (ratio of treatment benefit threshold to toxicity threshold) for CRRIs is much larger than for cancer treatment indications. CRRIs are intended for use by otherwise healthy, fully functional individuals with a higher cancer risk than the average population. Since such interventions are to be administered to otherwise healthy individuals, toxicities must be mild and not impact daily life activities. Second, the long latency for definition of risk to malignant transformation requires prolonged administration of an intervention with associated observation. Studies designed to define efficacy of a CRRI based upon a reduction of an incidence endpoint require enrollment and study of large numbers of subjects (thousands) over years. Such trials are not financially feasible and have limited progress in translation of molecular targeted therapeutics into human clinical trials.[33,34] Third, the cancer endpoint is a rare event. Despite the common diagnosis and high mortalities associated with cancer, the clinical appearance of a specific organ-based cancer event in a study population is a rare event. For example, in a colonoscopic screening population of otherwise healthy, average risk adults, colon cancer is incidentally found in 4 in 1,000 endoscopic screening procedures (0.4%).[35] Trials of CRRIs for the reversal or delay of colon cancer require approximately 13,000 subjects studied longitudinally over 3 years in order to have sufficient number of cancer events to statistically determine whether an intervention has cancer risk reductive efficacy. Fourth, nonadherence to recommended interventions biases data collected in large population-based cancer risk reductive clinical trials. Healthy individuals who do not consider themselves ill do not adhere to treatments intended to prevent a distant health event unless a clearly measurable biomarker defines benefit to risk reduction of a given disease. For example, hypertension, a surrogate biomarker for future vascular disease, is a definable, proximate biomarker that can be frequently measured and quantified.[36] For otherwise asymptomatic, healthy individuals, adherence to antihypertensive interventions, a vascular preventive intervention, is enhanced as "treatment" for hypertension, a surrogate biomarker.[36] For cancer, reliable, quantitative biomarkers will enhance interest and ultimately adherence to CRRIs. Fifth, risk assessment for common epithelial cancers, remains complex and as yet insufficiently developed to identify large populations outside of those individuals and families with highly penetrant but infrequent, inherited mutations. Currently available risk assessment tools consist of algorithms based upon epidemiologic associations with a given cancer such as an individual's personal and family history of cancers, environmental exposures, and lifestyle variables such as diet, exercise, and smoking, for example, the Gail scale for breast cancer risk.[37] While these are useful tools, they do not provide the simple, quantitative, continuous feedback that hypertension or lipid profiles provide for cardiovascular disease risk. Highly penetrant but infrequent, inherited genetic mutations predict for cancer risk for breast and colon cancers.[38,39] Genetic testing for such cancers (e.g., breast-ovarian cancer syndromes, familial adenomatous polyposis) identifies high-risk subjects who appear to benefit from CRRIs.[40–42]

V. Molecular Intermediates as Biomarkers for Cancer Risk Reductive Efficacy

A biomarker is defined as a characteristic that is measured and evaluated as an indicator of normal biologic processes, pathogenic processes, or pharmacologic responses to therapeutic interventions.[43] Motivations for the development of cancer-related biomarkers include detection of cancer at an early stage, identification of individuals at high risk for cellular transformation, cancer prognosis, and monitoring of efficacy of therapeutic or risk reductive interventions for patients. For CRRIs, biomarkers are crucial to enable reduction in the size and duration of an intervention trial by replacing a rare or distal end point, with a more frequent, proximate end point.[44]

Biomarkers may be derived from biologically derived products, such as a protein, gene, or quantitative cellular process used to predict cancer diagnosis or risk.[45] Hundt *et al.* group molecular markers into DNA, mRNA, and protein.[46] We group DNA and mRNA into a genetic material category.

A. Genetic Material

The direct detection of aberrant genes or genetic material specific to neoplasms, for example, colorectal neoplasm, includes mutated APC, β-catenin, K-ras, DCC, and p53 gene. Discovery has been limited by the technical challenge of DNA recovery, the large number of potential underlying genetic mutations, and by the limited sensitivity of any single genetic alteration due to the extremely low abundance gene mutations in circulating plasma or serum.[47–51] Genetic biomarkers, for example, K-ras mutations alone[52,53] or in combination with p16 methylation,[54] have shown promise for differentiating and risk-stratifying patients with colorectal adenocarcinoma. DNA hypermethylation affects identical residues in regulatory portions of genes and provides major advantages in designing biomarker assays.[50,55–57] New technologies that have improved the bisulfite conversion while minimizing bisulfite-associated DNA fragmentation[58] will permit cost-effective development of DNA hypermethylated gene biomarkers in the near future. Other biomarkers such as siRNAs are intriguing and under investigation.

B. Proteins

Hundt *et al.* categorize proteins into antigens, antibodies, cytokines, other proteins, and chromatographic/mass spectroscopy assays.[46]

1. Antigens

Since the discovery of carcinoembryonic antigen (CEA),[59,60] investigators have searched for immunological biomarkers that are both sensitive and specific. Measurement of serum prostate-specific antigen (PSA) and fecal immunochemical testing rely on the presence of particular molecules to indicate the presence of cancer. Both clearly can lead to the detection of tumors that would otherwise not be diagnosed.[61–64]

2. Antibodies

Serological screening of cDNA expression library (SEREX) or proteomics technologies have demonstrated that serum antibodies recognize antigens from common solid tumors. Such antibodies can be detected in serum, suggesting that those serum autoantibodies could potentially be exploited as diagnostic markers.[65–67]

3. Cytokines/Growth Factors

High serum concentrations of insulin-like growth factors (IGF) and low levels of their binding proteins have been shown to correlate with cancer risk in large cohort studies[68–71] and are associated with the development of common epithelial cancers. Other cytokines or angiogenesis factors, such as TGF-β1,[72–78] VEGF,[79,80] angiogenin,[81] endostatin,[82] and endothelins,[83,84] appear to have promise as biomarkers for the detection of common cancers. Matrix metalloproteinases,[85–87] for example, plasma TIMP1, has been found to be elevated in colorectal adenocarcinomas but has not had sufficient sensitivity in larger validation trials to merit development as a detection biomarker.[88] Other potential biomarkers include cell adhesion molecules.[89] Nuclear matrix proteins[90,91] have recently been published as having high sensitivity and specificity in convenience set data for urothelial tumors and GI tumors.

VI. Future Approaches to Molecular Biomarker Applications to CRRI

Systems biology strategies now permit interrogation of full pathways with analysis of pathway intermediates, and their role in the carcinogenesis process will identify important molecular biomarkers.[92,93] The markers may be useful as intermediate endpoints for the identification of efficacious risk reduction endpoints. Recently, stem cells in primary culture have been identified as useful biomarker endpoints for identification of potentially efficacious CRRIs[94,95] through inhibition of stem cell self-renewal and inhibition of key stem cell signaling pathways. For example, curcumin, a natural product with broad anticarcinogenesis mechanisms,[96,97] inhibits primary human breast stem cell and breast cancer stem cell self-renewal, at least in part, through inhibition of the Wnt signaling pathway.[94]

VII. Standards for Biomarkers as Endpoints for Cancer Risk Reductive Efficacy

A biomarker must have statistical accuracy, precision, and validity of results[98] that demonstrates prediction of a "hard" disease end point—a cancer incidence or mortality end point. Validation of a molecular biomarker for prediction of the hard cancer endpoint requires avoiding and accounting for overfitting and bias. To date, no biomarkers have met this standard. Nevertheless, the future of successful CRRI discovery and clinical validation rests upon the identification and validation of molecular biomarkers.

VIII. Examples of CRRIs and Their Molecular Targets

CRRIs in clinical use or development may be classified on the basis of molecular mechanisms as outlined in Table II. The major purified agents with known anticarcinogenesis activity that have been studied in humans or show great promise in preclinical studies are reviewed.

A. Nonsteroidal Anti-inflammatory Agents (NSAIDs)

NSAIDs, commonly used, over-the-counter drugs (e.g., aspirin, ibuprofen, naproxen), reduce cellular inflammation through inhibition of prostaglandin-H synthases 1 and 2 (PGHS) (commonly called cyclooxygenases-1 and -2).[99,100] PGHSs convert fatty acids presented from the lipid membrane to prostaglandins via two enzymatic reactions. The first, catalyzed by prostaglandin endoperoxide, produces PGH_2. PGH_2 is then isomerized to a "2-series" product—PGD_2, PGE_2,

TABLE II
REPRESENTATIVE MOLECULAR TARGETS AND AGENTS FOR CHEMOPREVENTION (ADAPTED FROM KELLOFF *ET AL.*[23])

Molecular target	Clinical target	Representative agents
Anti-inflammatory/antioxidant		
Eicosanoid pathways	Colon, esophagus, lung, breast, cervix, liver, bladder, head and neck	Non-steroidal anti-inflammatory drugs, lipoxygenase inhibitors (zileuton, zafirkulast), prostacyclin agonists (Iloprost)
NF-κB	Prostate, colon, head and neck, liver	Non-steroidal anti-inflammatory drugs, bortezomib, curcumin, tea polyphenols
Antioxidant response element (Nrf2)	Lung, liver, head and neck	Dithiolthiones, triterpenes
Epigenetic modulation		
Estrogen receptor-α	Breast, prostate, colon	Tamoxifen, raloxifene, arzoxifene, indole-3-carbinol, soy isoflavones
5α-steroid reductase	Prostate	Finasteride, dutasteride
Aromatase	Breast, prostate	Anastrozole, letrozole
DNA methylation	Prostate, lung, colon	Folic acid, azacytidine
Histone deacetylase	Breast, colon	Vorinostat (SAHA)
Retinoic acid receptor-β	Breast, colon, ovary, head and neck	Fenretinide, 9-*cis*-retinoic acid
Retinoid X receptor	Breast	Targretin
Vitamin D receptor	Colon, prostate	Vitamin D_3 analogs
Signal transduction modulation		
Ornithine decarboxylase	Colon, bladder, skin, esophagus	Difluoromethylornithine (DFMO)
HMGCoA reductase	Colon, melanocyte, breast, prostate	Statins
Ras	Colon, pancreas, lung	Perillyl alcohol (triterpenes)
mTOR	Prostate	RAD-001

$PGF_{2\alpha}$, PGI_2, or thromboxane A_2 (TxA_2) by specific synthases. Arachidonic acid also serves as a substrate for lipoxygenases with resultant conversion to leukotrienes. 5-lipoxygenase converts arachidonic acid to a leukotriene A_4 which is subsequently hydrolyzed to other downstream leukotrienes. Prostaglandins and leukotrienes bind to EP receptors, releasing coupled G-proteins to elicit responses in the same or neighboring cells.[101]

NSAID binding to the cycloxygenase domain of PGHS varies. Thus, each NSAID targets the eicosanoid enzymes with different binding affinities and ultimately different potencies.[102] For example, selective inhibitors of the Cox domain of PGHS-2 (Cox-2 inhibitor) were developed as less gastric toxic alternatives to nonselective NSAIDs.

In addition to targeting the Cox portion of PGHS, NSAIDs possess independent mechanisms of action. For example, some NSAIDs inhibit the transcription factor nuclear factor (NF)-κB at pharmacologic concentrations and key cellular proliferation signaling intermediates such as activator protein (AP)-1 and other intermediates of the mitogen-activated protein (MAP)-kinase pathway. PGHS-independent mechanisms of action of NSAIDs may, at least in part, explain NSAID preventive efficacy.[99–101]

NSAIDs (e.g., aspirin, indomethacin, piroxicam, sulindac) suppress chemically induced (1,2-dimethylhydrazine or its metabolites) or trangenically (Min^+)[103,104] of colorectal adenocarcinoma. They reduce carcinogenesis events in chemical carcinogenesis models of stomach, skin, breast, lung, prostate, and urinary bladder.[105]

Preclinical molecular targeting data have been verified in human clinical trials. Using adenoma recurrence as a pathologic endpoint, sulindac and the selective Cox-2 inhibitor, celecoxib, suppressed recurrence and caused regression of adenomatous polyps in patients with familial adenomatous polyposis.[41,106] In randomized, double-blinded placebo-controlled trial, aspirin reduces the recurrence of adenomas by 20%.[107,108] Neither sulindac nor piroxicam suppressed adenoma formation in high-risk, sporadic populations at tolerable doses.[109,110] While selective Cox-2 inhibitors (celecoxib, rofecoxib) demonstrate potent colorectal risk reduction,[111–113] their cardiovascular toxicity precludes their use in healthy, at-risk populations.

NSAIDs may reduce neoplastic progression in average risk populations, although the data are insufficient to recommend routine administration of NSAIDs, primarily aspirin below 100 mg daily, in routine risk populations.[114–116]

B. Posttranslational Molecular Targets for Cancer Risk Reductives

Posttranslational pathway targets remain a fertile source of cancer risk reductive strategies. The most prominent among these targets, the selective estrogen receptor modulators (SERMs), aromatase inhibitors, and 5α-steroid reductase inhibitors, have demonstrated substantial cancer risk reduction for breast and prostate cancers. Inhibition of the polyamine pathway in combination with inhibition of PGHS reduces colorectal adenoma recurrence.

1. Selective Estrogen Receptor Modulators

SERMs bind to and activate or repress estrogen receptors (ERs), depending upon the structure of the agent. Based upon the chemical structure, a SERM may block or expose coactivator recruitment into the ER-ligand binding pocket of the three-dimensional ER structure. A SERM may be specific or nonspecific for tissue ER receptor type (α or β). For example, tamoxifen functions as an ER agonist when binding to ERβ while functioning as an ER antagonist when binding to ERα.[117,118] SERM structures exploit differential binding to ER receptor types on the basis of tissue-specific distribution. Since the amount of ERα during carcinogenesis increases while the amount of ERβ decreases in breast tissues, tamoxifen antagonizes ER function and may serve as a breast CRRI.[117] Predominant ERβ expression in the uterus and liver accounts for tamoxifen's agonist activity at these organ sites and explains its toxicities of thromboembolism and endometrial cancer.

The clinical finding that tamoxifen reduces the incidence of contralateral second primary breast cancers during adjuvant treatment regimens catalyzed the push for its development as a CRRI.[119–121] Breast cancer risk reductive efficacy has been established by large, cancer incidence endpoint clinical trials.[122–125] These trials, recently updated with long-term follow-up ranging from 8 to 20 years, demonstrate a consistent reduction in the incidence of invasive, ER-positive breast cancer, ranging from 26% to 45%. The data to date demonstrate that tamoxifen prevents the development of breast cancer in women with increased risk. Despite the United States Food and Drug Administration approval of tamoxifen for the prevention of breast cancer, only 3–20% of eligible high-risk women take tamoxifen for primary prevention. Patients and their primary care physicians decline to use tamoxifen due to its rare but important toxicities (endometrial carcinoma and thromboembolism).[126,127] This demonstrates the need for less toxic, effective cancer risk reductive approaches.

Raloxifene's benzothiophene structure confers a different tissue-specific ER binding profile than tamoxifen. Its reduced ERβ affinity mitigates tamoxifen's hepatic and venous thrombotic toxicities, while its agonist activity to ERα enhances its estrogen agonist activity in the bone. This enhanced ERα agonist activity was exploited to reduce fractures in postmenopausal women[128] with no increased risk of thromboembolism or endometrial adenocarcinoma.[117,118,129] A prospective, cancer incidence endpoint, risk reductive clinical trial of raloxifene compared to tamoxifen identified raloxifene's cancer risk reductive efficacy as equivalent to tamoxifen, but with an improved safety profile.[130] Compared to tamoxifen, raloxifene reduces thromboembolic events but does not lower the risk for endometrial cancer.[130]

2. Aromatase Inhibitors

The family of aromatase inhibitors (anastrozole, exemestane, letrozole) enhances reduction of breast cancer recurrence in the contralateral breast compared to tamoxifen alone.[131] In a Phase I risk reduction trial, letrozole reduces the Ki-67 proliferation index of breast epithelial cells aspirated from high-risk women.[132] Two aromatase inhibitors are currently being studied with a breast cancer incidence endpoint in Canada (exemestane) and in the United Kingdom (anastrozole).

3. 5α-Steroid Reductase Inhibitors

Intracellular androgen receptors bind to DNA hormone response elements that control cellular proliferation and apoptotic responses in transformed prostate cells. Dihydrotestosterone controls androgen receptor action by binding to the intracellular receptor. The 5α-steroid reductase types 1 and 2 isozymes catalyze the synthesis of dihydrotestosterone from testosterone, thus controlling intracellular androgen reception function. Targeting 5α-steroid reductase with finasteride, a selective, competitive inhibitor of type 2 5α-steroid reductase,[133] deprives the transformed prostate cell of proliferation signaling. In the 3,2′-dimethyl-4-aminobiphenyl (DMAB), methylnitrosourea (MNU), and testosterone chemical carcinogenesis models in rats, finasteride reduces prostate tumor incidence by close to six fold. Finasteride appears to be more effective in the promotion phase of prostate carcinogenesis.[134] A randomized, placebo-controlled cancer incidence endpoint risk reductive clinical trial of finasteride demonstrated that the finasteride-treated arm reduced the prevalence of prostate cancer by 24.8%. The initial analysis suggested that tumors of high Gleason's grade[24–27] were higher in the finasteride arm (37%) compared to the placebo arm (22%); however, subsequent analyses have suggested that the increase in high-grade prostate cancer was due to PSA and DRE prompted biopsies. For men with the greatest exposure to the drug (those with an end of study biopsy), there was no significant increase in risk. Sexual function side effects (erectile dysfunction, loss of libido, gynecomastia) were more common in the finasteride treatment arm.[135,136] Recent analyses of longer-term follow-up from this trial demonstrate that finasteride significantly enhanced the ability of PSA to detect prostate cancer and high-grade prostate.[137] Dutasteride inhibits both 5α-steroid reductase inhibitor types 1 and 2 isoforms, whereas finasteride inhibits only the type-2 isoform. In a cancer randomized, placebo-controlled prostate cancer risk reduction trial, dutasteride reduced the relative risk of prostate cancer by 22% and the absolute risk reduction of 5.1%. Similar to the finasteride study, in the final 2 years of observation, more high-grade (Gleasons 8 to 10) tumors were observed in the dutaseride-treated arm than in the placebo arm.[138]

C. Signal Transduction Modifiers

Interventions aimed at modulating signal transduction pathways promise new approaches to interventions in the carcinogenesis process. New, small molecule or antibody-based targeted therapies inhibiting epidermal growth factor receptor function, components of angiogenesis, and enhancing p53 function in neoplastic cells, aimed primarily for the treatment of invasive neoplasms, may be useful as CRRIs. Because of the complexity of signaling systems, inhibition of single targets may not be effective or may cause unacceptable toxicity.

1. Difluoromethylornithine

Cellular growth requires polyamines, spermidine, spermine, and the diamine, putrescine.[139] Pharmacologic inhibition of polyamines stops cellular growth and proliferation.[139–141] Difluoromethylornithine (DFMO) is a potent, irreversible inhibitor of ornithine decarboxylase, inhibits polyamine metabolism, and prevents tumor promotion in a variety of systems—skin, mammary, colon, cervical, and bladder carcinogenesis models.[139] Synergistic or additive activity with retinoids, butylated hydroxyanisole, tamoxifen, piroxicam, and fish oil has been demonstrated with low concentrations of DFMO.[139,141] DFMO alone has not been proved safe and effective as a cancer risk reductive for common epithelial neoplasms, but in combination with sulindac, DFMO potently inhibits adenoma recurrence.[142]

IX. Nutritional Products

Many dietary polyphenolic compounds such as curcumin,[97,143] gingerols,[144,145] and resveratrol[146] have been studied in *in vitro* and *in vivo* models of carcinogenesis as well as tested in early phase human clinical trials. Resveratrol, for example (3,5,4′-trihydroxy-*trans*-stilbene), is a phytoalexin found in grapes, mulberries, peanuts, and *Cassia quinquangulata* plants that may help to protect against carcinogenesis and pathogenesis.[147] It has been reported to function as an antioxidant. Resveratrol in red wine has been suggested as one possible explanation for the "French paradox," that is, the finding that the incidence of coronary artery disease is low in southern France despite the high dietary saturated fat intake. Resveratrol has also been reported to possess a variety of anti-inflammatory, antiplatelet, and both pro- and anti-estrogenic effects.[147] Resveratrol administration to mice inhibits high caloric diet-induced body weight, increases life span, and reduces insulin-1 like growth factor 1 (IGF-1) concentrations.[148] Resveratrol's life-extending properties and possible anticarcinogenesis activity may be mediated through the activation of sirtuin proteins, a class of histone deacetylases involved in life span determination, although this mechanism remains controversial.[149,150] Other anticarcinogenesis mechanisms of action include inhibition of cyclooxygenases 1 and 2, 5-lipoxygenase, and many protein kinases, leading to inhibition of downstream pathways such as STAT-3, HER2/neu, MAP kinases, Akt, and NFκB release.[148,151,152] In rodent models, resveratrol inhibited the formation of AOM-induced aberrant crypt foci in rat colon,[153] decreased the number of adenomas in the small intestine, and suppressed tumor formation in the colon of *Min*$^+$ mice.[154] It also reduced mammary tumor formation in NMU-treated rats when given at a relatively high dose (100 mg/kg).[155] Data in humans to date are limited, suggesting poor bioavailability when measured in human plasma,[146] similar to other nutritional polyphenols.[143,145] Ten-fold higher concentrations in human colon tissue as compared to plasma tissue suggest that resveratrol has sufficient bioavailability in human tissue to be considered for further CRRI investigation.[156] Nutritional polyphenols may represent a category of cancer risk reductive agents with low toxicity and multimechanistic anticarcinogenesis potential if the universal problem of limited bioavailability in humans can be overcome with liposomal formulations, nutritional extract combinations, or nanopackaging strategies.

X. Multiagent CRRIs

The likelihood of successful cancer risk reduction through inhibition of a single molecular intermediate within complex signaling pathways is low. Both preclinical and recent clinical data demonstrate the feasibility and high potential of successful cancer risk reduction based upon a multiple targeted molecular pathway modulation approach as opposed to targeting a single intermediate. For example, interactive signaling of epidermal growth factor receptors and cyclooxygenase-2[157] experiments rodent model systems[158] demonstrated synergistic reduction of colonic neoplastic events. Statin inhibition of HMG CoA combined with predominant cyclooxgygenase-1 inhibitor (aspirin) or with a selective cyclooxygenase-2 inhibitor (celecoxib) enhances the inhibition of azoxymethane-induced colon carcinogenesis in F344 rats and reduces the dose of the combined drugs required to achieve reduction of colon carcinogenesis.[159] Based upon coordinate inhibition of the polyamine and eicosanoid metabolic pathways, Meyskens *et al.* demonstrated synergistic cancer risk reductive effect in humans.[142] As more data accumulate from *in vivo* models, combined drugs aimed as specific targets in coordinated signaling pathways will enter clinical biomarker-based trials.

XI. Molecular Viral Targets for Cancer Risk Reduction

A. Human Papilloma Virus

The human papilloma virus, a causal event in the carcinogenesis process of cervical and some upper airway epithelial squamous cell neoplasms,[160,161] provides a molecular basis to reduce cancer risk through the development and deployment of vaccines toward the antigenic, single virion capsid protein, L1. The vaccines are based upon data demonstrating (1) that L1 has the intrinsic ability to self-assemble into a virus-like particle that can induce high concentrations of neutralizing antibodies due to the repetitive arrangement of the epitopes found on the virus-like particle[162,163] and (2) that in animal models of papilloma virus infection, vaccination with L1 virus-like particles protects from virus infectious challenge, while denatured L1 or L1 virus-like particles are not protective.[163]

Two vaccines, one produced by GlaxoSmithKine (GSK) (Cervarix) and the second produced by Merck (Gardasil), have notable differences. The GSK vaccine is bivalent, containing virus-like particles from HPV16 and HPV18, the two types found in 70% of cervix cancers worldwide. The vaccine uses a proprietary adjuvant, AS04 consisting of aluminum salts monophosphoryl lipid A. Merck's vaccine is quadravalent, containing virus-like proteins from HPV types 6, 11, 16, and 18, and uses a simple aluminum salt adjuvant. HPVs 6 and 11 cause 90% of cutaneous genital warts. For this reason, the quadravalent vaccine targets two distinct hyperproliferative diseases.[162,164,165] Both vaccines are remarkably effective in preventing persistent HPV infections and low-(Grade 1) and high (Grades 2 and 3)-grade cervical intraepithelial neoplasia.[166–169] In women with no evidence of vaccine HPV type (by HPV PCR and serology) who received all the three doses of the vaccine, the tetravalent vaccine has high efficacy (96.9%; 95% CI 81.3%-99.9%) against incident infection with HPV-16 and HPV endpoints, sustained for up to 4.5 years.[166] The vaccine also has efficacy against cervical intraepithelial neoplasia lesions (100%; 95% CI 42.4–100%) associated with vaccine type. The vaccine does not protect against persistent infection, cervical intraepithelial neoplasia, or genital warts in females who were infected at the time of vaccination.

B. Hepatitis B

The hepatitis B virus, a 42 nm double-stranded, partially circular, 3,200 base pair DNA virus, causes a chronic inflammatory state in the liver with resultant cirrhosis and hepatocellular carcinoma. The viral envelope, coded for by the S gene, consists of three components, large, middle, and major proteins. The hepatitis B surface antigen, HBsAg, consists of a small sphere and rods with an average diameter of 22 nm and can be found in the circulation in large quantities. The virus replicates in the liver, and there is no evidence indicating that the virus replicates at mucosal surfaces.[170] Infection may be asymptomatic or symptomatic. Worldwide, 50–55% of all hepatocellular carcinomas are attributable to persistent viral infection with hepatitis B virus.

HBsAg is used for hepatitis vaccine. The HBsAg vaccine has a protective efficacy for hepatitis B infection of 90–95%.[171] Due to a comprehensive vaccination strategy in the United States that included universal vaccination for infants and children, a 78% decline in hepatitis B infections occurred in the United States from 1990 to 2005.[172,173] Nevertheless, the incidence of hepatocellular carcinoma is increasing in the United States and worldwide.[174,175] The increased incidence of hepatocellular carcinoma is linked to a cohort effect linked to a rising incidence of hepatitis C and the impact hepatitis B exposures in the past. The rising incidence of hepatocellular carcinoma in developed countries is attributed to hepatitis C infections, while the cases of hepatocellular carcinoma in Africa remain associated with hepatitis B as well as hepatitis C.

References

1. Sporn M. Approaches to prevention of epithelial cancer during the preneoplastic period. *Cancer Res* 1976;**36**:2699–702.
2. Wattenberg L. Chemoprevention of cancer. *Cancer Res* 1985;**45**:1–8.
3. Greenwald P, Kelloff G. The role of chemoprevention in cancer control. *IARC Sci Publ* 1996;**139**:13–22.
4. Brenner DE. Principles of chemoprevention. In: DeVita V, Lawrence T, Rosenberg S, editors. *Cancer: Principles and Practice of Oncology*. 8th ed. USA: Lippincott, Williams and Wilkins; 2008. p. 630–5.
5. Brenner DE, Gescher A. Cancer chemoprevention: lessons learned and future directions. *Br J Cancer* 2005;**93**:735–9.
6. Hanahan D, Weinberg RA. The hallmarks of cancer. *Cell* 2000;**100**:57–70.
7. Nelson WG, De Marzo AM, Isaacs WB. Prostate cancer. *N Engl J Med* 2003;**349**:366–81.
8. von Knobloch R, Konrad L, Barth PJ, Brandt H, Wille S, Heidenreich A, et al. Genetic pathways and new progression markers for prostate cancer suggested by microsatellite allelotyping. *Clin Cancer Res* 2004;**10**:1064–73.
9. Palapattu GS, Sutcliffe S, Bastian PJ, Platz EA, De Marzo AM, Isaacs WB, et al. Prostate carcinogenesis and inflammation: emerging insights. *Carcinogenesis* 2005;**26**:1170–81.
10. Fearon ER, Vogelstein B. A genetic model for colorectal tumorigenesis. *Cell* 1990;**61**:759–67.
11. Barrett MT, Sanchez CA, Prevo LJ, Wong DJ, Galipeau PC, Paulson TG, et al. Evolution of neoplastic cell lineages in Barrett oesophagus. *Nat Genet* 1999;**22**:106–9.
12. Dontu G, Liu S, Wicha MS. Stem cells in mammary development and carcinogenesis: implications for prevention and treatment. *Stem Cell Rev* 2005;**1**:207–13.
13. Liu S, Dontu G, Mantle ID, Patel S, Ahn NS, Jackson KW, et al. Hedgehog signaling and Bmi-1 regulate self-renewal of normal and malignant human mammary stem cells. *Cancer Res* 2006;**66**:6063–71.
14. Wistuba II, Lam S, Behrens C, Virmani AK, Fong KM, LeRiche J, et al. Molecular damage in the bronchial epithelium of current and former smokers. *J Natl Cancer Inst* 1997;**89**:1366–73.
15. Massion PP, Carbone DP. The molecular basis of lung cancer: molecular abnormalities and therapeutic implications. *Respir Res* 2003;**4**:12.
16. Mao L, Lee JS, Kurie JM, Fan YH, Lippman SM, Lee JJ, et al. Clonal genetic alterations in the lungs of current and former smokers. *J Natl Cancer Inst* 1997;**89**:857–62.
17. Ha PK, Benoit NE, Yochem R, Sciubba J, Zahurak M, Sidransky D, et al. A transcriptional progression model for head and neck cancer. *Clin Cancer Res* 2003;**9**:3058–64.

18. Califano J, Westra WH, Meininger G, Corio R, Koch WM, Sidransky D. Genetic progression and clonal relationship of recurrent premalignant head and neck lesions. *Clin Cancer Res* 2000;**6**:347–52.
19. Califano J, van der Riet P, Westra W, Nawroz H, Clayman G, Piantadosi S, et al. Genetic progression model for head and neck cancer: implications for field cancerization. *Cancer Res* 1996;**56**:2488–92.
20. Braakhuis BJ, Tabor MP, Kummer JA, Leemans CR, Brakenhoff RH. A genetic explanation of Slaughter's concept of field cancerization: evidence and clinical implications. *Cancer Res* 2003;**63**:1727–30.
21. Reid BJ, Levine DS, Longton G, Blount PL, Rabinovitch PS. Predictors of progression to cancer in Barrett's esophagus: baseline histology and flow cytometry identify low- and high-risk patient subsets. *Am J Gastroenterol* 2000;**95**:1669–76.
22. Thorgeirsson SS, Grisham JW. Molecular pathogenesis of human hepatocellular carcinoma. *Nat Genet* 2002;**31**:339–46.
23. Kelloff GJ, Lippman SM, Dannenberg AJ, Sigman CC, Pearce HL, Reid BJ, et al. Progress in chemoprevention drug development: the promise of molecular biomarkers for prevention of intraepithelial neoplasia and cancer—a plan to move forward. *Clin Cancer Res* 2006;**12**:3661–97.
24. Boland CR, Koi M, Chang DK, Carethers JM. The biochemical basis of microsatellite instability and abnormal immunohistochemistry and clinical behavior in Lynch syndrome: from bench to bedside. *Fam Cancer* 2008;**7**:41–52.
25. Boland CR. The molecular biology of gastrointestinal cancer: implications for diagnosis and therapy. *Gastrointest Endosc Clin N Am* 2008;**18**:401–13, vii.
26. D'Andrea AD. Susceptibility pathways in Fanconi's anemia and breast cancer. *N Engl J Med* 2010;**362**:1909–19.
27. Gudmundsdottir K, Ashworth A. The roles of BRCA1 and BRCA2 and associated proteins in the maintenance of genomic stability. *Oncogene* 2006;**25**:5864–74.
28. Farmer H, McCabe N, Lord CJ, Tutt AN, Johnson DA, Richardson TB, et al. Targeting the DNA repair defect in BRCA mutant cells as a therapeutic strategy. *Nature* 2005;**434**:917–21.
29. Dantzer F, de La Rubia G, Menissier-De Murcia J, Hostomsky Z, de Murcia G, Schreiber V. Base excision repair is impaired in mammalian cells lacking Poly(ADP-ribose) polymerase-1. *Biochemistry* 2000;**39**:7559–69.
30. Dantzer F, Schreiber V, Niedergang C, Trucco C, Flatter E, De La Rubia G, et al. Involvement of poly(ADP-ribose) polymerase in base excision repair. *Biochimie* 1999;**81**:69–75.
31. Fong PC, Boss DS, Yap TA, Tutt A, Wu P, Mergui-Roelvink M, et al. Inhibition of poly(ADP-ribose) polymerase in tumors from BRCA mutation carriers. *N Engl J Med* 2009;**361**:123–34.
32. Fong PC, Yap TA, Boss DS, Carden CP, Mergui-Roelvink M, Gourley C, et al. Poly(ADP)-ribose polymerase inhibition: frequent durable responses in BRCA carrier ovarian cancer correlating with platinum-free interval. *J Clin Oncol* 2010;**28**:2512–9.
33. Shureiqi I, Reddy P, Brenner DE. Chemoprevention: general perspective. *Crit Rev Oncol Hematol* 2000;**33**:157–67.
34. Kelloff GJ, Bast Jr. RC, Coffey DS, D'Amico AV, Kerbel RS, Park JW, et al. Biomarkers, surrogate end points, and the acceleration of drug development for cancer prevention and treatment: an update prologue. *Clin Cancer Res* 2004;**10**:3881–4.
35. Lieberman DA. Clinical practice. Screening for colorectal cancer. *N Engl J Med* 2009;**361**:1179–87.
36. Stamler J, Stamler R, Neaton J. Blood pressure, systolic and diastolic, and cardiovascular risks. US population data. *Arch Intern Med* 1993;**153**:598–615.

37. Costantino JP, Gail MH, Pee D, Anderson S, Redmond CK, Benichou J, et al. Validation studies for models projecting the risk of invasive and total breast cancer incidence. *J Natl Cancer Inst* 1999;**91**:1541–8.
38. Miki Y, Swensen J, Shattuck-Eidens D, Futreal PA, Harshman K, Tavtigian S, et al. A strong candidate for the breast and ovarian cancer susceptibility gene BRCA1. *Science* 1994;**266**:66–71.
39. Powell SM, Petersen GM, Krush AJ, Booker S, Jen J, Giardiello FM, et al. Molecular diagnosis of familial adenomatous polyposis. *N Engl J Med* 1993;**329**:1982–7.
40. Giardiello FM, Yang VW, Hylind LM, Krush AJ, Petersen GM, Trimbath JD, et al. Primary chemoprevention of familial adenomatous polyposis with sulindac. *N Engl J Med* 2002;**346**:1054–9.
41. Steinbach G, Lynch PM, Phillips RK. The effect of celecoxib, a cyclooxygenase-2 inhibitor, in familial adenomatous polyposis. *N Engl J Med* 2000;**342**:1946–52.
42. Gronwald J, Tung N, Foulkes WD, Offit K, Gershoni R, Daly M, et al. Tamoxifen and contralateral breast cancer in BRCA1 and BRCA2 carriers: an update. *Int J Cancer* 2006;**118**:2281–4.
43. Definitions Working Group , editor. *Biomarkers and Surrogate Endpoints*. Bethesda, MD: National Institutes of Health, Food and Drug Administration; 1999.
44. Prentice R. Surrogate endpoints in clinical trials: definition and operational criteria. *Stat Med* 1989;**8**:431–40.
45. Sidransky D. Emerging molecular markers of cancer. *Nat Rev Cancer* 2002;**2**:210–9.
46. Hundt S, Haug U, Brenner H. Blood markers for early detection of colorectal cancer: a systematic review. *Cancer Epidemiol Biomarkers Prev* 2007;**16**:1935–53.
47. Hibi K, Robinson CR, Booker S, Wu L, Hamilton SR, Sidransky D, et al. Molecular detection of genetic alterations in the serum of colorectal cancer patients. *Cancer Res* 1998;**58**:1405–7.
48. Vlems FA, Diepstra JH, Cornelissen IM, Ligtenberg MJ, Wobbes T, Punt CJ, et al. Investigations for a multi-marker RT-PCR to improve sensitivity of disseminated tumor cell detection. *Anticancer Res* 2003;**23**:179–86.
49. Yamaguchi K, Takagi Y, Aoki S, Futamura M, Saji S. Significant detection of circulating cancer cells in the blood by reverse transcriptase-polymerase chain reaction during colorectal cancer resection. *Ann Surg* 2000;**232**:58–65.
50. Zou H, Yu B, Wang Z, Sun J, Cang H, Gao F, et al. Detection of aberrant p16 methylation in the serum of colorectal cancer patients. *Clin Cancer Res* 2002;**8**:188–91.
51. Noh YH, Im G, Ku JH, Lee YS, Ahn MJ. Detection of tumor cell contamination in peripheral blood by RT-PCR in gastrointestinal cancer patients. *J Korean Med Sci* 1999;**14**:623–8.
52. Borchers R, Heinzlmann M, Zahn R, Witter K, Martin K, Loeschke K, et al. K-ras mutations in sera of patients with colorectal neoplasia and long-standing inflammatory bowel disease. *Scand J Gastroenterol* 2002;**37**:715–8.
53. Salbe C, Trevisiol C, Ferruzzi E, Mancuso T, Nascimbeni R, Di Fabio F, et al. Molecular detection of codon 12 K-RAS mutations in circulating DNA from serum of colorectal cancer patients. *Int J Biol Markers* 2000;**15**:300–7.
54. Lecomte T, Berger A, Zinzindohoue F, Micard S, Landi B, Blons H, et al. Detection of free-circulating tumor-associated DNA in plasma of colorectal cancer patients and its association with prognosis. *Int J Cancer* 2002;**100**:542–8.
55. Grady WM, Rajput A, Lutterbaugh JD, Markowitz SD. Detection of aberrantly methylated hMLH1 promoter DNA in the serum of patients with microsatellite unstable colon cancer. *Cancer Res* 2001;**61**:900–2.
56. Nakayama H, Hibi K, Taguchi M, Takase T, Yamazaki T, Kasai Y, et al. Molecular detection of p16 promoter methylation in the serum of colorectal cancer patients. *Cancer Lett* 2002;**188**:115–9.

57. Verma M, Srivastava S. Epigenetics in cancer: implications for early detection and prevention. *Lancet Oncol* 2002;**3**:755–63.
58. Li M, Chen WD, Papadopoulos N, Goodman SN, Bjerregaard NC, Laurberg S, et al. Sensitive digital quantification of DNA methylation in clinical samples. *Nat Biotechnol* 2009;**27**:858–63.
59. Gold P, Freedman SO. Specific carcinoembryonic antigens of the human digestive system. *J Exp Med* 1965;**122**:467–81.
60. Gold P, Freedman SO. Demonstration of tumor specific antigens in human colonic carcinoma by immunologic tolerance and absorption techniques. *J Exp Med* 1965;**121**:439.
61. Allison JE, Sakoda LC, Levin TR, Tucker JP, Tekawa IS, Cuff T, et al. Screening for colorectal neoplasms with new fecal occult blood tests: update on performance characteristics. *J Natl Cancer Inst* 2007;**99**:1462–70.
62. Shastri YM, Stein J. Quantitative immunochemical fecal occult blood test for diagnosing colorectal neoplasia. *Ann Intern Med* 2007;**147**:522–3, author reply 3.
63. van Rossum LG, van Rijn AF, Laheij RJ, van Oijen MG, Fockens P, van Krieken HH, et al. Random comparison of guaiac and immunochemical fecal occult blood tests for colorectal cancer in a screening population. *Gastroenterology* 2008;**135**:82–90.
64. Walsh PC. Acceptability and complications of prostate biopsy in population-based PSA screening versus routine clinical practice: a prospective, controlled study. *J Urol* 2003;**170**:315.
65. Scanlan MJ, Welt S, Gordon CM, Chen YT, Gure AO, Stockert E, et al. Cancer-related serological recognition of human colon cancer: identification of potential diagnostic and immunotherapeutic targets. *Cancer Res* 2002;**62**:4041–7.
66. Lu H, Goodell V, Disis ML. Targeting serum antibody for cancer diagnosis: a focus on colorectal cancer. *Expert Opin Ther Targets* 2007;**11**:235–44.
67. Nam MJ, Madoz-Gurpide J, Wang H, Lescure P, Schmalbach CE, Zhao R, et al. Molecular profiling of the immune response in colon cancer using protein microarrays: occurrence of autoantibodies to ubiquitin C-terminal hydrolase L3. *Proteomics* 2003;**3**:2108–15.
68. Giovannucci E. Insulin-like growth factor-1 and binding protein-3 and risk of cancer. *Horm Res* 1999;**51**:34–41.
69. Giovannucci E, Pollak MN, Platz EA, Willet WC, Stampfer MJ, Majeed N, et al. A prospective study of plasma insulin-like growth factor and binding protein-3 and risk of colorectal neoplasia in women. *Cancer Epidemiol Biomarkers Prev* 2000;**9**:345–9.
70. Giovannucci E, Pollak M, Platz EA, Willet WC, Stampfer MJ, Majeed N, et al. Insulin-like growth factor I (IGF-I), IGF-binding protein-3 and the risk of colorectal adenoma and cancer in the Nurses' Health Study. *Growth Horm IGF Res* 2000;**10**(Suppl. A):S30–1.
71. Palmquist R, Stattin P, Rinaldi S, Biessy C, Stenling R, Riboli E, et al. Plasma insulin, IGF-binding proteins-1 and -2 and risk of colorectal cancer: a prospective study in northern Sweden. *Int J Cancer* 2003;**107**:89–93.
72. Cruz-Correa M, Cui H, Giardiello FM, Powe NR, Hylind L, Robinson A, et al. Loss of imprinting of insulin growth factor II gene: a potential heritable biomarker for colon neoplasia predisposition. *Gastroenterology* 2004;**126**:964–70.
73. Cui H, Onyango P, Brandenburg S, Wu Y, Hsieh CL, Feinberg AP. Loss of imprinting in colorectal cancer linked to hypomethylation of H19 and IGF2. *Cancer Res* 2002;**62**:6442–6.
74. Cui H, Cruz-Correa M, Giardiello FM, Hutcheon DF, Kafonek DR, Brandenburg S, et al. Loss of IGF2 imprinting: a potential marker of colorectal cancer risk. *Science* 2003;**299**:1753–5.
75. Woodson K, Flood A, Green L, Tangrea JA, Hanson J, Cash B, et al. Loss of insulin-like growth factor-II imprinting and the presence of screen-detected colorectal adenomas in women. *J Natl Cancer Inst* 2004;**96**:407–10.

76. Tsushima H, Kawata S, Tamura S, Ito N, Shirai Y, Kiso S, et al. High levels of transforming growth factor beta 1 in patients with colorectal cancer: association with disease progression. *Gastroenterology* 1996;**110**:375–82.
77. Tsushima H, Ito N, Tamura S, Matsuda Y, Inada M, Yabuuchi I, et al. Circulating transforming growth factor beta-1 as a predictor of liver metastasis after resection in colorectal cancer. *Clin Cancer Res* 2001;**7**:1258–62.
78. Narai S, Watanabe M, Hasegawa H, Nishibori H, Endo T, Kubota T, et al. Significance of transforming growth factor beta 1 as a new tumor marker for colorectal cancer. *Int J Cancer* 2002;**97**:508–11.
79. Broll R, Erdmann H, Duchrow M, Oevermann E, Schwandner O, Markert U, et al. Vascular endothelial growth factor (VEGF)—a valuable serum tumour marker in patients with colorectal cancer? *Eur J Surg Oncol* 2001;**27**:37–42.
80. Takeda A, Shimada H, Imaseki H, Okazumi S, Natsume T, Suzuki T, et al. Clinical significance of serum vascular endothelial growth factor in colorectal cancer patients: correlation with clinicopathological factors and tumor markers. *Oncol Rep* 2000;**7**:333–8.
81. Shimoyama S, Yamasaki K, Kawahara M, Kaminishi M. Increased serum angiogenin concentration in colorectal cancer is correlated with cancer progression. *Clin Cancer Res* 1999;**5**:1125–30.
82. Feldman AL, Alexander Jr HR, Bartlett DL, Kranda KC, Miller MS, Costouros NG, et al. A prospective analysis of plasma endostatin levels in colorectal cancer patients with liver metastases. *Ann Surg Oncol* 2001;**8**:741–5.
83. Simpson RA, Dickinson T, Porter KE, London NJ, Hemingway DM. Raised levels of plasma big endothelin 1 in patients with colorectal cancer. *Br J Surg* 2000;**87**:1409–13.
84. Peeters CF, Thomas CM, Sweep FC, Span PN, Wobbes T, Ruers TM. Elevated serum endothelin-1 levels in patients with colorectal cancer; relevance for prognosis. *Int J Biol Markers* 2000;**15**:288–93.
85. Pellegrini P, Contasta I, Berghella AM, Gargano E, Mammarella C, Adorno D. Simultaneous measurement of soluble carcinoembryonic antigen and the tissue inhibitor of metalloproteinase TIMP 1 serum levels for use as markers of pre-invasive to invasive colorectal cancer. *Cancer Immunol Immunother* 2000;**49**:388–94.
86. Yukawa N, Yoshikawa T, Akaike M, Sugimasa Y, Takemiya S, Yanoma S, et al. Plasma concentration of tissue inhibitor of matrix metalloproteinase 1 in patients with colorectal carcinoma. *Br J Surg* 2001;**88**:1596–601.
87. Barozzi C, Ravaioli M, D'Errico A, Grazi GL, Poggioli G, Cavrini G, et al. Relevance of biologic markers in colorectal carcinoma: a comparative study of a broad panel. *Cancer* 2002;**94**:647–57.
88. Holten-Anderson MN, Christensen IJ, Nielsen HJ, Stephens RW, Jensen V, Nielsen OH, et al. Total levels of tissue inhibitor of metalloproteinases 1 in plasma yield high diagnostic sensitivity and specificity in patients with colon cancer. *Clin Cancer Res* 2002;**8**:156–64.
89. Alexiou D, Karayiannakis AJ, Syrigos KN, Zbar A, Kremmyda A, Bramis I, et al. Serum levels of E-selectin, ICAM-1 and VCAM-1 in colorectal cancer patients: correlations with clinicopathological features, patient survival and tumour surgery. *Eur J Cancer* 2001;**37**:2392–7.
90. Leman ES, Schoen RE, Magheli A, Sokoll LJ, Chan DW, Getzenberg RH. Evaluation of colon cancer-specific antigen 2 as a potential serum marker for colorectal cancer. *Clin Cancer Res* 2008;**14**:1349–54.
91. Leman ES, Schoen RE, Weissfeld JL, Cannon GW, Sokoll LJ, Chan DW, et al. Initial analyses of colon cancer-specific antigen (CCSA)-3 and CCSA-4 as colorectal cancer-associated serum markers. *Cancer Res* 2007;**67**:5600–5.
92. Gonzalez-Angulo AM, Hennessy BT, Mills GB. Future of personalized medicine in oncology: a systems biology approach. *J Clin Oncol* 2010;**28**:2777–83.

93. Stemke-Hale K, Gonzalez-Angulo AM, Lluch A, Neve RM, Kuo WL, Davies M, et al. An integrative genomic and proteomic analysis of PIK3CA, PTEN, and AKT mutations in breast cancer. *Cancer Res* 2008;**68**:6084–91.
94. Kakarala M, Brenner DE, Korkaya H, Cheng C, Tazi K, Ginestier C, et al. Targeting breast stem cells with the cancer preventive compounds curcumin and piperine. *Breast Cancer Res Treat* 2010;**122**:777–85.
95. Li Y, Zhang T, Korkaya H, Liu S, Lee HF, Newman B, et al. Sulforaphane, a dietary component of broccoli/broccoli sprouts, inhibits breast cancer stem cells. *Clin Cancer Res* 2010;**16**:2580–90.
96. Sharma RA, Gescher AJ, Steward WP. Curcumin: the story so far. *Eur J Cancer* 2005;**41**:1955–68.
97. Aggarwal BB, Sung B. Pharmacological basis for the role of curcumin in chronic diseases: an age-old spice with modern targets. *Trends Pharmacol Sci* 2009;**30**:85–94.
98. Ransohoff DF. Rules of evidence for cancer molecular-marker discovery and validation. *Nat Rev Cancer* 2004;**4**:309–14.
99. Thun MJ, Henley SJ, Patrono C. Nonsteroidal anti-inflammatory drugs as anticancer agents: mechanistic, pharmacologic, and clinical issues. *J Natl Cancer Inst* 2002;**94**:252–66.
100. Chan TA. Cyclooxygenase inhibition and mechanisms of colorectal cancer prevention. *Curr Cancer Drug Targets* 2003;**3**:455–63.
101. Wang D, Mann JR, DuBois RN. The role of prostaglandins and other eicosanoids in the gastrointestinal tract. *Gastroenterology* 2005;**128**:1445–61.
102. Meade EA, Smith WL, DeWitt DL. Differential inhibition of prostaglandin endoperoxide synthase (cyclooxygenase) isozymes by aspirin and other non-steroidal anti-inflammatory drugs. *J Biol Chem* 1993;**268**:6610–4.
103. Pollard M, Luckert PH. Effect of indomethacin on intestinal tumor induced in rats by the acetate derivative of dimethylnitrosamine. *Science* 1981;**214**:558–9.
104. Jacoby RF, Marshall DJ, Newton MA, Novakovic K, Tutsch K, Cole CE, et al. Chemoprevention of spontaneous intestinal adenomas in the Apc Min mouse model by the nonsteroidal anti-inflammatory drug piroxicam. *Cancer Res* 1996;**56**:710–4.
105. Anderson WF, Umar A, Viner JL, Hawk ET. The role of cyclooxygenase inhibitors in cancer prevention. *Curr Pharm Des* 2002;**8**:1035–62.
106. Giardiello FM, Hamilton SR, Krush AJ, Piantadosi S, Hylind LM, Celano P, et al. Treatment of colonic and rectal adenomas with sulindac in familial adenomatous polyposis. *N Engl J Med* 1993;**328**:1313–6.
107. Baron JA, Cole BF, Sandler RS, Haile RW, Ahnen D, Bresalier R, et al. A randomized trial of aspirin to prevent colorectal adenomas. *N Engl J Med* 2003;**348**:891–9.
108. Sandler RS, Halabi S, Baron JA, Budinger S, Paskett E, Keresztes R, et al. A randomized trial of aspirin to prevent colorectal adenomas in patients with previous colorectal cancer. *N Engl J Med* 2003;**348**:883–90.
109. Ladenheim J, Garcia G, Titzer D, Herzenberg H, Lavori P, Edson R, et al. Effects of sulindac on sporadic colonic polyps. *Gastroenterology* 1995;**108**:1083–7.
110. Calaluce R, Earnest DL, Heddens D, Einspahr JG, Roe D, Bogert CL, et al. Effects of piroxicam on prostaglandin E2 levels in rectal mucosa of adenomatous polyp patients: a randomized phase IIb trial. *Cancer Epidemiol Biomarkers Prev* 2000;**9**:1287–92.
111. Arber N, Eagle CJ, Spicak J, Racz I, Dite P, Hajer J, et al. Celecoxib for the prevention of colorectal adenomatous polyps. *N Engl J Med* 2006;**355**:885–95.
112. Bertagnolli MM, Eagle CJ, Zauber AG, Redston M, Solomon SD, Kim K, et al. Celecoxib for the prevention of sporadic colorectal adenomas. *N Engl J Med* 2006;**355**:873–84.
113. Baron JA, Sandler RS, Bresalier RS, Quan H, Riddell R, Lanas A, et al. A randomized trial of rofecoxib for the chemoprevention of colorectal adenomas. *Gastroenterology* 2006;**131**:1674–82.

114. Cook NR, Lee IM, Gaziano JM, Gordon D, Ridker PM, Manson JE, et al. Low-dose aspirin in the primary prevention of cancer: the Women's Health Study: a randomized controlled trial. *JAMA* 2005;**294**:47–55.
115. Gann PH, Manson JE, Glynn RJ, Buring JE, Hennekens CH. Low-dose aspirin and incidence of colorectal tumors in a randomized trial. *J Natl Cancer Inst* 1993;**85**:1220–4.
116. U.S. Preventive Services Task Force. Routine aspirin or nonsteroidal anti-inflammatory drugs for the primary prevention of colorectal cancer: U.S. Preventive Services Task Force recommendation statement. *Ann Intern Med* 2007;**146**:361–4.
117. Fabian CJ, Kimler BF. Selective estrogen-receptor modulators for primary prevention of breast cancer. *J Clin Oncol* 2005;**23**:1644–55.
118. Jordan VC. Chemoprevention of breast cancer with selective oestrogen-receptor modulators. *Nat Rev Cancer* 2007;**7**:46–53.
119. Cuzick J, Baum M. Tamoxifen and contralateral breast cancer. *Lancet* 1985;**2**:282.
120. Rutqvist LE, Cedermark B, Glas U, Mattsson A, Skoog L, Somell A, et al. Contralateral primary tumors in breast cancer patients in a randomized trial of adjuvant tamoxifen therapy. *J Natl Cancer Inst* 1991;**83**:1299–306.
121. Fisher B, Redmond C. New perspective on cancer of the contralateral breast: a marker for assessing tamoxifen as a preventive agent. *J Natl Cancer Inst* 1991;**83**:1278–80.
122. Fisher B, Costantino J, Wickerham D, Redmond C, Kavanah M, Cronin W, et al. Tamoxifen for prevention of breast cancer: report of the National Surgical Adjuvant Breast and Bowel Project P-1 study. *J Natl Cancer Inst* 1998;**90**:1371–88.
123. Cuzick J, Forbes JF, Sestak I, Cawthorn S, Hamed H, Holli K, et al. Long-term results of tamoxifen prophylaxis for breast cancer—96-month follow-up of the randomized IBIS-I trial. *J Natl Cancer Inst* 2007;**99**:272–82.
124. Powles TJ, Ashley S, Tidy A, Smith IE, Dowsett M. Twenty-year follow-up of the Royal Marsden randomized, double-blinded tamoxifen breast cancer prevention trial. *J Natl Cancer Inst* 2007;**99**:283–90.
125. Veronesi U, Maisonneuve P, Rotmensz N, Bonanni B, Boyle P, Viale G, et al. Tamoxifen for the prevention of breast cancer: Late results of the Italian randomzied tamoxifen prevention trial among women with hysterectomy. *JNCI* 2007;**99**:727–37.
126. Tchou J, Hou N, Rademaker A, Jordan VC, Morrow M. Acceptance of tamoxifen chemoprevention by physicians and women at risk. *Cancer* 2004;**100**:1800–6.
127. Melnikow J, Paterniti D, Azari R, Kuenneth C, Birch S, Kuppermann M, et al. Preferences of Women Evaluating Risks of Tamoxifen (POWER) study of preferences for tamoxifen for breast cancer risk reduction. *Cancer* 2005;**103**:1996–2005.
128. Ettinger B, Black DM, Mitlak BH, Knickerbocker RK, Nickelsen T, Genant HK, et al. Reduction of vertebral fracture risk in postmenopausal women with osteoporosis treated with raloxifene: results from a 3-year randomized clinical trial. Multiple Outcomes of Raloxifene Evaluation (MORE) Investigators. *JAMA* 1999;**282**:637–45.
129. Cauley JA, Norton L, Lippman ME, Eckert S, Krueger KA, Purdie DW, et al. Continued breast cancer risk reduction in postmenopausal women treated with raloxifene: 4-year results from the MORE trial. Multiple outcomes of raloxifene evaluation. *Breast Cancer Res Treat* 2001;**65**:125–34.
130. Vogel VG, Costantino JP, Wickerham DL, Cronin WM, Cecchini RS, Atkins JN, et al. Effects of tamoxifen vs raloxifene on the risk of developing invasive breast cancer and other disease outcomes: the NSABP Study of Tamoxifen and Raloxifene (STAR) P-2 trial. *JAMA* 2006;**295**:2727–41.
131. Goss PE, Ingle JN, Martino S, Robert NJ, Muss HB, Piccart MJ, et al. A randomized trial of letrozole in postmenopausal women after five years of tamoxifen therapy for early-stage breast cancer. *N Engl J Med* 2003;**349**:1793–802.

132. Fabian CJ, Kimler BF, Zalles CM, Khan QJ, Mayo MS, Phillips TA, et al. Reduction in proliferation with six months of letrozole in women on hormone replacement therapy. *Breast Cancer Res Treat* 2007;**106**:75–84.
133. Hess-Wilson JK, Knudsen KE. Endocrine disrupting compounds and prostate cancer. *Cancer Lett* 2006;**241**:1–12.
134. Andriole G, Bostwick D, Civantos F, Epstein J, Lucia MS, McConnell J, et al. The effects of 5alpha-reductase inhibitors on the natural history, detection and grading of prostate cancer: current state of knowledge. *J Urol* 2005;**174**:2098–104.
135. Thompson IM, Goodman PJ, Tangen CM, Lucia MS, Miller GJ, Ford LG, et al. The influence of finasteride on the development of prostate cancer. *N Engl J Med* 2003;**349**:215–24.
136. Thompson IM, Tangen CM, Goodman PJ, Lucia MS, Klein EA. Chemoprevention of prostate cancer. *J Urol* 2009;**182**:499–507, discussion 8.
137. Thompson IM, Chi C, Ankerst DP, Goodman PJ, Tangen CM, Lippman SM, et al. Effect of finasteride on the sensitivity of PSA for detecting prostate cancer. *J Natl Cancer Inst* 2006;**98**:1128–33.
138. Andriole GL, Bostwick DG, Brawley OW, Gomella LG, Marberger M, Montorsi F, et al. Effect of dutasteride on the risk of prostate cancer. *N Engl J Med* 2010;**362**:1192–202.
139. Gerner EW, Meyskens Jr. FL. Polyamines and cancer: old molecules, new understanding. *Nat Rev Cancer* 2004;**4**:781–92.
140. Pegg A. Polyamine metabolism and its importance in neoplastic growth and as a target for chemotherapy. *Cancer Res* 1988;**48**:759–74.
141. Meyskens Jr. FL, Gerner EW. Development of difluoromethylornithine (DFMO) as a chemoprevention agent. *Clin Cancer Res* 1999;**5**:945–51.
142. Meyskens Jr. FL, McLaren CE, Pelot D, Fujikawa-Brooks S, Carpenter PM, Hawk E, et al. Difluoromethylornithine plus sulindac for the prevention of sporadic colorectal adenomas: a randomized placebo-controlled, double-blind trial. *Cancer Prev Res (Phila Pa)* 2008;**1**:32–8.
143. Vareed SK, Kakarala M, Ruffin MT, Crowell JA, Normolle DP, Djuric Z, et al. Pharmacokinetics of curcumin conjugate metabolites in healthy human subjects. *Cancer Epidemiol Biomarkers Prev* 2008;**17**:1411–7.
144. Zick S, Ruffin M, Lee J, Normolle D, Siden R, Alrawi S, et al. Phase II trial of encapsulated ginger as a treatment for chemotherapy-induced nausea and vomiting. *Support Care Cancer* 2009;**17**:563–72.
145. Zick SM, Djuric Z, Ruffin MT, Litzinger AJ, Normolle DP, Alrawi S, et al. Pharmacokinetics of 6-gingerol, 8-gingerol, 10-gingerol, and 6-shogaol and conjugate metabolites in healthy human subjects. *Cancer Epidemiol Biomarkers Prev* 2008;**17**:1930–6.
146. Boocock DJ, Faust GE, Patel KR, Schinas AM, Brown VA, Ducharme MP, et al. Phase I dose escalation pharmacokinetic study in healthy volunteers of resveratrol, a potential cancer chemopreventive agent. *Cancer Epidemiol Biomarkers Prev* 2007;**16**:1246–52.
147. Jang M, Cai L, Udeani GO, Slowing KV, Thomas CF, Beecher CW, et al. Cancer chemopreventive activity of resveratrol, a natural product derived from grapes. *Science* 1997;**275**:218–20.
148. Pirola L, Frojdo S. Resveratrol: one molecule, many targets. *IUBMB Life* 2008;**60**:323–32.
149. Baur JA, Pearson KJ, Price NL, Jamieson HA, Lerin C, Kalra A, et al. Resveratrol improves health and survival of mice on a high-calorie diet. *Nature* 2006;**444**:337–42.
150. Baur JA, Sinclair DA. Therapeutic potential of resveratrol: the in vivo evidence. *Nat Rev Drug Discov* 2006;**5**:493–506.
151. Aggarwal BB, Bhardwaj A, Aggarwal RS, Seeram NP, Shishodia S, Takada Y. Role of resveratrol in prevention and therapy of cancer: preclinical and clinical studies. *Anticancer Res* 2004;**24**:2783–840.

152. Aggarwal BB, Shishodia S, Sandur SK, Pandey MK, Sethi G. Inflammation and cancer: how hot is the link? *Biochem Pharmacol* 2006;**72**:1605–21.
153. Tessitore L, Davit A, Sarotto I, Caderni G. Resveratrol depresses the growth of colorectal aberrant crypt foci by affecting bax and p21(CIP) expression. *Carcinogenesis* 2000;**21**:1619–22.
154. Schneider Y, Duranton B, Gosse F, Schleiffer R, Seiler N, Raul F. Resveratrol inhibits intestinal tumorigenesis and modulates host-defense-related gene expression in an animal model of human familial adenomatous polyposis. *Nutr Cancer* 2001;**39**:102–7.
155. Bhat KP, Lantvit D, Christov K, Mehta RG, Moon RC, Pezzuto JM. Estrogenic and anti-estrogenic properties of resveratrol in mammary tumor models. *Cancer Res* 2001;**61**:7456–63.
156. Brown V, Patel K, Hemingway D, Miller A, West K, Jones D, et al. Pilot study of the putative cancer chemopreventive agent resveratrol in colorectal cancer patients: concentration of resveratrol and its metabolites in colorectal tissue. *Cancer Res* 2010; In press.
157. Coffey RJ, Hawkey CJ, Damstrup L, Graves-Deal R, Daniel VC, Dempsey PJ, et al. Epidermal growth factor receptor activation induces nuclear targeting of cyclooxygenase-2, basolateral release of prostaglandins, and mitogenesis in polarizing colon cancer cells. *Proc Natl Acad Sci USA* 1997;**94**:657–62.
158. Torrance CJ, Jackson PE, Montgomery E, Kinzler KW, Vogelstein B, Wissner A, et al. Combinatorial chemoprevention of intestinal neoplasia. *Nat Med* 2000;**6**:1024–8.
159. Reddy BS, Wang CX, Kong AN, Khor TO, Zheng X, Steele VE, et al. Prevention of azoxymethane-induced colon cancer by combination of low doses of atorvastatin, aspirin, and celecoxib in F 344 rats. *Cancer Res* 2006;**66**:4542–6.
160. Schiffman MH, Bauer HM, Hoover RN, Glass AG, Cadell DM, Rush BB, et al. Epidemiologic evidence showing that human papillomavirus infection causes most cervical intraepithelial neoplasia. *J Natl Cancer Inst* 1993;**85**:958–64.
161. Lowy DR, Sedman SA, Cohen BD, Schiller JT. Human papillomaviruses and cancer. *Recent Results Cancer Res* 1993;**128**:351–2.
162. Schiller JT, Lowy DR. Prospects for cervical cancer prevention by human papillomavirus vaccination. *Cancer Res* 2006;**66**:10229–32.
163. Lowy DR, Schiller JT. Prophylactic human papillomavirus vaccines. *J Clin Invest* 2006;**116**:1167–73.
164. Saslow D, Castle PE, Cox JT, Davey DD, Einstein MH, Ferris DG, et al. American Cancer Society Guideline for human papillomavirus (HPV) vaccine use to prevent cervical cancer and its precursors. *CA Cancer J Clin* 2007;**57**:7–28.
165. Centers for Disease Control and Prevention . Quadravalent human papilloma virus vaccine: recommendations of the Advisory Committee on Immunization Practices. *MMWR Morb Mortal Wkly Rep* 2007;**56**:1–24.
166. Harper DM, Franco EL, Wheeler CM, Moscicki AB, Romanowski B, Roteli-Martins CM, et al. Sustained efficacy up to 4.5 years of a bivalent L1 virus-like particle vaccine against human papillomavirus types 16 and 18: follow-up from a randomised control trial. *Lancet* 2006;**367**:1247–55.
167. Villa LL, Costa RL, Petta CA, Andrade RP, Paavonen J, Iversen OE, et al. High sustained efficacy of a prophylactic quadrivalent human papillomavirus types 6/11/16/18 L1 virus-like particle vaccine through 5 years of follow-up. *Br J Cancer* 2006;**95**:1459–66.
168. Villa LL, Ault KA, Giuliano AR, Costa RL, Petta CA, Andrade RP, et al. Immunologic responses following administration of a vaccine targeting human papillomavirus Types 6, 11, 16, and 18. *Vaccine* 2006;**24**:5571–83.
169. Block SL, Nolan T, Sattler C, Barr E, Giacoletti KE, Marchant CD, et al. Comparison of the immunogenicity and reactogenicity of a prophylactic quadrivalent human papillomavirus (types 6, 11, 16, and 18) L1 virus-like particle vaccine in male and female adolescents and young adult women. *Pediatrics* 2006;**118**:2135–45.

170. Kao JH, Chen DS. Global control of hepatitis B virus infection. *Lancet Infect Dis* 2002;**2**:395–403.
171. Ni YH, Huang LM, Chang MH, Yen CJ, Lu CY, You SL, et al. Two decades of universal hepatitis B vaccination in Taiwan: impact and implication for future strategies. *Gastroenterology* 2007;**132**:1287–93.
172. Centers for Disease Control and Prevention. A comprehensive immunization strategy to eliminate transmission of hepatitis B virus infection in the United States. Part 1: immunization of infants, children, and adolescents. *MMWR Morb Mortal Wkly Rep* 2005;**54**:1–32.
173. Centers for Disease Control and Prevention. A comprehensive immunization strategy to eliminate transmission of hepatitis B virus infection in the United States. Part 2: immunization of adults. *MMWR Morb Mortal Wkly Rep* 2006;**55**:1–33.
174. Bosch FX, Ribes J, Diaz M, Cleries R. Primary liver cancer: worldwide incidence and trends. *Gastroenterology* 2004;**127**:S5–S16.
175. Seeff LB, Hoofnagle JH. Epidemiology of hepatocellular carcinoma in areas of low hepatitis B and hepatitis C endemicity. *Oncogene* 2006;**25**:3771–7.

Index

A

B

C

D

E

F

G

N

T

V

W

FSC
www.fsc.org
MIX
Papier | Fördert
gute Waldnutzung
FSC® C083411

Kolibri 360 GmbH
Ferdinand-Jühlke-Straße 7
99095 Erfurt, Deutschland
produktsicherheit@kolibri360.de